Product Design for Manufacture and Assembly

MANUFACTURING ENGINEERING AND MATERIALS PROCESSING
A Series of Reference Books and Textbooks

EDITOR

Ioan Marinescu
University of Toledo
Toledo, Ohio

FOUNDING EDITOR

Geoffrey Boothroyd
Boothroyd Dewhurst, Inc.
Wakefield, Rhode Island

Additional Volumes in Preparation

Product Design for Manufacture and Assembly

Second Edition
Revised and Expanded

Geoffrey Boothroyd

Boothroyd Dewhurst, Inc.
Wakefield, Rhode Island

Peter Dewhurst

Winston Knight

University of Rhode Island
Kingston, Rhode Island

MARCEL DEKKER, INC.   NEW YORK · BASEL

ISBN: 0-8247-0584-X

This book is printed on acid-free paper.

Headquarters
Marcel Dekker, Inc.
270 Madison Avenue, New York, NY 10016
tel: 212-696-9000; fax: 212-685-4540

Eastern Hemisphere Distribution
Marcel Dekker AG
Hutgasse 4, Postfach 812, CH-4001 Basel, Switzerland
tel: 41-61-261-8482; fax: 41-61-261-8896

World Wide Web
http://www.dekker.com

The publisher offers discounts on this book when ordered in bulk quantities. For more information, write to Special Sales/Professional Marketing at the headquarters address above.

Current printing (last digit):
10 9 8 7 6 5 4 3 2

PRINTED IN THE UNITED STATES OF AMERICA

Preface to the Second Edition

This second edition of *Product Design for Manufacture and Assembly* includes three new chapters, describing the processes of sand casting, investment casting, and hot forging. These chapters, combined with the chapters describing design for machining, injection molding, sheet metalworking, die casting, and powder metals, cover a wide range of the most basic forming processes used in industry.

In addition, substantial material has been added to the introductory chapter illustrating the effects that the application of design for manufacture and assembly (DFMA) has had on U.S. industry as a whole. Chapter 2, dealing with the selection of materials and processes for manufacture, now includes further material describing material selection specifically and the economic ranking of processes using a new software tool.

Chapter 3, dealing with product design for manual assembly, includes an updated special section dealing with the effect of design on product quality. Finally, additional material has been added to Chapter 15 discussing links between computer-aided design (CAD) solid models and design analysis tools.

As with the previous edition, we thank the various companies who have supported research on DFMA at the University of Rhode Island and the graduate students who have contributed to the research. We particularly acknowledge the help of Allyn Mackay, on whose work the new chapter on investment casting is largely based.

Finally, thanks are due to Shirley Boothroyd for typing much of the new material and to Kenneth Fournier for preparing some of the additional artwork.

Geoffrey Boothroyd
Peter Dewhurst
Winston Knight

Preface to the First Edition

We have been working in the area of product design for manufacture and assembly (DFMA) for over twenty years. The methods that have been developed have found wide application in industry—particularly U.S. industry. In fact, it can be said that the availability of these methods has created a revolution in the product design business and has helped to break down the barriers between design and manufacture; it has also allowed the development of concurrent or simultaneous engineering.

This book not only summarizes much of our work on DFMA, but also provides the details of DFMA methods for practicing and student engineers.

Much of the methodology involves analytical tools that allow designers and manufacturing engineers to estimate the manufacturing and assembly costs of a proposed product before detailed design has taken place. Unlike other texts on the subject, which are generally descriptive, this text provides the basic equations and data that allow manufacturing and assembly cost estimates to be made. Thus, for a limited range of materials and processes the engineer or student can make cost estimates for real parts and assemblies and, therefore, become familiar with the details of the methods employed and the assumptions made.

For practicing manufacturing engineers and designers, this book is not meant as a replacement for the DFMA software developed by Boothroyd Dewhurst, Inc., which contains more elaborate databases and algorithms, but rather provides a useful companion, allowing an understanding of the methods involved.

For engineering students, this book is suitable as a text on product design for manufacture and assembly and, in fact, is partially based on notes for a two-course sequence developed by the authors at the University of Rhode Island.

The original work on design for assembly was funded at the University of Massachusetts by the National Science Foundation. Professor K. G. Swift and Dr. A. H. Redford of the Universities of Hull and Salford, respectively, collaborated with G. Boothroyd in this early work and were supported by the British Science Research Council.

The research continued at the University of Rhode Island and was supported mainly by U.S. industry. We thank the following companies for their past and, in some cases, continuing support of the work: Allied, AMP, Digital Equipment, DuPont, EDS, Ford, General Electric, General Motors, Gillette, IBM, Instron, Loctite, Motorola, Navistar, Westinghouse, and Xerox.

We also thank all the graduate assistants and research scholars who over the years have contributed to the research, including: N. Abbatiello, A. Abbot, A. Anderson, J. Anderson, T. Andes, D. Archer, G. Bakker, T. Becker, C. Blum, T. Bassinger, K. P. Brindamour, R. C. Burlingame, T. Bushman, J. P. Cafone, A. Carnevale, M. Caulfield, H. Connelly, T. J. Consunji, C. Donovan, J. R. Donovan, W. A. Dvorak, C. Elko, B. Ellison, M. C. Fairfield, J. Farris, T. J. Feenstra, M. B. Fein, R. P. Field, T. Fujita, A. Fumo, A. Girard, T. S. Hammer, P. Hardro, Y. S. Ho, L. Ho, L. S. Hu, G. D. Jackson, J. John II, B. Johnson, G. Johnson, K. Ketelsleger, G. Kobrak, D. Kuppurajan, A. Lee, C. C. Lennartz, H. C. Ma, D. Marlowe, S. Naviroj, N. S. Ong, C. A. Porter, P. Radovanovic, S. C. Ramamurthy, B. Rapoza, B. Raucent, M. Roe, L. Rosario, M. Schladenhauffen, B. Seth, C. Shea, T. Shinohara, J. Singh, R. Stanton, M. Stanziano, G. Stevens, A. Subramani, B. Sullivan, J. H. Timmins, E. Trolio, R. Turner, S. C. Yang, Z. Yoosufani, J. Young, J. C. Woschenko, D. Zenger, and Y. Zhang.

We would also like to thank our colleagues, the late Professor C. Reynolds, who collaborated in the area of early cost estimating for manufactured parts, and Professor G. A. Russell, who collaborated in the area of printed circuit board assembly.

Finally, thanks are due to Kenneth Fournier for preparing much of the artwork.

Geoffrey Boothroyd
Peter Dewhurst
Winston Knight

Contents

Contents

Contents

Product Design for Manufacture and Assembly

1
Introduction

1.1 WHAT IS DESIGN FOR MANUFACTURE AND ASSEMBLY?

In this text we shall assume that "to manufacture" refers to the manufacturing of the individual component parts of a product or assembly and that "to assemble" refers to the addition or joining of parts to form the completed product. This means that for the purposes of this text, assembly will not be considered a manufacturing process in the same sense that machining, molding, etc., are manufacturing processes. Hence, the term "design for manufacture" (or DFM) means the design for ease of manufacture of the collection of parts that will form the product after assembly and "design for assembly" (or DFA) means the design of the product for ease of assembly. Thus, "design for manufacture and assembly" (DFMA) is a combination of DFA and DFM.

DFMA is used for three main activities:

1. As the basis for concurrent engineering studies to provide guidance to the design team in simplifying the product structure, to reduce manufacturing and assembly costs, and to quantify the improvements.
2. As a benchmarking tool to study competitors' products and quantify manufacturing and assembly difficulties.
3. As a should-cost tool to help negotiate suppliers contracts.

The development of the original DFA method stemmed from earlier work in the 1960s on automatic handling [1]. A group technology classification system was developed to catalogue automatic handling solutions for small parts [2]. It became apparent that the classification system could also help designers to design parts that would be easy to handle automatically.

1

In the mid-1970s the U.S. National Science Foundation (NSF) awarded a substantial grant to extend this approach to the general areas of DFM and DFA. Essentially, this meant classifying product design features that significantly effect assembly times and manufacturing costs and quantifying these effects. At the same time, the University of Salford in England was awarded a government grant to study product design for automatic assembly. As part of the study, various designs of domestic gas flow meters were compared. These meters all worked on the same principal and had the same basic components. However, it was found that their manufacturability varied widely and that the least manufacturable design had six times the labor content of the best design.

Figure 1.1 shows five different solutions for the same attachment problem taken from the gas flow meters studied. It can be seen that, on the left, the simplest method for securing the housing consisted of a simple snap fit. In the examples on the right, not only does the assembly time increase, but both the number and cost of parts increases. This illustrates the two basic principles of design for ease of assembly of a product: reduce the number of assembly operations by reducing the number of parts and make the assembly operations easier to perform.

The DFA time standards for small mechanical products resulting from the NSF-supported research were first published in handbook form in the late 1970s, and the first successes resulting from the application of DFA in industry were reported in an article in *Assembly Engineering* [3] . In the article, Sidney Liebson, corporate director of manufacturing for Xerox and a long-time supporter of our research, suggested that "DFA would save his company hundreds of millions of dollars over the next ten years." The article generated intense interest in U.S. industry.

At that time, microcomputers were coming onto the market. A version of DFA, running on an Apple II plus computer proved attractive to those wishing to obtain the reported benefits of DFA applications. It appeared that, unlike their European or Japanese counterparts, U.S. designers preferred to use the new computers rather than perform hand calculations to analyze their designs for ease of

FIG. 1.1   Examples of design features affecting assembly.

assembly. As a result, engineers at IBM and Digital funded the development of versions of the DFA software to run on their own company products.

A major breakthrough in DFA implementation was made in 1988 when Ford Motor Company reported that our DFA software had helped them save billions of dollars on their Taurus line of automobiles. Later, it was reported [4] that General Motors (GM) made comparisons between its assembly plant at Fairfax, Kansas, which made the Pontiac Grand Prix, and Ford's assembly plant for its Taurus and Mercury Sable models near Atlanta. GM found a large productivity gap and concluded that 41% of the gap could be traced to the manufacturability of the two designs. For example, the Ford car had fewer parts—10 in its front bumper compared with 100 in the GM Pontiac—and the Ford parts fit together more easily.

Not surprisingly, GM has now become one of the leading users of DFMA. In fact, a GM executive has stated that:

DFM/DFA is a primary driver of quality and cost improvement.
It impacts every system of the vehicle.
It is an integral part of engineering and manufacturing employee training.
It provides knowledge and capabilities for individuals and organizations.
It provides technical improvements to both product and process.
It's not an option—it's a requirement.

In the 1960s there was much talk about designing products so they could be manufactured more easily. Recommendations commonly known as producibility guidelines were developed. Figure 1.2 shows a typical design guideline published in 1971 that emphasized simplifying the individual parts [5]. The authors of this guideline mistakenly assumed that several simple-shaped parts are inherently less expensive to manufacture than a single complex part and that any assembly costs are more than offset by the savings in part costs. They were wrong on both counts, as the results in Table 1.1 show. Even ignoring assembly costs, the two

FIG. 1.2 Misleading producibility guideline for the design of sheet metal parts.

TABLE 1.1 Estimated Costs in Dollars for the Two Examples in Fig. 1.2 if 100,000 Are Made

	Wrong	Right
Setup	0.015	0.023
Process	0.535	0.683
Material	0.036	0.025
Piece part	0.586	0.731
Tooling	0.092	0.119
Total manufacture	0.678	0.850
Assembly	0.000	0.200
Total	0.678	1.050

parts in the "right" design are significantly more expensive than the single part in the "wrong" design—even the piece part costs (neglecting tooling costs) are more expensive. Taking assembly costs into account and ignoring storage, handling, quality, and paperwork costs, the "right" design is 50% more costly than the "wrong" design!

Once methods for analyzing assembly difficulties were developed in the 1970s it became recognized that there was a conflict between producibility and assembly. It was found that the simplification of products by reducing the number of separate parts through DFA—on the order of 50% on average—could easily achieve substantial reductions in assembly costs. Much more important, however, was the fact that even greater savings could be achieved in the cost of the parts. The ability to estimate both assembly and part manufacturing costs at the earliest stages of product design is the essence of DFMA. The authors of this text have carried out numerous research programs over the past two decades on the subject of DFMA. A primary objective of this work has been to develop economic models of manufacturing processes, based on product design information, and which require a minimum of manufacturing knowledge [6,7,8].

The simple example in Fig. 1.2 and Table 1.1 illustrates this. If the "right" design were subject to a DFA analysis, the designer would be challenged as to why the subassembly could not be manufactured as a single part thereby eliminating an assembly cost of $0.20. Further analysis would show an additional saving of $0.17 in part costs.

That designers should give more attention to possible manufacturing problems has been advocated for many years. Traditionally, it was expected that engineering students should take "shop" courses in addition to courses in machine design. The idea was that a competent designer should be familiar with manufacturing processes to avoid adding unnecessarily to manufacturing costs during design.

Unfortunately, in the 1960s shop courses disappeared from university curricula in the United States; they were not considered suitable for academic credit by the new breed of engineering theoreticians. In fact, a career in design was not generally considered appropriate for one with an engineering degree. Of course, the word "design" has many different meanings. To some it means the aesthetic design of a product such as the external shape of a car or the color, texture, and shape of the casing of a can opener. In fact, in some university curricula this is what would be meant by a course in "product design."

On the other hand, design can mean establishing the basic parameters of a system. For example, before considering any details, the design of a power plant might mean establishing the characteristics of the various units such as generators, pumps, boilers, connecting pipes, etc.

Yet another interpretation of the word "design" would be the detailing of the materials, shapes, and tolerance of the individual parts of a product. This is the aspect of product design mainly considered in this text. It is an activity that starts with sketches of parts and assemblies; it then progresses to the CAD workstation, where assembly drawings and detailed part drawings are produced. These drawings are then passed to the manufacturing and assembly engineers whose job it is to optimize the processes used to produce the final product. Frequently, it is at this stage that manufacturing and assembly problems are encountered and requests are made for design changes. Sometimes these design changes are large in number and result in considerable delays in the final product release. In addition, the later in the product design and development cycle the changes occur, the more expensive they become. Therefore, not only is it important to take manufacture and assembly into account during product design, but also these considerations must occur as early as possible in the design cycle.

This is illustrated qualitatively by the chart in Fig. 1.3 showing that extra time spent early in the design process is more than compensated for by savings in time when prototyping takes place. Thus, in addition to reducing product costs, the application of design for manufacture and assembly (DFMA) shortens the time to bring the product to market. As an example, Ingersoll-Rand Company reported [9] that the use of DFMA software from Boothroyd Dewhurst, Inc., slashed product development time from two years to one. In addition, the simultaneous engineering team reduced the number of parts in a portable compressor radiator and oil-cooler assembly from 80 to 29, decreased the number of fasteners from 38 to 20, trimmed the number of assembly operations from 159 to 40 and reduced assembly time from 18.5 to 6.5 min. Developed in June 1989, the new design went into full production in February, 1990.

Another reason why careful consideration of manufacture and assembly should be considered early in the design cycle is because it is now widely accepted that over 70% of final product costs are determined during design [10]. This is illustrated in Fig. 1.4.

FIG. 1.3 DFMA shortens the design process. (From *Plastics Design Forum*, October 1993.)

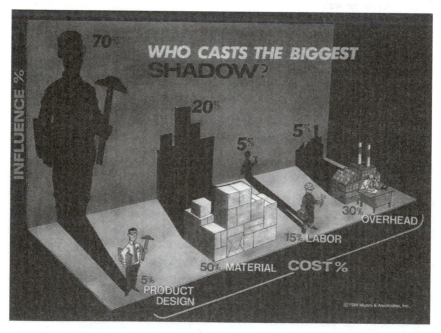

FIG. 1.4 Who casts the biggest shadow? (From Ref. 10.)

FIG. 1.5 "Over the wall" design, historically the way of doing business. (From Ref. 10.)

Traditionally, the attitude of designers has been "we design it, you build it."
This has now been termed the "over-the-wall approach" where the designer is
sitting on one side of the wall and throwing designs over the wall (Fig. 1.5) to the
manufacturing engineers, who then have to deal with the various manufacturing
problems arising because they were not involved in the design effort. One means
of overcoming this problem is to consult the manufacturing engineers at the
design stage. The resulting teamwork avoids many problems. However, these
teams, now called simultaneous engineering or concurrent engineering teams,
require analysis tools to help them study proposed designs and evaluate them
from the point of view of manufacturing difficulty and cost.

By way of illustration we see that DFMA efforts at Hewlett Packard Loveland
[11] started in the mid-1980s with redesign of existing products and continued
with application to new product design. During these studies, which proved
increasingly successful, product development involved one to three manufacturing engineers interacting frequently with the R&D team members. Eventually, by
1992, HP Loveland had incorporated DFMA into a formal concurrent engineering approach. The gradual improvements in their product manufacturing and
assembly costs are shown in Fig. 1.6.

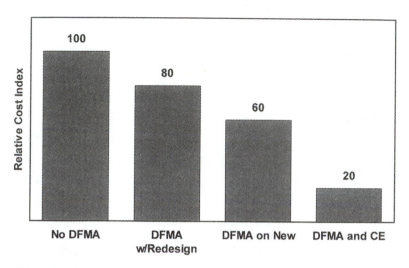

FIG. 1.6 Effects of DFMA and CE on product cost at Hewlett Packard. (Adapted from Ref. 11.)

1.2 HOW DOES DFMA WORK?

Let's follow an example from the conceptual design stage. Figure 1.7 represents a motor drive assembly that is required to sense and control its position on two steel guide rails. The motor must be fully enclosed for aesthetic reasons and have a removable cover to provide access to adjustment of the position sensor. The principal requirements are a rigid base designed to slide up and down with guide rails that will both support the motor and locate the sensor. The motor and sensor have wires connecting to a power supply and control unit, respectively.

A proposed solution is shown in Fig. 1.8, where the base is provided with two bushings to provide suitable friction and wear characteristics. The motor is secured to the base with two screws and a hole accepts the cylindrical sensor, which is held in place with a set screw. The motor base and sensor are the only items necessary for operation of the device. To provide the required covers, an end plate is screwed to two standoffs, which are screwed into the base. This end plate is fitted with a plastic bushing through which the connecting wires pass. Finally, a box-shaped cover slides over the whole assembly from below the base and is held in place by four screws, two passing into the base and two into the end cover.

There are two subassemblies, the motor and the sensor, which are required items, and, in this initial design, there are eight additional main parts and nine screws making a total of nineteen items to be assembled.

FIG. 1.7 Configuration of required motor drive assembly.

When DFA began to be taken seriously in the early 1980s and the consequent benefits were appreciated, it became apparent that the greatest improvements arose from simplification of the product by reducing the number of separate parts. In order to give guidance to the designer in reducing the part count, the DFA methodology [12] provides three criteria against which each part must be examined as it is added to the product during assembly.

1. During operation of the product, does the part move relative to all other parts already assembled. Only gross motion should be considered—small motions that can be accommodated by integral elastic elements, for example, are not sufficient for a positive answer.
2. Must the part be of a different material than or be isolated from all other parts already assembled? Only fundamental reasons concerned with material properties are acceptable.
3. Must the part be separate from all other parts already assembled because otherwise necessary assembly or disassembly of other separate parts would be impossible.

Application of these criteria to the proposed design (Fig. 1.8) during assembly would proceed as follows:

1. Base: Since this is the first part to be assembled, there are no other parts with which it can be combined, so it is a theoretically necessary part.

FIG. 1.8 Proposed design of motor drive assembly (dimensions in inches).

2. Bushings (2): These do not satisfy the criteria because, theoretically, the base and bushings could be of the same material.
3. Motor: The motor is a standard subassembly of parts that, in this case, is purchased from a supplier. Thus, the criteria cannot be applied and the motor is a necessary separate item.
4. Motor screws (2): Invariably, separate fasteners do not meet the criteria because an integral fastening arrangement is always theoretically possible.
5. Sensor: This is another standard subassembly and will be considered a necessary separate item.
6. Set screw: Theoretically not necessary.
7. Standoffs (2): These do not meet the criteria; they could be incorporated into the base.
8. End plate: Must be separate for reasons of assembly of necessary items.
9. End plate screws (2): Theoretically not necessary.
10. Plastic bushing: Could be of the same material as, and therefore combined with, the end plate.
11. Cover: Could be combined with the end plate.
12. Cover screws (4): Theoretically not necessary.

From this analysis it can be seen that if the motor and sensor subassemblies could be arranged to snap or screw into the base and a plastic cover designed to snap on, only four separate items would be needed instead of 19. These four items represent the theoretical minimum number needed to satisfy the requirements of the product design without considering practical limitations.

It is now necessary for the designer or design team to justify the existence of those parts that did not satisfy the criteria. Justification may arise from practical or technical considerations or from economic considerations. In this example, it could be argued that two screws are needed to secure the motor and one set screw is needed to hold the sensor because any alternatives would be impractical for a low-volume product such as this. However, the design of these screws could be improved by providing them with pilot points to facilitate assembly.

It could be argued that the two powder metal bushings are unnecessary because the part could be machined from an alternative material, such as nylon, having the necessary frictional characteristics. Finally, it is difficult to justify the separate standoffs, end plate, cover, plastic bushing, and six screws.

Now, before an alternative design can be considered, it is necessary to have estimates of the assembly times and costs so that any possible savings can be taken into account when considering design alternatives. Using the techniques described in this text, it is possible to make estimates of assembly costs, and later estimate the cost of the parts and associated tooling, without having final detail drawings of the part available.

Table 1.2 presents the results of an assembly analysis for the original motor drive assembly where it can be seen that an assembly design index of 7.5%

TABLE 1.2 Results of Design for Assembly (DFA) Analysis for the Motor Drive Assembly Proposed Design (Fig. 1.8)

	No.	Theoretical part count	Assembly time (s)	Assembly cost (¢)[a]
Base	1	1	3.5	2.9
Bushing	2	0	12.3	10.2
Motor subassembly	1	1	9.5	7.9
Motor screw	2	0	21.0	17.5
Sensor subassembly	1	1	8.5	7.1
Set screw	1	0	10.6	8.8
Standoff	2	0	16.0	13.3
End plate	1	1	8.4	7.0
End plate screw	2	0	16.6	13.8
Plastic bushing	1	0	3.5	2.9
Thread leads	—	—	5.0	4.2
Reorient	—	—	4.5	3.8
Cover	1	0	9.4	7.9
Cover screw	4	0	31.2	26.0
Totals	19	4	160.0	133

$$\text{Design efficiency} = \frac{4 \times 3}{160} = 7.5\%$$

[a] For a labor rate of $30/h.

given. This figure is obtained by comparing the estimated assembly time of 160 s with a theoretical minimum time obtained by multiplying the theoretical minimum part count of four by a minimum time of assembly for each part of 3 s. It should be noted that for this analysis standard subassemblies are counted as parts.

Considering first the parts with zeros in the theoretical part count column, it can be seen that those parts that did not meet the criteria for minimum part count involved a total assembly time of 120.6 s. This figure should be compared with the total assembly time for all 19 parts of 160 s. It can also be seen that parts involving screw-fastening operations resulted in the largest assembly times. It has already been suggested that the elimination of the motor screws and the set screw would probably be impractical. However, elimination of the remaining parts not meeting the criteria would result in the design concept shown in Fig. 1.9 where the bushings are combined with the base and the standoffs, end plate, cover, plastic bushing, and six screws are replaced by one snap-on plastic cover. The eliminated items involved an assembly time of 97.4 s. The new cover would take only 4 s to assemble and would avoid the need for a reorientation. In addition,

FIG. 1.9 Redesign of motor drive assembly following design for assembly (DFA) analysis.

screws with pilot points would be used and the base redesigned so that the motor was self-aligning.

Table 1.3 presents the results of an assembly analysis of the new design where it can be seen that the new assembly time is only 46 s—less than one-third of the original assembly time. The assembly index is now 26%, a figure that approaches

TABLE 1.3 Results of Design for Assembly (DFA) Analysis for the Motor Drive Assembly Redesign (Fig. 1.9)

	No.	Theoretical part count	Assembly time (s)	Assembly cost (¢)[a]
Base	1	1	3.5	2.9
Motor subassembly	1	1	4.5	3.8
Motor screw	2	0	12.0	10.0
Sensor subassembly	1	1	8.5	7.1
Set screw	1	0	8.5	7.1
Thread leads	—	—	5.0	4.2
Plastic cover	1	1	4.0	3.3
Totals	6	4	46.0	38.4

$$\text{Design efficiency} = \frac{4 \times 3}{46.0} = 26\%$$

[a] For a labor rate of $30/h.

the range found from experience to be representative of good designs of small electromechanical devices produced in relatively low volume.

Table 1.4 compares the cost of the parts for the two designs showing a savings of $15 in parts cost. However, the tooling for the new cover is estimated to be $6307—an investment that would have to be made at the outset. The parts cost and tooling cost estimates were made using the techniques described in this text.

Thus, the outcome of this study is a second design concept representing a total savings of $15.95, of which only 95 cents represents the savings in assembly time. In addition, the design index has been improved by about 250%.

It is interesting to note that the redesign suggestions arose through the application of the minimum part count criteria during the DFA analysis—the final cost comparison being made after assembly cost and parts cost estimates were considered.

The second step in an analysis is Design for Manufacture (DFM). This means estimating the cost of the manufactured parts in order to quantify the effects of any design improvements suggested by the initial Design for Assembly (DFA) analysis. In the present example the DFM analysis of the base revealed the cost of providing each set of features by machining. Interestingly, it was found that

TABLE 1.4 Comparison of Parts Cost for the Motor Drive Assembly Proposed Design and Redesign (purchased motor and sensor subassemblies not included)

(a) Proposed design		(b) Redesign	
Item	Cost ($)	Item	Cost ($)
Base (aluminum)	12.91	Base (nylon)	13.43
Bushing (2)	2.40[a]	Motor screw (2)	0.20[a]
Motor screw (2)	0.20[a]	Set screw	0.10[a]
Set screw	0.10[a]	Plastic cover	6.71
Standoff (2)	5.19	(includes tooling)	
End plate	5.89		
End plate screw (2)	0.20[a]	Total	20.44
Plastic bushing	0.10[a]		
Cover	8.05	Tooling cost for plastic cover, $6307	
Cover screw (4)	0.40[a]		
Total	35.44		

[a] Purchased in quantity.

elimination of the two drilled and tapped screw holes in the side of the base and the two drilled and tapped holes provided for the standoffs would reduce the total machining cost by $1.14. Thus, these changes would save more than the total possible savings in assembly cost of 95 cents. This is an indication that it is important not only to know the total estimated manufacturing cost of an item but, more importantly, to know the cost of providing the various features. This case study is typical in the sense that although DFA means design for assembly, the results of improving assemblability usually manifest themselves in significant reductions in part manufacturing costs.

Figure 1.10 summarizes the steps taken when using DFMA during design. The DFA analysis is first conducted leading to a simplification of the product structure. Then, using DFM, early cost estimates for the parts are obtained for both the original design and the new design in order to make trade-off decisions. During this process the best materials and processes to be used for the various parts are considered. In the example, would it be better to manufacture the cover in the new design from sheet metal? Once final selection of materials and processes has occurred, a more thorough analysis for DFM can be carried out for the detail design of parts. All of these steps are considered in the following chapters.

FIG. 1.10 Typical steps taken in a DFMA study using DFMA software.

1.3 REASONS FOR NOT IMPLEMENTING DFMA

No Time

In making presentations and conducting workshops on DFMA, the authors have found that the most common complaint among designers is that they are not allowed sufficient time to carry out their work. Designers are usually constrained by the urgent need to minimize the design-to-manufacture time for a new product. Unfortunately, as was illustrated earlier (Fig. 1.3), more time spent in the initial stages of design will reap benefits later in terms of reduced engineering changes after the design has been released to manufacturing. Company executives and managers must be made to realize that the early stages of design are critical in determining not only manufacturing costs, but also the overall design-to-manufacturing cycle time.

Not Invented Here

Enormous resistance can be encountered when new techniques are proposed to designers. Ideally, any proposal to implement DFMA should come from the designers themselves. However, more frequently it is the managers or executives who have heard of the successes resulting from DFMA and who wish their own designers to implement the philosophy. Under these circumstances, great care must be taken to involve the designers in the decision to implement these new techniques. Only then will the designers support the idea of applying DFMA. If they don't support DFMA, it won't be successfully applied.

The Ugly Baby Syndrome

Even greater difficulties can exist when an outside group or a separate group within the company undertakes to analyze current designs for ease of manufacture and assembly. Commonly, it will be found that significant improvements could be made to the current design, and when these improvements are brought to the attention of those who produced the design this can result in extreme resistance. Telling a designer that their designs could be improved is much like telling a mother that her baby is ugly! (Fig. 1.11). It is important, therefore, to involve the designers in the analysis and provide them with the incentive to produce better designs. If they perform the analysis, they are less likely to take as criticism any problems that may be highlighted.

Low Assembly Costs

The earlier description of the application of DFMA showed that the first step is a DFA analysis of the product or subassembly. Quite frequently it will be suggested that since assembly costs for a particular product form only a small proportion of the total manufacturing costs, there is no point in performing a DFA analysis. Figure 1.12 shows the results of an analysis where the assembly costs were

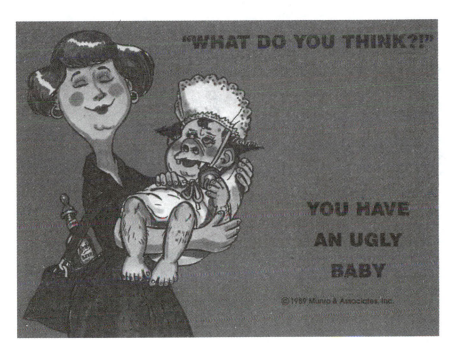

FIG. 1.11 Would you tell this mother that her baby is ugly? (From Ref. 10.)

extremely small compared with material and manufacturing costs. However, DFA analysis would suggest replacement of the complete assembly with, say, a machined casting. This would reduce total manufacturing costs by at least 50%.

Low Volume

The view is often expressed that DFMA is worthwhile only when the product is manufactured in large quantities. It could be argued, though, that use of the DFMA philosophy is even more important when the production quantities are small. This is because, commonly, reconsideration of an initial design is usually not carried out for low-volume production. Figure 1.12 is an example of this where the assembly was designed to be built from items machined from stock as if the product were one-of-a-kind. The prototype then became the production model. Applying the philosophy "do it right the first time" becomes even more important, therefore, when production quantities are small. In fact, the opportunities for parts consolidation are usually greater under these circumstances because it is not usually a consideration during design.

(a)

(b)

FIG. 1.12 DFA analysis can reduce total costs significantly even though assembly costs are small.

The Database Doesn't Apply to Our Products

Everyone seems to think that their own company is unique and, therefore, in need of unique databases. However, when one design is rated better than another using the DFA database it would almost certainly be rated in the same way using a customized database. Because DFMA should be applied at the early design stage before detailed design has taken place, there is a need for a generalized database for this purpose. Later, when more accurate estimates are desired, then the user can employ a customized database if necessary.

We've Been Doing It for Years

When this claim is made, it usually means that some procedure for "design for producibility" has been in use in the company. However, design for producibility usually means detailed design of individual parts for ease of manufacture. It was made clear earlier that such a process should occur only at the end of the design cycle; it can be regarded as a "fine tuning" of the design. The important decisions affecting total manufacturing costs will already have been made. In fact, there is a great danger in implementing design for producibility in this way. It has been found that the design of individual parts for ease of manufacture can mean, for example, limiting the number of bends in a sheet metal part. Again, experience has shown that it is important to combine as many features in one part as possible. In this way, full use is made of the abilities of the various manufacturing processes.

It's Only Value Analysis

It is true that the objectives of DFMA and value analysis are the same. However, it should be realized that DFMA is meant to be applied early in the design cycle and that value analysis does not give proper attention to the structure of the product and its possible simplification. DFMA has the advantage that it is a systematic step-by-step procedure that can be applied at all stages of design and that challenges the designer or design team to justify the existence of all the parts and to consider alternative designs. Experience has shown that DFMA still makes significant improvements of existing products even after value analysis has been carried out.

DFMA Is Only One Among Many Techniques

Since the introduction of DFMA, many other techniques have been proposed, for example, design for quality (DFQ), design for competitiveness (DFC), design for reliability, and many more. Many have even suggested that design for performance is just as important as DFMA. One cannot argue with this. However, DFMA is the subject that has been neglected over the years, while adequate consideration has always been given to the design of products for performance, appearance, etc.

FIG. 1.13   Improved assembly design efficiency results in increased reliability. (Courtesy of Motorola Inc.)

The other factors, such as quality, reliability, etc., will follow when proper consideration is given to the manufacture and assembly of the product. In fact, Fig. 1.13 shows a relationship between the quality of a design measured by the design efficiency obtained during DFA and the resulting product quality measured in defects per million parts assembled. Each data point on this graph represents a different product designed and manufactured by Motorola. It clearly shows that if design for assembly is carried out leading to improved design efficiencies, then improved quality will follow.

DFMA Leads to Products That Are More Difficult to Service

This is absolute nonsense. Experience shows that a product that is easy to assemble is usually easier to disassemble and reassemble. In fact, products that need continual service involving the removal of inspection covers and the replacement of various items should have DFMA applied even more rigorously during the design stage. How many times have we seen an inspection cover fitted with numerous screws, only to find that after the first inspection only two are replaced?

I Prefer Design Rules

There is a danger in using design rules because they can guide the designer in the wrong direction. Generally, rules attempt to force the designer to think of simpler-shaped parts that are easier to manufacture. In an earlier example, it was pointed out that this can lead to more complicated product structures and a resulting increase in total product costs. In addition, in considering novel designs of parts that perform several functions, the designer needs to know what penalties are associated when the rules are not followed. For these reasons the systematic procedures used in DFMA that guide the designer to simpler product structures and provide quantitative data on the effect of any design changes or suggestions are found to be the best approach.

I Refuse to Use DFMA

Although a designer will not say this, if the individual does not have the incentive to adopt this philosophy and use the tools available, then no matter how useful the tools or how simple they are to apply, the individual will see to it that they do not work. Therefore, it is imperative that the designer or the design team be given the incentive and the necessary facilities to incorporate considerations of assembly and manufacture during design.

1.4 WHAT ARE THE ADVANTAGES OF APPLYING DFMA DURING PRODUCT DESIGN?

Surveys taken at engineering design shows reveal, somewhat surprisingly, that reduction in product manufacturing cost is not necessarily considered to be the most desired outcome of redesign efforts. The example in Fig. 1.14 shows that reduced time to market and improved quality were thought to be more important than cost reduction.

Another advantage is that DFMA provides a systematic procedure for analyzing a proposed design from the point of view of assembly and manufacture.

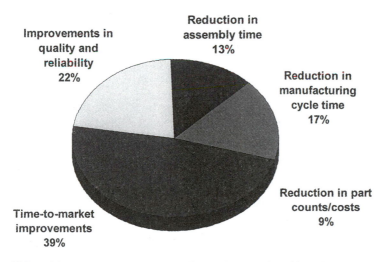

FIG. 1.14   Survey on importance of reductions produced by DFMA. (From Reader poll, *Computer-Aided Engineering*, June 1993.)

This procedure results in simpler and more reliable products that are less expensive to assemble and manufacture. In addition, any reduction in the number of parts in an assembly produces a snowball effect on cost reduction because of the drawings and specifications that are no longer needed, the vendors that are no longer needed, and the inventory that is eliminated. All of these factors have an important effect on overheads, which, in many cases, form the largest proportion of the total cost of the product.

DFMA tools also encourage dialogue between designers and the manufacturing engineers and any other individuals who play a part in determining final product costs during the early stages of design. This means that teamwork is encouraged and the benefits of simultaneous or concurrent engineering can be achieved.

The savings in manufacturing costs obtained by many companies who have implemented DFMA are astounding. As mentioned earlier, Ford Motor Company has reported savings in the billions of dollars as a result of applying DFMA to the original Ford Taurus line of automobiles. NCR anticipated savings in the millions of dollars as a result of applying DFMA to their new point-of-sales terminals. These are high-volume products. At the other end of the spectrum, where production quantities are low, Brown & Sharpe were able, through DFMA, to introduce their revolutionary coordinate-measuring machine, the MicroVal, at half the cost of their competitors, resulting in a multimillion dollar business for the company. These are but a few of the examples that show that DFMA really works.

1.5 TYPICAL DFMA CASE STUDIES

1.5.1 Defense Industry

Defense contractors have special difficulties in applying design for manufacture and assembly. Often, the designers do not know who will be manufacturing the product they are designing because the design will eventually go out for bid after it is fully detailed. Under these circumstances, communications between design and manufacturing are not possible. In addition, until recently, defense contractors have not had the normal incentives with regard to minimizing the final product cost. We have all heard horror stories regarding the ridiculously high cost of seemingly simple items such as toilet seats and door latches used by the military. This means that the defense industry in general is a very fertile area for the successful application of DFMA.

Figure 1.15 shows the original design of a reticle assembly for a thermal gunsight used in a ground-based armored vehicle [13]. The thermal gunsight is manufactured by the Defense Systems and Electronics Group of Texas Instruments (now Raytheon Systems). It is used to track and sight targets at night,

BEFORE

FIG. 1.15 Reticle assembly—original design. (Courtesy Texas Instruments, Inc.)

under adverse battlefield conditions, and to align the video portion of the system with the trajectory path of the vehicle's weapon to ensure accurate remote-controlled aiming. It makes steady, precise adjustments of a critical optical element, while handling ballistic shock from the vehicle's weapon systems and mechanical vibrations generated by the vehicle's engine and rough terrain. It must also be lightweight, as this is a major consideration for all such systems.

The results of a DFA analysis showed that fasteners and reorientations of the assembly were the two main contributors to the assembly time. Special operations for drilling and pinning couplers and applying adhesive to screws were also major contributors. The main objective during the redesign was to reduce hardware, eliminate unnecessary parts, standardize the remainder, and reduce or eliminate reorientations. Once the analysis had begun, several design alternatives were proposed within a matter of hours. Eventually, the best features of the alternative proposals were combined to produce a new design (Fig. 1.16).

The new design was analyzed using the design for assembly procedure, and Table 1.5 presents the results for the original design and for the redesign. It can be seen that impressive results were obtained in all aspects of the manufacture of this assembly. In the original design there were 24 different parts and in the new design only eight. This means that the documentation, acquisition, and inventory of 16 part types has been eliminated.

The U.S. Army's long-range advanced scout surveillance system is used to pinpoint far target locations, explains Paul Zimmermann [14], Principal Mechanical Engineer at Raytheon Systems Company, which was commissioned to build the unit. It uses an I-safe laser range finder, a global positioning system, forward-looking infrared sensors, a video camera, and a classified computer system to gather and display information on far-away targets during the day or night. According to Zimmermann, DFMA analysis simplified the final design through parts reduction and also quantified assembly times and costs. Raytheon also used an in-house six-sigma process to improve quality by avoiding defects.

"The major benefit of using DFMA with Six Sigma for this project was reducing the time and cost associated with repair and re-work" Zimmermann says. He estimates the method saved more than $2 million during the design phase.

1.5.2 Aerospace

David Eakin, DFMA specialist, has explained [15] that one of the most successful projects conducted by Bombardier Aerospace was the redesign of part of the engine nacelle on their Canadair Regional Jet series 200. A multidisciplinary DFMA team implemented a new design for one of the subassemblies on the torque box. They reduced the number of major parts from 110 to 86, the number

AFTER

SPRING
(2X)

CAM

RETICLE
CARRIAGE
SUBASSY

BUSHING
(3X)

SHAFT
(2X)

HOUSING

FIG. 1.16 Reticle assembly—new design. (Courtesy Texas Instruments, Inc.)

TABLE 1.5 Comparison of Original and New Designs of the Retical Assembly

	Original design	Redesign	Improvement (%)
Assembly time (h)	2.15	0.33	84.7
Number of different parts	24	8	66.7
Total number of parts	47	12	74.5
Total number of operations	58	13	77.6
Metal fabrication time (h)	12.63	3.65	71.1
Weight (lb)	0.48	0.26	45.8

Source: Texas Instruments, Inc. (Raytheon Systems).

of fasteners from 1090 to 916, and the assembly time from 126 to 96 h. The result will be recurring cost savings in excess of $450,000.

In the last few years military aircraft manufacturers have reported amazing successes with the application of DFMA. For example, McDonnell Douglas used DFMA software to help in the redesign of the F/A Hornet jet fighter [16]. The new aircraft is 25% larger than its predecessor, contains 42% fewer parts, and is 1000 lb below specification.

More recently, we have had reports of substantial improvements with the Apache Longbow helicopter [17]. In one of the several individual studies made during the redesign of the helicopter, high-speed machined parts were used to replace sheet metal angles and extruded stiffeners attached with rivets. In the new design the number of parts and fasteners was down from 74 to 9, the weight from 3 to 2.74 kg, fabrication time from 305 to 20 h, and manufacture/assembly time from 697 to 181 h. The cost savings were $43,000 per item and additional advantages of the new design were greater rigidity, greater stability, easier alignment, and reduced installation and inspection time.

1.5.3 Manufacturing Equipment

The need for a low-cost, high-accuracy coordinate measuring machine (CMM) was the impetus behind the development of the MicroVal personal CMM by Brown & Sharpe [18]. The primary design consideration was to produce a CMM that would sell for one-half the price of the existing product. The CMM was to compete with low-priced imports that had penetrated the CMM market to an even greater extent than imports had in the automotive industry. Since the CMM customer is not driven by price alone, the new CMM would have to be more accurate than the current design, while also being easier to install, use, maintain, and repair.

Brown & Sharpe started with a clean sheet of paper. Instead of designing the basic elements of the machine and then adding on parts that would perform specific functions required for the operation of the machine, it was decided to build as many functions into the required elements as feasible. This concept was called integrated construction. However, until the DFA methodology was applied, the cost objectives could not be met with the original design proposal. After DFA, for example, the shape of the Z-rail was changed to an elongated hexagonal, thus providing the necessary antirotation function. As a result, the number of parts required to provide the antirotation function was reduced from 57 to 4. In addition, the time required to assemble and align the antirotation rail was eliminated. Similar savings were made in other areas such as the linear displacement measuring system and the Z-rail counterbalance system. Upon introduction at the Quality Show in Chicago in 1988 the machine became an instant success, setting new industry standards for price and ease of operation.

1.5.4 Computers

Engineers from Dell Computer Corporation have described how they used several tools, including DFMA, to help with the design and manufacture of a new desktop computer chassis [19]. They achieved a 32% reduction in assembly time, a 44% reduction in service time, a part count reduction of 50%, and a direct labor cost reduction of 80%. All this translates into estimated savings close to $15 million.

Following a year-long competition for the nation's "outstanding example of applied assembly technology and thinking," *Assembly Engineering* magazine selected Bill Sprague of NCR Corporation, Cambridge, Ohio, as the PAT (productivity through technology) recipient [20]. Sprague, a senior advanced manufacturing engineer, was recognized for his contribution in designing a new point-of-sales terminal called the 2760. The DFA methodology, used in conjunction with solid modeling, assisted NCR engineers in making significant changes over the previous design. Those changes translated into dramatic reductions and savings as follows:

65% fewer suppliers
75% less assembly time
100% reduction in assembly tools
85% fewer parts

A total lifetime manufacturing cost reduction of 44% translated into millions of dollars. Indeed, Sprague estimated that the removal of one single screw from the original design would reduce lifetime product costs by as much as $12,500.

A multifunctional design team at Digital Equipments Corporation redesigned the company's computer mouse [21]. They began with the competitive benchmarking of Digital's products and mice made by other companies. They used DFMA software to compare such figures as assembly times, part counts, assembly operations, labor costs, and total costs of the products. They also consulted with hourly people who actually assembled the mice. Gordon Lewis, the DFMA coordinator and team leader, stated that DFMA gives the design team a "focal point so that they can pinpoint the problems from a manufacturing perspective and a design perspective." "It's the 80/20 rule," says Mr. Lewis. "You spend 80% of your time on 20% of your problems. DFMA is one of the tools that helps design teams identify the right 20% of the problems to work on," he says.

Figure 1.17 show the old and new mice. In the new DFMA design, 130 s of assembly time for a ball cage device has been reduced to 15 s for the device that replaced it. Other changes to the product structure also brought cost savings. For instance, the average of seven screws in the original mouse was reduced to zero by the use of snap fits. The new mouse also required no assembly adjustments, whereas the average number for previous designs was eight. The total number of

BEFORE   **AFTER**

FIG. 1.17   Digital's mouse.

assembly operations went from 83 in the old product down to 54 in the new mouse. All these improvements added up to a mouse that could be assembled in 277 s, rather than 592 s for the conventional one. Cycle time, too, was reduced by DFMA. A second development project that adhered to the new methodology was finished in 18 weeks, including the hard-tooling cycle. "That's unbelievable," admits Mr. Lewis, "normally it takes 18 weeks to do hard tooling alone."

1.5.5   Telecommunications

Design for assembly methods have been used at Motorola to simplify products and reduce assembly costs [22]. As part of the commitment to total customer satisfaction, Motorola has embraced the six-sigma philosophy for product design and manufacturing. It seemed obvious that simpler assembly should result in improved assembly quality. With these precepts in mind, they set about designing the new generation of vehicular adaptors.

The portable products division of Motorola designs and manufactures portable two-way Handi-Talkie radios for the landmobile radio market. This includes such users as police, firemen, and other public safety services, in addition to construction and utility fields. These radios are battery operated and carried about by the user.

The design team embraced the idea that designing a product with a high assembly efficiency would result in lower manufacturing costs and provide the high assembly quality desired. They also considered that an important part of any design process is to benchmark competitors' products as well as their own. At the time, Motorola produced two types of vehicular adaptors called Convert-a-Com (CVC) for different radio products. Several of their competitors also offered similar units for their radio products. The results of the redesign efforts are summarized in Table 1.6. Encouraged by this result, Motorola surveyed several products that had been designed using the DFA methodology to see if there might be a general correlation of assembly efficiency to manufacturing quality. Figure 1.13, introduced earlier, shows what they found. The defect levels are reported as defects per million parts assembled, which allows a quality evaluation independent of the number of parts in the assembly. Motorola's six-sigma quality goal is *3.4 defects per million parts assembled*. Each result in Fig. 1.13 represents a product having an analyzed assembly efficiency and a reported quality level. The results represent assembly efficiency improvements for a sequence of products over a period of several years. The vehicular adaptor represented by the symbol in the lower right of Fig. 1.13 was the first product to meet Motorola's six-sigma quality goal.

TABLE 1.6   Motorola's Redesign of Vehicular Adaptor

	Old product	New product	Improvement
DFA assembly efficiency (%)	4	36	800
Assembly time (s)	2742	354	87
Assembly count	217	47	78
Fasteners	72	0	100

Nortel product engineers in Electronic Systems Packaging utilized DFMA to redesign two products in the company's broadband portfolio, the S/DMS TransportNode OC-3 Express and the S/DMS TransportNode OC-192 ("OC" is "optical carrier"). The function of the TransportNode family of products is to carry voice, video, and data transmissions along an interconnected synchronous optical network (SONET) [23].

The original OC-3 Express was a successful design, but it did not reach the market soon enough. It offered more features and flexibility than the competitor's version and promised broader marker penetration, but the competitor soon lowered the price on its more mature product. In response, Nortel's product designers began to look for ways to reduce costs to match the competitor's tactic.

Dean Flockton, mechanical systems design engineer, explained that a DFA analysis of the original design showed that the hinged aluminum front cover of the box alone consisted of 53 parts and cost $78 to make. A majority of the parts were fasteners. The redesigned plastic cover with snaps consisted of only 17 parts and took only 95 s to assemble, versus 378 s for the original cover. The total parts cost for the new cover was $26, a savings of $52 per unit.

DFMA-guided redesign of the OC-3 Express shelf system resulted in a total cost of $136, a considerable reduction as compared to the original cost of $276. The expected savings to Nortel in assembly and manufacturing costs are estimated at $700,800 annually.

The TransportNode OC-192 consists of several shelves of components that provide device control, manage the fiber optics, and contain fans for cooling the unit. Installed on the transport shelf are a variety of circuit packs. Each transport shelf has enough space for ten circuit packs, which fit, like books into a bookshelf, into vertical slots about 2 in. wide, 12 in. tall, and 10 in. deep. Depending on the customer's requirements, fewer than ten active circuit packs may be installed. Filler packs are inserted into the vertical slots unoccupied by active circuit packs.

A filler pack contains a printed circuit board that plugs into the backplane of the unit, signaling to the operating system that a pack is in place and blocking electromagnetic interference that would otherwise radiate out the opening. Using the DFMA method, the design team was able to reduce the cost of a filler pack from $410 to $65 (U.S. dollar amounts). The total number of parts was reduced from 59 to 32, and the time to assemble each filler pack was cut by two-thirds, from 15 min to 5 min. The entire redesign process, from defining the functional requirements to producing the part, took only 10 months. The annual expected cost savings to Nortel was estimated to be $3.45 million.

1.5.6 Medical Equipment

When Ciba Corning Diagnostics Corp. set out to upgrade its line of blood-gas analyzers, it established aggressive cost and high-quality targets [24]. The new family of analyzers had to share a single base platform that could be upgraded at the customer's site to change with clinical demands. The analyzers also had to be durable, easy to assemble, and simple to use. Says David Yeo, new product manufacturing engineer at Ciba Corning, "Each design engineer was teamed with a manufacturing engineer throughout the project to focus on manufacturing and assembly methods. We set up an area where those who had input on the design could work together."

All assemblies were run through a DFMA analysis that prompted discussions about material selection and part fit. Some of the most significant changes made to the new model were in the hydraulic system in the upper caseworks. For example, the reagent manifold was made from a solid block of acrylic, eliminating the need for numerous tubes. The redesign reduced hydraulics parts content by 80%.

The upper caseworks presented a particular challenge. From the solid-model geometry, stereolithography was used to develop the first model, followed by rubber and cast urethane, and finally structural foam. "DFMA reduced total parts count, but the product still had to be highly functional with numerous features. Packing more functions into a single component presented a challenge to conventional prototyping methods. Three-dimensional solid modeling and stereolithography for rapid prototyping were able to meet that challenge," said Yeo.

DFMA analysis helped reduce the overall number of parts by 48%, fluidic connections by 60% and the cost by 22%. The designers also found trade-offs in molding snap fits and holes for screws or fasteners that helped reduce cost. The difference between the old 200 series and the new 800 series is dramatic. The new model contains fewer components, most of which are self-aligning. "We paid special attention to self-locating features. We didn't want to reorient any part more than we needed. Now, when assembling the 800 series, the pins match up to specific holes and parts fit mating parts," says Yeo.

1.5.7 Transportation

In the automotive industry the pressure to reduce prices is present at all levels, from the vehicle manufacturer through to the lowest level of the supply base. Most first- and second-tier suppliers are faced with contractual performance or productivity clauses that require annual downward price adjustments of 3–5% on the products they provide to the manufacturer.

Faced with these pricing constraints and continuing cost pressures, the suppliers and manufacturers alike have resorted to a multitude of formalized techniques designed to meet the challenge. These include GFD, Total Quality Management (TQM), Value Engineering (VE), Value Analysis (VA), DFM, and DFA. Some of these techniques have met with success while others have been tried briefly and dropped. Richard Johnson, Manufacturing Engineer, explains [25]:

at the Magna Interior Systems Seating Group, we have found the most powerful combination has been the melding of the disciplines of DFA and Value Engineering/Value Analysis.

At Magna Seating, we have used the DFA technique to strongly enhance our cost reduction efforts, not only in the area of metal parts, but also in the analysis of cut & sew products such as seat and armrest covers. We believe this second application of DFA is an industry first. We have developed our own formulas for each sewing operation and have verified the accuracy of the outcomes by comparing them with actual observations during visits to the various facilities.

An example that demonstrated the power of DFA to reduce parts and operations costs was a project to tool and build a jump seat for use in a major automotive manufacturer's pickup. A DFA analysis on the jump seat pointed out assembly problems and opportunities for change. The results were rather startling; the assembly consisted of 105 separate parts made of four basic materials types, formed using several different manufacturing techniques and assembled by welding, riveting, screwing, snapping, and swaging. Many of the parts were hidden from view until the assembly was actually cut apart revealing multiple layers of material (i.e., tubes inserted within tubes, multipiece plastic subassemblies, etc.). The resultant changes reduced part count from 105 to 19; major subassemblies were reduced to 5 and assembly time from 1445 s to 258 s.

Ford trained nearly 10,000 engineers in the DFA methodology and has contributed heavily to new research programs and to expanding existing DFMA tools. James Cnossen, Ford manager of manufacturing systems and operations research, has concluded that [26] "now it's part of the very fabric of Ford Motor Co." This is not surprising when Ford reports savings of over $1 billion as a result of applying DFA to the Taurus line of cars.

The Transmission and Chassis Division (T&C) is responsible for the design and manufacture of automatic transmissions of Ford vehicles. The transmission is a complex product, with approximately 500 parts and 15 model variations. The following describes the introduction and implementation of DFA.

1. Provide DFA overview for senior management.
2. Choose DFA champion/coordinator.
3. Define objectives.
4. Choose pilot program.
5. Choose test case.
6. Identify team structure.
7. Identify team members.
8. Coordinate training.
9. Have first workshop.

During the workshop:
1. Review the parts list and processes.
2. Break up into teams.
3. Analyze the existing design for manual assembly.
4. Analyze the teams' redesigns for manual assembly.
5. Teams present results of original design analysis versus redesign analysis.
6. Prioritize ideas, A, B, C,..., etc.
7. Incorporate all "A" and "B" ideas into one analysis.
8. Assign responsibilities and timing.

Results: The combined results of all of the workshops indicated a potential total transmission assembly savings as follows:

Labor minutes—29%
Number of parts—20%
Number of operations—23%

The cost benefits that have been gained since introduction of the DFA methodology in the T&C division are nothing less than staggering. Even more importantly, the changes resulting from DFA have brought substantial quality improvements. Moreover, the design lead time was reduced by one-half and would soon be halved again.

1.5.8 Consumer Products

Stefan Wohnhas, a DFA champion at Whirlpool notes [27] "It was important for us to develop a DFMA implementation strategy that would train people in using DFMA as a standard tool in the product development process." His objective was to focus the design teams on existing products and ongoing design projects so that participants would see measurable benefits from learning and applying the DFA methodology and the DFMA software.

Fortunately, an ideal in-house project presented both a new design and a benchmark. Whirlpool Sweden was planning the introduction of a new oven. Since the new oven would replace the then current model, the VIP 20, Whirlpool decided to benchmark the VIP 20 and use DFA analysis to create the new design for the new oven.

The VIP 20 is a 900-W microwave oven with the cavity-sealing quartz grill option for grilling meat. Controls are mechanical, a style preferred in Europe over touchpads on digital displays. Controls include seven presets, a 24-hour clock, and a 90-minute timer.

"The early involvement of the production engineers was new to us and very valuable" says Johan Dahm, a mechanical engineer in the Development Department. "Before their involvement, we discovered that we risked designing assembly problems into the oven in the concept phase. Now, the production engineers help to eliminate assembly problems before the design is finalized."

The design for the new oven is leaner than that of the VIP 20 despite adding forced convection. The model's original design encompassed 150 parts while the new oven is down to 106. This parts reduction coupled with an open assembly strategy and elimination of reorientations lowered assembly time by an estimated 26%.

Improved assembly methods and parts reduction have produced savings greatly exceeding expectations. Payback time for the entire DFA training project at Whirlpool was six months.

1.6   OVERALL IMPACT OF DFMA ON U.S. INDUSTRY

The preceding case studies describe but a few of the successes resulting from the application of DFMA software. A summary of results of case studies from various companies is presented in Fig. 1.18, which shows that the average part reduction is around 50% with some studies resulting in reductions in the range of 81 to 90%! Table 1.7 shows other improvements due to DFMA applications mentioned in the case studies. For example, 12 studies reported an average 37% reduction in product cost.

According to our records, over 800 different U.S. companies have implemented DFMA software since 1990. Of course, this figure does not indicate the

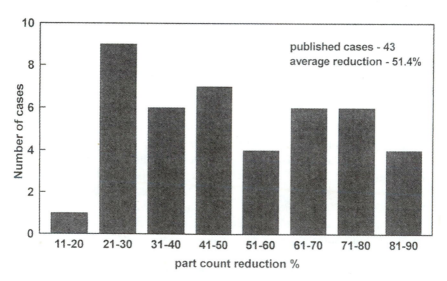

FIG. 1.18   Part count reductions from 43 published case studies where DFMA methods were used.

TABLE 1.7 DFMA Software Results from 117 Published Case Studies[a]

Category	No. of cases	Average reduction (%)
Part count	100	54
Assembly time	65	60
Product cost	31	50
Assembly cost	20	45
Assembly operations	23	53
Separate fasteners	20	60
Labor costs	8	42
Manufacturing cycle	7	63
Weight	11	22
Assembly tools	6	73
Part cost	8	52
Unique parts	8	45
Material cost	4	32
Manufacturing process steps	3	45
Number of suppliers	4	51
Assembly defects	3	68
Annual cost savings	11	$1,417,091

[a] From 56 companies as of April 1999.

number of individual users. The largest company, GM, reportedly has 1500 users, whereas others have only one. No other similar design methodology has had a significant impact on U.S. industry as a whole. In the early 1980s, the Hitachi Assembly Evaluation Method (AEM) was licensed by a few major manufacturers; almost all now use DFMA instead.

Of the Fortune 500 companies (those with annual revenues greater than about $3 billion) that are manufacturers, 60% have divisions or own companies that have licensed DFMA software. These represent about 38% of licensed manufacturers (Fig. 1.19).

The incentive to implement DFMA or concurrent engineering often arises because a company is under pressure to reduce manufacturing costs, improve time to market, etc., because of competitive pressures. This was certainly true of the U.S. computer equipment and transportation industries 20 years ago. More recently, the U.S. aerospace and defense industries have been under pressure to become more competitive. It is not surprising that these developments are reflected in the industries using DFMA software (Fig. 1.20).

Figure 1.21 shows the distribution of the case studies according to the type of product. These results give an indication of the level of activity in DFMA use

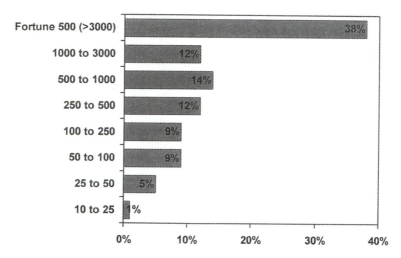

FIG. 1.19   Percent distribution of DFMA software users by company annual revenue.

during the past 15 years or so. Of course, not all of the successful applications of DFMA are publicized. Companies are often reluctant to reveal their developments—certainly not before the product in question appears on the market. Thus, the published case studies will form only a small proportion of the successes resulting from DFMA applications.

According to a recent survey conducted by the Society of Manufacturing Engineers [28], 92% of employers in mid-size manufacturing companies

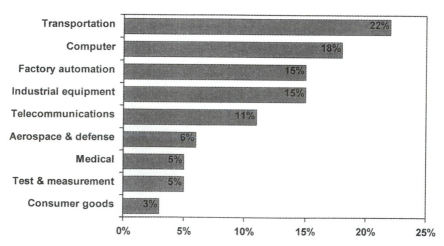

FIG. 1.20   Percent distribution of DFMA software users by industry.

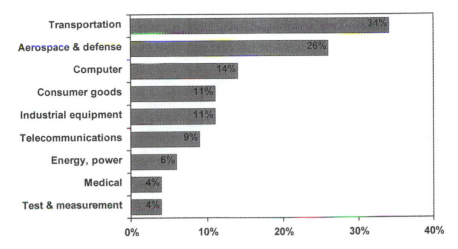

FIG. 1.21 Breakdown of published case studies where DFMA software was used.

($20–200 million annual sales) thought employees' skills were not adequate in design and concurrent engineering, whereas only 62% of employees thought there was a need for improvement. Incidentally, both groups agreed there was a significant need for improvement in employees' knowledge of computers and information technology. The same study investigated the make-up of design teams. Figure 1.22 shows that manufacturing is nearly always represented.

We have made attempts to investigate how the benefits of DFMA can be realized by small manufacturers. According to estimates there are more than 380,000 manufacturing companies with fewer than 500 workers. These companies employ more than 12 million people, over 65% of the U.S. manufacturing workforce, and account for more than $185 billion in payroll. Small manufacturers produce over half of all U.S.-manufactured goods. Many of these need assistance in upgrading their capabilities to meet the needs of global competition.

In 1992, U.S. Undersecretary of Commerce, Robert White, announced a Department of Commerce initiative—called The Analysis and Educational Tool Project—which would rely on our DFMA software tools to benchmark the way small companies make existing products. If analysis showed that a company's manufacturing process needed upgrading, the program would help the firm locate needed equipment and necessary education/training resources.

Accordingly, we conducted the pilot study with the objective of testing whether DFMA tools could be applied in small businesses through regional technology centers. We first provided training for representatives of eight centers, who were then to carry out the following:

**The midsize manufacturers Grant Thornton surveyed pull many areas into
their design teams, with the folks from manufacturing almost always on board.**

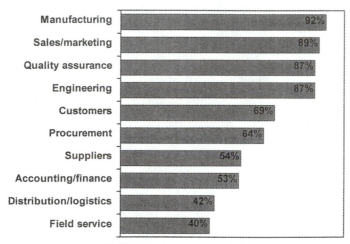

Multiple responses accepted

FIG. 1.22   Who's on the design team?

1. Test the utility of the DFMA tools in providing a "should cost analysis" for
    specimen components produced by the company and thereby measuring the
    efficiency of the company's manufacturing process.
2. Test the use of the DFMA tools as diagnostics to indicate whether the
    company has a problem.
3. Test whether DFMA can guide the company in acquiring the equipment that
    will improve their competitiveness.
4. Increase awareness of the importance of the design of a product in
    determining its final cost.

The general conclusion from these studies was that the DFMA programs were
very applicable to the needs of small manufacturers. With appropriate tailoring of
the machine databases the programs can provide good estimates of time and cost
for a manufacturing job. The greatest benefit resulting from application of these
programs appeared to be the introduction of a systematic approach for analysis of
the design and manufacturing process. Unfortunately, after the change in
administration at the White House, the program was discontinued.

Also in 1992 another pilot study was carried out by the Diecasting Development Council, who used DFMA to provide a free "costing service" that gave
designers and companies an estimate of the cost of using diecasting to make parts
that are currently being made of sheet metal or molded plastic. To help this study

the Diecasting Development Council offered to provide free in-booth diecast conversion computer cost analysis at the National Design Engineering Show in Chicago. This initiative also failed, and the problem of how to help small manufacturers reap the benefits of DFMA still remains.

It was mentioned earlier that there was initially great interest in software versions of DFMA. It was also stated that U.S. engineers preferred to use computers for DFMA analyses. Experience now shows that use of software was the right approach. DFMA is not only a procedure, it is knowledge—so much knowledge that it cannot be satisfactorily presented in handbook form. Those who decided that they could manage with a handbook and thought they did not need the more sophisticated assistance offered by computer software have generally not succeeded. It is worth pointing out that the numerous impressive published case studies have arisen from those who have implemented the software version of DFMA. Again, the objective is to put manufacturing knowledge in the hands of the concurrent engineering design team, and the most efficient way to do this is with software.

1.7 CONCLUSIONS

It is clear that the use of DFMA software has a tremendous impact where properly applied in a concurrent engineering environment. Our experience in running DFMA design workshops has shown that significant ideas for design improvements can invariably be made in the space of only a few hours. Unfortunately, small manufacturers have not been able to take advantage of the huge potential of DFMA; they do not have the manpower or the necessary experience. Even with large companies, proven successes in one division do not necessarily spread to other divisions without management support and commitment. However, we remain encouraged by the continuing reports of substantial product improvements and often staggering cost savings obtained by both high- and low-volume manufacturers in the United States.

It should be noted that in all of the case studies mentioned, a systematic step-by-step DFMA analysis and quantification procedure was used. However, as pointed out earlier, it is still claimed by some that design rules or guidelines (sometimes called producibility rules) by themselves can give similar results. This is not so. In fact, guidelines or qualitative procedures can lead to increased product complexity because they are usually aimed at simplifying the individual component parts, resulting in a design that has a large number of parts and poor quality and involves greater overheads due to larger inventory, more suppliers, and more record keeping. Rather, the objective should be to utilize the capabilities of the individual manufacturing processes to the fullest extent in order to keep the product structure as simple as possible.

In spite of all the success stories, the major barrier to DFMA implementation continues to be human nature. People resist new ideas and unfamiliar tools, or claim that they have always taken manufacturing into consideration during design. The DFMA methodology challenges the conventional product design hierarchy. It reorders the implementation sequence of other valuable manufacturing tools, such as SPC and Taguchi methods. Designers are traditionally under great pressure to produce results as quickly as possible and often perceive DFMA as yet another time delay.

In fact, as the preceding case studies have shown, the overall design development cycle is shortened through use of early manufacturing analysis tools, because designers can receive rapid feedback on the consequences of their design decisions where it counts—at the conceptual stage.

One hears a lot these days about concurrent or simultaneous engineering. In some people's minds, simultaneous engineering means gathering together designers, manufacturing engineers, process monitors, marketing personnel, and the outside "X factor" person. Working with teams at the predesign stage is a laudable practice and should be undertaken in every company. But unless one can provide a basis for discussion grounded in quantified cost data and systematic design evaluation, directions will often be dictated by the most forceful individual in the group, rather than being guided by a knowledge of the downstream results. The Portable Compressor Division of Ingersoll-Rand has used various aspects of simultaneous engineering for the past 10 years. However, the introduction of DFMA in 1989 as a simultaneous engineering tool served as a catalyst that provided dramatic increases in productivity and reduced new development time from 2 years to 12 months.

In conclusion, it appears that in order to remain competitive in the future, almost every manufacturing organization will have to adopt the DFMA philosophy and apply cost quantification tools at the early stages of product design.

REFERENCES

1. Boothroyd, G., Automatic Handling of Small Parts, Proc. International Production Engineering Conference, Tokyo, 1974, p. 475.
2. Boothroyd, G., and Ho, C., Coding System for Small Parts for Automatic Handling, SME paper ADR76-13, presented at Assemblex III, Chicago, October 1976.
3. Boothroyd, G., Design for Producibility: The Road to Higher Productivity, Assembly Eng., March 1992.
4. Womak, J.P., Jones, D.T., and Roos, D., The Machine that Changed the World, Macmillan, New York, 1990.
5. Pahl, G., and Beitz, W., Engineering Design, English Edition, The Design Council, London, 1984.
6. Dewhurst, P., and Boothroyd, G., Early Cost Estimating in Product Design, J. Manuf. Syst., Vol. 7, No. 3, 1988.

7. Boothroyd, G., and Dewhurst, P., Computer-Aided Design for Manufacture, Manuf. Eng., April 1988.
8. Boothroyd, G., Dewhurst, P., and Knight, W.A., Research Program on the Selection of Materials and Processes for Component Parts, Int. J. Adv. Manuf. Technol., Vol. 6, 1991.
9. Bauer, L., Team Design Cuts Time, Cost, Welding Des. Fabricat., September, 1990, p. 35.
10. Munro and Associates, Inc., 911 West Big Beaver Road, Troy, MI 48084.
11. Williams, R.A., Concurrent Engineering Delivers on Its Promise: Hewlett Packard's 34401A Multimeter. In S.G. Shina, ed., Successful Implementation of Engineering Products and Processes, Van Nostrand, New York, 1994.
12. Boothroyd, G., and Dewhurst, P., Product Design for Assembly, Boothroyd Dewhurst, Inc., Wakefield, RI, 1990.
13. DFMA Optimizes Reticle Assembly, DFMA Insight, Vol. 1, No. 3, Boothroyd Dewhurst, Inc.
14. Vermillion, L., Out of Harm's Way, Prod. Des. Develop., April 1999.
15. Product Design for Manufacture and Assembly, Engineering, January 1998.
16. Flannery, W., Assembly System May Set Standard for Lowest Cost, St. Louis Post-Dispatch, October 1994.
17. Parker, M., Mission: Possible—Improving Design and Reducing Costs at Boeing Helicopter, IIE Solutions, December 1997.
18. McCabe, W.J., Maximizing Design Efficiencies for a Coordinate Measuring Machine, Des. Insight, Vol. 1, No. 1, Boothroyd Dewhurst, Inc., Wakefield, RI, 1988.
19. Dewey, D.M., and Keup, B.S., DFX: A Dell Success Story, International Forum on DFMA, Newport, RI, June 1999.
20. NCR's Bill Spague Lands PAT Design Award, Des. Insight, Vol. 1, No. 2, Boothroyd Dewhurst, Inc., Wakefield, RI, 1988.
21. Digital Builds a Better Mouse Trap, DFMA Insight, Vol. 1, No. 5, Boothroyd Dewhurst, Inc., Wakefield, RI, 1990.
22. Branan, B., Six-Sigma Quality and DFMA, DFMA Insight, Vol. 2, No. 1, Boothroyd Dewhurst, Inc., Wakefield, RI, 1991.
23. Flockton, D., It's All in the Packaging, Mech. Eng., February 1999.
24. Development of Blood-Gas Analyzers, Benefits from DFMA Software, Med. Prod. Manuf. News, March 1997.
25. Design for Manufacture and Assembly: Auto Suppliers Use DFMA to Impact Product Cost and Quality and Gain a Competitive Edge, Automotive Eng., December 1997.
26. Burke, G.J., and Carlson, J.B., DFA at Ford Motor Company, DFMA Insight, Vol. 1, No. 4, Boothroyd Dewhurst, Inc., Wakefield, RI, 1989.
27. Trained Teams/Lean Design, Appl. Manuf., July 1999.
28. Owen, J.V., Best Practices, Best Plants, Manuf. Eng., SME, Vol. 122, No. 4, 1999.

2

Selection of Materials and Processes

2.1 INTRODUCTION

An integral part of design for manufacture is the systematic early selection of material and process combinations for the manufacture of parts, which can then be ranked according to various criteria. Unfortunately, designers tend to conceive parts in terms of the processes and materials with which they are most familiar, and they may, as a consequence, exclude from consideration processes and process/material combinations that might have proved to be more economic. Opportunities for major manufacturing improvements may be lost through such limited selections of manufacturing processes and the associated materials in the early stages of product design. This can be well illustrated by the results of a survey of designers' knowledge of manufacturing processes and materials carried out in Britain [1]. This survey covered a wide range of design offices in various sectors of industry. For manufacturing processes (Fig. 2.1), more than half of those surveyed professed little or no knowledge of metal extrusion, two-thirds knew little about glass-reinforced molding, and over three-quarters were uninformed about plastic extrusion, technical blow molding, and sintering. For less common processes, such as hot isostatic pressing, outsert molding, and superplastic forming, the percentage of designers claiming some process knowledge was only 6, 7, and 8, respectively. Similar results were found for materials, and Fig. 2.2 illustrates designers' knowledge about a range of polymeric materials. This again shows a surprising lack of familiarity with some commonly used materials. The overall implication of these findings is that because material and process combinations are likely to be chosen from those with which designers are

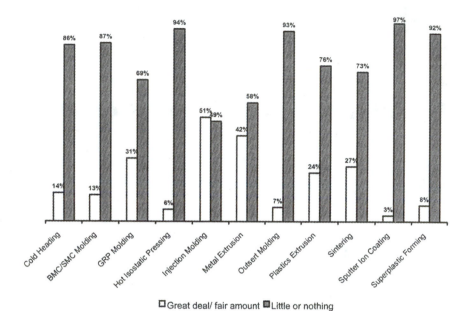

FIG. 2.1 Survey of designers' knowledge of manufacturing processes. (Adapted from Ref. 1.)

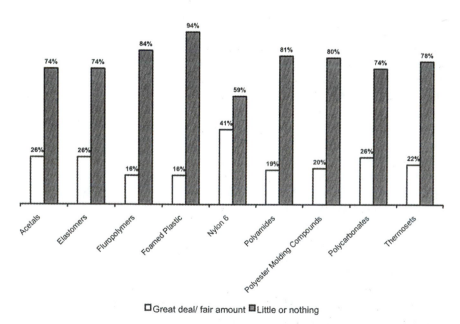

FIG. 2.2 Survey of designers' knowledge of polymer materials. (Adapted from Ref. 1.)

most comfortable, the possibilities of using other processes that may be much more cost effective may be missed.

2.2 GENERAL REQUIREMENTS FOR EARLY MATERIALS AND PROCESS SELECTION

In order to be of real design value, the information on which the initial selection of material/process combinations and their ranking is to be based should be available at the early concept design stage of a new product. Such information might include, for example:

Product life volume
Permissible tooling expenditure levels
Possible part shape categories and complexity levels
Service or environment requirements
Appearance factors
Accuracy factors

It is important to realize that for many processes the product and process are so intimately related that the product design must use an anticipated process as a starting point. In other words, many design details of a part cannot be defined without a consideration of processing. For this reason, it is crucial that an economic evaluation of competing processes be performed while the product is still at the conceptual stage. Such an early evaluation will ensure that every economically feasible process is investigated further before the product design evolves to a level where it becomes process specific.

As a design progresses from the conceptual stage to production, different methods can be used to perform cost modeling of the product. At the conceptual stage, rough comparisons of the costs of products of similar size and complexity may be sufficient. While this procedure contains a certain degree of uncertainty, it only requires conceptual design information and is useful for the purpose of early economic comparison. As the design progresses and specific materials and processes are selected, more advanced cost modeling methods may be employed. These may be particularly useful in establishing the relationship between design features and manufacturing costs for the chosen process. The basis of several cost estimation procedures for different processes is outlined in later chapters.

2.2.1 Relationship to Process and Operations Planning

There is an obvious relationship between the initial selection of process/material combinations and process planning. During process planning the detailed elements of the sequence of manufacturing operations and machines are determined. It is at this stage that the final detailed cost estimates for the manufacture of the part are determined. Considerable work has been done in the area of computer-aided process planning (CAPP) systems [2–4], although closer examination shows that the majority of this work has been devoted to machining processes only. These systems are utilized after detailed design of the part has been carried out, with an implied manufacture process. The initial decision on the material and process combination to be used for the part is most important, as this will determine the majority of subsequent manufacturing costs. The goal of systematic early material and process selection is to influence this initial decision on which combination to use, before detailed design of the part has been carried out and before detailed process planning is attempted.

2.3 SELECTION OF MANUFACTURING PROCESSES

The selection of appropriate processes for the manufacture of a particular part is based upon a matching of the required attributes of the part and the various process capabilities. Once the overall function of a part is determined, a list can be formulated giving the essential geometrical features, material properties, and other attributes that are required. This represents a "shopping list" that must be filled by the material properties and process capabilities. The attributes on the "shopping list" are related to the final function of the part and are determined by geometric and service conditions.

Most component parts are not produced by a single process, but require a sequence of different processes to achieve all the required attributes of the final part. This is particularly the case when forming or shaping processes are used as the initial process, and then material removal and finishing processes are required to produce some or all of the final part features. Combinations of many processes are used, and this is necessary because a single process cannot in general provide all of the attributes of the finished part. However, one of the goals of DFMA analysis is product structure simplification and parts consolidation. Experience shows that it is generally most economical to make the best use of the capabilities of the initial manufacturing process in order to provide as many of the required attributes of a part as possible.

There are hundreds of processes and thousands of individual materials. Moreover, new processes and materials are being developed continually. Fortunately, the following observations help to simplify the overall selection problem:

1. Many combinations of processes and materials are not possible. Figure 2.3 shows a compatibility matrix for a selected range of processes and material types.
2. Many combinations of processes are not possible and, therefore, do not appear in any processing sequences.

FIG. 2.3 Compatibility between processes and materials.

3. Some processes affect only one attribute of the part, particularly surface treatment and heat treatment processes.
4. Sequences of processes have a natural order of shape generation, followed by feature addition or refinement by material removal and then material property enhancement.

Processes can be categorized as:

Primary processes
Primary/secondary processes
Tertiary processes

Some texts refer to primary processes as those used for producing the raw materials for manufacturing such as flat rolling, tube sinking, and wire drawing. In the context of producing component parts in this text, the term primary process will refer to the main shape-generating process. Such processes should be selected to produce as many of the required attributes of the part as possible and usually appear first in a sequence of operations. Casting, forging, and injection molding are examples of primary shape-generating processes.

Primary/secondary processes, on the other hand, can generate the main shape of the part, form features on the part, or refine features on the part. These processes appear at the start or later in a sequence of processes. This category includes material removal and other processes such as machining, grinding, and broaching.

Tertiary processes do not affect the geometry of the part and always appear after primary and primary/secondary processes. This category consists of finishing processes such as surface treatments and heat treatments. Selection of tertiary processes is simplified, because many tertiary processes only affect a single attribute of the part. For instance, lapping is employed to achieve a very good surface finish, and plating is often used to improve appearance or corrosion resistance.

2.4 PROCESS CAPABILITIES

A great deal of general information is available on manufacturing processes in a wide range of textbooks, handbooks, and so on. Each process can be analyzed to determine the range of its capabilities in terms of attributes of the parts that can be produced. Included in these capabilities are shape features that can be produced, natural tolerance ranges, surface roughness capabilities, and so on. These capabilities determine whether a process can be used to produce the corresponding part attributes. Table 2.1 shows some of the general capabilities of a range of commonly used processes that can be used as a guide to selection.

The shape-generating capabilities of the range of the processes characterized in Table 2.1 are shown in Table 2.2, with the various terms used in this table defined as follows.

General Shape Attributes

Depressions (Depress): The ability to form recesses or grooves in the surfaces of the part. The first column entry refers to the possibility of forming depressions in a single direction, while the second entry refers to the possibility of forming depressions in more than one direction. These two entries refer to depressions in the direction of tooling motion and those in other directions. The following are some examples of tooling motion directions. Processes with split molds—the direction of mold opening.

Processes that generate continuous profiles—normal to the direction of extrusion or normal to the axis of the cutting medium.

Forging (impact) processes—the direction of impact of the tooling onto the part.

Uniform wall (UniWall): Uniform wall thickness. Any nonuniformity arising from the natural tendency of the process, such as material stretching or build-up behind projections in centrifugal processes is ignored, and the wall is still considered uniform.

Uniform cross section (UniSect): Parts where any cross sections normal to a part axis are identical, excluding draft.

Axis of rotation (AxisRot): Parts whose shape can be generated by rotation about a single axis: a solid of revolution.

Regular cross section (RegXSec): Cross sections normal to the part's axis contain a regular pattern (e.g., a hexagonal or splined shaft). Changes in shape that maintain a regular pattern are permissible (e.g., splined shaft with a hexagonal head).

Captured cavities (CaptCav): The ability to form cavities with reentrant surfaces (e.g., a bottle).

Enclosed (Enclosed): Parts that are hollow and completely enclosed.

Draft-free surfaces (NoDraft): The capability of producing constant cross sections in the direction of tooling motion. Many processes can approach this capability when less than ideal draft allowances are specified, but this designation is reserved for processes where this capability is a basic characteristic and no draft can be obtained without cost penalty.

DFA Compatibility Attributes

Manufacturing processes have varying levels of compatibility with the basic goals of DFA of simplified product structure and ease of assembly. This relative compatibility in Table 2.2 is measured in the following key areas.

TABLE 2.1 Capabilities of a Range of Manufacturing Processes

Process	Part size	Tolerances[a]	Surface finish	Shapes produced competitively[b]
Sand casting	Weight: 0.2 lb–450 ton Min. wall: 0.125 in.	General: ± 0.02 (1 in.), ± 0.1 (24 in.) For dimensions across parting line add ± 0.03 (50 in.^{2}), ± 0.04 (200 in.^{2})	500–1000 μin.	Large parts with walls and internal passages of complex geometry requiring good vibration damping characteristics
Investment casting	Weight: 1 oz–110 lb Major dimension: to 50 in. Min. wall: 0.025 (ferrous), 0.060 (nonferrous)	General: ± 0.002 (1 in.) ± 0.004 (6 in.)	63–25 μin.	Small intricate parts requiring good finish, good dimensional control, and high strength
Die casting	Min. wall (in.): 0.025 (Zn), 0.05 (Al, Mg) Min. hole dia. (in.): 0.04 (Zn), 0.08 (Mg), 0.1 (Al) Max. weight (lb): 35 (Zn), 20 (Al), 10 (Mg)	General: ± 0.002 (1 in.) ± 0.005 (6 in.) (Zinc) ± 0.003 (1 in.), ± 0.006 (6 in.) (Alum, Mg) Add ± 0.004 across parting line or moving core	32–85 μin.	Similar to injection molding
Injection molding (thermoplastics)	Envelope: 0.01 in.^{3}–80 ft^3 Wall: 0.03–0.250 in.	General: ± 0.003 (1 in.), ± 0.008 (6 in.) Hole dia.: ± 0.001 (1), ± 0.002 (1 dia.) Flatness: ± 0.002 in./in. Increase tol. 5% for each additional mold cavity Increase tolerance: ± 0.004 for dimensions across parting line	8–25 μin.	Small-to-medium sized parts with intricate detail and good surface finish
Structural foam molding	Weight: 25–50 lb Wall: 0.09–2.0 in.	Approximately that of injection molding	Poor; paint generally required properties	Large, somewhat intricate parts, requiring high stiffness and/or thermal or acoustical insulating properties
Blow molding (extrusion and injection)	Envelope: Up to 800 gal containers (105 ft^3) Wall: 0.015–0.125 in.	General: ± 0.02 (1 in.), ± 0.04 (6 in.) Wall: $\pm 50\%$ of nominal wall Neck: ± 0.004 (injection only)	250–500 μin.	Hollow, well-rounded thin-walled parts with low degree of asymmetry
Rotational molding	Envelope: Up to 5000 gal containers (670 ft^3) Wall: 0.06–0.40 in.	General: ± 0.025 (1 in.) ± 0.05 (6 in.), ± 0.01 (24) Wall: ± 0.015	Poor; parts generally textured	Large containers with minimal detail
Impact extrusion (forward and backward)	Dia.: 0.075–2.5 in. Length: 3–24 in.	O.D.: ± 0.002 (0.5 in.) I.D.: ± 0.003 (5 in.) Bottom dia: ± 0.005 (5 in.) Tolerances approximately 50% greater for rectangular parts	20–63 μin.	Approx. 1–2 in. dia. part with a closed end thicker than side walls (backward extrusion) Headed parts with large L/D ratio and zero draft (forward extrusion) Comb. of forward/backward common
Cold heading	Shank dia.: 0.03–2.0 in. Length: 0.6–9.0 in.	Head height: ± 0.006 (0.025 shank dia.), ± 0.008 (0.50 shank dia.) Head dia. ± 0.01 (0.25 shank dia., ± 0.018 (0.50 shank dia.) Length: ± 0.03 (1 in.)	32–85 μin.	Small symmetrical, or near symmetrical, headed cylindrical parts, with shank length greater than shank dia.
Hot forging (closed die)	Weight: 0.1–500 lb	Perpendicular to die motion: $\pm 0.7\%$ of dimension Parallel to die motion: ± 0.03 (10 in.^{2} area), ± 0.12 (100 in.^{2} area)	125–250 μin.	Parts of moderate complexity, in a wide range of sizes, whose failure in service would be catastrophic

TABLE 2.1 Continued

Process limitations	Typical application	Mat'ls[c]	Comments
Secondary machining usually required Production rates often lower than that for other casting processes Tolerances, surface finish coarser than other casting processes Requires generous draft (approx. 3 deg) and radii (approx. equal to thickness)	Engine blocks Engine manifolds Machine bases Gears Pulleys	1, 2, 3 4, 5, 6 7[d], 8, 12	Very flexible manufacturing process in terms of possible geometries, part size, and possible materials Pattern in reusable and mold expendable
Most investment castings are less than 12 in. long and less than 10 lb L/D ratio of through or blind holes less than 4 : 1 and 1 : 1, respectively Tooling cost and lead time generally greater than for other casting processes except die casting	Turbine blades Burner nozzles Armament components Lock components Sewing machine components Industrial handtools bodies	2, 3, 4[d] 5, 6, 8 12	Expendable pattern and mold Greater flexibility in material choices or part geometry than die casting, but much higher production costs Less susceptible to porosity than most casting process Multiple parts may be cast simultaneously around central sprue
Trimming operations required for flash and overflow removal Porosity can be present Die life limited to approximately 200 k shots in Al or Mg or 1 million in Zn	Similar to injection molding in part geometry, but particularly suited where higher mechanical properties or the absence of creep are required	5, 6[d], 7 8	Produces thinnest walls of all casting processes Production rate approximately 100 parts/h in alum and approximately 200 parts/h in zinc Tooling cost and lead time similar to that for injection molding but trimming and surface treatment can make process less economic
Tooling is costly and requires greater lead time than most alternative processes Poor design can result in high levels of molded-in stress, resulting in warpage or failure	Numerous applications, often replacing die casting or sheet metal assemblies	10, 11	Typical cycle time 20–40 s Details such as living hinges, insert molding and snap features allow significant opportunity for part consolidation Injection molding of thermoset materials also possible: longer cycle time, no reprocessing of waste, generally harder, more brittle, but more stable material which can be used at higher service temperatures
Details as sharp as those of injection molding not possible Cycle time is long (2–3 min)	Pallets, housing, drawers, TV cabinets, fan shrouds	10	Tooling approximately 20% less than for injection molding Solid skin approximately 0.03–0.8 in. thick; entire wall cross section has densities between 50% and 90% of solid weight Process generates a low level of internal stress RIM is a similar foaming process utilizing thermosets (generally polyurethane)
With extrusion blow molding, somne geometries produce a high level of material scrap Integral handles possible with extrusion blow molding only Poor control of wall thickness	Most polymer containers to 5 gal Toys Auto heater ducting	10	Injection blow molding: smaller parts, more accurate necks Extrusion blow molding: more asymmetrical parts, less costly tooling High production rates, particular for injection blow molding (as low as 10 s per cycle)
Abrupt wall changes, long, thin projections, and small separations between opposing part surfaces not possible	Toys Containers	10	Cycle time 8–20 min Inserts for securing or stiffening are possible Less detail possible than with blow molding
Flat inner bottom requires additional operation Tooling costs are high Maximum L/D ratio for backward extrusion is 10 (in some aluminum alloys) L/D ratio almost unlimited in forward extrusion Tolerances not as good as machining	Fasteners Sockets for socket wrench Gear blanks with shank	2, 3[d], 5 6, 7, 8	Generally chosen over screw machined part if material savings are significant (approximately 25% or more) Significant improvement in mechanical properties due to cold working, allowing further material reduction Limited asymmetry possible
Seldom used for diameters greater than 1.25 in. Must allow much more generous radii than with machining Significant asymmetry difficult	Nails Fasteners Spark plug pot Ball joint Shafts	2, 3, 4[d] 5, 6, 12[d]	Minimization of shank diameter and upset volume important Production rates 35–120 parts/min Process can also be carried out warm (800–1200°F)
Holes may not be produced directly Flash must be removed and secondary machining is often required Die wear and die mismatch can be significant Generous draft angles and radii are suggested	Crankshafts Airframe components Tools Nuclear components Agricultural components	2, 3, 4 5, 6, 8 9, 12[d]	By controlling material flow, grain structure may be applied with the direction of principal stress Closed die forgings nearly always pass through series of impressions before completion In decreasing order of forgability: Al, Mg, steel, St steel, titanium, high-temperature

TABLE 2.1 Continued

Process	Part size	Tolerances[a]	Surface finish	Shaped produced competitively[b]
Pressing and sintering (power metal parts)	Min. wall: 0.06 in. Min. hole dia. 0.06 in. Max. length (in direction of press): 4.0 in. Max. projected area: 40 in.^{2}	Perpendicular to press direction: ±0.15% of dimension (±0.05% if repressed) Parallel to press direction: ±0.30% of dimension	8–50 μin.	Small parts of uniform height with parallel, but fairly intricate walls
Rotary swaging	Dia.: 0.01–5.0 in. (bar), 14 in. (tubing)	Dia.: ±0.003 (1 in.)	20% of original stock finish (5-fold improvement)	Tapered cylindrical rod or tubing
Hot extrusion	Cross-sectional area: 0.1–225 in.^{2} (Alum), 0.5–4.0 in.^{2} (LC steel) Min. wall: 1.5% of circumscribed dia.	General: ±0.01 (1 in.) ±0.03 (6 in.) (±0.005 if cold drawn after extrusion) Angles ±2 deg. Twist: 1 deg. per foot for width less than 2 in. Flatness: 0.004 in./in.	63 μin. (Alum), 125 micro- includes (LC steel)	Straight part with constant cross section that is fairly complex, but balanced, without extreme change in wall thickness
Machining (from stock)	Limited only by machine capability	Turning ±0.001, boring ±0.0005, Milling ±0.002, Drilling ±0.008–0.002, Broaching ±0.005, Grinding + 0.002 (dia.); ±0.008 (surface), Reaming, ±0.001 (all for dimensions of 1 in.)	Turning 63–125 Boring 32–125 Milling 63–125 Drilling 63–250 Grinding 8–32 Reaming 63	Rotational: Axisymmetrical part with L/D ratio of 3 or less and major dia. of 2 in. or less Nonrotational: Rectangular part with all feature parallel and open in the same direction
Electrochemical machining (ECM)	Min. hole diam.: 0.01 in. Max. hole depth: 50 × dia.	General: ±0.001	8–63 μin.	Highly accurate complex, or finely detailed shapes in hardened materials or those susceptible to damage due to heat build-up Production of high aspect or burr-free holes and processing of flimsy materials
Electrical discharge machining (EDM)	Min. hole dia.: 0.002 Min. slot width: 0.002 in.	General: ±0.001	8–250 μin. (dependent on removal rate)	Same as ECM
Sheet metal stamping/ bending	Material thickness 0.001–0.75 in. (normally 0.050–0.375 in.) Area: 80 ft^2 with turret press and press brake, 10 ft^2 with die sets	Punching or stamping: ±10% of mat'l. thickness (2.0 in.) Press brake: ±2 deg. on bend, ±0.015 in. hole-to-bend	For cold rolled sheet or coil: 32–125 μin.	Moderate complexity parts of constant material thickness with flanges in a single direction
Thermoforming	Area: 1 in.^{2}–300 ft^2	General: ±0.05% of dimension Wall: ±20% of nominal	60–120 μin.	Large, shallow, thin wall parts with generous radii
Metal spinning	Dia.: 25 in.–26 ft Mat'l. thickness: (Alum), 0.004–1.5 (LC steel), (0.025–0.05 in. most common)	Dia.: ±0.01 (1 in.), ±0.03 (24 in.) Angle: ±3 deg	32–65 μin.	Thin-walled conical shape with diameter greater than twice depth

[a] Limits shown represent fine tolerances. More stringent requirements will significantly increase cost. [b] Part types that can be produced cost effectively in comparison to other processes. [c] Materials. [d] Used on a limited basis: 1, cast iron; 2, carbon steel; 3, alloy steel; 4, stainless steel;

TABLE 2.1 Continued

Process limitations	Typical application	Mat'ls.[c]	Comments
Generally lower mechanical properties than wrought metals Undercuts, off-axis holes, and threads cannot be produced directly Thin sections and feature edges should be avoided Max. L/D ratio approximately 3	Small gears Lock mechanisms components Small arms parts Filters Bearings	1, 2, 3 4, 5, 6 9[d], 12[d]	Production rates approximately 700 parts/h Impregnation with lubricants gives self-lubrication properties Density range 75%–95% (compared to raw material) Maximum compression ratio (powder volume before and after pressing and sintering) approximately 2.5 : 1
Taper should be 6 degrees or less included angle for manual feeding and up to 14 degrees for power feed Shoulders perpendicular to part axis not possible	Tube: Gold club shafts, table legs, exhaust pipes Bar: Punches, screwdriver blades	2, 3, 4 5, 6, 7[d] 8[d], 12	Tooling costs are generally less than those for cold extrusion or cold heading Noncylindrical part can be swaged in stationary die machines Production rates can range from 100–3000 parts/h Shapes like splines can be produced by swaging tubing over an internal former called a mandrel
Dimensional accuracy and part-to-part consistency generally not as high as competing processes. Warp and twist can be troublesome Use of materials other than aluminum and copper alloys can cause some shape restrictions Avoid knife edges and long, unsupported projections	Heatsinks Structural corner and edge members Decorative trim	2, 3[d], 4[d] 5, 6, 7[d] 8, 9[d]	Plastic working produces favorable grain structure Maximum extrusion ratios are 40 : 1 (Alum), 5 : 1 (LC steel) Shorter setup time than rolling, but a lower production rate (1–8 ips) crossover point at approximately 50,000 ft Low tooling costs, therefore short runs can often be justified if part consolidation and integral fastening is considered.
Little opportunity for part consolidation Most parts produced by a sequence of several operations and machines Need for multiple operations can impact part quality Tool wear is significant	Widely varied applications	1, 2, 3 4, 5, 6 7[d], 8, 9[d] 10[d], 11[d] 12[d]	Closer to true CAD/CAM link than most other processes Most flexible of manufacturing processes
Some taper of walls Minimum radius of 0.002 all around Material most be electrically conductive	Various jet engine parts	1[d], 3, 6[d] 9, 12	Material removal rates much greater than EDM (approximately 5 in.³/min.) although tooling, equipment and energy costs are much higher Surface finish not nearly as closely tied to removal rates as with EDM Generally more cost-effective than precision machining and grinding for all but the most easily machined materials
Electrode wear impacts accuracy and requires periodic replacement Material removal rate is extremely slow (0.01–0.5 in.³/h) Additional limitations identical to ECM	Due to low production rates, EDM is generally used in toolmaking rather than part production, or for deburring, where other methods aren't satisfactory	2[d], 3[d] 5[d], 6, 9[d] 12[d]	A very different variation of conventional EDM, wire EDM is used to cut highly accurate, and sometimes complex profiles in hardened materials up to 6 in. thick These components are often used in drawing, extruding, or stamping dies
Holes with diameter lest than stock thickness need to be drilled Since 1/2–2/3 of material thickness is fractured, rather than sheared, secondary operations or fineblanking is needed for good edge finish or parallel sides Finishing and material scrap costs are often substantial	Numerous consumer and industrial applications	2, 3, 4 5, 6, 7[d] 8[d], 12[d]	Mechanical reciprocating presses operate at 35–500 strokes/min. CNC Turret presses achive 55–265 hits/min. at 1 in. centers Often when the cost of dies exceeds the total costs of parts, die sets are no longer cost effective (approximately 20,000 pcs for common geometries) Progressive dies can often be justified if they can save two or more secondary operations on individual die sets
Low degree of part complexity Low dimensional accuracy Minimal opportunity for integral fasteners or attachment points	Various consumer packaging Bus, aircraft interior panels Refrigerator linings Signs Boat hulls	10	Tooling less expensive than other plastic processing, methods High production rates possible (drinking cups: 2000–3000 pcs.min.) Material properties can be improved due to molecular orientation Reinforcing fibers may also be added to improve strength Of several processes available (vacuum, pressure, drape), vacuum is most popular
Stiffening beads should be formed externally rather than internally Cylindrical sections and reentrant angles are possible but more costly Minimal radius 1.5 × thickness Maximum thickness for hand spinning: 0.25 in. (Al), 0.187 (LC steel), 0.125 (S steel)	Cooking utensils Lamp bases Nose cones Reflectors	2, 3[d], 4 5, 6, 7[d] 8[d], 12[d]	Conventional spinning and displacement spinning differ in that displacement spinning moves material back along forming member refining grain structure in direction of flow Tooling costs are much less than for stamping or deep-drawing, very small quantities may be economically produced Tube spinning reduced I.D., O.D. or lengthens tubes or preforms

5, aluminum and alloys; 6, copper and alloys; 7, zinc and alloys; 8, magnesium and alloys; 9, titanium; 10, thermoplastics; 11, thermosets; 12, nickel and alloys. *Source*: Boothroyd Dewhurst, Inc., Wakefield, RI.

TABLE 2.2 Shape Generation Capabilities of Processes

	Depress	UniWall	UniSect	AxisRot	RegXSec	CaptCav	Enclosed	NoDraft	PConsol	Alignmt	IntFast	
Sand casting	Y	Y	Y	Y	Y	Y	N	N	4	3	1	Solidification
Investment casting	Y	Y	Y	Y	Y	Y	N	N	5	5	2	processes
Die casting	Y[a]	Y	Y	Y	Y	N	N	N	4	5	3	
Injection molding	Y[a]	Y	Y	Y	Y	N[b]	N	N	5	5	5	
Structural foam	Y[a]	Y	Y	Y	Y	N	N	N	4	4	3	
Blow molding (extr)	Y[a]	M	N	Y	Y	M	Y	N	3	4	3	
Blow molding (inj)	Y[a]	M	N	Y	Y	M	N	N	3	4	3	
Rotational molding	Y[a]	M	N	Y	Y	N	M	N	2	2	1	
Impact extrusion	N	Y	N	Y	Y	N	N	Y	3	3	1	Bulk
Cold heading	N	Y	N	Y	Y	N	N	Y	3	3	1	deformation
Closed die forging	Y[a]	Y	Y	Y	Y	N	N	N	3	2	1	processes
Power metal parts	N	Y	Y	Y	Y	N	N	Y	3	3	1	
Hot extrusion	Y[d]	Y	M	Y	Y	N	N	Y	2	2	3	
Rotary swaging	N[c]	N	N	M	N[c]	N	N	N	1	1	1	
Machining (from stock)	Y	Y	Y	Y	Y	Y	N	Y	2	3	2	Material
ECM	Y[c]	Y	Y	Y	Y	N	N	N	3	4	1	removal
EDM	Y[c]	Y	Y	Y	Y	N	N	N	3	4	1	processes
Wire-EDM	Y[d]	Y	Y	Y	Y	N	N	Y	2	2	3	Profile generating processes
Sheetmetal stamp/bend	Y	M	Y	Y	Y	N	N	N	4	3	4	Sheet
Thermoforming	Y[a]	M	N	Y	Y	N	N	N	3	3	3	forming
Metal spinning	N	M	N	M	N	Y	N	N	1	1	1	processes

[a] Possible at higher cost.
[b] Shallow undercuts are possible without significant cost penalty.
[c] Possible with more specialized machine and tooling.
[d] Only continuous, open-ended possible.
Y, Process is capable of producing parts with this characteristic. N, Process is not capable of producing parts with this characteristic. M, Parts produced with this process must have this characteristic. An underlined entry indicates that parts using this process are easier to form with this characteristic.
The last three columns refer to DFA guidelines and are rates on a scale of 1 to 5, with 5 assigned to processes most capable of incorporating the respective guideline.

Part consolidation (PConsol): The ability to incorporate several functional requirements into a single piece, eliminating the need for multipart assemblies.

Alignment features (Alignmt): The ease of incorporating in the part positive alignment or location features that will aid in the assembly of mating parts.

Integral fasteners (IntFast): The cost-effectiveness and scope of fastening elements that can be designed into the part. The ability to incorporate features such as threads, which generally involve separate fasteners, is not given as much consideration as elements such as snap features.

2.5 SELECTION OF MATERIALS

The systematic selection of specific materials to meet required properties has been given considerable attention. Numerous textbooks and handbooks have been devoted to this topic [5–7]. Comprehensive procedures have been developed for material selection, such as, for example, the detailed handbook system of the Fulmer Research Institute in the United Kingdom [8]. Similarly, software systems based upon comprehensive databases of material properties are available, such as the Mat.DB system from the American Society of Metals [9] and more recently the Cambridge Material Selector [10]. While these procedures are a valuable contribution to the systematic selection of materials, their usefulness at the very early stages of product design, when initial decisions on materials and processes are made, is restricted for several reasons, including:

1. These procedures are aimed at the selection of specific materials based on detailed material property specifications that may not be available early in the design process. At this stage only general ranges of properties may have been decided upon.

2. The material selection is considered independently from the manufacturing processes that may be used, whereas compatibility between processes and materials is important. Several approaches can be adopted to rationalize the search for suitable materials for application during early product design.

2.5.1 Grouping of Materials into Process Compatible Classes

Rather than using a single comprehensive materials database, it is preferable to divide the material databases into classes related to the principal shape-generating processes used in discrete parts manufacture. This is necessary because of incompatibility between some processes and materials and because, generally, the selection of processes and materials must be considered together. The processes used for producing material forms for other shape-generating processes need not be included, since discrete parts manufacture starts with materials that have already undergone this primary processing. Thus, the separate material databases should include, for example, standard metal stockforms (wire, rod, etc.), sand and permanent mold-casting alloys, die-casting alloys, metal powders, thermoplastics granules, thermoplastic sheet and extruded stockforms, and so on.

During an initial search phase when a rapid response to changes in input is essential, it is perhaps inappropriate to search extensive material databases in order to identify precise metal alloys, polymer specifications, powder mixes, etc. This leads to unacceptably slow search procedures and provides information largely irrelevant to early process/material decision making. For example, listing all the thermoplastic resins that would satisfy the specified performance requirements would clearly be premature in early discussions of the relative merits of alternative processes, their required tooling investments, the likely size and shape capabilities, etc. A more efficient procedure is to have, for each process, an associated supermaterial specification that comprises the best attainable properties of all the materials in the corresponding category. If a new alloy is added, say, to the die-casting material database, its properties are compared with the die-casting supermaterial specification, which is then updated as necessary.

The approach to preliminary process selection through supermaterial specification is compatible with the tradeoff and compromise decisions that are part of early design work. A typical scenario might involve the specification of possible shape attributes, size, and one or more production and performance parameters. The next step would be to change the input specifications or add to the specification list, or to investigate a process further for acceptable associated materials.

2.5.2 Material Selection by Membership Function Modification

One challenge of designing a system so that appropriate materials are chosen at the early stages of design lies in modeling ambiguous or vague material constraints. For instance, a designer may want to use a material with a yield stress of "about" 2000 psi and a service temperature "in the neighborhood of" 90°C. A conventional database search for materials with properties greater than those specified would unnecessarily exclude materials with properties close to the desired values, but not in the range specified.

Some material selection systems have attempted to model vague designer specifications by breaking material property values into discrete ranges. However, an alternative approach is to model such vague qualifiers as "about" and "in the neighborhood of" using aspects of fuzzy logic. Fuzzy logic relies on the concept of a membership function to determine how well an object fits into a defined set.

Ambiguity in the material constraints specified by the designer is modeled by providing the designer with different levels of accuracy to further describe the material constraints specified. These levels could correspond, for example, to the qualifiers "approximately," "close to," and "more or less." These levels of precision are illustrated in Fig. 2.4. The ability to assign different levels of accuracy or precision to each constraint is an advantage of fuzzy logic.

A simple example may help to illustrate the advantages and flexibility of this approach (Fig. 2.5). For instance, if pressing and sintering has been selected as a candidate primary process and the user has restricted the material to one with an ultimate tensile strength between 25 and 30 kpsi, then a conventional search of a small database that contains 102 entries would yield 15 candidate materials [11]. A fuzzy search with the qualifier "close to" would yield 29 candidate materials with ultimate tensile strengths between 21 to 29 kpsi. The qualifier "approximately" produces 38 materials with ultimate tensile strengths from 19 to 36 kpsi. Seventeen additional materials with tensile strengths between 16 and 39 kpsi are chosen when the qualifier "more or less" is used. The additional materials selected by the modified membership function may become increasingly important as other material constraints eliminate many materials from consideration.

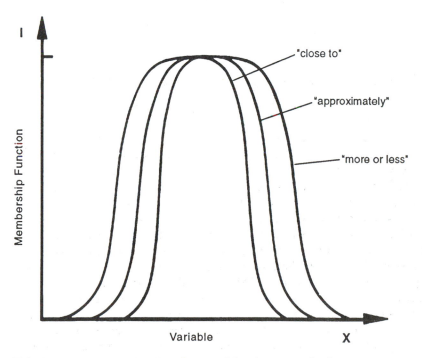

FIG. 2.4 Membership functions for material and process selection.

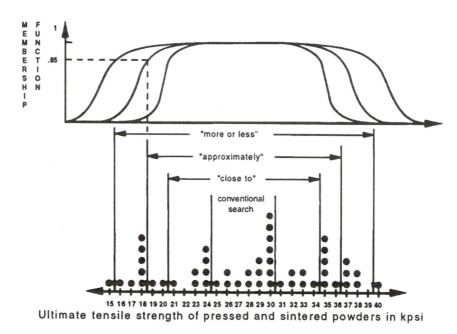

Ultimate tensile strength of pressed and sintered powders in kpsi

FIG. 2.5   Selection of sintered powder materials by membership function modification.
(From Ref. 19.)

2.5.3   Material Selection by Dimensionless Ranking

An aspect of material selection, which is a great source of difficulty, is the
distinction between the fundamental material properties, which are given in
material databases, and the actual design requirements, which are usually based
on a combination of different property values. For the present purposes, material
cost per unit weight will be included as a property of the material, so that
economic constraints on design can be considered in just the same manner as
weight constraints, strength constraints, and so on. Thus, for a structural member
in an aerospace product the designers may be interested in maximum stiffness per
unit weight, while for a high-volume consumer product, maximum stiffness per
unit cost may be more important. In the first case, the materials would be
compared on the basis of a function of Young's modulus divided by density, and
in the second case a combination of Young's modulus, density, and cost per unit
weight would be used for comparison purposes. Some derived parameters, which
are commonly used in mechanical design, have been established in the literature
[12,13]. In this section we are concerned with establishing a simple procedure for
comparing materials based on either single fundamental properties, or the general
form of derived parameters. Such material comparisons may typically be required

on the basis of total performance, best performance per unit weight, or best performance per unit cost. A procedure will be established in this section for making these comparisons on a dimensionless scale from 0 to 100.

The authors, in their work on computer-aided material and process selection, have observed that material properties tend to be distributed approximately uniformly when presented on logarithmic scales. This fact can be seen clearly in the work by Ashby [13], in which material properties are plotted on a variety of combinations of logarithmic scales, such as yield strength plotted against density, coefficient of expansion against thermal conductivity, and so on. In all cases the properties of groups of material represented as bubbles are seen to spread out in an approximately uniform manner across the logarithmic scales. For example, Fig. 2.6 shows the spread of the property of elastic modulus for the different classes of materials when presented on a linear scale. It can be seen that there is a crowding of materials at the beginning of the scale with poor discrimination between material properties in that region because of the coarse scale. Figure 2.7 shows the same data represented on a logarithmic scale, which spreads out the

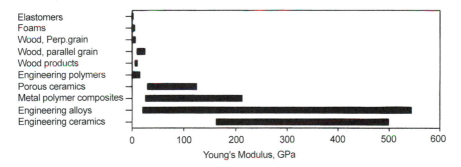

FIG. 2.6 Elastic modulus for classes of materials plotted on linear scales. (After Ref. 13.)

FIG. 2.7 Elastic modulus for classes of materials plotted on logarithmic scales. (After Ref. 13.)

material classes across the wide range of property values. Similar approximately uniform spread of materials across logarithmic property scales is found to apply also to derived material properties. This can be seen by viewing the spread of materials along the inclined lines in the Ashby plots [13], which represent a wide variety of derived parameters.

Property distributions of the form shown in Fig. 2.6 can be changed to a log-linear scale through a transformation of the form

$$P = \alpha 10^{\beta N} \tag{2.1}$$

where

$P =$ the actual property value
$N =$ the log-linear scale value
$\alpha, \beta =$ constants for a particular material property.

We are interested in using Eq. (2.1) to develop look-up tables or spreadsheets in which fixed upper and lower values provide benchmarks for comparison of material properties. These fixed values, for the material properties, will be set at zero for the least value in the material database and 100 for the highest value. The database will then be updated automatically if a new material is introduced with a higher or lower property value than any of the existing materials. However, these extreme materials are in most cases unlikely to be exceeded.

Consider property P and let P_{max}, P_{min} be the highest and least values in the database. Thus from Eq. (2.1)

$$\alpha = P_{min} \tag{2.2}$$
$$\beta = \log(P_{max}/P_{min})/100 \tag{2.3}$$

Substituting Eqs. (2.2) and (2.3) into (2.1) gives

$$N = 100 \log(P/P_{min})/\log(P_{max}/P_{min}) \tag{2.4}$$

For example, for Young's modulus, E, the largest value in the database is likely to be the value for diamond, which gives

$$E_{max} = 1.03 \times 10^6 \text{ MN/m}^2$$

while the least value may be the value for natural rubber which gives

$$E_{min} = 4.59 \text{ MN/m}^2$$

With these values, the "100-scale" for Young's modulus is given from Eq. (2.4) as

$$N = 18.68 \log(0.218E) \tag{2.5}$$

where E has units of MN/m^2.

TABLE 2.3 "100 Scale" Values for Young's Modulus

Material name	N
Diamond	100
Tungsten carbide	95
Steel	87
Magnesium	75
Polycarbonate (33% glass)	64
Pine (parallel to grain)	61
Particle board	51
High-density polyethylene	42
Urethane foam	26
Cork	12
Natural rubber	0

Table 2.3 gives N values for Young's modulus for a small range of commonly used materials. It can be seen that the values appear to represent an engineer's perception of material stiffness. In particular values greater than 50 apply to materials that are found in structural applications.

Largest and least values for a range of principal fundamental material properties are given in Table 2.4. A small material database is given in Table 2.5 that includes representative materials from metal alloys, polymers, rubbers, foams, ceramics, and natural materials. The last column in the table gives "100-scale" values for the derived property in the adjacent column. These will be discussed below.

TABLE 2.4 Largest and Least Material Property Values

Property	Material name	Largest value	Least value	Units
Tensile yield strength	Alloy steel	1375	—	MN/m^2
	Cork	—	1.0	
Compressive strength	Tungsten carbide	4950	—	MN/m^2
	Cork	—	1.0	
Young's modulus	Diamond	1.0×10^6	—	MN/m^2
	Rubber	—	4.6	
Density	Tungsten carbide	13,300	—	MN/m^3
	Cork	—	140	
Cost	Industrial diamond	725	—	$/kg
	Concrete	—	0.13	

TABLE 2.5 Material Database and Derived Parameter Ranking

Generic material	Cost, $/kg	Tensile yield str. MN/m²	Elastic modulus, MN/m²	Compressive yield str. MN/m²	Density, kg/m³	Derived parameter	N
Gray cast iron	2.86E−01	2.93E+02	1.34E+05	2.93E+02	7.21E+03	7.08E−03	41
Ductile iron	3.52E−01	4.48E+02	1.65E+05	3.10E+02	7.13E+03	7.67E−03	43
Malleable iron	4.18E−01	3.45E+02	1.60E+05	3.45E+02	7.38E+03	7.33E−03	42
Mild steel (c.q. annealed)	9.90E−01	2.62E+02	2.07E+05	2.62E+02	7.77E+03	7.58E−03	43
Alloy steel (high-strength)	6.16E+00	1.38E+03	2.07E+05	1.38E+03	7.85E+03	7.50E−03	42
Stainless steel	2.75E+00	2.48E+02	1.93E+05	2.48E+02	8.04E+03	7.16E−03	41
Aluminum alloy (high-strength)	5.28E+00	1.93E+02	7.10E+04	1.93E+02	2.75E+03	1.50E−02	62
Beryllium copper	3.85E+01	1.10E+03	1.28E+05	1.10E+03	8.27E+03	6.07E−03	36
Copper, hard	2.86E+00	3.10E+02	1.17E+05	3.10E+02	8.96E+03	5.44E−03	33
Magnesium	7.70E+00	2.34E+02	4.48E+04	2.34E+02	1.80E+03	1.96E−02	70
Titanium	2.68E+01	9.45E+02	1.13E+05	9.45E+02	4.74E+03	1.02E−02	51
Lead	2.86E+00	2.00E+01	1.52E+04	2.00E+01	1.14E+04	2.17E−03	7
Epoxy (glass-reinforced)	5.28E+00	6.55E+01	3.10E+03	2.48E+02	1.91E+03	7.60E−03	43
Polyethylene (high-density)	7.48E−01	2.48E+01	8.27E+02	2.48E+01	9.71E+02	9.65E−03	50
Polycarbonate (glass-reinforced)	3.52E+00	1.59E+02	1.16E+04	1.45E+02	1.53E+03	1.48E−02	62
Rubber (isoprene)	3.48E+00	2.76E+01	4.59E+00	2.76E+01	9.71E+02	1.71E−03	0
Polyurethane foam	1.76E+00	1.52E+01	1.08E+02	1.72E+01	4.99E+02	9.51E−03	49
Particle board (med. density)	3.52E−01	1.55E+01	2.93E+03	1.45E+01	6.10E+02	2.34E−02	75
Pine	2.05E+00	7.93E+01	8.27E+03	3.31E+01	3.61E+02	5.59E−02	100
Diamond	7.26E+02	2.69E+02	1.03E+06	4.00E+03	3.52E+03	2.86E−02	81
Silicon carbide (sintered)	6.60E+01	6.90E+01	3.31E+05	1.03E+03	2.97E+03	2.32E−02	75
Tungsten carbide	2.64E+02	8.96E+02	5.39E+05	4.95E+03	1.33E+04	6.09E−03	36
Glass (soda lime, gen. purpose)	3.30E−01	9.17E+01	7.31E+04	1.38E+03	2.47E+03	1.69E−02	66
Pottery	6.60E−01	3.31E+01	7.03E+04	5.00E+02	2.22E+03	1.85E−02	68
Concrete	1.32E−01	1.65E+00	3.00E+04	2.48E+01	2.50E+03	1.24E−02	57
Cork	1.50E+00	1.00E+00	2.00E+01	1.00E+00	1.39E+02	1.96E−02	70
Index values	0.000	0.000	0.333	0.000	−1.000		

TABLE 2.6 Derived Parameter D for Best Performance

To obtain:	Maximum performance	Minimum weight	Minimum cost
Strongest tension member	Y_t	Y_t/ρ	$Y_t/\rho C_m$
Strongest compression member	Y_c	Y_c/ρ	$Y_c/\rho C_m$
Strongest beam or plate	Y_t	$Y_t^{0.5}/\rho$	$Y_t^{0.5}/\rho C_m$
Stiffest structural beam	E	$E^{1/3}/\rho$	$E^{1/3}/\rho C_m$
Best coil or tension spring	Y_t^2/E	$Y_t^2/(E\rho)$	$Y_t^2/(E\rho C_m)$
Best diaphragm spring	$Y_t^{1.5}/E$	$Y_t^{1.5}/(E\rho)$	$Y_t^{1.5}/(E\rho C_m)$

Y_t = tensile yield stress, Y_c = compressive yield stress, E = Young's modulus, ρ = density, C_m = material cost/weight.

A number of derived parameters of importance in mechanical design [6,11] are given in Table 2.6. It can be seen that they are represented by the general form

$$D = P_1^{m_1} P_2^{m_2} P_3^{m_3} \cdots \qquad (2.6)$$

For example, if $P_1 = Y_t$, $P_2 = E$, $P_3 = \rho$, $m_1 = 2$, $m_2 = -1$ and $m_3 = -1$, then D is the derived parameter for best spring performance per weight as given in Table 2.6.

Let the log-linear relationships for $P_1, P_2, P_3, \ldots$, be

$$P_1 = {}_1 10^{{}_1 N_1}$$
$$P_2 = {}_2 10^{{}_2 N_2} \qquad (2.7)$$
$$P_3 = {}_3 10^{{}_3 N_3}$$

The general form of a derived parameter then becomes

$$D = (\alpha_1^{m_1} \alpha_2^{m_2} \alpha_3^{m_3} \cdots) 10^{(m_1\beta_1 N_1 + m_2\beta_2 N_2 + m_3\beta_3 N_3 + \cdots)} \qquad (2.8)$$

and we require that D be represented by

$$D = \alpha 10^{\beta N} \qquad \text{where } 0 \le N \le 100 \qquad (2.9)$$

Thus, from Eq. (2.4), the "100-scale" value for the derived parameter is given by

$$N = 100 \log(D/D_{\min})/\log(D_{\max}/D_{\min}) \qquad (2.10)$$

Equation (2.10) can be simplified further by recognizing that the factor $(\alpha_1^{m_1} \alpha_2^{m_2}, \ldots)$ will cancel in the argument of both logarithmic expressions. Thus, define parameter W as

$$W = m_1\beta_1 N_1 + m_2\beta_2 N_2 + \cdots \qquad (2.11)$$

Substituting W into Eq. (2.10) gives

$$N = 100 \log(10^{W-W_{min}})/\log(10^{W_{max}-W_{min}}) \tag{2.12}$$

$$= 100(W - W_{min})/(W_{max} - W_{min}) \tag{2.13}$$

This transformation from "100-scale" values for individual parameters to the "100-scale" value for any derived parameter is easily accomplished on a spreadsheet using Eqs. (2.11) and (2.13). Table 2.5 is the printout of such a spreadsheet written with Microsoft Excel. The bottom row of the spreadsheet contains values for m, the index values of the derived parameter. The last two columns contain the values for W and the "100-scale" N values for the derived parameter, respectively.

Note that the index values entered into the last row are those for beam stiffness for minimum weight. It can be seen that for this application (and no other design constraints), pine is the best choice ($N = 100$) and rubber is the worst choice ($N = 0$). The 100 score for pine indicates why straight grain wood is still a material of choice for small aerobatic aircraft structures. Note that manufacturing feasibility is not a part of this selection process. Thus while diamond scores a credible 81, its use would obviously be restricted to very small and very expensive devices. If we change the index for cost from 0 to -1, then the derived parameter changes to represent beam stiffness for minimum cost. The best choice then changes to concrete. Particle board is a close second with a score of 96, diamond becomes 6, and tungsten carbide is zero because of its combination of high cost and high density. The high scores for concrete and particleboard explains their use for low-cost beams and floor structures, respectively. The main purpose of the "100-scale" method is for such easy visualization of the relative merits of materials for different applications. The method can be extended to include combinations of two or more derived parameters. For example, the primary requirement for an automobile panel may be bending stiffness for least cost. However, diaphragm spring quality is a valuable additional material property since this makes the material more dent resistant. The 100-scale method can be expanded to cover such situations by using a weighted geometric mean of the two derived parameters [14].

Sometimes the choice of a material is based on the inverse of one of the fundamental properties. Examples would include specific volume representing lightness instead of density representing heaviness, flexibility instead of stiffness, softness instead of compressive strength, and so on. Assume we are interested in inverse property $(1/P)$, where P is represented by Eq. (2.1). Let

$$(1/P) = \alpha 10^{\beta M} \tag{2.14}$$

From Eqs. (2.2) and (2.3), the values of α and β for the inverse property are given by

$$\alpha = 1/P_{max} \tag{2.15}$$
$$\beta = \log[(1/P_{min})/(1/P_{max})]/100 \tag{2.16}$$
$$= \log(P_{max}/P_{min})/100 \tag{2.17}$$

and so the value for M becomes [refer to Eq. (2.4)],

$$M = 100 \log[(1/P)/(1/P_{max})]/\log(P_{max}/P_{min})$$
$$= 100\{[\log(P_{max}/P_{min}) - \log(P/P_{min})]\}/\log(P_{max}/P_{min}) \tag{2.18}$$
$$= 100 - N$$

This result simply stems from the "100-scale" span from minimum to maximum values, which exchange places when the inverse property is considered. However, the fact that not just "100" and "0" change places, but that "95" becomes "5", "90" becomes "10," and so on, is intuitively satisfying.

Problem: Construct the Excel spreadsheet illustrated in Table 2.5. Use this to explore the best material choices for each of the 18 criteria given in Table 2.6.

2.6 PRIMARY PROCESS/MATERIAL SELECTION

Systematic procedures can be developed for the selection of primary process/material combinations. Such procedures operate by eliminating processes and materials as more detailed specification of the required part's attributes occurs. The elements of such a selection procedure can be illustrated by considering, as an example, the part shown in Fig. 2.8, which is to be used as an oven bracket. The example has been used previously by Wilson et al. [15,16] and is used again here as an illustrative example. In terms of the shape-producing capabilities listed in Table 2.2, this part is specified as follows.

Shape Attributes

1. Depressions   Yes
2. Uniform wall   Yes
3. Uniform cross section Yes
4. Axis of rotation   No
5. Regular cross section   No
6. Captured cavity   No
7. Enclosed cavity   No
8. No draft   Yes

Material Requirements

A. Maximum temperature of 500°C
B. Excellent corrosion resistance to weak acids and alkalis

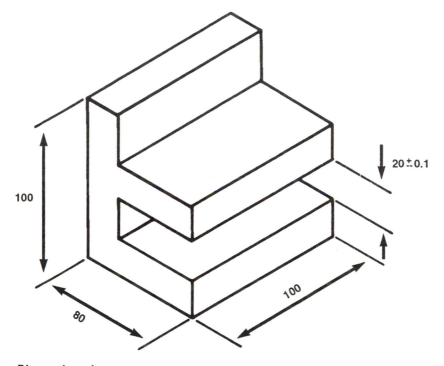

Dimensions in mm

FIG. 2.8 Oven bracket part.

In this list, the shape attributes with a "Yes" will eliminate those processes that are not capable of producing these features. Those features with a "No" will eliminate those processes that are only capable to producing parts with these features present. Applying these requirements progressively to the basic process/material compatibility matrix shown in Fig. 2.3 produces the results shown in Figs. 2.9 to 2.11. Figure 2.9 shows the processes eliminated by the first four shape attributes listed above and Fig. 2.10 shows the processes eliminated by the other four shape attributes. Combining these together results in four selected processes (Fig. 2.11): powder metal parts, hot extrusion, machining from stock, and wire EDM. Finally, imposing the material requirements results in the final selection of processes and materials shown in Fig. 2.12. These selected combinations can then be ranked by other criteria, such as estimates of manufacturing costs.

1. Depression ....Yes
2. Uniform Wall ....Yes
3. Uniform Cross Section ....Yes
4. Axis of Rotation ....No

Column headers (left to right): Cast Iron, Carbon Steel, Alloy Steel, Stainless Steel, Aluminum and Alloys, Copper and Alloys, Zinc and Alloys, Magnesium and Alloys, Titanium and Alloys, Nickel and Alloys, Refractory Metals, Thermoplastics, Thermosets

Solidification Processes
- Sand Casting
- Investment Casting
- Die Casting
- Injection Molding
- Structural Foam Molding
- 3 Blow Molding(ext.)
- 3 Blow Molding(inj.)
- 3 Rotational Molding

Bulk Deformation Processes
- 3 Impact Extrusion
- 3 Cold Heading
- Closed Die Forging
- Powder Metal Processing
- Hot Extrusion
- 1, 2, 3, 4 Rotary Swaging

Material Removal Processes
- Machining (from stock)
- ECM
- EDM

Profiling
- Wire EDM

Sheet Forming Processes
- Sheet Metal(stamp/bend)
- 3 Thermoforming
- 1, 3, 4 Metal Spinning

Legend: ☐ Normal Practice ■ Not Applicable ▦ Less common

FIG. 2.9 Process elimination based on four geometric attributes of the part in Fig. 2.8.

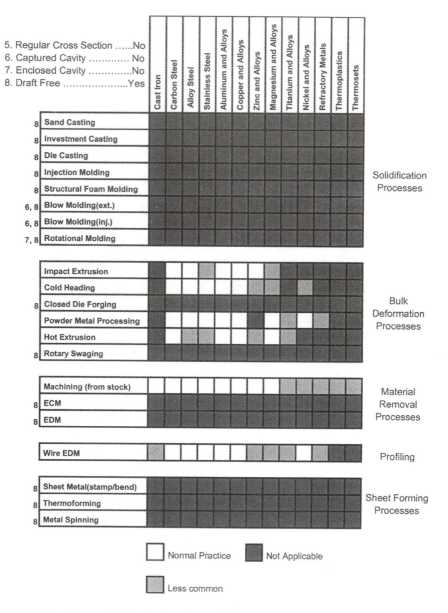

FIG. 2.10   Process elimination based on a further four attributes of the part in Fig. 2.8.

FIG. 2.11   Final process selection based on geometric attributes of the part in Fig. 2.8.

FIG. 2.12 Final selection based on process/material combinations of the part shown in Fig. 2.8.

2.7 SYSTEMATIC SELECTION OF PROCESSES AND MATERIALS

The development of computer-based procedures for process/material selection from general part attributes can have a significant impact on early product design, and several approaches to this problem have been made.

2.7.1 Computer-Based Primary Process/Material Selection

An initial research program has been carried out in the area of combined material/process selection [15,16]. This work by Wilson and co-workers resulted in the development of a Fortran-based computer program given the acronym MAPS. A more recently developed primary material/process selector uses a commercially available relational database system. This selector has the acronym CAMPS, for computer-aided material and process selection [17]. In the selector, inputs made under the headings of part shape, size, and production parameters are used to search a comprehensive process database to identify processing possibilities. However, process selection completely independent of material performance requirements would not be satisfactory, and for this reason, required performance parameters can also be specified by making selections under the general categories of mechanical properties, thermal properties, electrical properties, and physical properties. As many selections as required can be made, and at each stage the candidate processes are presented to the system user. Processes may be eliminated directly because of shape or size, or when performance selections eliminate all of the materials associated with a particular process, in a similar manner to the charts shown in Figs. 2.9 to 2.12. For the materials in the CAMPS system for each process, the type of supermaterial specification described above is utilized. The supermaterial specifications are maintained automatically by the program.

The CAMPS system also classifies all possible selections into ranges labeled A through F. This is intended to ensure ease of use in the early stages of design when precise numerical values for many of the parameters would not be known. For example, for a structural part, yield strength will clearly be an important requirement. However, the minimum allowable yield stress value will depend on the part wall thickness, which will in turn depend on the process/material combination to be used.

A simplified version of this approach to process and material selection has been incorporated into a cost-estimating tool used at the early stages of product design [18]. This program contains definable process limits, such as maximum dimensions, minimum wall thickness, and so on. The selection procedure indicates any combinations of process and material that are not suitable or for which the part geometry may be outside normal processing limits. Figure 2.13

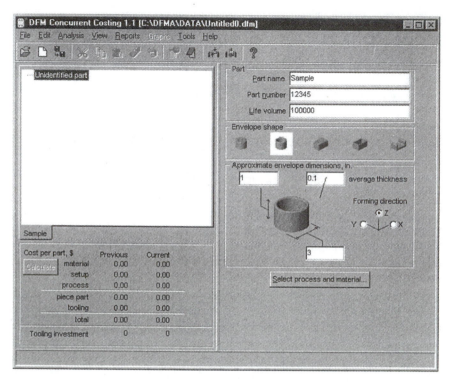

FIG. 2.13 General description of proposed part.

shows an initial part description where the general part type and overall dimensions, including wall thickness, are defined. Following this a process must be selected as indicated in Fig. 2.14, and then the compatible materials are indicated (red for incompatible, green for compatible, and yellow for compatible materials, but exceeding a normal processing limit). In the case of Fig. 2.14 cold chamber die casting has been selected and the indicated compatible processes are copper, magnesium, aluminum, and zinc alloys.

2.7.2 Expert Processing Sequence Selector

An approach to the preliminary selection of materials and processes has been described above. While this approach may generally result in selection of appropriate combinations of materials and primary processes, in some cases matching of the material and primary process alone to the finished part attributes, without considering viable sequences of operations, may lead to the omission of some appropriate combinations of primary processes and materials. An expert

FIG. 2.14 Material classes compatible with cold-chamber die casting.

processing sequence generator has been investigated to enhance this aspect of material and process selection [19,20].

With this procedure the user classifies the geometry and specifies the material constraints for the part. The result is a list of viable sequences of processes and compatible materials. The procedure is divided into four steps: geometry input, process selection, material selection, and system update. The geometry of the part is first classified according to its size, shape, cross section, and features. Using pattern-matching rules, processes are then selected that would form the geometry of the part. Material selection uses fuzzy set theory materials, as described earlier.

The geometrical classification of a part is concerned with the following characteristics:

1. The overall size
2. The basic shape
3. The accuracy and surface finish
4. The cross section
5. Functional features—projections, depressions, etc.

As described earlier, processes are classified as either primary, primary/secondary, or tertiary to take advantage of the natural order of processes in a sequence. Rules, formulated from knowledge about processes and materials, are used to select sequences of processes and materials for part manufacture. Processes are selected using a pattern-matching expert system and rules of the form. If . . .

(condition 1)
(condition 2)
(condition 3)

Then . . .

(action 1)

For primary process selection, the conditions are restrictions on the size of the enclosing envelope, the size and shape of the fundamental envelope, and the cross-sectional description of the part. The action is the selection of a candidate primary process. If a part satisfies the restrictions, then the process is chosen as a candidate primary process. Other rules of the same form then assess which features of the part can be formed by the primary process. The conditions for these rules are restrictions on the descriptors of the features and the action is to conclude that the primary process can form the feature.

The boundaries of a processes' capabilities, for inclusion in the selection rules, are not well defined. Parts with requirements that are near the boundaries of the capabilities of a process are more difficult to produce than parts that fall well within these boundaries. Therefore, the process selection rules are better formulated with fuzzy logic membership functions to model the progressive transition from "easy" to "difficult or impossible" to manufacture by the selected process. For example, Fig. 2.15 shows a typical membership function for primary process selection for the attribute part size for, say, die casting. Similar fuzzy selection rules can be applied to other part attributes. This process also enables a preliminary ranking of selections to be made based on the values of the membership functions obtained.

Next, the material database is searched for the primary process selected and uses the fuzzy logic approach described earlier to choose candidate materials. Since the properties of a material are related to how the material is processed, each process has its own material database. Materials are selected by mapping the user's input onto the material properties. Material properties that can be affected by tertiary manufacturing processes are not used to exclude materials from consideration, at this stage. For example, corrosion resistance could be achieved by plating an otherwise unacceptable material.

Primary/secondary processes are selected in a similar manner to form any features of the part that cannot be formed by the primary process. Similarly,

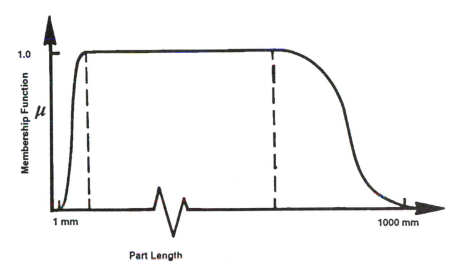

FIG. 2.15   Example of membership function for process selection rules.

tertiary processes are selected to fulfill material requirements that the candidate material could not fulfill. A viable sequence of processes is found when all of the geometrical and material goals specified by the user are satisfied. Figure 2.16 shows this process graphically. Here circles represent the goals and processes. Satisfied goals are indicated by filled-in circles with arrows pointing to the material or process that satisfied the goal.

If a suitable process or material cannot be found to form required features or satisfy material requirements, then the procedure backtracks to solve the impasse. For instance, if a suitable material cannot be found, then the procedure backtracks to choose another primary process. Similarly, if a tertiary process cannot be found to satisfy a material requirement, then the procedure backtracks to choose an alternative material.

A characteristic of this approach to material and process selection is that as the list of part attributes to be fulfilled grows, the number of possible sequences may also increase. This differs from the procedure for selecting primary process/material combinations in which the list of possible combinations generally decreases, as the specification of the part becomes more precise. For example, the addition of a surface finish tolerance to the attribute list will introduce secondary processes into sequences that could produce this requirement. For this reason it is important that consideration be given to the economic ranking of the processing sequences generated.

SEQUENCE GENERATION

| PART GOALS | PRIMARY PROCESS CHOICE | MATERIAL CHOICE | SECONDARY PROCESS CHOICE | TERTIARY PROCESS CHOICE |

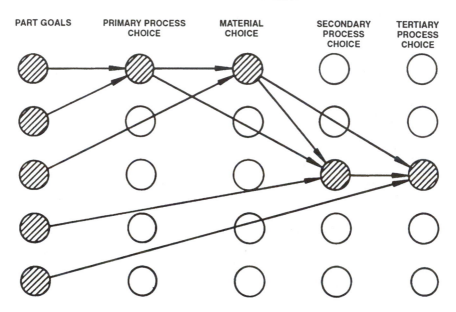

FIG. 2.16 Procedure for processing sequence selection.

2.7.3 Economic Ranking of Processes

The viable material/process combinations determined by the selection procedures described above require evaluation as to which is the most suitable, usually by estimating which will be the most economic. This requires the availability of procedures for realistically evaluating manufacturing costs early in the design process. Several of the later chapters deal with simplified cost-estimating procedures for various processes. However, at the very early design stages even simpler methods for cost assessment can be used for the ranking of alternative material/process combinations.

As an example of how early cost estimates can be made for a particular process, machining will be considered. Further details on cost estimating for design for machining are contained in Chapter 7. From a detailed analysis of cost estimating for machined parts [21,22] it can be concluded that, in general, the time to remove a given volume of material in rough machining is determined mainly by the specific cutting energy (or unit power) of the material and the power available for machining. For finish-machining a given surface area, the recommended speeds and feeds for minimum machining cost could be used. Also, it is possible to make appropriate allowances for tool replacement costs.

Further research [23] has shown that a large amount of statistical data is available on the shapes and sizes of machined components and the amount of machining carried out on them. Statistical data is also available on the sizes of the machine tools relative to the sizes of the workpieces machined. Combining this data with information gathered on machine costs and power availability, it can be shown that estimates of machined component costs can be made based on the minimum of design information, such as might be readily available early in the design process [24,25].

The information required can be divided into three areas:

1. Workpiece and production data
2. Factors affecting nonproductive costs
3. Factors affecting machining times and costs

The first item under the heading of workpiece and production data describes the shape category of the workpiece. It was found in previous studies [26] that common workpieces can be classified into seven basic categories, as illustrated in Fig. 2.17. Other items under this first heading include: the material, the form of the material (standard stock or near-net shape), dimensions of the workpiece, cost per unit weight, average machine and operator rate, and batch size per setup. A

FIG. 2.17 Seven basic categories of machines component parts. (From Ref. 26.)

knowledge of the workpiece and production data not only allows the cost of the workpiece to be estimated, but also allows predictions to be made of the probable magnitudes of the remaining items necessary for estimates of nonproductive costs and machining costs.

For example, for the workpiece shown in Fig. 2.18, the total cost of the finished component was estimated to be $24.32—a figure obtained from knowledge of the work material, its general shape category and size, and its cost per unit volume. A cost estimate for this component based on its actual machined features and using the approximate equations developed in [23] gives a total cost of $22.83, which is within 6%. A more detailed estimate obtained using more traditional cost-estimating methods gives a total cost of $22.95.

Comparison of these three estimates together with comparisons for several other workpieces are presented in Fig. 2.19, where it can be seen that the approximate method using actual data gives figures surprisingly close to the accurate estimate. Also, the initial crude estimate based on typical workpieces is quite accurate and probably sufficient for the purposes of early cost estimating

aluminum shaft

Machine	Feature	Operations
Horizontal band saw	—	Cut off workpiece
CNC lathe	1	Finish face
	2	Center drill, drill, tap
	3	Finish turn
	4	Groove
	—	Reclamp
	5	Rough and finish turn
	6	Finish face
	7	Finish face
Vertical miller	8	End mill keyway

FIG. 2.18   Category 2 part—rotational with secondary features.

FIG. 2.19 Comparison of machining cost estimates. The cost indicated next to each part drawing is the material cost for the part, ▨, detailed analysis; ☐, estimate; ▨, initial estimate.

when various material and process combinations are being considered and before the component has been designed. However, these considerations should preferably occur after the proposed product has been simplified as much as possible, through design for assembly analysis, as will be described in more detail in the next chapter.

This example for machining illustrates that it is possible to obtain reliable cost estimates based on some initial general design information and that such cost estimates can be refined as more detailed design information becomes available. The cost models for the common primary shape-generating processes detailed in the later chapters have been combined into a software tool [18] that enables a cost comparison to be obtained for a proposed part using different manufacturing processes and materials. The estimation procedures are structured in such a way that an initial cost estimate is determined from the general shape, size, and material selected for the part, using appropriate default values for various parameters. These initial cost estimates are then refined as more details of the part geometry and other factors are entered. The geometric features such as part volume, perimeter, and projected area can be ontained from an in-built geometry calculator or from a computer-aided design (CAD) model of the part.

As an illustration of the use of such a tool to compare material and process selections consider the following example. Figure 2.20 shows a small connecting-rod shape that is being considered for manufacture. This is a symmetrical part with a flat parting plane that divides the shape into two mirror images. It will be assumed that this part is to be made from brass. The type of process selection procedures described earlier, as illustrated in Figs. 2.13 and 2.14, show that the following compatible processes are suitable for manufacture of the part:

Die casting
Hot forging
Investment casting
Automatic sand casting
Powder metal processing

Some of the features, in particular the horizontal holes, will need to be produced by secondary machining for some of the processes, as illustrated in the figure.

Figure 2.21 compares the estimated cost for the selected processes for different production volumes. These cost estimates include the costs of tooling and any secondary operations required. These curves show that powder metallurgy would be the least expensive process for life volumes greater than about 20,000 and that hot forging would be more economical for smaller life volumes. However, it should be realized that for very small quantities machining from the solid would undoubtedly be the most economical procedure, but this has not been considered here.

For sand casting, hot forging or powder metal processes these holes will be drilled. For investment casting, ceramic cores might be used and die casting will require a side pull.

Parting plane ————

FIG. 2.20 Connecting rod.

Figure 2.22 shows a more detailed breakdown of the cost contributions for a production volume of 100,000 parts for each process. The material costs for investment casting and hot forging are higher than the costs for other processes. In investment casting the pattern costs and ceramic molding costs are added to the metal costs and in hot forging where four forgings per cycle were assumed there is a large amount of scrap generated in the form of flash. The high cost of setup for automatic sand casting would be expected. The lowest process cost is for die casting, which again would be expected, since this is a highly efficient operation once the tooling has been manufactured. However, the high cost of tooling for this process is evident in the bar chart and is partly due to the short cavity life with the selected material. Powder metal processing is clearly the preferred process for this life volume of 100,000 (Fig. 2.23).

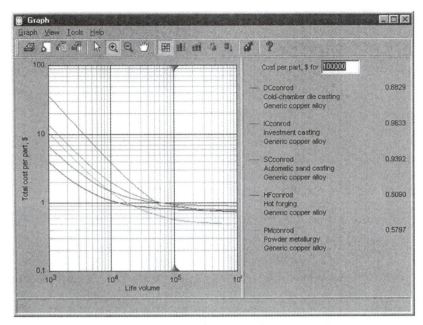

FIG. 2.21   Connecting rod costs for different processes and production volumes.

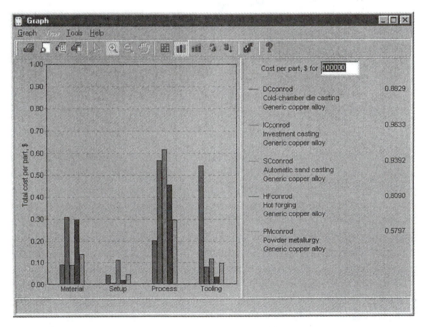

FIG. 2.22   Cost breakdown for production volume of 100,000.

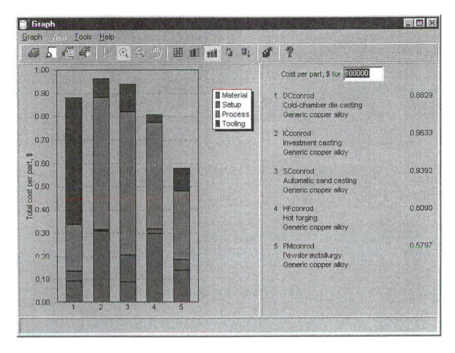

FIG. 2.23   Cost comparisons for production volume of 100,000.

REFERENCES

1. Bishop, R., Huge Gaps in Designers' Knowledge Revealed, Eureka, October 1985.
2. Chang, F.C., and Wysk, R.A., An Itroduction to Automated Process Planning Systems, Prentice-Hall, Englewood Cliffs, NJ, 1985.
3. Ham, I., and Lu, C.Y., Computer-Aided Process Planning: The Present and the Future, Annals CIRP, Vol. 37, No. 2, 1988, p. 591.
4. Bedworth, D.B., Henderson, M.R., and Wolfe, P.M., Computer Integrated Design and Manufacturing, McGraw-Hill, New York, 1991.
5. Farag, M.M., Materials and Process Selection in Engineering, Applied Science Publishers, Barting, United Kingdom, 1979.
6. Crane, F.A.A., and Charles, J.A., Selection and Use of Engineering Materials, Butterworths, London, 1984.
7. Hamley, D.P., Introduction to the Selection of Engineering Materials, Van Nostrand Reinhold, New York, 1980.
8. Fulmer Institute, Fulmer Materials Optimiser, Fulmer Institute, Stoke Poges, United Kingdom, 1975.
9. American Society of Metals, Mat.DB User's Manual, ASM International, Cleveland, Ohio, 1990.
10. Cambridge Materials Selector Software, Granta Design Limited, Cambridge, United Kingdom, 1998.

11. Kalpakjian, S., Manufacturing Processes for Engineering Materials, 1st ed., Addison-Wesley, Reading, MA, 1984.
12. Dieter, G.E., Engineering Design: A Materials and Processing Approach, McGraw-Hill, London, 1983.
13. Ashby, M.F., Materials Selection in Mechanical Design, Pergamon Press, New York, 1992.
14. Dewhurst, P., and Reynolds, C.R., A Novel Procedure for the Selection of Materials in Concept Design, J. Mater. Eng. Perform., Vol. 6, No. 3, June 1997.
15. Dargie, P.P., A System for Material and Manufacturing Process Selection (MAPS), M.S. Project Report, Department of Mechanical Engineering, University of Massachusetts, Amherst, May 1980.
16. Dargie, P.P., Parmeshwar, K., and Wilson, W.R.D., MAPS-1: Computer-Aided Design System for Preliminary Material and Manufacturing Process Selection, ASME Trans., Vol. 104, January 1982, pp. 126–136.
17. Shea, C., and Dewhurst, P., Computer-Aided Materials and Process Selection, Proc. 4th lnt. Conf. on Design for Manufacture and Assembly, Newport, RI, June 1989.
18. Concurrent Costing Software, Boothroyd Dewhurst, Inc., Wakefield, RI, 1999.
19. Farris, J., and Knight, W.A., Selecting Sequences of Processes and Material Combinations for Part Manufacture, Proc. Int. Forum of Design for Manufacture and Assembly, Newport, RI, June 10–11 1991.
20. Farris, J., Selection of Processing Sequences and Materials During Early Product Design, Ph.D. Thesis, University of Rhode Island, 1992.
21. Boothroyd, G., Cost Estimating for Machined Components, Report 15, Department of Industrial and Manufacturing Engineering, University of Rhode Island, 1987.
22. Boothroyd, G., Grinding Cost Estimating, Report 16, Department of Industrial and Manufacturing Engineering, University of Rhode Island, 1987.
23. Boothroyd, G., and Radovanovik, P., Estimating the Cost of Machined Components During the Conceptual Design of a Product, CIRP Annals, 1989, Vol. 38, No. 1, pp. 157–160.
24. Boothroyd, G., and Schorr-Kon, T., Power Availability and Cost of Machine Tools, Report 18, Department of Industrial and Manufacturing Engineering, University of Rhode Island, 1987.
25. Boothroyd, G., and Reynolds, C., Approximate Machining Cost Estimates, Report 17, Department of Industrial and Manufacturing Engineering, University of Rhode Island, 1987.
26. PERA, Survey of Machining Requirements in Industry, PERA, Melton Mowbray, United Kingdom.

3

Product Design for Manual Assembly

3.1 INTRODUCTION

Design for assembly (DFA) should be considered at all stages of the design process, but especially the early stages. As the design team conceptualizes alternative solutions, it should give serious consideration to the ease of assembly of the product or subassembly. The team needs a DFA tool to effectively analyze the ease of assembly of the products or subassemblies it designs. The design tool should provide quick results and be simple and easy to use. It should ensure consistency and completeness in its evaluation of product assemblability. It should also eliminate subjective judgment from design assessment, allow free association of ideas, enable easy comparison of alternative designs, ensure that solutions are evaluated logically, identify assembly problem areas, and suggest alternative approaches for simplifying the product structure—thereby reducing manufacturing and assembly costs.

By applying a DFA tool, communication between manufacturing and design engineering is improved, and ideas, reasoning, and decisions made during the design process become well documented for future reference.

The *Product Design for Assembly* handbook [1] originally developed as a result of extensive university research, and, more recently, expanded versions in software form provide systematic procedures for evaluating and improving product design for ease of assembly. This goal is achieved by providing assembly information at the conceptualization stage of the design process in a logical and organized fashion. This approach also offers a clearly defined procedure for evaluating a design with respect to its ease of assembly. In this manner a feedback loop is provided to aid designers in measuring improvements resulting from

85

specific design changes. This procedure also functions as a tool for motivating designers; through this approach they can evaluate their own designs and, if possible, improve them. In both cases, the design can be studied and improved at the conceptual stage when it can be simply and inexpensively changed. The DFA method accomplishes these objectives by:

1. Providing a tool for the designer or design team which assures that considerations of product complexity and assembly take place at the earliest design stage. This eliminates the danger of focusing exclusively during early design on product function with inadequate regard for product cost and competitiveness.
2. Guiding the designer or design team to simplify the product so that savings in both assembly costs and piece parts can be realized.
3. Gathering information normally possessed by the experienced design engineer and arranging it conveniently for use by less-experienced designers.
4. Establishing a database that consists of assembly times and cost factors for various design situations and production conditions.

The analysis of a product design for ease of assembly depends to a large extent on whether the product is to be assembled manually, with special-purpose automation, with general-purpose automation (robots), or a combination of these. For example, the criteria for ease of automatic feeding and orienting are much more stringent than those for manual handling of parts. In this chapter we shall introduce design for manual assembly, since it is always necessary to use manual assembly costs as a basis for comparison. In addition, even when automation is being seriously considered, some operations may have to be carried out manually, and it is necessary to include the cost of these in the analysis.

3.2 GENERAL DESIGN GUIDELINES FOR MANUAL ASSEMBLY

As a result of experience in applying DFA it has been possible to develop general design guidelines that attempt to consolidate manufacturing knowledge and present them to the designer in the form of simple rules to be followed when creating a design. The process of manual assembly can be divided naturally into two separate areas, handling (acquiring, orienting and moving the parts) and insertion and fastening (mating a part to another part or group of parts). The following design for manual assembly guidelines specifically address each of these areas.

3.2.1 Design Guidelines for Part Handling

In general, for ease of part handling, a designer should attempt to:

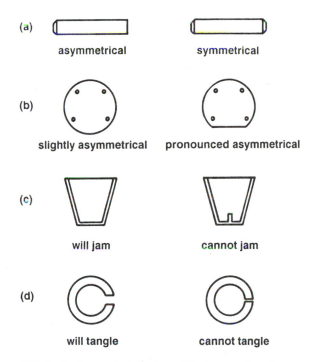

FIG. 3.1 Geometrical features affecting part handling.

1. Design parts that have end-to-end symmetry and rotational symmetry about the axis of insertion. If this cannot be achieved, try to design parts having the maximum possible symmetry (see Fig. 3.1a).
2. Design parts that, in those instances where the part cannot be made symmetric, are obviously asymmetric (see Fig. 3.1b).
3. Provide features that will prevent jamming of parts that tend to nest or stack when stored in bulk (see Fig. 3.1c).
4. Avoid features that will allow tangling of parts when stored in bulk (see Fig. 3.1d).
5. Avoid parts that stick together or are slippery, delicate, flexible, very small, or very large or that are hazardous to the handler (i.e., parts that are sharp, splinter easily, etc.) (see Fig. 3.2).

3.2.2 Design Guidelines for Insertion and Fastening

For ease of insertion a designer should attempt to:

1. Design so that there is little or no resistance to insertion and provide chamfers to guide insertion of two mating parts. Generous clearance

FIG. 3.2   Some other features affecting part handling.

should be provided, but care must be taken to avoid clearances that will result in a tendency for parts to jam or hang-up during insertion (see Figs. 3.3 to 3.6).

2. Standardize by using common parts, processes, and methods across all models and even across product lines to permit the use of higher volume processes that normally result in lower product cost (see Fig. 3.7).

3. Use pyramid assembly—provide for progressive assembly about one axis of reference. In general, it is best to assemble from above (see Fig. 3.8).

4. Avoid, where possible, the necessity for holding parts down to maintain their orientation during manipulation of the subassembly or during the placement of another part (see Fig. 3.9). If holding down is required, then try to design so that the part is secured as soon as possible after it has been inserted.

5. Design so that a part is located before it is released. A potential source of problems arises from a part being placed where, due to design constraints, it

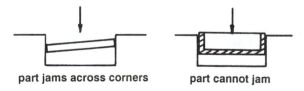

FIG. 3.3   Incorrect geometry can allow part to jam during insertion.

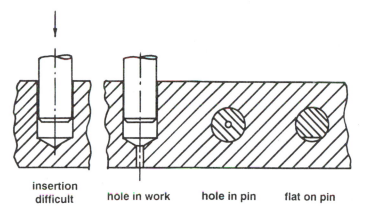

FIG. 3.4 Provision of air-relief passages to improve insertion into blind holes.

must be released before it is positively located in the assembly. Under these circumstances, reliance is placed on the trajectory of the part being sufficiently repeatable to locate it consistently (see Fig. 3.10).

6. When common mechanical fasteners are used the following sequence indicates the relative cost of different fastening processes, listed in order of increasing manual assembly cost (Fig. 3.11).

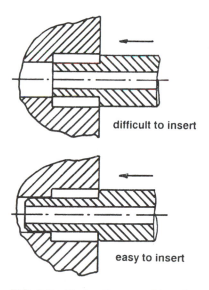

FIG. 3.5 Design for ease of insertion—assembly of long stepped bushing into counterbored hole.

part can hang-up   **part falls into place**

FIG. 3.6   Provision of chamfers to allow easy insertion.

old design

new design

FIG. 3.7   Standardize parts.

    a. Snap fitting
    b. Plastic bending
    c. Riveting
    d. Screw fastening

7. Avoid the need to reposition the partially completed assembly in the fixture (see Fig. 3.12).

Although functioning well as general rules to follow when design for assembly is carried out, guidelines are insufficient in themselves for a number of reasons.

FIG. 3.8 Single-axis pyramid assembly.

self-locating

**holding down and alignment
required for subsequent operation**

FIG. 3.9 Provision of self-locating features to avoid holding down and alignment.

**part must be released
before it is located**

part located before release

FIG. 3.10 Design to aid insertion.

FIG. 3.11   Common fastening methods.

First, guidelines provide no means by which to evaluate a design quantitatively for its ease of assembly. Second, there is no relative ranking of all the guidelines that can be used by the designer to indicate which guidelines result in the greatest improvements in handling, insertion, and fastening; there is no way to estimate the improvement due to the elimination of a part, or due to the redesign of a part for handling, etc. It is, then, impossible for the designer to know which guidelines to emphasize during the design of a product.

Finally, these guidelines are simply a set of rules that, when viewed as a whole, provide the designer with suitable background information to be used to develop a design that will be more easily assembled than a design developed without such a background. An approach must be used that provides the designer with an organized method that encourages the design of a product that is easy to assemble. The method must also provide an estimate of how much easier it is to assemble one design, with certain features, than to assemble another design with different features. The following discussion describes the DFA methodology, which provides the means of quantifying assembly difficulty.

FIG. 3.12   Insertion from opposite direction requires repositioning of assembly.

3.3   DEVELOPMENT OF THE SYSTEMATIC DFA METHODOLOGY

Starting in 1977, analytical methods were developed [2] for determining the most economical assembly process for a product and for analyzing ease of manual, automatic, and robot assembly. Experimental studies were performed [3–5] to measure the effects of symmetry, size, weight, thickness, and flexibility on manual handling time. Additional experiments were conducted [6] to quantify the effect of part thickness on the grasping and manipulation of a part using tweezers, the effect of spring geometry on the handling time of helical compression springs, and the effect of weight on handling time for parts requiring two hands for grasping and manipulation.

Regarding the design of parts for ease of manual insertion and fastening, experimental and theoretical analyses were performed [7–11] on the effect of chamfer design on manual insertion time, the design of parts to avoid jamming during assembly, the effect of part geometry on insertion time, and the effects of obstructed access and restricted vision on assembly operations.

A classification and coding system for manual handling, insertion, and fastening processes, based on the results of these studies, was presented in the form of a time standard system for designers to use in estimating manual assembly times [12,13]. To evaluate the effectiveness of this DFA method the ease of assembly of a two-speed reciprocating power saw and an impact wrench were analyzed and the products were then redesigned for easier assembly [14]. The initial design of the power saw (Fig. 3.13) had 41 parts and an estimated assembly time of 6.37 min. The redesign (Fig. 3.14) had 29 parts for a 29% reduction in part count, and an estimated assembly time of 2.58 min for a 59% reduction in assembly time. The outcome of further analyses [14] was a more than 50% savings in assembly time, a significant reduction in parts count and an anticipated improvement in product performance.

3.4   ASSEMBLY EFFICIENCY

An essential ingredient of the DFA method is the use of a measure of the DFA index or "assembly efficiency" of a proposed design. In general, the two main factors that influence the assembly cost of a product or subassembly are

The number of parts in a product.
The ease of handling, insertion, and fastening of the parts.

The DFA index is a figure obtained by dividing the theoretical minimum assembly time by the actual assembly time. The equation for calculating the DFA index E_{ma} is

$$E_{ma} = N_{min}t_a/t_{ma} \tag{3.1}$$

FIG. 3.13   Power saw (initial design—41 parts, 6.37 min assembly time). (After Ref. 14.)

where N_{min} is the theoretical minimum number of parts, t_a is the basic assembly time for one part, and t_{ma} is the estimated time to complete the assembly of the product. The basic assembly time is the average time for a part that presents no handling, insertion, or fastening difficulties (about 3 s).

The figure for the theoretical minimum number of parts represents an ideal situation where separate parts are combined into a single part unless, as each part is added to the assembly, one of the following criteria is met:

1. During the normal operating mode of the product, the part moves relative to all other parts already assembled. (Small motions do not qualify if they can be obtained through the use of elastic hinges.)

FIG. 3.14   Power saw (new design—29 parts, 2.58 min assembly time). (After Ref. 14.)

2.   The part must be of a different material than, or must be isolated from, all other parts assembled (for insulation, electrical isolation, vibration damping, etc.).

3.   The part must be separate from all other assembled parts; otherwise the assembly of parts meeting one of the preceding criteria would be prevented.

It should be pointed out that these criteria are to be applied without taking into account general design or service requirements. For example, separate fasteners will not generally meet any of the preceding criteria and should always be considered for elimination. To be more specific, the designer considering the design of an automobile engine may feel that the bolts holding the cylinder head onto the engine block are necessary separate parts. However, they could be eliminated by combining the cylinder head with the block—an approach that has proved practical in certain circumstances.

If applied properly, these criteria require the designer to consider means whereby the product can be simplified, and it is through this process that enormous improvements in assemblability and manufacturing costs are often achieved. However, it is also necessary to be able to quantify the effects of changes in design schemes. For this purpose the DFA method incorporates a system for estimating assembly cost which, together with estimates of parts cost, will give the designer the information needed to make appropriate trade-off decisions.

3.5 CLASSIFICATION SYSTEMS

The classification system for assembly processes is a systematic arrangement of part features that affect acquisition, movement, orientation, insertion, and fastening of the part together with some operations that are not associated with specific parts such as turning the assembly over.

Selected portions of the complete classification system, its associated definitions, and the corresponding time standards are presented in tables in Figs. 3.15 to 3.17. It can be seen that the classification numbers consist of two digits; the first digit identifies the row and the second digit identifies the column in the table.

The portion of the classification system for manual insertion and fastening processes is concerned with the interaction between mating parts as they are assembled. Manual insertion and fastening consists of a finite variety of basic assembly tasks (peg-in-hole, screw, weld, rivet, press-fit, etc.) that are common to most manufactured products.

It can be seen that for each two-digit code number, an average time is given. Thus, we have a set of time standards that can be used to estimate manual assembly times. These time standards were obtained from numerous experiments, some of which will now be described.

3.6 EFFECT OF PART SYMMETRY ON HANDLING TIME

One of the principal geometrical design features that affects the times required to grasp and orient a part is its symmetry. Assembly operations always involve at least two component parts: the part to be inserted and the part or assembly (receptacle) into which the part is inserted [15]. Orientation involves the proper alignment of the part to be inserted relative to the corresponding receptacle and can always be divided into two distinct operations: (1) alignment of the axis of the

for parts that can be grasped and manipulated with one hand without the aid of grasping tools

sym (deg) = (alpha+ beta)		no handling difficulties			part nests or tangles		
		thickness > 2mm		< 2mm	thickness > 2mm		< 2mm
		size > 15mm	6mm < size < 15mm	size > 6mm	size > 15mm	6mm < size < 15mm	size > 6mm
		0	1	2	3	4	5
sym < 360	0	1.13	1.43	1.69	1.84	2.17	2.45
360 <= sym < 540	1	1.5	1.8	2.06	2.25	2.57	3.0
540 <= sym < 720	2	1.8	2.1	2.36	2.57	2.9	3.18
sym = 720	3	1.95	2.25	2.51	2.73	3.06	3.34

for parts that can be lifted with one hand but require two hands because they severely nest or tangle, are flexible or require forming etc.

	alpha <= 180		alpha = 360
	size > 15mm	6mm < size <15mm	size > 6mm
	0	1	2
4	4.1	4.5	5.6

FIG. 3.15 Selected manual handling time standards, seconds (parts are within easy reach, are no smaller than 6 mm, do not stick together, and are not fragile or sharp). (Copyright 1999 Boothroyd Dewhurst, Inc.)

part that corresponds to the axis of insertion, and (2) rotation of the part about this axis.

It is therefore convenient to define two kinds of symmetry for a part:

1. *Alpha symmetry*: depends on the angle through which a part must be rotated about an axis perpendicular to the axis of insertion to repeat its orientation.

part inserted but not secured immediately or secured by snap fit

		secured by separate operation or part				secured on insertion by snap fit	
		no holding down required		holding down required			
		easy to align	not easy to align	easy to align	not easy to align	easy to align	not easy to align
		0	1	2	3	4	5
no access or vision difficulties	0	1.5	3.0	2.6	5.2	1.8	3.3
obstructed access or restricted vision	1	3.7	5.2	4.8	7.4	4.0	5.5
obstructed access and restricted vision	2	5.9	7.4	7.0	9.6	7.7	7.7

part inserted and secured immediately by screw fastening with power tool
(times are for 5 revs or less and do not include a tool acquisition time of 2.9s)

		easy to align	not easy to align
		0	1
no access or vision difficulties	3	3.6	5.3
restricted vision only	4	6.3	8.0
obstructed access only	5	9.0	10.7

FIG. 3.16 Selected manual insertion time standards, seconds (parts are small and there is no resistance to insertion). (Copyright 1999 Boothroyd Dewhurst, Inc.)

2. *Beta symmetry*: depends on the angle through which a part must be rotated about the axis of insertion to repeat its orientation.

For example, a plain square prism that is to be inserted into a square hole would first have to be rotated about an axis perpendicular to the insertion axis. Since, with such a rotation, the prism will repeat its orientation every 180°, it can be

	screw tighten with power tool	manipulation, reorientation or adjustment	addition of non solids
	0	1	2
6	5.2	4.5	7

FIG. 3.17 Selected separate operation times, seconds (solid parts already in place). (Copyright 1999 Boothroyd Dewhurst, Inc.)

Definitions:

For Fig. 3.15
Alpha is the rotational symmetry of a part about an axis perpendicular to its axis of insertion. For parts with one axis of insertion, end-to-end orientation is necessary when alpha equals 360 degrees, otherwise alpha equals 180 degrees.

Beta is the rotational symmetry of a part about its axis of insertion. The magnitude of rotational symmetry is the smallest angle through which the part can be rotated and repeat its orientation. For a cylinder inserted into a circular hole, beta equals zero.

Thickness is the length of the shortest side of the smallest rectangular prism that encloses the part. However, if the part is cylindrical, or has a regular polygonal cross-section with five or more sides, and the diameter is less than the length, then thickness is defined as the radius of the smallest cylinder which can enclose the part.

Size is the length of the longest side of the smallest rectangular prism that can enclose the part.

For Fig. 3.16
Holding down required means that the part will require gripping, realignment, or holding down before it is finally secured.

Easy to align and position means that insertion is facilitated by well designed chamfers or similar features.

Obstructed access means that the space available for the assembly operation causes a significant increase in the assembly time.

Restricted vision means that the operator has to rely mainly on tactile sensing during the assembly process.

termed 180° alpha symmetry. The square prism would then have to be rotated about the axis of insertion, and since the orientation of the prism about this axis would repeat every 90°, this implies a 90° beta symmetry. However, if the square prism were to be inserted in a circular hole, it would have 180° alpha symmetry

α	0	180	180	90	360	360
β	0	0	90	180	0	360

FIG. 3.18 Alpha and beta rotational symmetries for various parts.

and 0° beta symmetry. Figure 3.18 gives examples of the symmetry of simple-shaped parts.

A variety of predetermined time standard systems are presently used to establish assembly times in industry. In the development of these systems, several different approaches have been employed to determine relationships between the amount of rotation required to orient a part and the time required to perform that rotation. Two of the most commonly used systems are the methods time measurement (MTM) and work factor (WF) systems.

In the MTM system, the "maximum possible orientation" is employed, which is one-half the beta rotational symmetry of a part defined above [16]. The effect of alpha symmetry is not considered in this system. For practical purposes, the MTM system classifies the maximum possible orientation into three groups, namely, (1) symmetric, (2) semisymmetric, and (3) nonsymmetric [3], Again, these terms refer only to the beta symmetry of a part.

In the WF system, the symmetry of a part is classified by the ratio of the number of ways the part can be inserted to the number of ways the part can be grasped preparatory to insertion [17]. In the example of a square prism to be inserted into a square hole, one particular end first, it can be inserted in four ways out of the eight ways it could be suitably grasped. Hence, on the average, one-half of the parts grasped would require orientation, and this is defined in the WF system as a situation requiring 50% orientation [17]. Thus, in this system, account is taken of alpha symmetry, and some account is taken of beta symmetry. Unfortunately, these effects are combined in such a way that the classification can only be applied to a limited range of part shapes.

Numerous attempts were made to find a single parameter that would give a satisfactory relation between the symmetry of a part and the time required for orientation. It was found that the simplest and most useful parameter was the sum

FIG. 3.19 Effect of symmetry on the time required for part handling. Times are average for two individuals and shaded areas represent nonexistent values of the total angle of symmetry.

of the alpha and beta symmetries [5]. This parameter, which will be termed the total angle of symmetry, is therefore given by

$$\text{Total angle of symmetry} = \text{alpha} + \text{beta} \tag{3.2}$$

The effect of the total angle of symmetry on the time required to handle (grasp, move, orient, and place) a part is shown in Fig. 3.19. In addition, the shaded areas indicate the values of the total angle of symmetry that cannot exist. It is evident from these results that the symmetry of a part can be conveniently classified into five groups. However, the first group, which represents a sphere, is not generally of practical interest; therefore, four groups are suggested that are employed in the coding system for part handling (Fig. 3.15). Comparison of these experimental results with the MTM and WF orientation parameters showed that these parameters do not account properly for the symmetry of a part [5].

3.7 EFFECT OF PART THICKNESS AND SIZE ON HANDLING TIME

Two other major factors that affect the time required for handling during manual assembly are the thickness and the size of the part. The thickness and size of a part are defined in a convenient way in the WF system, and these definitions have

FIG. 3.20   Effect of part thickness on handling time.

been adopted for the DFA method. The thickness of a "cylindrical" part is defined as its radius, whereas for noncylindrical parts the thickness is defined as the maximum height of the part with its smallest dimension extending from a flat surface (Fig. 3.20). Cylindrical parts are defined as parts having cylindrical or other regular cross sections with five or more sides. When the diameter of such a part is greater than or equal to its length, the part is treated as noncylindrical. The reason for this distinction between cylindrical and noncylindrical parts when defining thickness is illustrated by the experimental curves shown in Fig. 3.20. It can be seen that parts with a "thickness" greater than 2 mm present no grasping or handling problems. However, for long cylindrical parts this critical value would have occurred at a value of 4 mm if the diameter had been used for the "thickness". Intuitively, we see that grasping a long cylinder 4 mm in diameter is equivalent to grasping a rectangular part 2 mm thick if each is placed on a flat surface.

The size (also called the major dimension) of a part is defined as the largest nondiagonal dimension of the part's outline when projected on a flat surface. It is normally the length of the part. The effects of part size on handling time are shown in Fig. 3.21. Parts can be divided into four size categories as illustrated. Large parts involve little or no variation in handling time with changes in their size; the handling time for medium and small parts displays progressively greater sensitivity with respect to part size. Since the time penalty involved in handling very small parts is large and very sensitive to decreasing part size, tweezers will usually be required to manipulate such parts. In general, tweezers can be assumed to be necessary when size is less than 2 mm.

FIG. 3.21 Effect of part size on handling time.

3.8 EFFECT OF WEIGHT ON HANDLING TIME

Work has been carried out [18] on the effects of weight on the grasping, controlling, and moving of parts. The effect of increasing weight on grasping and controlling is found to be an additive time penalty and the effect on moving is found to be a proportional increase of the basic time. For the effect of weight on a part handled using one hand, the total adjustment t_{pw} to handling time can be represented by the following equation [3]:

$$t_{pw} = 0.0125W + 0.011Wt_h \qquad (3.3)$$

where W (lb) is the weight of the part and t_h (s) is the basic time for handling a "light" part when no orientation is needed and when it is to be moved a short distance. An average value for t_h is 1.13, and therefore the total time penalty due to weight would be approximately $0.025W$.

If we assume that the maximum weight of a part to be handled using one hand is around 10–20 lb, the maximum penalty for weight is 0.25–0.5 s and is a fairly small correction. It should be noted, however, that Eq. (3.3) does not take into account the fact that larger parts will usually be moved greater distances, resulting in more significant time penalties. These factors will be discussed later.

3.9 PARTS REQUIRING TWO HANDS FOR MANIPULATION

A part may require two hands for manipulation when:

The part is heavy.
Very precise or careful handling is required.
The part is large or flexible.
The part does not possess holding features, thus making one-hand grasp difficult.

Under these circumstances, a penalty is applied because the second hand could be engaged in another operation—perhaps grasping another part. Experience shows that a penalty factor of 1.5 should be applied in these cases.

3.10 EFFECTS OF COMBINATIONS OF FACTORS

In the previous sections, various factors that affect manual handling times have been considered. However, it is important to realize that the penalties associated with each individual factor are not necessarily additive. For example, if a part requires additional time to move it from A to B, it can probably be oriented during the move. Therefore, it may be wrong to add the extra time for part size and an extra time for orientation to the basic handling time. The following gives some examples of results obtained when multiple factors are present.

3.11 EFFECT OF SYMMETRY FOR PARTS THAT SEVERELY NEST OR TANGLE AND MAY REQUIRE TWEEZERS FOR GRASPING AND MANIPULATION

A part may require tweezers when (Fig. 3.22):

Its thickness is so small that finger-grasp is difficult.
Vision is obscured and prepositioning is difficult because of its small size.
Touching it is undesirable, because of high temperature, for example.
Fingers cannot access the desired location.

A part is considered to nest or tangle severely when an additional handling time of 1.5 s or greater is required due to these factors. In general, two hands will be required to separate severely nested or tangled parts. Helical springs with open ends and widely spaced coils are examples of parts that severely nest or tangle. Figure 3.23 shows how the time required for orientation is affected by the alpha and beta angles of symmetry for parts that nest or tangle severely and may require tweezers for handling.

In general, orientation using hands results in a smaller time penalty than orientation using tweezers, therefore factors necessitating the use of tweezers should be avoided if possible.

**thickness so small
that finger grasp
is difficult**

**vision is obscured and pre-position-
ing is difficult because of small size**

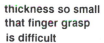 HOT

**fingers cannot
access desired
location**

undesirable to touch the part

FIG. 3.22 Examples of parts that may require tweezers for handling.

FIG. 3.23 Effect of symmetry on handling time when parts nest or tangle severely. (Disentangling time is not included.)

3.12 EFFECT OF CHAMFER DESIGN ON INSERTION OPERATIONS

Two common assembly operations are the insertion of a peg (or shaft) into a hole and the placement of a part with a hole onto a peg. The geometries of traditional conical chamfer designs are shown in Fig. 3.24. In Fig. 3.24a, which shows the

(a) Geometry of Peg

(b) Geometry of Hole

FIG. 3.24   Geometries of peg and hole.

design of a chamfered peg, d is the diameter of the peg, w_1 is the width of the chamfer, and θ_1 is the semiconical angle of the chamfer. In Fig. 3.24b, which shows the design of a chamfered hole, D is the diameter of the hole, w_2 is the width of the chamfer, and θ_2 is the semiconical angle of the chamfer. The dimensionless diametral clearance c between the peg and the hole is defined by

$$(D-d)/D \tag{3.4}$$

A typical set of results [9] showing the effects of various chamfer designs on the time taken to insert a peg in a hole are presented in Fig. 3.25. From these and other results, the following conclusions have been drawn:

1.   For a given clearance, the difference in the insertion time for two different chamfer designs is always a constant.
2.   A chamfer on the peg is more effective in reducing insertion time than the same chamfer on the hole.
3.   The maximum width of the chamfer that is effective in reducing the insertion time for both the peg and the hole is approximately $0.1D$.
4.   For conical chamfers, the most effective design provides chamfers on both the peg and the hole, with $w_1 = w_2 = 0.1D$ and $\theta_1 = \theta_2 < 45$.
5.   The manual insertion time is not sensitive to variations in the angle of the chamfer for the range $10 < \theta < 50$.

6. A radiused or curved chamfer can have advantages over a conical chamfer for small clearances.

It was learned from the peg insertion experiments [9] that the long manual insertion time for the peg and hole with a small clearance is probably due to the type of engagement occurring between the peg and the hole during the initial stages of insertion. Figure 3.26 shows two possible situations that will cause difficulties. In Fig. 3.26a, the two points of contact arising on the same circular cross section of the peg give rise to forces resisting the insertion. In Fig. 3.26b, the peg has become jammed at the entrance of the hole. An analysis was carried out to find a geometry that would avoid these unwanted situations. It showed that a chamfer conforming to a body of constant width (Fig. 3.27) is one of the designs having the desired properties. It was found that for such a chamfer, the insertion time is independent of the dimensionless clearance c in the range $c > 0.001$. Therefore, the curved chamfer is the optimum design for peg-in-hole insertion operations (Fig. 3.25). However, since the manufacturing costs for curved chamfers would normally be greater than for conical chamfers, the modified chamfer would only be worthy of consideration for very small values of clearance when the significant reductions in insertion time might compensate for the higher cost. An interesting example of a curved chamfer is the geometry of a bullet. Its design not only has aerodynamic advantages but is also ideal for ease of insertion.

FIG. 3.25   Effect of clearance on insertion time. (After Ref. 9.) (For clarity, experimental results are shown for only one case.)

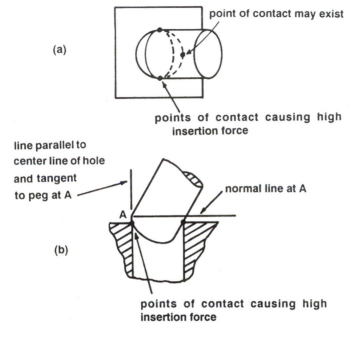

FIG. 3.26   Points of contact on chamfer and hole.

3.13 ESTIMATION OF INSERTION TIME

Empirical equations have been derived [9] to estimate the manual insertion time t_i for both conical chamfers and curved chamfers. For conical chamfers (Fig. 3.24), where the width of 45° chamfers is $0.1d$, the manual insertion time for a plain cylindrical peg t_i is given by

$$t_i = -70 \ln c + f(\text{chamfers}) + 3.7L + 0.75d \text{ ms} \tag{3.5}$$

or

$$t_i = 1.4L + 15 \text{ ms} \tag{3.6}$$

whichever is larger, and where

$$
\begin{aligned}
f(\text{chamfers}) = &-100 \text{ (no chamfer)}\\
&-220 \text{ (chamfer on hole)}\\
&-250 \text{ (chamfer on peg)}\\
&-370 \text{ (chamfer on peg and hole)}
\end{aligned}
\tag{3.7}
$$

d < r < D

FIG. 3.27 Chamfer of constant width.

For modified curved chamfers (Fig. 3.27) the insertion time is given by

$$t_i = 1.4L + 15 \tag{3.8}$$

Example: $D = 20$ mm, $d = 19.5$ mm, and $L = 75$ mm. There are chamfers on both peg and hole. From Eq. (3.4):

$$c = (20 - 19.5)/20 = 0.025$$

From Eq. (3.5):

$$t_i = -70 \ln(0.025) - 370 + 3.7(75) + 0.75(19.5)$$
$$= 181 \text{ ms}$$

From Eq. (3.6):

$$t_i = 120 \text{ ms}$$

3.14 AVOIDING JAMS DURING ASSEMBLY

Parts with holes that must be assembled onto a peg can easily jam if they are not dimensioned carefully. This problem is typical of assembling a washer on a bolt. In analyzing a part assembled on a peg [7] the hole diameter can be taken to be one unit; all other length dimensions are then measured relative to this unit and are dimensionless (Fig. 3.28). The peg diameter is $1 - c$, where c is the dimensionless diametral clearance between the two mating parts. The resultant force applied to the part during the assembly operation is denoted by P. The line of action of P intercepts the x axis at e, 0. If the following equation is satisfied, the part will slide freely down the peg:

$$P \cos \theta > \mu(N_1 + N_2) \tag{3.9}$$

By resolving forces horizontally

$$P \sin \theta + N_2 - N_1 = 0 \tag{3.10}$$

(a) Part Assembled on Peg   (b) Part Wedged on Peg

FIG. 3.28   Geometry of part and peg.

and by taking moments about $(0,0)$

$$\{[1 + L^2 - (1 - c)^2]^{1/2} + \mu(1 - c)\}N_2 - eP\cos\theta = 0 \qquad (3.11)$$

From Eqs. (3.9), (3.10), and (3.11):

$$(2\mu e/q - 1)\cos\theta + \mu\sin\theta < 0 \qquad (3.12)$$

where

$$q = [1 + L^2 - (1 - c)^2]^{1/2} + \mu(1 - c)$$

Thus, when $e = 0$ and $\cos\theta > 0$, the condition

$$\tan\theta < 1/\mu \qquad (3.13)$$

ensures free sliding. If $e = 0$ and $\cos\theta$ is less than 0, then the condition becomes

$$\tan\theta > 1/\mu \qquad (3.14)$$

In the case when $\theta = 0$ (the assembly force is applied vertically), Eq. (3.12) yield

$$2\mu e < q \qquad (3.15)$$

or

$$e = m(1 - c)/2 \qquad (3.16)$$

where m is a positive number. Substituting Eq. (3.16) into Eq. (3.15) gives

$$1 + L^2 > (1 - c)^2[\mu^2(m - 1)^2 + 1] \qquad (3.17)$$

When $m = 1$, the force is applied along the axis of the peg. Because $(1 + L^2)$ must always be larger than $(1 - c)^2$, the parts will never jam under these circumstances.

Even if the part jams, a change in the line of action of the force applied will free the part. However, it is also necessary to consider whether the part can be rotated and wedged on the peg. If the net moment of the reaction forces at the contact points is in the direction that rotates the part from the wedged position, then the part will free itself. Thus, for the part to free itself when released:

$$1 + L^2 > (1 - c)^2(\mu^2 + 1) \tag{3.18}$$

Comparing Eq. (3.17) with Eq. (3.18), shows that the condition for the part to wedge without freeing itself occurs when $m = 2$ in Eq. (3.17).

3.15 REDUCING DISC-ASSEMBLY PROBLEMS

When an assembly operation calls for the insertion of a disc-shaped part into a hole, jamming or hang-up is a common problem. Special handling equipment can prevent jams but a simpler, less-costly solution is to analyze all part dimensions carefully before production begins.

Again, the diameter of the hole is one unit; all other dimensions are measured relative to this unit and are dimensionless (Fig. 3.29). The disc diameter is $1 - c$, where c is the dimensionless diametral clearance between the mating parts, P is the resultant force in the assembly operation, and μ is the coefficient of friction. When a disc with no chamfer is inserted into a hole, the condition for free sliding can be determined by

$$L^2 > \mu^2 + 2c - c^2 \tag{3.19}$$

If c is very small, then Eq. (3.19) can be expressed as

$$L > \mu + c/\mu \tag{3.20}$$

If the disc is very thin, that is, if

$$(1 - c)^2 + L^2 < 1 \tag{3.21}$$

the disc can be inserted into the hole by keeping its circular cross section parallel to the wall of the hole and reorienting it when it reaches the bottom of the hole.

FIG. 3.29   Geometry of disc and hole.

3.16   EFFECTS OF OBSTRUCTED ACCESS AND RESTRICTED VISION ON INSERTION OF THREADED FASTENERS OF VARIOUS DESIGNS

Considerable experimental work has been conducted on the time taken to insert threaded fasteners of different types under a variety of conditions. Considering first the time taken to insert a machine screw and engage the threads, Fig. 3.30a shows the effects of the shape of the screw point and hole entrance, when the assembly worker cannot see the operation and when various levels of obstruction are present. When the distance from the obstructing surface to the hole center was greater than 16 mm, the surface had no effect on the manipulations and the restriction of vision was the only factor. Under these circumstances, the standard screw inserted into a recessed hole gave the shortest time. For a standard screw with a standard hole an additional 2.5 s was required. When the hole was closer to the wall, thereby inhibiting the manipulations, a further time of 2 or 3 s was necessary.

Figure 3.30b shows the results obtained under similar conditions but when vision was not restricted. Comparison with the previous results indicates that restriction of vision had little effect when access was obstructed. This was because the proximity of the obstructing surface allowed tactile sensing to take the place of sight. However, when the obstruction was removed, restricted vision could account for up to 1.5 s additional time.

Once the screw threads are engaged, the assembly worker must grasp the necessary tool, engage it with the screw, and perform sufficient rotations to tighten the screw. Figure 3.31 shows the total time for these operations for a variety of screw head designs and for both hand-operated and power tools. There was no restriction on tool operation for any of these situations. Finally, Fig. 3.32 shows the time to turn down a nut using a variety of hand-operated tools and

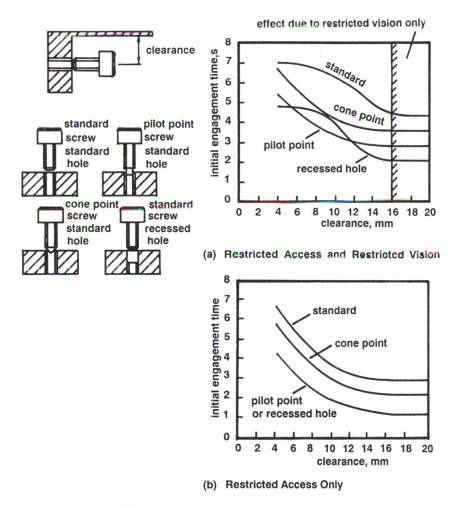

FIG. 3.30 Effects of restricted access and restricted vision on initial engagement of screws.

where the operation of the tools was obstructed to various degrees. It can be seen that the penalties for a box-end wrench are as high as 4 s per revolution when obstructions are present. However, when considering the design of a new product, the designer will not normally consider the type of tool used and can reasonably expect that the best tool for the job will be selected. In the present case this would be either the nut driver or the socket ratchet wrench.

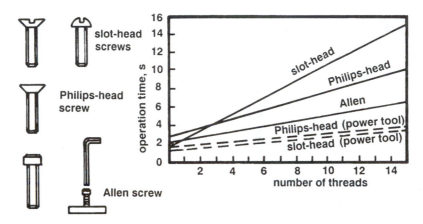

FIG. 3.31 Effect of number of threads on time to pick up the tool, engage the screw, tighten the screw, and replace the tool.

FIG. 3.32 Effect of obstructed access on time to tighten a nut.

3.17   EFFECTS OF OBSTRUCTED ACCESS AND RESTRICTED VISION ON POP-RIVETING OPERATIONS

Figure 3.33 summarizes the results of experiments [10] on the time taken to perform pop-riveting operations. In the experiments, the average time taken to pick up the tool, change the rivet, move the tool to the correct location, insert the rivet and return the tool to its original location was 7.3 s. In Fig. 3.33a the combined effects of obstructed access and restricted vision are summarized, and Fig. 3.33b shows the effects of obstructed access alone. In the latter case, time penalties of up to 1 s can be incurred although, unless the clearances are quite small, the penalties are negligible. With restricted vision present, much higher penalties, on the order of 2 to 3 s, were obtained.

3.18   EFFECTS OF HOLDING DOWN

Holding down is required when parts are unstable after insertion or during subsequent operations. It is defined as a process that, if necessary, it maintains the position and orientation of parts already in place prior to or during subsequent operations. The time taken to insert a peg vertically through holes in two or more stacked parts can be expressed as the sum of a basic time t_b and a time penalty t_p. The basic time is the time to insert the peg when the parts are prealigned and self-locating, as shown in Fig. 3.34a and can be expressed [11] as:

$$t_b = -0.07 \ln c - 0.1 + 3.7L + 0.75d_g \tag{3.22}$$

where
  $c = (D - d)/D$ and is the dimensionless clearance $(0.1 \geq c \geq 0.0001)$
  $L =$ the insertion depth in meters
  $d_g =$ the grip size in meters $(0.1 \text{ m} \geq d_g \geq 0.01 \text{ m})$.

Example: $D = 20$ mm, $d = 19.6$ mm, $c = (D - d)/D = (20 - 19.6)/20 = 0.02$, $L = 100$ mm $= 0.10$ m, $d_g = 40$ mm $= 0.04$ m then

$$t_b = -0.07 \ln c - 0.1 + 3.7L + 0.75d_g$$
$$= -0.07 \ln 0.02 - 0.1 + 3.7 \times 0.10 + 0.75 \times 0.04$$
$$= 0.27 - 0.1 + 0.37 + 0.03$$
$$= 0.57 \text{ s}$$

The graphs presented in Figs. 3.34 and 3.35 will allow the time penalty t_p to be determined for three conditions:

When easy-to-align parts have been aligned and require holding down (Fig. 3.34b)

View from above

View from side

(a) vision restricted   (b) vision unrestricted

FIG. 3.33 Effects of obstructed access and restricted vision on the time to insert a pop rivet. (After Ref. 10.)

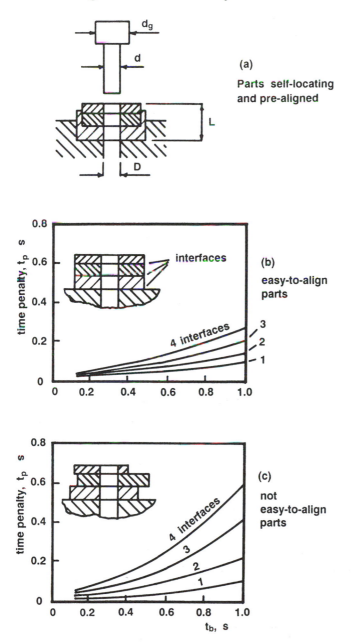

FIG. 3.34 Effects of holding down on insertion time. (After Ref. 11.)

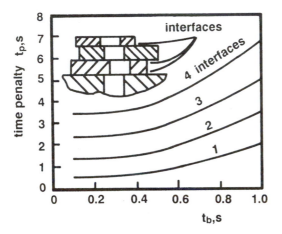

FIG. 3.35 Effects of holding down and realignment on insertion time for difficult-to-align parts. (After Ref. 11.)

When difficult-to-align parts have been aligned and require holding down (Fig. 3.34c)

When difficult-to-align parts require alignment and holding down (Fig. 3.35).

For the example given above where $t_b = 0.57$ s, the time penalty $t_p = 0.1$ s for the conditions of Fig. 3.34b, the time penalty $t_p = 0.15$ s for the conditions of Fig. 3.34c, and $t_p = 3$ s for the conditions of Fig. 3.35.

3.19 MANUAL ASSEMBLY DATABASE AND DESIGN DATA SHEETS

The preceding sections have presented a selection of the results of some of the analyses and experiments conducted during the development of the DFA method. For the development of the classification schemes and time standards presented earlier it was necessary to obtain an estimate of the average time, in seconds, to complete the operation for all the parts falling within each classification or category. For example, the uppermost left-hand box in Fig. 3.15 (code 00) gives 1.13 for the average time to grasp, orient, and move a part

That can be grasped and manipulated with one hand
Has a total symmetry angle of less than 360° (a plain cylinder, for example)
Is larger than 15 mm
Has a thickness greater than 2 mm
Has no handling difficulties, such as flexibility, tendency to tangle or nest, etc.

Clearly, a wide range of parts will fall within this category and their handling times will vary somewhat. The figure presented is only an average time for the range of parts.

To illustrate the type of problem that can arise through the use of the group technology coding or classification scheme employed in the DFA method, we can consider the assembly of a part having a thickness of 1.9 mm. We shall assume that, except for its thickness of less than 2 mm, the part would be classified as code 00 (Fig. 3.15). However, because of the part's thickness, the appropriate code would be 02 and the estimated handling time would be 1.69 instead of 1.13 s, representing a time penalty of 0.56 s. Turning now to the results of experiments for the effect of thickness (Fig. 3.20), it can be seen that for a cylindrical part the actual time penalty is on the order of only 0.01 to 0.02 s. We would therefore expect an error in our results of about 50%. Under normal circumstances, experience has shown that these errors tend to cancel—with some parts the error results in an overestimate of time and with some an underestimate. However, if an assembly contains a large number of identical parts, care must be taken to check whether the part characteristics fall close to the limits of the classification; if they do, then the detailed results presented above should be consulted.

3.20   APPLICATION OF THE DFA METHODOLOGY

To illustrate how DFA is applied in practice, we shall consider the controller assembly shown in Fig. 3.36. The assembly of this product first involves securing a series of assemblies to the metal frame using screws, connecting these assemblies together in various ways and then securing the resulting assembly into the plastic cover, again using screws. An undesirable feature of the design of the plastic cover is that the small subassemblies must be fastened to the metal frame before the metal frame can be secured to the plastic cover.

Figure 3.37 shows a completed worksheet analysis for the controller in the form of a tabulated list of operations and the corresponding assembly times and costs. Each assembly operation is divided into handling and insertion, and the corresponding times and two-digit code numbers for each process are given. Assembly starts by placing the pressure regulator (a purchased item) upside-down into a fixture. The metal frame is placed onto the projecting spindle of the pressure regulator and secured with the nut. The resulting assembly is then turned over in the fixture to allow for the addition of other items to the metal frame.

Next the sensor and the strap are placed and held in position while two screws are installed. Clearly, the holding of these two parts and the difficulty of the screw insertions will impose time penalties on the assembly process.

After tape is applied to the thread on the sensor, the adaptor nut can be screwed into place. Then one end of the tube assembly is screwed to the threaded

FIG. 3.36   Controller assembly.

extension on the pressure regulator and the other end to the adaptor nut. Clearly, both of these are difficult and time-consuming operations.

The printed circuit board (PCB) assembly is now positioned and held in place while two screws are installed, after which its connector is snapped into the sensor and the earth lead is snapped into place.

The whole assembly must be turned over once again to allow for the positioning and holding of the knob assembly while the screw-fastening operation can be carried out. Finally, the plastic cover is placed in position and the entire

	No. of items RP	Tool aquire time TA	Hand-ling code	Hand-ling time TH	Insert-ion code	Insert-ion time TI	Total time TA+RP* (TH+TI)	Mini-mum part count	
1. Pressure regulator	1	-	30	1.95	00	1.5	3.45	1	Place in fixture
2. Metal frame	1	-	30	1.95	02	2.6	4.55	1	Add
3. Nut	1	2.9	00	1.13	31	5.3	9.33	0	Add and screw fasten
4. Reorientation	1	-	-	-	61	4.5	4.50	-	Reorient and adjust
5. Sensor	1	-	30	1.95	03	5.2	7.15	1	Add
6. Strap	1	-	20	1.80	03	5.2	7.00	0	Add and hold down
7. Screw	2	2.9	11	1.80	31	5.3	17.10	0	Add and screw fasten
8. Apply tape	1	-	-	-	62	7.0	7.00	-	Special operation
9. Adapter nut	1	2.9	10	1.50	51	10.7	15.10	0	Add and screw fasten
10. Tube assembly	1	-	42	5.60	10	3.7	9.30	0	Add and screw fasten
11. Screw fastening	1	2.9	-	-	60	5.2	8.10	-	Standard operation
12. PCB assembly	1	-	42	5.60	03	5.2	10.80	1	Add and hold down
13. Screw	2	2.9	11	1.80	31	5.3	17.10	0	Add and screw fasten
14. Connector	1	-	30	1.95	05	3.3	5.25	0	Add and snap fit
15. Earth lead	1	-	42	5.60	05	3.3	8.90	0	Add and snap fit
16. Reorientation	1	-	-	-	61	4.5	4.50	-	Reorient and adjust
17. Knob assembly	1	-	30	1.95	03	5.2	7.15	1	Add and hold down
18. Screw fastening	1	2.9	-	-	60	5.2	8.10	-	Standard operation
19. Plastic cover	1	-	30	1.95	03	5.2	7.15	0	Add and hold down
20. Reorientation	1	-	-	-	61	4.5	4.50	-	Reorient and adjust
21. Screw	3	2.9	11	1.80	51	10.7	40.1	0	Add and screw fasten
							206.43	5	Totals

FIG. 3.37 Completed worksheet analysis for the controller assembly.

assembly is turned over for the third time to allow the three screws to be inserted. It should be noted that access for the insertion of these screws is very restricted.

It is clear from this description of the assembly sequence that many aspects of the design could be improved. However, a step-by-step analysis of each operation is necessary before changes to simplify the product structure and reduce assembly difficulties can be identified and quantified. First we shall look at how the handling and insertion times are established. The addition of the strap will be considered by way of example. This operation is the sixth item on the worksheet and the line of information is completed as follows:

NUMBER OF ITEMS, RP

There is one strap.

HANDLING CODE

The insertion axis for the strap is horizontal in Fig. 3.36 and the strap can only be inserted one way along this axis, so the alpha angle of symmetry is 360°. If the strap is rotated about the axis of insertion, it will repeat its orientation every 180°, which is, therefore, the beta angle of symmetry. Thus, the total angle of symmetry is 540°. Referring to the database for handling time (Fig. 3.15), since the strap can be grasped and manipulated using one hand without the aid of tools and alpha plus beta is 540°, the first digit of the handling code is 2. The strap presents no handling difficulties (can be grasped and separated from bulk easily), its thickness is greater than 2 mm, and its size is greater than 15 mm; therefore, the second digit is 0 giving a handling code of 20.

HANDLING TIME PER ITEM, TH

A handling time of 1.8 s corresponds to a handling code of 20 (Fig. 3.15).

INSERTION CODE

The strap is not secured as part of the insertion process and since there is no restriction to access or vision, the first digit of the insertion code is 0 (Fig. 3.16). Holding down is necessary while subsequent operations are carried out, and the strap is not easy to align because no features are provided to facilitate alignment of the screw holes. Therefore, the second digit will be 3, giving an insertion code of 03.

INSERTION TIME PER ITEM, TI

An insertion time of 5.2 s corresponds to an insertion code of 03 (Fig. 3.16).

TOTAL OPERATION TIME

This is the sum of the handling and insertion times multiplied by the number of items plus tool acquisition time if necessary i.e., TA + RP(TH + TI). For the strap the total operation time is therefore 7.0 s.

FIGURES FOR MINIMUM PARTS

As explained earlier, the establishment of a theoretical minimum part count is a powerful way to identify possible simplifications in the product structure. For the strap the three criteria for separate parts are applied after the pressure regulator, the metal frame, the nut, and the sensor have been assembled.

1. The strap does not move relative to these parts and so it could theoretically be combined with any of them.
2. The strap does not have to be of a different material—in fact it could be of the same plastic material as the body of the sensor and therefore take the form of two lugs with holes projecting from the body. At this point in the analysis the designer would probably determine that since the sensor is a purchased stock item, its design could not be changed. However, it is important to ignore these economic considerations at this stage and consider only theoretical possibilities.
3. The strap clearly does not have to be separate from the sensor in order to allow assembly of the sensor, and therefore none of the three criteria are met and the strap becomes a candidate for elimination. For the strap a zero is placed in the column for minimum parts.

3.20.1 Results of the Analysis

Once the analysis is complete for all operations the appropriate columns can be summed. Thus, for the controller, the total number of parts and subassemblies is 19 and there are six additional operations. The total assembly time is 206.43 s, and, for an assembly worker's rate of $30/hr, the corresponding assembly cost is $1.72. The theoretical minimum number of parts is five.

A DFA index is now obtained using Eq. (3.1). In this equation t_a is the basic assembly (handling and insertion) time for one part and can be taken as 3 s on average. Thus, the DFA index is

$$5 \times 3/206.43 = 0.07 \text{ or } 7\%$$

The high-cost processes should now be identified—especially those associated with the installation of parts that do not meet any of the criteria for separate parts. From the worksheet results (Fig. 3.37) it can be seen that attention should clearly be paid to combining the plastic cover with the metal frame. This would eliminate the assembly operation for the cover, the three screws, and the reorientation operation—representing a total time saving of 52.05 s—a figure that forms 25% of the total assembly time. Of course, the designer must check that the cost of the combined plastic cover and frame is less than the total cost of the individual items.

A summary of the items that can be identified for elimination or combination and the appropriate assembly time savings are presented in Table 3.1.

TABLE 3.1 Design Changes and Associated Savings

	Design change	Items	Time saving (s)
1.	Combine plastic cover with frame, eliminate three screws and a reorientation	19, 20, 21	52.05
2.	Eliminate strap and two screws (provide snaps in plastic frame to hold sensor if necessary)	6, 7	24.1
3.	Eliminate screws holding PCB assembly (provide snaps in plastic frame)	13	17.1
4.	Eliminate two reorientations	4, 16	9.0
5.	Eliminate tube assembly and two screw fastening operations (screw adaptor nut and sensor direct to the pressure regulator)	10, 11	17.4
6.	Eliminate earth lead (not required with plastic frame)	15	8.7
7.	Eliminate connector (plug sensor into PCB)	14	5.25

We have now identified design changes that could result in savings of 133.8 s of assembly time, which forms 65% of the total. In addition, several items of hardware would be eliminated, resulting in reduced part costs. Figure 3.38 shows a conceptual redesign of the controller in which all the proposed design changes have been made, and Fig. 3.39 presents the corresponding revised worksheet. The total assembly time is now 77.93 s and the assembly efficiency is increased to 19%—a fairly respectable figure for this type of assembly. Of course, the designer or design team must now consider the technical and economic consequences of the proposed designs.

First there is the effect on the cost of the parts. However, experience shows, and this example would be no exception, that the savings from parts cost reduction would be greater than the savings in assembly costs, which in this case is $1.07. It should be realized that the documented savings in materials,

FIG. 3.38   Conceptual redesign of the controller assembly.

	No. of items RP	Tool aquire time TA	Hand -ling code	Hand- ling time TH	Insert -ion code	Insert -ion time TI	Total time TA+RP* (TH+TI)	Mini- mum part count	
1. Pressure regulator	1	-	30	1.95	00	1.5	3.45	1	Place in fixture
2. Plastic cover	1	-	30	1.95	03	5.2	7.15	1	Add and hold down
3. Nut	1	2.9	00	1.13	31	5.3	9.33	0	Add and screw fasten
4. Knob assembly	1	-	30	1.95	03	5.2	7.15	1	Add and hold down
5. Screw fastening	1	2.9	-	-	60	5.2	8.10	-	Standard operation
6. Reorientation	1	-	-	-	61	4.5	4.50	-	Reorient and adjust
7. Apply tape	1	-	-	-	62	7.0	7.00	-	Special operation
8. Adapter nut	1	2.9	10	1.50	51	10.7	15.10	0	Add and screw fasten
9. Sensor	1	-	30	1.95	31	5.3	7.25	1	Add and screw fasten
10. PCB assembly	1	-	42	5.60	05	3.3	8.9	1	Add and snap fit
							77.93	5	Totals

FIG. 3.39 Completed analysis for the controller assembly redesign.

manufacturing, and assembly represent direct costs. To obtain a true picture, overheads must be added and these can often amount to 200% or more. In addition, there are other savings more difficult to quantify. For example, when a part such as the metal frame is eliminated, all associated documentation— including part drawings—is also eliminated. Also, the part cannot be misassembled or fail in service—factors that lead to improved reliability, maintainability and quality of the product. It is not surprising, therefore, that many U.S. companies have been able to report annual savings measured in millions of dollars as a result of the application of the DFA analysis method described here.

3.21 FURTHER DESIGN GUIDELINES

Some guidelines or design rules for the manual handling and insertion of parts were listed earlier. However, it is possible to identify a few more general guidelines that arise particularly from the application of the minimum parts criteria, many of which found application in the analysis of the controller.

1. *Avoid connections*: If the only purpose of a part or assembly is to connect A to B, then try to locate A and B at the same point. Figure 3.40 illustrates this guideline. Here the two connected assemblies are rearranged to provide increasing assembly and manufacturing efficiency. Also, two practical examples occurred during the analysis of the controller, when it was found that the entire tube assembly could be eliminated and that the wires from the PCB assembly to the connector were not necessary (Fig. 3.36).

2. *Design so that access for assembly operations is not restricted*: Figure 3.41 shows two alternative design concepts for a small assembly. In the first concept the installation of the screws would be very difficult because of the restricted access within the box-shaped base part. In the second concept access is relatively unrestricted because the assembly is built up on the flat base part. An example of

Increased   Efficiency

FIG. 3.40   Rearrangement of connected items to improve assembly efficiency and reduce costs.

this type of problem occurred in the controller analysis when the screws securing the metal frame to the plastic cover were installed (item 21—Fig. 3.37).

3. *Avoid adjustments*: Figure 3.42 shows two parts of different materials secured by two screws in such a way that adjustment of the overall length of the assembly is necessary. If the assembly were replaced by one part manufactured

Restricted access for assembly of screws

FIG. 3.41   Design concept to provide easier access during assembly.

Stainless Steel Fingers

**Low Carbon Steel Bracket
adjustment required**

**Stainless Steel Bracket
no adjustment**

FIG. 3.42　Design to avoid adjustment during assembly.

from the more expensive material, difficult and costly operations would be avoided. These savings would probably more than offset the increase in material costs.

4. *Use kinematic design principles*: There are many ways in which the application of kinematic design principles can reduce manufacturing and assembly cost. Invariably, when located parts are overconstrained, it is necessary either to provide a means of adjustment of the constraining items or to employ more accurate machining operations. Figure 3.43 shows an example where to locate the square block in the plane of the page, six point constraints are used, each one requiring adjustment. According to kinematic design principles only three point constraints are needed together with closing forces. Clearly, the redesign shown in Fig. 3.43 is simpler, requiring fewer parts, fewer assembly operations, and less adjustment. In many circumstances designs where overconstraint is involved result in redundant parts. In the design involving overconstraint in Fig. 3.44 one

Overconstrained Design　　　**Sound Kinematic Design**

FIG. 3.43　Showing how overconstraint leads to unnecessary complexity in product design.

Over-constrained   **Kinematically sound**

FIG. 3.44   Showing how overconstraint leads to redundancy of parts.

of the pins is redundant. However, application of the minimum parts criteria to the design with a single pin would suggest combining the pin with one of the major parts and combining the washer with the nut.

3.22 LARGE ASSEMBLIES

In the original DFA (design for assembly) method estimates of assembly time were based on a group technology approach in which design features of parts and products were classified into broad categories and, for each category, average handling and insertion times were established. Clearly, for any particular operation, these average times can be considerably higher or lower than the actual times. However, for assemblies containing a significant number of parts, the differences tend to cancel so that the total time will be reasonably accurate. In fact, application of the DFA method in practice has shown that assembly time estimates are reasonably accurate for small assemblies in low-volume production where all the parts are within easy arm reach of the assembly worker.

Clearly, with large assemblies, the acquisition of the individual parts from their storage locations in the assembly area will involve significant additional time. Also, in mass production transfer-line situations, the data for low-volume production will overestimate these times. Obviously, one database of assembly times cannot be accurate for all situations.

Let us take one example. From the DFA databases in Figs. 3.15 to 3.17, the time for acquiring and inserting a standard screw, which is not easy to align, is 8.2 s. This time includes acquisition of the screw, placing it in the assembly manually with a couple of turns, acquiring the power tool, operating the tool, and then replacing it. However, in high-volume production situations, the screws are often automatically fed, and so the time is reduced to about 3.6 s per screw or, for well-designed screws, the time per screw can be less than 2 s.

The DFA method was extended to allow for these possibilities, and more accurate estimates of assembly times are obtained. However, this can reduce the effectiveness of the method. In the preceding example, an analysis using the shorter time for screw insertion would indicate that eliminating screws would not be so advantageous in reducing the assembly time. However, it is known that simplifying the product by combining parts and eliminating separate fasteners has the greatest benefit through reductions in parts cost rather than through reductions in assembly cost; yet the suggestions for these improvements arise from analyses of the assembly of the product. Hence, separate fasteners should, perhaps, be severely penalized even if they take little time to install. In fact, it can be argued that in the preceding example, where screws could be inserted quickly, special equipment was being used to solve problems arising from poor design. This is a good argument for suggesting that early DFA analyses should be carried out assuming that only standard equipment is available. Perhaps later, at the detailed-design stage, attempts can be made to improve the assembly time estimates. Clearly, accurate estimates cannot be made unless detailed descriptions of manufacturing and assembly procedures are available—a situation not present during the early stages of design when the possibilities for cost savings through improved product design are at their greatest. On the other hand, a database of assembly times suitable for small assemblies measuring only a few inches cannot be expected to give even approximate estimates for assemblies containing large parts measuring several feet. It is desirable, therefore, to have databases appropriate to those situations where the size of the product and the production conditions differ significantly. Again, it should be realized that great detail regarding the assembly work area will not generally be available to the designer during the conceptual stages of design.

With these points in mind, the following sections describe an approach to the development of databases that are used to estimate acquisition and insertion times for parts assembled into large products.

3.23   TYPES OF MANUAL ASSEMBLY METHODS

Part acquisition time is highly dependent on the nature of the layout of the assembly area and the method of assembly. For small parts placed within easy reach of the assembly worker, the handling times given in Fig. 3.15 are adequate if bench assembly (Fig. 3.45) or multistation assembly (Fig. 3.46) is employed. It is assumed in both cases that major body motions by the assembly worker are not required.

For volumes that do not justify transfer systems and if the assembly contains several parts that weigh more than about 5 lb or that are over 12 in. in size, it will not be possible to place an adequate supply of parts within easy arm's reach of the assembly worker. In this case, provided the largest part is less than 35 in. in size and no part weighs more than 30 lb, the modular assembly center might be used. This is an arrangement of workbench and storage shelves where the parts are situated as conveniently for the assembly worker as possible (Fig. 3.47). However, because turning, bending, or walking may be necessary for acquisition of some of the parts, the handling times will be increased. It is convenient to identify three modular work centers to accommodate assemblies falling within three size categories where the largest part in the assembly is less than 15 in., from 15 to 25 in., and from 25 to 35 in., respectively.

FIG. 3.45   Bench assembly.

FIG. 3.46   Multistation assembly.

FIG. 3.47   Modular assembly center.

For products with even larger parts, the custom assembly layout can be used. Here the product is assembled on a worktable or on the floor and the various storage shelves and auxiliary equipment are arranged around the periphery of the assembly area (Fig. 3.48). The total working area is larger than that for the modular assembly center and depends on the size category of the largest parts in the assembly. Three subcategories of the custom assembly layout are employed: for assemblies whose largest parts are from 35 to 50 in., from 50 to 65 in., and larger than 65 in.

Also, for large products a more flexible arrangement can be used; this is called the flexible assembly layout. The layout (Fig. 3.49) would be similar in size to the custom assembly layout and the same three subcategories would be employed according to the size of the largest part. However, the use of mobile storage carts and tool carts can make assembly more efficient.

In both the custom assembly layout and the flexible assembly layout, the possibility arises that mechanical assistance in the form of cranes or hand trucks may be needed. In these cases, the working areas may need to be increased in order to accommodate the additional equipment.

For high-volume assembly of products containing large parts (such as in the automobile industry) transfer lines moving past manual assembly stations would be employed (Fig. 3.50).

Two other manual assembly situations exist. The first is assembly of small products with very low volumes—perhaps in a clean room. This would include the assembly of intricate and sensitive devices such as the fuel control valves for an aircraft where instructions must be read for each step and where the worker is near the beginning of the learning curve. The second is where assembly of large products is mainly carried out on site. This type of assembly is usually termed

FIG. 3.48   Custom assembly layout.

installation, and an example would be the assembly and installation of a passenger elevator in a multistory building.

In any assembly situation, special equipment may be needed. For example, a positioning device is sometimes needed for positioning and aligning the part—especially prior to welding operations. In these cases, the device must be brought from storage within the assembly area, and then returned after the part has been positioned and perhaps secured. Thus, the total handling time for the device will be roughly twice the handling time for the part and must be taken into account if the volume to be produced is small.

Figure 3.51 summarizes the basic types of manual assembly methods described above. It can be seen that the first three methods assume only small parts are being assembled. In these cases it can be assumed that the parts are all placed close to hand and will be acquired one-at-a-time. Therefore if, say, six screws are to be inserted, there is no advantage in collecting the six screws simultaneously. However, with the assembly of products containing large parts, where small items such as fasteners may not be located within easy reach or where the assembly worker must move to various locations for the small items, there may be considerable advantage in acquiring multiple parts when needed.

FIG. 3.49 Flexible assembly layout.

3.24 EFFECT OF ASSEMBLY LAYOUT ON ACQUISITION TIMES

For assembly method categories, 4, 5, and 6, Fig. 3.52 presents a summary of the results obtained from a thorough study of typical assembly layouts of various sizes. For each of the nine subcategories described above, a typical layout was designed using standard items such as worktables and storage racks. Then the various sizes and weights of parts were assumed to be stored at the most suitable locations. An example for the custom assembly layout is shown in Fig. 3.53. Using MTM time standards [19], the times for the retrieval of parts within the various size and weight categories were then estimated [20]. Finally, the results were averaged to give the data in Fig. 3.52. In addition, since it was found that the times for the custom assembly and flexible assembly layout were similar, these were combined and averaged. Thus, for example, the basic part retrieval or acquisition time for the mid-sized custom assembly layout or the mid-sized flexible assembly layout (largest part 50 to 65 in.) was determined to be 11.61 s.

For the effect of part weight, a correction factor can be applied, as described in an earlier section. However, the resulting correction is quite small and, therefore, it would seem feasible to divide parts into broad weight categories of, say, 0 to

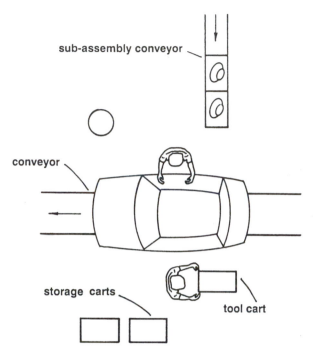

sub-assembly conveyor

conveyor

storage carts

tool cart

FIG. 3.50   Multistation assembly of large products.

30 lb and over 30 lb. The reason for the last category is that such parts would normally require two persons or lifting equipment for handling. Corrected values of handling time for the first weight category (0–30 lb) were obtained by assuming an average weight of 15 lb.

For the second weight category, the part requires two persons to handle. The figures for this category were obtained by estimating the time for two persons to acquire a part weighing 45 lb, doubling this time, and multiplying the result by a factor of 1.5. This factor allows for the fact that two persons working together will typically only manage to work in coordination for 67% of their time.

For the third weight category, where lifting equipment is needed, allowance must be provided for the time taken for the worker to acquire the equipment, use it to acquire the part, move the part to the assembly, release the part, and finally return the lifting equipment to its original location. Figure 3.52 gives the estimated times for these operations, assuming that no increase in the size of the assembly area was required in order to accommodate the lifting equipment. These estimates were obtained using the MOST time standard system [21] and included the times required to acquire the equipment, move it to the parts'

FIG. 3.51   Manual assembly methods.

location, hook the part, transport it to the assembly, unhook the part, and, finally, return the equipment to its original location.

For small parts, where several are required and can be grasped in one hand, it will usually be advantageous to acquire all the needed parts in one trip to the storage location. Figure 3.54 presents the results of an experiment where the effect of the distance traveled by the assembly worker on the acquisition and handling time per part was studied. It can be seen that when the parts are stored out of easy arm reach, it is preferable to acquire all the parts needed in the one trip to the storage location. From results such as these it is possible to estimate acquisition times for the multiple acquisition of parts stored away from the assembly fixture. Figure 3.52 presents these times for each of the product size categories.

In the work [20] leading to the development of Fig. 3.52, the size of the largest part in the product was assumed to determine the size and nature of the assembly area layout. However, to provide an alternative method of identifying the layout size, for each layout an average distance from the assembly to the storage location of the major parts was determined. These averages are listed in the second column in Fig. 3.52 and provide an alternative method for the determination of the appropriate layout.

Product Design for Manual Assembly

average distance to location of parts (ft.)	size of largest part in assembly (in.)		one item (small or large) or multiple small parts				small part – jumbled can be grasped in multiples additional time per part	
			weight < 30 lbs.		weight > 30 lbs.			
			easy to grasp	difficult to grasp (1)	two persons	manual crane	easy to grasp	difficult to grasp (1)
			0	1	2	3	4	5
< 4	< 15	0	2.54	4.54	8.82	18.42	0.84	1.11
4 to 7	15 to 25	1	4.25	6.25	14.34	27.10	0.84	1.11
7 to 10	25 to 35	2	5.54	7.54	18.54	31.22	0.84	1.11
10 to 13	35 to 50	3	9.93	11.93	32.76	39.50	0.84	1.11
13 to 16	50 to 65	4	11.61	13.61	36.75	44.93	0.84	1.11
> 16	> 65	5	12.41	14.41	40.80	50.07	0.84	1.11

factory assembled large products (2)

Notes: 1) For large items, no features to allow easy grasping (eg. no finger hold).
For small items, those that are slippery, nested, tangled, or stuck together require careful handling are difficult to grasp.
2) Times are for acquisition only. Multiply by 2 if replacement time is to be included (eg. fixture).

FIG. 3.52 Acquisition times (s) for items not stored within easy reach of the assembly worker. (Copyright Boothroyd Dewhurst, Inc. 1991.)

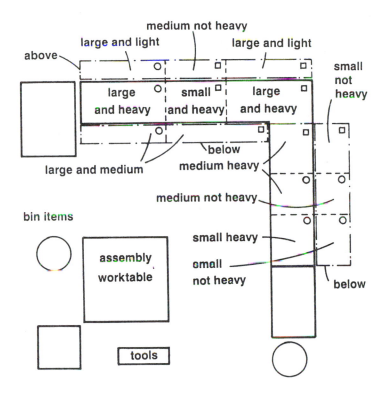

FIG. 3.53   Assumed distribution of parts in custom assembly layout: □, low turnover; ○, high turnover.

3.25 ASSEMBLY QUALITY

The design for assembly procedures described in this chapter have been used by Motorola, Inc., since the mid-1980s. In 1991 Motorola reported the results of a DFA redesign of the motor vehicle adapter for their family of two-way professional hand-held radios [22]. Their benchmarking of competitors' electronic products indicated a best-in-class manual assembly efficiency (DFA index), as given by Eq. (3.1), of 50%, and they evaluated many different concepts to reach that goal. The final design had 78% fewer parts than their previous vehicle adapter and an 87% reduction in assembly time. They also measured the assembly defect rates of the new design in production and compared the results to defect rates for the old design. The result was a 95.6% reduction in assembly defects per product. Encouraged by this result, the Motorola engineers surveyed a number of products that had been analyzed using DFA and produced a relationship between assembly defects per part and the DFA assembly efficiency values.

FIG. 3.54 Effects of distance to storage location on the acquisition and assembly time for small parts.

This relationship was discussed in Chapter 1, and as illustrated in Fig. 1.10 shows a strong relationship between assembly quality and the assembly efficiency value of a product.

These Motorola data were subsequently analyzed independently by other researchers [23] to produce an even more powerful relationship for use in early design evaluation. These researchers postulated that since DFA assembly time values are related to the difficulty of assembly operations, then the probability of an assembly error may also be a function of predicted assembly operation times. In the study it was reported that 50 combinations of defect rates versus assembly characteristics were tested for meaningful correlation. Of these, the variation of average assembly defect rate per operation with average DFA time estimate per operation showed the strongest linear correlation, with correlation coefficient $r = 0.94$. The actual data are shown in Fig. 3.55. The equation of the regression line is given by

$$D_i = 0.0001(t_i - 3.3)$$ (3.23)

where D_i = average probability of assembly defect per operation
        t_i = average assembly time per operation

As discussed earlier, the average assembly time predicted by the DFA time standard database, for parts which present no assembly difficulties, is approxi-

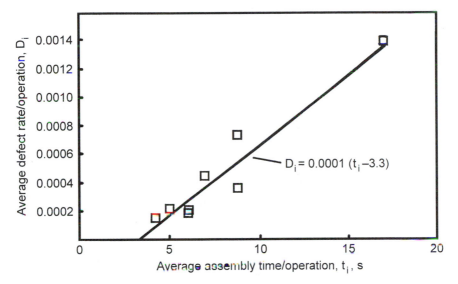

FIG. 3.55 Relation between assembly defect rate and average DFA assembly time. (After Ref. 23.)

mately 3 s. Thus Eq. (3.23) can be interpreted as an estimated assembly defect rate of 0.0001, or 1 in 10,000, for every second of extra time associated with difficulties of assembly. In fact, if the regression line in Fig. 3.55 is constrained to pass through $t_i = 3$, then the correlation coefficient is still 0.94 to two decimal places. For this reason we will use 3.0 instead of 3.3 in the expressions and calculations below.

For a product requiring n assembly operations, the probability of a defective product, containing one or more assembly errors, is therefore approximately

$$D_a = 1 - [1 - 0.0001(t_i - 3.0)]^n \qquad (3.24)$$

Alternatively, the expected number of assembly errors in one of these products is given by

$$N_d = 0.0001(t_i - 3.0)n \qquad (3.25)$$

These relationships can be applied very easily in the early stages of design to compare the likely assembly error rates of alternative design concepts for small products assembled in large quantities. This can provide powerful directional guidance for product quality improvements, since it is becoming widely accepted that faulty assembly steps, rather than defective components, are more often the reason for production quality problems [24].

For the controller assembly example discussed earlier, the existing design requires 25 operations for final assembly and has a total estimated assembly time, obtained from the DFA analysis, of 206.43 s; see Fig. 3.37. The average time per operation is thus 206.43/25, equal to 8.26 s. Applying Eq. (3.24) then gives

$$D_a = 1 - [1 - 0.0001(8.26 - 3.0)]^{25} = 0.0131 \tag{3.26}$$

Thus, the estimated probability of a defective assembly is 0.0131, or 1.31%.

For the redesigned controller assembly, the number of final assembly operations has been reduced to 10, and the estimated assembly time is 77.93 s; see Fig. 3.39. The average time per operation is now 7.8 s, and the likely number of defective assemblies is given by

$$D_a = 1 - [1 - 0.0001(7.8 - 3.0)]^{10} = 0.0048 \tag{3.27}$$

Thus the predicted assembly defect rate is about one-third of that for the original design. This defect rate could be reduced further, by considering detail design improvements to further decrease the average operation time of 7.8 s.

The expression $(t_i - 3.0)n$ in Eq. (3.25) can be interpreted as the total time penalty associated with the assembly of a product. For example, for the original design of the controller assembly, the 25 assembly operations would take only 75.0 s if there were no assembly difficulties. The total assembly time penalty is $(8.25 - 3.0)\,25$, equal to 131.4 s. Equation (3.25) thus predicts that the assembly defects per unit will be proportional to the total assembly time penalty. This interpretation is supported by another set of industrial quality data illustrated in Fig. 3.56. These data were obtained from a disk drive manufacturer [25], and it

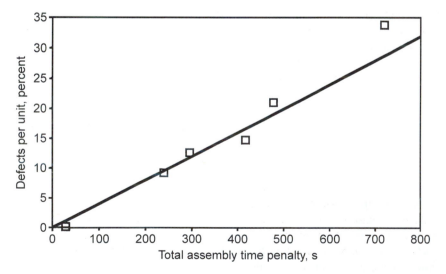

FIG. 3.56 Relation between total DFA assembly time penalty and assembly defect rates per unit. (After Ref. 26.)

should be noted that one outlier with a defect rate of approximately 65% has been omitted from the figure. It should also be noted that the slope of the regression line in Fig. 3.56 is approximately 0.0004. This suggests that the assembly workers are making mistakes at four times the rate of those involved in building the Motorola products. This may be the result of the delicate nature of disk drive assemblies and the need to maintain very small clearances in these devices.

3.26 APPLYING LEARNING CURVES TO THE DFA TIMES

Learning-curve theory applied to industrial production has its roots in early aerospace manufacturing. A now-famous paper by T. P. Wright [26] resulted from the study of small aircraft construction in the 1920s and 1930s. Wright noted that as repetitive tasks are performed, improvement occurs at a diminishing rate. In particular, he noted that twice as many repetitions are required to achieve each successive constant incremental improvement. This constant improvement with doubling is embodied in a simple power law, referred to as the Wright model, which is expressed as

$$T_{1,x} = T_1 x^b \tag{3.28}$$

where
$T_{1,x}$ = average time of production for x units
T_1 = time of production for the first unit
x = number of identical units
b = the reduction exponent

The reduction exponent can, in turn, take the general form

$$b = \log r / \log f \tag{3.29}$$

where
r = the average time for a factor increase in output divided by the time for the first output expressed as a percentage
f = the factor increase in output

Normally a doubling of output is the basis for learning curve analyses, and the improvement is given as r percent. For this case Eq. (3.29) becomes

$$b = \log(r/100) / \log 2 \tag{3.30}$$

For example, the often-quoted 90% learning curve represents the situation where the average time to produce $2x$ units will be 90% of the average time taken to produce the first x of those units. From Eq. (3.30) this gives

$$b = \log 0.9 / \log 2 = -0.152$$

Note that if b is known, then the learning curve percent value is given by the inverse of Eq. (3.30)

$$r = 2^b 100\% \tag{3.31}$$

Other forms of learning curve have been proposed and used since the work of Wright; see Refs. [27] and [28]. In particular, in cases where the times for producing successive individual units are of primary interest a more appropriate model may be represented by

$$T_x = T_1 x^c \tag{3.32}$$

where
$\quad T_x =$ time to produce the xth unit
$\quad x =$ number of identical units
$\quad c =$ "individual" reduction exponent

This alternative model is attributed to Crawford [29], whose work was carried out for Lockheed Aircraft Corporation. For batch assembly work the Wright model is more appropriate and will be used below to calculate learning-curve adjustments to predicted assembly times.

The reduction exponent represents the improvements that would be expected as the worker "learns" to operate more efficiently through repetitive identical tasks. However, Eq. (3.28) can be viewed in a much broader sense as a progress function that models progress accomplished through improved process techniques, tools, and training methods, as well as the learning associated with repetitions of the same task [30]. For these situations a learning-curve percentage value of 85 is more appropriate than the 90% value typically used to model worker performance improvement only [27].

The learning-curve equations described above can be used to adjust DFA times for situations in which only a few assemblies are to be produced. To carry out this adjustment, the number of task repetitions, for which the DFA times apply, must first be established. Since the handling and insertion experiments used to obtain the DFA times were mainly based on 100 task repetitions, this number will be used as the basis for the example calculations below. In other words, it will be assumed that DFA times are equivalent to $T_{1,100}$ times on the appropriate learning curve. We can thus make learning-curve transformations, using the Wright model, as follows:

$$T_{1,100} = T_1 100^b \tag{3.33}$$

or

$$T_1 = T_{1,100}/100^b \tag{3.34}$$

Thus for the first assembly of a small batch B of products, the average time will be given by the Wright model as

$$T_{1,B} = T_1 B^b \tag{3.35}$$

Finally, substituting from Eq.(3.34) for T_1 gives

$$T_{1,B} = T_{1,100}(B/100)^b \tag{3.36}$$

where

b = reduction exponent
$T_{1,B}$ = adjusted DFA time for batch size B
B = total batch to be assembled
$T_{1,100}$ = assumed basic DFA time value

Example: A set of only five measuring instruments is to be produced. A DFA analysis shows an estimated assembly time given by T_A. Using a 90% learning curve, provide an estimate for the average time to build the first two units. Use this value to determine the average time to assemble the next three units. Assume that the DFA time estimate would apply well for the average assembly time of 100 units.

(a) Substituting T_A for $T_{1,100}$ in Eq. (3.36) and using $b = -0.152$ for a 90% learning curve gives

$$T_{1,2} = (2/100)^{-0.152} T_A = 1.81 T_A \tag{3.37}$$

(b) The predicted assembly time for all five units is given by

$$5 \times T_{1,2} = 5(5/100)^{-0.152} T_A = 7.88 T_A \tag{3.38}$$

Similarly the total time for the first two units is

$$2 \times T_{1,2} = 2(2/100)^{-0.152} T_A = 3.62 T_A \tag{3.39}$$

Thus, the average time for units 3, 4, and 5 can be given by

$$T_{3,5} = (7.88 - 3.62) T_A / 3 = 1.42 T_A \tag{3.40}$$

REFERENCES

1. Boothroyd, G., Product Design for Assembly, Department of Mechanical Engineering, University of Massachusetts, Amherst, 1980.
2. Boothroyd, G., Design for Economic Manufacture, Annals CIRP, Vol. 28, No. 1, 1979, p. 345.
3. Yoosufani, Z., and Boothroyd, G., Design of Parts for Ease of Handling, Report No. 2, Department of Mechanical Engineering, University of Massachusetts, September 1978.

4. Boothroyd, G., Design for Manual Handling and Assembly, Report No. 4, Department of Mechanical Engineering, University of Massachusetts, Amherst, September 1979.

5. Yoosufani, Z., Ruddy, M., and Boothroyd, G., Effect of Part Symmetry on Manual Assembly Times, J. Manuf. Syst., Vol. 2, No. 2, 1983, pp. 189–195.

6. Seth, B., and Boothroyd, G., Design for Manual Handling, Report No. 9, Department of Mechanical Engineering, University of Massachusetts, January 1979.

7. Ho, C., and Boothroyd, G., Avoiding Jams During Assembly, Machine Des., Tech. Brief, January 25, 1979.

8. Ho, C., and Boothroyd, G., Reducing Disc-Assembly Problems, Machine Des., Tech. Brief, March 8, 1979.

9. Ho, C., and Boothroyd, G., Design of Chamfers for Ease of Assembly, Proc. 7th North American Metalworking Conference, May 1979, p. 345.

10. Fujita, T., and Boothroyd, G., Data Sheets and Case Study for Manual Assembly, Report No. 16, Department of Mechanical Engineering, University of Massachusetts, Amherst, April 1982.

11. Yang, S.C., and Boothroyd, G., Data Sheets and Case Study for Manual Assembly, Report No. 15, Department of Mechanical Engineering, University of Massachusetts, Amherst, December 1981.

12. Dvorak, W.A., and Boothroyd, G., Analysis of Product Designs for Ease of Manual Assembly: A Systematic Approach, Report No. 11, Department of Mechanical Engineering, University of Massachusetts, Amherst, May 1982.

13. De Lisser, W.A., and Boothroyd, G., Analysis of Product Designs for Ease of Manual Assembly: A Systematic Approach, Report No. 17, Department of Mechanical Engineering, University of Massachusetts, Amherst, May 1982.

14. Ellison, B., and Boothroyd, G., Applying Design for Assembly Handbook to Reciprocating Power Saw and Impact Wrench, Report No. 10, Department of Mechanical Engineering, University of Massachusetts, Amherst, August 1980.

15. Karger, W., and Bayha, F.H., Engineered Work Measurement, Industrial Press, Inc., New York, 1966.

16. Raphael, David L., A Study of Positioning Movements, Research Report No. 109, MTM Association of Standards and Research, Ann Arbor, Michigan, 1957 and 1967.

17. Quick, Joseph H., Work Factor Time Standards, McGraw-Hill, New York, 1962.

18. Raphael, D.L., A Study of Arm Movements Involving Weight, Research Report No. 108, MTM Association of Standards and Research, Ann Arbor, Michigan 1955 and 1968.

19. Karge, D.W., and Hancock, W.M., Advanced Work Measurement, H.B. Maynard & Co., Pittsburgh, 1982.

20. Fairfield, M.C., and Boothroyd, G., Part Acquisition Time During Assembly of Large Products, Report No. 44, Department of Industrial and Manufacturing Engineering, University of Rhode Island, 1991.

21. Zandin, K.B., MOST Work Measurement Systems, 2nd ed., Marcel Dekker, New York, 1989.

22. Branan, W., DFA Cuts Assembly Defects by 80%, Appliance Manufacturer, November 1991.

23. Barkan, P., and Hinckley, C.M., The Benefits and Limitations of Structured Design Methodologies, Manufacturing Rev., Vol. 6, No. 3, September 1993.

24. Hinckley, C.M., The Quality Question, Assembly, November 1997.

25. Hinckley, C.M., Global Conformance Quality Model, Sandia National Laboratory Report SAND94-8451, 1994.

26. Wright, T.P., Factors Affecting the Cost of Aeroplanes, J. Aeronaut. Sci., Vol. 3, 49–73, 1936.
27. Teplitz, C.J., The Learning Curve Deskbook, Quorum Books, Westport, CT, 1991.
28. Belkaoni, A., The Learning Curve: A Management Accounting Tool, Greenwood Press, Westport, CT, 1986.
29. Crawford, J.R., Learning Curve, Ship Curve, Ratios, Related Data, Lockheed Aircraft Corporation, Burbank, CA.
30. Kniep, J.G., The Maintenance Program Function, J. Ind. Eng., November–December, 1965.

4
Electrical Connections and Wire Harness Assembly

4.1 INTRODUCTION

The design for manual assembly procedures described in Chapter 3 have been successfully applied to mechanical products. However, when products contain a significant number of electrical connections, the labor involved in their assembly often far outweighs the labor involved in the assembly of the mechanical parts and associated fasteners.

For example, Fig. 4.1 shows the potential reduction in assembly time through the redesign of a control unit [1]. It can be seen that in the original design, where the total assembly time was 260 min (4.3 h), about half was devoted to the hand soldering of wires and a further 31% was devoted to the mechanical fastening of wires. A proposed redesign using no hand soldering and eliminating many of the connections would reduce the total assembly time to only 33 min (0.55 h).

A further example is that of a descrambler for satellite TV reception. This product, about the size of a VCR, contained ten printed circuit boards and, as can be seen in Table 4.1, 51 connecting wires or cables and 31 circuit board jumper wires, making a total of 164 connections. Table 4.1 shows that these items involved an assembly time of 7236 s (2.01 h) which formed 68% of the total assembly time for the product of 10,613 s (2.95 h). Table 4.2 presents a summary of the opportunities for savings through redesign. For example, if the 10 circuit boards could be combined into one board thereby eliminating the interconnections and if the jumper wires could be avoided, we might be able to reduce assembly costs by 61%.

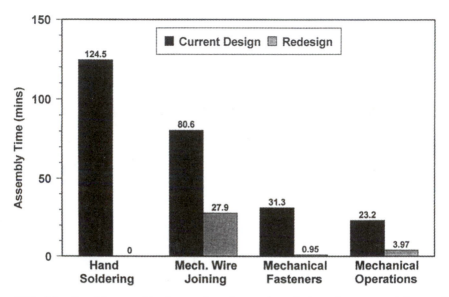

FIG. 4.1   Possible assembly time savings due to the redesign of a commercial control unit.

TABLE 4.1   Labor Content of a Descrambler

Number	Item	Assembly time (s)
51	Connecting wires or cables	5,768
31	Circuit board jumper wires	1,468
138	Mechanical fasteners	1,143
80	Electronic components manually inserted	796
112	Mechanical operations	753
20	Hand solder of components	403
33	Other parts identified as candidates for elimination	282
	Totals	10,613

TABLE 4.2   Possible Savings Through Descrambler Redesign

Design change	Saving ($)	Percent
Eliminate cabling	48.06	40
Combine boards	15.00	11
Eliminate jumper wires	12.23	10
Eliminate mechanical fasteners	9.53	8
Total	84.82	69

> The Personal System/2 Model 50 is the first computer to come with a built-in party game.
> The object of this game is to disassemble and reassemble the machine as quickly as possible. Screwdrivers are prohibited. Sounds impossible? With a little practice, you might be able to clock in at under a minute.

FIG. 4.2 Advertisement in *PC Magazine*, May 26, 1987.

Figure 4.2 shows one of the first advertisements for the IBM PS2 computer. This amazing achievement in ease of assembly was mainly brought about by the elimination of all internal cabling.

It is interesting to note that if the minimum parts criteria described in Chapter 3 are applied to connections of any kind, these connections can never be theoretically justified. In fact, if the sole purpose of an item is to connect A to B, then the item can be eliminated by arranging for A and B to be adjacent to one another, as was illustrated in Fig. 3.40. Often, unfortunately, other constraints dictate that A and B must be positioned at different locations and that connections are necessary. This means that it is desirable in DFMA analysis to be able to estimate the penalties resulting from such constraints. In other words, we must quantify the cost of interconnections.

When several electrical interconnections are necessary, the cables or wires can be installed separately in the product and then tied together and secured as appropriate. Alternatively, they can be assembled and tied together prior to installation in the product. This latter assembly of wires or cables is called a wire or cable harness assembly.

4.2 WIRE OR CABLE HARNESS ASSEMBLY

A completed harness assembly usually consists of a main trunk, where multiple wires or cables are bundled and tied together, with individual wires or smaller bundles of wires leaving the main trunk at various points known as breakouts. Figure 4.3 diagrammatically illustrates possible arrangements and terminology in wire harness assembly. Figure 4.4 shows the principal operations involved: starting with wire preparation, followed by harness assembly, and ending with installation in the product. Harnesses are usually constructed by manually laying out the individual wires or cables on a board that has a full-size schematic drawing of the harness mounted on its surface to guide the assembly worker. Also, pegs or nails are positioned in such a way that the wires are constrained to run along the required paths. During construction, the ends of the wires must be held in position. If wires are to be terminated in a connector, then the connector may previously have been inserted into a receptacle mounted on the board in the correct position. The ends of the terminated wires are then inserted into the back

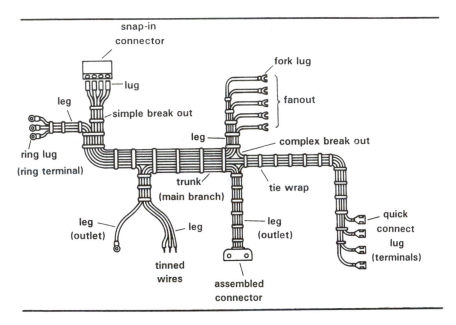

FIG. 4.3   Terminology.

of the connector. Sometimes, if the wires are sufficiently rigid, they are simply inserted into the connector, which does not need positive location on the board. Wire ends that are to be connected during installation in the product are retained temporarily on the board. This is often accomplished by the use of a helical spring mounted on the board with its axis horizontal and perpendicular to the direction of the wire. The assembly worker simply presses the wire end between two coils of the spring.

Once all the wires or cables have been laid (sometimes referred to as laid-in), they are bundled together using tie wraps, lacing, or split conduit tubing or they are bound with electrical tape. Usually, one assembly worker completes all of the operations involved in the construction phase. However, wire preparation usually requires special tools and is carried out in a separate earlier phase. In high-volume situations such as in the automobile industry, assembly of cable harnesses may be carried out in assembly line fashion. Here, the boards will be transferred from station to station with one or two assembly workers completing a few operations at each station.

When complex harnesses are assembled, the connector receptacles are wired to a computer that continually tests whether the wires have been inserted properly. Such on-line testing is typically carried out in the automobile and defense industries, for example.

FIG. 4.4   Principal operations.

Usually, each end of a wire must eventually be connected. This is often accomplished using an intermediate item such as a connector or wire termination to which the wire is directly connected. The final connection of the connector or terminal is made during installation in the product. Alternatively, wires may be connected directly to an item such as a printed wiring board, a switch, or a terminal block.

At various stages of manufacture, each wire or cable must be cut to length, terminated, installed, and finally connected. The stage at which these operations are carried out is generally determined by the complexity of the wiring installation or wire harness assembly. Figure 4.5 illustrates the sequence of operations for low, normal, and high complexities. For low complexity, where only a few wires are necessary, the wires (or cables) would first be cut to length and terminated. During final assembly they would be installed separately into the product and connected.

For normal complexity, typically occurring in the automobile, TV, and computer industries, the wires would first be cut and terminated, then laid on a board (laid-in) and tied, laced, or wrapped. After delivery to the product assembly station, they would be routed, connected, and finally secured.

High complexity refers to the type of harness used in the aerospace or defense industry. Here, the wire is arranged in position and then cut to length and terminated on the lay-in board.

The degree of automation used in wire harness assembly depends on the complexity of the harness and the quantity to be produced. However, automation is generally restricted to wire or cable preparation. Fully automatic machines are available that will cut wire to length and then strip and terminate both ends. Such machines are typically employed in the automobile industry but are finding increasing applications in lower volume production situations. More common is the semiautomatic procedure. Here the assembly worker transfers the wire to a series of special machines where individual operations such as cut, strip, and/or crimp are performed. All the wire handling is performed manually. Although some robotic cells have been developed for harness assembly, this operation is usually performed manually even in high-volume situations such as those in the automobile industry. Finally, installation is performed manually.

4.3 TYPES OF ELECTRICAL CONNECTIONS

Single conductors and cables are normally connected to terminals or connectors. The types of connections can be broadly classified into three categories, as shown in Fig. 4.6.

4.3.1 Solder Connections

Although relatively expensive, soldering is one of the most versatile and reliable methods of joining wires to pins, terminals, etc., and the main function of soldered wire connections is to provide a mechanical joint and an electrically conductive path. Wires may be secured mechanically to terminals prior to soldering, depending on the criticality of the connections. Mechanically secured soldered wires involve additional assembly time because the assembly worker

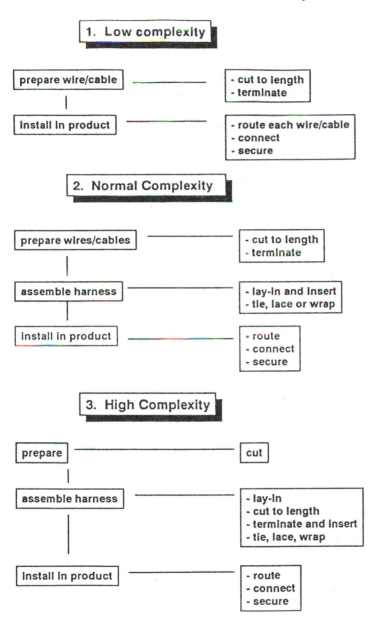

FIG. 4.5 Classification of electrical interconnections.

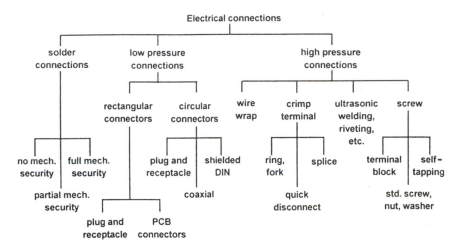

FIG. 4.6 Types of electrical connections.

needs to acquire a hand tool, move the tool to the wire, secure the wire to the terminal, and put aside the tool in addition to the subsequent soldering operation. A long-nose plier is often used for temporarily securing wires which are wrapped 180 degrees in electronic wiring. For full mechanical security, a 360-degree wrap can be used.

When mechanical strength requirements are not important, wires can be soldered without prior securing, as with solder lugs. For this type of connection, the diameter of the wire should be slightly less than that of the hole in the lug.

4.3.2 Low-Pressure Connections

Low-pressure connections can be separated without the aid of a tool. Electrical and electronic connectors fall into this category. Besides providing a good solid connection, connectors are particularly convenient when frequent disconnection and reconnection are required; they also tend to minimize errors during servicing because the wires remain in the correct sequence when disconnected. The general expectations from low-pressure connections are as follows [2]:

Sufficient contact force for good conduction
Good cleaning action during assembly
Low resistance to mating of parts
Low wear on parts
Long life
Ease in connecting and disconnecting

Some low-pressure connections have difficulty in meeting all the criteria stated above. For reliability, the higher the contact force, the better the connection. On the other hand, higher forces make a multicontact connector more difficult to connect and disconnect. Therefore a compromise has to be maintained among the preceding criteria.

Low-pressure connections can be grouped by their construction (e.g., rectangular and circular connectors). One-piece card-edge connectors and two-piece plug and receptacle connectors such as socket and PCB connectors are classified as rectangular connectors. With a one-piece card-edge connector, the PCB foils extend to an edge of the board, which is inserted into the connector. Two-piece connectors come in the form of male and female halves. The male half consist of pins (male contacts) and the female part consist of sockets (female contacts). Coaxial, shielded DIN, and other circular plug and receptacle connectors are classified as circular connectors.

Rectangular and circular connectors have a variety of contact styles and the frequency of mating will suggest the types of contacts to be used. The contacts can be machined or stamped, and they come in many shapes and sizes. Figure 4.7 shows pin and socket contacts, a combination commonly used because of its low cost and its ready availability. The disadvantages of these types of contacts are high insertion forces and susceptibility to damage. Other types of contacts include blade and fork contacts and bellows contacts [2]. Some high-priced systems can provide as many as 100,000 high-compliance contacts without failure [3]. They are generally found on two-piece connectors. The points to consider when choosing a connector are the types of conductors to be terminated,

FIG. 4.7 Pin and socket contacts.

the assemblies to be connected, the number of contacts required, the final product function, and the cost.

There are three methods used to connect wires to the connector contacts. In the first method the pins or sockets could be crimped onto wires and then poked into the connector body with a cylindrical-shaped hand tool. The second method involves soldering the wires to the contacts. Soldering generally requires a skilled worker and has high labor cost. The third method is the insulation displacement technique, where the insulation of the wire is pierced and displaced from the conductor. The conductor is then forced into the U-shaped contact in one operation, greatly reducing the time needed to connect a wire to the contact. This method is explained later when we consider multiwire flat cables that are terminated by insulation displacement

Cable connectors that are used for high-frequency signals must be positively and firmly coupled. Coupling devices for both male and female connectors have been designed for use in applications where there is not enough access space to turn a coupling nut (see Fig. 4.8). Other considerations involve the ability of the connector to stay coupled during shock and vibration conditions and to resist deterioration when used in a variety of environments. The commonly used methods of coupling circular connectors are screw-thread, push-on, or friction type and bayonet devices. The coupling devices used on rectangular connectors are latches or snap-on clips, locking clips, spring clips, or screws (Fig. 4.8).

4.3.3 High-Pressure Connections

The high-pressure connection is one that requires a tool to create a metal-to-metal contact under pressure and/or plastic deformation. The most commonly used methods under this category are wire-wrap, crimp, and screw connections. Other methods are ultrasonic welding, riveting, etc.

Wire wrap is the most reliable method for making point-to-point electrical connections between wires and terminals. It consists of wrapping the bare end of an insulated solid wire in a helix, tightly around a long terminal post having a square or rectangular cross section. The resulting high-pressure metal-to-metal contacts produce a gas-tight connection that has a large contact area with low resistance. With time, the molecules of the post and wire start to diffuse at the joint and the bond strength of the joint increases to nearly that of a welded joint. Stranded wire cannot be used for this joining method. Wire wrap can be performed manually using a pneumatic- or electric-powered rotary tool. Alternatively, semiautomatic or automatic methods are available. Semiautomatic wire wrap can connect over 300 wires/h with an error rate 10 to 20 times less than that of the manual method [2]. In the case of automatic wrapping, wrapping rates of 1000 wires/h with a reject rate of 0.05% are achievable. Figure 4.9 shows the standard wire-wrap connection.

FIG. 4.8 Coupling devices used on circular and rectangular connectors.

A crimped terminal is a permanent connection and it is used to connect individual wires to a terminal block with screws or it can be inserted into a connector block. Crimping can be done with hand tools or by machines. During crimping, pressure is applied to compress the terminal onto the wire to a critical depth. This depth is different for every terminal size and wire gauge. Two common configurations of the crimped terminal are the ring tongue and the fork tongue styles (see Fig. 4.10). The fork tongue style can be connected and

FIG. 4.9 Wire-wrap termination.

disconnected quickly and is used where mechanical security is not a major requirement.

The screw connection, probably the oldest type of mechanical connection, involves screwing down a metal screw or nut and clamping the wire under the screw head or nut. The wire is usually wrapped one turn around the screw or screwed post in a clockwise direction (Fig. 4.10). This method is particularly advantageous where field installation requires fast and simple connections. However, screw connections may deteriorate under conditions of corrosion or severe vibration, and they have voltage and electrical frequency limitations.

Terminal blocks, with screws attached to each connecting point and separated by insulated partitions, are used to connect discrete wires. The wires may be connected directly onto the terminal block using the bare wire ends, or the wires may be provided with lugs. The type of lug may influence the assembly time. For example, the use of ring lugs would require the operator to totally unscrew the connecting point whereas fork lugs eliminate this requirement (see Fig. 4.10).

FIG. 4.10 Attachment of wire, fork lug, and ring lug.

4.4 TYPES OF WIRES AND CABLES

Wire conductors are used to interconnect electrical and electronic systems to transmit current or signal and are selected for their current-carrying capacity, mechanical strength, type of insulation, and cost. Cables generally consist of two or more conductors within a common covering and are terminated by one or more connectors. Proper matching of connector and cable is important to ensure reliable field performance. Electrical conductors that are commonly used include single solid or stranded wire, twisted pair and trio, multiconductor cables, coaxial cable, ribbon cable and flexible flat cable. Soft, annealed copper is normally used for wires because of its high conductivity, ductility, and solderability.

Solid wire does not possess the flexibility characteristics nor the fatigue life of stranded wire and tends to fracture even under mild flexing. Solid wire may be used for such applications as jumpers on printed circuit boards where the leads are fixed securely and not subjected to vibration. The advantages of solid wires are its rigidity as well as its efficiency at high frequencies. Solid and stranded wire is commonly designated by American Wire Gauge (AWG) number, diameter of the wire in mils (thousandths of an inch), or cross section in circular mils (square of the diameter expressed in mils).

A twisted pair consists of two stranded conductor insulated wires twisted together. The number of twists per inch follows the engineering specification used in signal applications. The twist will reduce noise, similar to the hum in a radio. A twisted trio is the same as a twisted pair except that it consists of three conductors.

Multiconductor cable consists of two or more color-coded rubber- or PVC-insulated conductors. Hemp cord may be used as a filler to add strength. The outer insulation or covering is made of neoprene rubber or PVC. Cable of this type is used to carry power from a source to required areas within the unit or system.

Coaxial cable is used in radio-frequency circuits where the distributed capacity must be constant over the length of the cable. It consists of an insulated length of conductor enclosed in a conductive envelope of braided wire shield and an outer insulating jacket isolating the shield from ground.

Ribbon cable, sometimes called flat cable, consists of numerous conductors of the same gauge held side by side in flexible strips of insulation. The growing demand for the use of ribbon cables has resulted from the advantages of faster insulation stripping and mass termination at low cost. Mass termination is the process of simultaneously connecting a number of conductors to the same number of U-contacts of the connectors. During termination, using either manual or automatic equipment, the U-contact displaces the insulation and each conductor is then forced and wedged within a U-contact. The result is a high-pressure, gas-tight, solderless connection. It can be placed in narrow rectangular openings and used where great flexibility in one plane is needed. Wiring errors are eliminated, and the harness assembly is simplified when point-to-point wiring is required. Also, the ribbon cable has excellent characteristics for conveying high-speed digital signals and a more precise and fixed capacitance between conductors [3]. These characteristics are maintained in service because the separation between cable conductors remains constant.

Flexible circuitry, which includes flexible cable, can be bent, rolled, or folded many times, depending on the material used. Flexible circuits are becoming widely used as a form of interconnection in applications requiring size and weight reduction, controlled impedance, reduced labor, and ease of assembly [4]. They can be used as one-to-one connectors or as complex harnesses, allowing break-outs and special routing [5]. Replacing discrete wiring with flexible circuits can reduce assembly time considerably. Other benefits are reliability and service-ability of the circuits. A number of companies have converted from the use of conventional backplanes, motherboards, and wire harnessing to flexible and rigid/flexible circuits [4].

4.5 PREPARATION AND ASSEMBLY TIMES

In a study of preparation and assembly times [6] a variety of industrial time standards and published data were compared with on-site industrial studies and laboratory studies. The published data were taken from Refs. 7–10. Information was obtained from two companies, referred to as Company 1 and Company 2 in the text and as Co. 1 and Co. 2 in the figures.

4.5.1 Preparation

Figure 4.11 presents the times for manually and semiautomatically stripping one end of a wire. The manual operation involves inserting the wire to the correct stripping length and into the proper station of a stripping tool. The tool is then closed thereby severing the insulation, which is removed by the tool. An average experimental time of 7.0 s was obtained for stripping one end of a wire and this agreed closely with published times [8,10] and those observed in Company 2 (see Fig. 4.11). Machine stripping of wires involves grasping the wire and inserting the end into a stripping machine. This triggers the machine to strip off the insulation to the correct length. The stripping time obtained from published figures [10] suggests that an approximate time of 3 s can be assumed.

Figure 4.12 presents the times for tinning one bare end of a wire. In an experiment, the wire was held by a fixture and the bare end tinned using a soldering iron held in one hand and solder held in the other hand. An experimental time of 9 s was obtained and is comparable to that given in [8] and observed in Company 2. If a solder pot is used for tinning the wire end, an

FIG. 4.11 Results for stripping one wire end. (From Ref. 6.)

FIG. 4.12 Results for tinning one wire end. (From Ref. 6.)

average time of 7.2 s obtained from [8] and observed in Company 2 is recommended.

Figure 4.13 presents the times for crimping one terminal to the bare end of a wire. For manual crimping, a terminal is inserted into a color-coded notch of a crimping tool and the bare end of the wire is placed into the barrel of the terminal. The tool is then closed to compress the barrel so that it clamps tightly onto the wire. The experimental time for the manual operation was found to be 13.9 s. This time compared well; with [7,10] and with Company 2 but differed from that found from [8,9]. Crimp terminals come in various shapes and sizes and this may explain the wide variation in the times. The average experimental time of 13.9 s would seem to be a reasonable figure.

In the semiautomatic process of crimping terminals, the operation involves placing the bare end of the wire into the die set. This action triggers the crimping machine. After crimping, a new terminal advances automatically into the die set for the next crimping operation. It can be seen that the values obtained from various sources for the semiautomatic crimping operation agree closely, with the exception of the time observed in Company 2. A time of 3 s is suggested for this operation.

Semiautomatic machines are available that can strip the wire and crimp a terminal after insertion of unstripped wire into the machine. The insertion of wire is carried out manually, and from observations of machines that carried out these operations separately, a total time of 3.6 s is suggested for the combined operations. There are also machines capable of automatically cutting, stripping,

FIG. 4.13   Results for crimping a terminal to one wire end. (From Ref. 6.)

and crimping terminals to wires in one operation. The production rates vary, depending on the make and model of the machine used. This automatic operation can be as fast as 1.8 s per wire end for a length of 10 ft (3 m) [11].

For multiconductor cable, the experimental time for stripping the outer insulation and then the inner insulation of all conductors within the cable is shown in Fig. 4.14. It can be seen that the stripping times vary linearly with the number of conductors. A time of 30 s is required for stripping a multiconductor having two wires, and for each additional wire, a time of 6 s should be added. Also, the results indicate that the manual stripping of the outer insulation would take 18 s.

Figure 4.15 presents the manual assembly times for soldering the contacts of circular or rectangular connectors. The operation involves placing the connector into a fixture and filling all the solder cups. A handful of wires is then grasped using one hand, with the other hand holding the soldering iron. A solder cup is then reheated and simultaneously a bare wire is inserted. The wire is held in position until the solder solidifies. It can be seen that the times for filling solder cups and soldering wires to contacts vary linearly with the number of contacts to be soldered. The total experimental time t_s for assembling a rectangular or

FIG. 4.14 Experimental results for stripping a multiconductor cable: □, experimental. (From Ref. 6.)

FIG. 4.15   Manual assembly time for connector with solder contacts: ○, (7); ◤, (9); □, experimental (total); ▲, Co. 2. (From Ref. 6.)

circular connector is the summation of the times for filling solder cups $(6.8 + 3.1N_c)$ and soldering wires $(0.2 + 8.3N_c)$ and was found to be:

$$t_s = 7 + 11.4N_c \tag{4.1}$$

where N_c is the number of contacts.

The total time given by this equation does not include cutting and stripping the wires or assembling the connector.

It was noted that the design of connectors varied with different suppliers. The assembly of small mechanical parts is covered adequately by the methods in Chapter 3.

The total experimental time given by Eq. (4.1) was found to be quite close to the time given in [9,10] and the time observed in Company 2.

Figure 4.16 gives the manual assembly times for crimping the contacts of circular or rectangular connectors. A contact is first inserted into a crimping tool

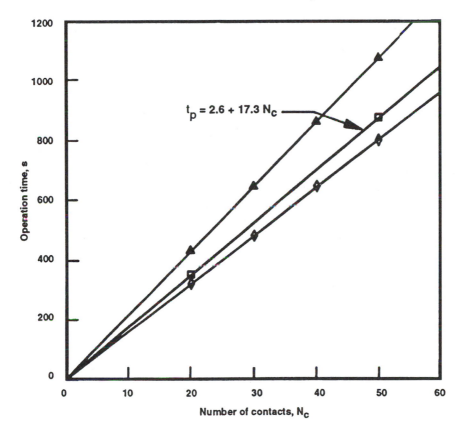

FIG. 4.16 Manual assembly time for connector with crimp contacts: □, experimental (total); ◇, Co. 1; ▲, Co. 2. (From Ref. 6.)

and the bare end of the wire is then placed into the contact. The tool is squeezed so that the barrel of the contact grips the bare end of the wire as well as the insulation on the wire. The operation is repeated for each contact. With the connector clamped in a fixture, the wire contacts are then inserted one by one, using an insertion tool. The times for crimping contacts to wires $(1.5 + 12.4N_c)$ and inserting wire contacts $(1.1 + 4.9N_c)$ vary linearly with the number of contacts. The total experimental time t_p for assembling wires to a circular or rectangular connector is then given by

$$t_p = 2.6 + 17.3N_c \tag{4.2}$$

where N_c is the number of contacts.

The total time given by this equation does not include the cutting and stripping of the wires and the assembly of the connector.

For semiautomatic crimping of contacts to wires (3 s per contact) and manually inserting the crimped contacts into the connector $(1.1 + 4.9N_c)$, a total time of 9 s for one contact is suggested. For installing more than one contact, an additional time of 7.9 s per contact should be added to the basic time.

Figure 4.17 presents the manual assembly times for one coaxial connector termination. The coaxial cable is cut to length, the outer insulation is stripped, and the polyethylene dielectric and shield braid are cut to the correct dimensions. The shield braid is then folded back smoothly and a plastic grommet assembled onto the cable. The contact is crimped to the conductor after it is assembled, flush against the dielectric. The connector body is then assembled, the braid clamp crimped, and the grommet pushed flush against the body. The experimental time for assembling a coaxial connector was 152 s. This time was very much shorter than the figure of 290 s given by [9] (which included tinning of the center conductor) and longer than the 115 s given by [10] (which does not include measuring and cutting the cable). The large difference in assembly time could probably be clue to differences in connector construction.

Figure 4.18 presents the manual assembly time for mass termination of a flat cable. This operation first involves placing the body of the connector into the locator plate of a press. The flat cable is then positioned into the body of the connector and the mating connector cover is placed over the assembly. The U-

FIG. 4.17 Manual assembly time for coaxial connector. (From Ref. 6.)

FIG. 4.18 Manual assembly time for mass termination of a flat cable.

contacts of the connector are forced into the flat cable conductors by operating the press. Finally, the cable connector is removed from the locator plate. The experimental time for flat cable connector termination was 30 s. The time obtained from the observations in Company 1 agreed fairly closely with the experimental time. However, the times obtained from [7,10] were considerably lower and higher respectively than the experimental time. For flat cable connector termination using a press machine, a time of 2 s is suggested.

Insulation displacement connection methods involve placing a connector into a semiautomatic machine, spreading the wires, and simultaneously inserting one pair at a time into the connector's contacts [7]. The machine is then triggered and the pair of wires are pierced and cut. An operation time of 7.3 s for each pair of wires is given in [7].

4.5.2 Assembly and Installation

Figure 4.19 shows the times for point-to-point wiring (direct wiring) of the items in a chassis. After one end of the wire is attached, it is dressed neatly to follow the contour of the equipment chassis before the other end of the wire is attached. The dressing time does not include cutting and stripping the ends of the wire, adding the terminations, or attaching the terminated wire. The experimental times are lower than those observed in Company 2. The differences could be due to the

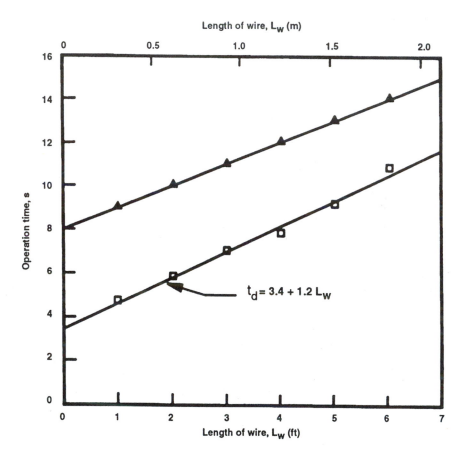

FIG. 4.19   Results for point-to-point (direct) wiring: □, experimental; ▲, Co. 2. (From Ref. 6.)

degree of obstruction encountered while dressing the wire. The experimental time t_d for dressing the wire is given by:

$$t_d = 3.4 + 1.2L_w$$

where L_w is the length of wire measured in feet.

or

$$t_d = 3.4 + 3.94L_w \tag{4.3}$$

where L_w is measured in meters.

Thus, for direct wiring, a time of 4.6 s is required for the first foot and an additional time of 1.2 s per ft (0.3 m) is added for a wire of length greater than 1 ft (0.3 m).

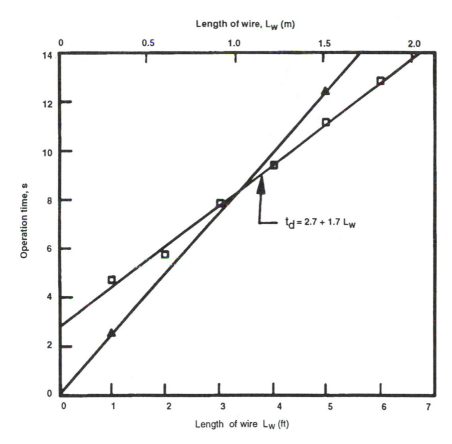

FIG. 4.20 Results for dressing a wire into a U-channel: ▲, (9); □, experimental. (From Ref. 6.)

Figure 4.20 shows the times for dressing a wire in a U-channel. After the wire is attached at one end, it is inserted (dressed) into the U-channel before the other end of the wire is attached. The dressing time does not include cutting and stripping the ends of the wire, adding the terminations, or attaching the terminated wire. For a wire of length 3 to 4 ft, the experimental time obtained is comparable to that given in [9]. A dressing time of 4.4 s is recommended for a wire 1 ft long, and an additional time of 1.7 s per ft should be added for a longer wire.

Figure 4.21 presents the times for laying a flat cable directly into the equipment chassis. The operation involves dressing the flat cable after the flat cable connector is attached onto its mating part. The experimental times obtained agreed fairly closely with those observed in Company 1. A time of 7.7 s is

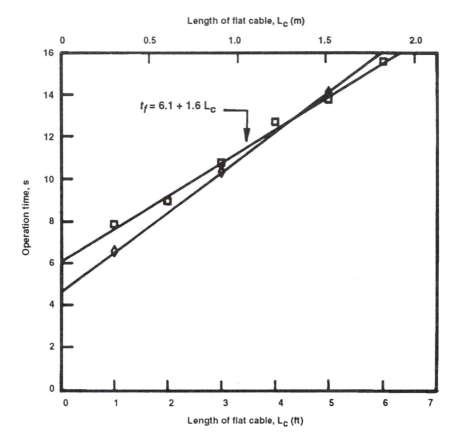

FIG. 4.21 Results for laying a flat cable: □, experimental; ◇, Company 1. (From Ref. 6.)

suggested for laying a flat cable 1 ft (0.3 m) long. An additional time of 1.6 s per ft should be added for a longer cable. The time for laying a flat cable does not include the time for bending and pressing of the cable so that it stays bent during laying. This operation is usually done before laying the flat cable, and a time of 15.1 s per bend should be added to the time for laying a flat cable.

Figures 4.22 and 4.23 present the times for laying wires onto a harness jig. A wire is first grasped and one end is attached to a holding device. The wire is then laid on the board according to the wiring layout and finished, with the other end of the wire being attached to another holding device. Multiple wires can be laid simultaneously if the wires start and end at the same breakout. Figure 4.22 shows the comparisons of published results with experimental times for laying one wire or six wires simultaneously onto the harness jig.

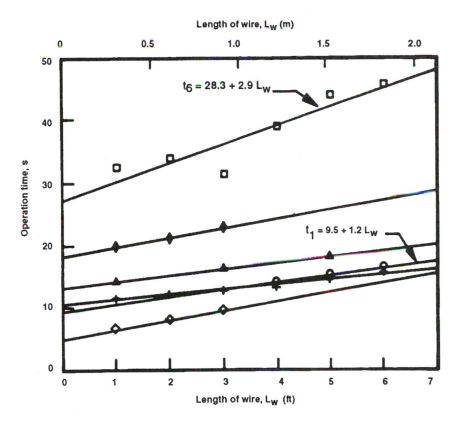

FIG. 4.22 Results for laying a wire or six wires simultaneously onto a harness jig: ○, 1 wire (experimental); □, 6 wires (experimental); +, 1 (7); ◆, 6 (8); ◇, 1 (8); ▲, 1 (Co. 2.)

Figure 4.23 shows further experimental results for laying one to six wires simultaneously onto a harness jig. From these results a general equation for the time t_n to assemble N_w wires simultaneously is

$$t_n = 6.4 + 3.8 N_w + (0.5 + 0.4 N_w) L_w \qquad (4.4)$$

Figures 4.24 and 4.25 show the times for laying wires already connected to a connector onto a harness jig. The operations involve the selection of the wire(s) after the cable connector is installed on the harness jig, laying the wire(s) according to the wiring layout, and attaching the ends to a holding device. The times obtained do not include attaching the cable connector to its mating part on the harness jig. Figure 4.24 shows the comparison between the times obtained experimentally and those given by [8]. It was found that the experimental times

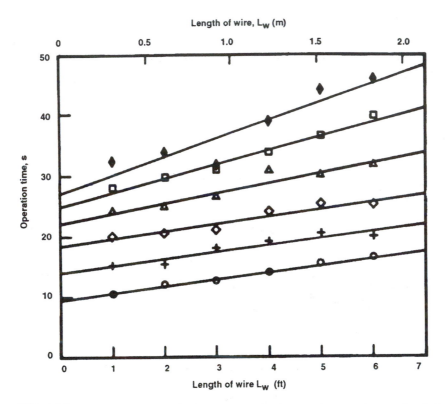

FIG. 4.23 Experimental times for laying wires simultaneously onto a harness jig; $\bigcirc$, lay 1 wire; $+$, 2; $\Diamond$, 3; $\triangle$, 4; $\square$, 5; $\blacklozenge$, 6. (From Ref. 6.)

were higher. Further experimental results are shown in Fig. 4.25 and a general equation for the time t_m to assembly N_w wires simultaneously from a connector is:

$$t_m = 6.9 + 2N_w + (0.5 + 0.3N_w)L_w \tag{4.5}$$

4.5.3 Securing

Figure 4.26 presents the times for acquiring a tie cord, spot tying a bundle of wires, and cutting the excess cord with a pair of scissors. The experimental time of 16.6 s is comparable with times obtained from other sources.

Figure 4.27 shows the times for tying a cable tie or strap onto a bundle of wires. The operation involved acquiring a cable tie, looping the tapered end of the strap around the harness, and inserting it through the strap eyelet. The strap is pulled tightly and the excess is cut off with a tool. The experimental time of 14.4 s agreed closely with the times from all other sources with the exception of [7].

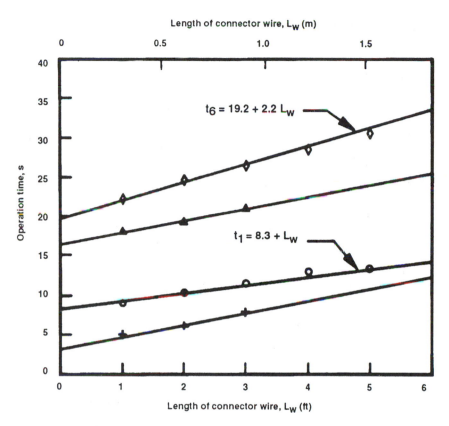

FIG. 4.24 Results for laying wires simultaneously (from a connector) onto a harness jig; ○, 1 wire (experimental); ◇, 6 wires (experimental); +, 1 wire (8); ▲, 6 wires (8). (From Ref. 6.)

Figure 4.28 shows the times for lacing a wire harness. The operations involve acquiring a lacing cord and making an initial stitch (girth stitch) at one end of the trunk. To aid in lacing, a lacing bobbin (tool) may be used. The girth stitch is formed by wrapping one end of the lacing cord into a double loop. The free end of the cord and the bobbin are then passed through this loop. Tension is applied to the free ends of the cord to dress the girth stitch firmly. Additional stitches are formed and spaced uniformly at distances approximately equal to the diameter of the harness (but never less than 0.5 in. apart). The completion of the lacing terminates when the end of the cord is secured and trimmed. The experimental time t_{st} was found to be:

$$t_{st} = 11.5 + 7.6 N_{st}$$

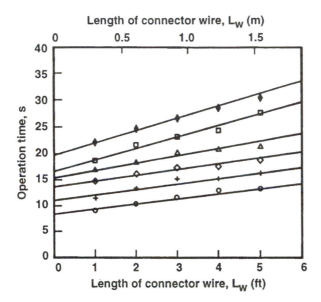

FIG. 4.25 Experimental times for laying wires simultaneously (from a connector) onto a harness jig: ○, lay 1 wire; +, 2; ◇, 3; △, 4; □, 5; ◆, 6. (From Ref. 6.)

FIG. 4.26 Results for spot tying a harness trunk/branch. (From Ref. 6.)

FIG. 4.27 Results for tying a cable tie onto a harness trunk/branch. (From Ref. 6.)

where N_{st} is the number of stitches formed on the harness. Comparing the experimental results with those obtained from other sources, the variation in the lacing times arose at the first stitch. The first stitch took 69.6 s [8]. This large time probably included cutting a sufficient length of lacing cord and winding it onto the bobbin before the lacing operation.

FIG. 4.28 Results for lacing a harness trunk/branch. (From Ref. 6.)

FIG. 4.29   Results for taping a bundle of wires. (From Ref. 6.)

Figure 4.29 gives the times for taping a bundle of wires. The operation involves obtaining a roll of tape, taping and wrapping to a specific location, and cutting the tape. For a tape 1 in. wide, the experimental time for taping 1 in. (25 mm) was 13.8 s for the first wrap and agreed closely with the data obtained from all the other sources. The taping of the second inch would involve more than one wrap of tape due to the overlapping nature of the taping process. The wide variation in times given for taping an additional inch could be due to the number of wraps involved. In the experiment, three wraps of tape constituted about 1 in. (25 mm) and a time of 7 s per additional inch (25 mm) was considered acceptable.

Figure 4.30 presents the times for inserting a precut tube or sleeve over a bundle of insulated wires. The operation involved acquiring and arranging a bundle of wires and inserting the bundle into a tube or sleeve which is grasped with the free hand. The experimental time differed greatly from the times given by other sources. The ease of inserting the wires into the tube depends very much on the amount of clearance between the inner wall of the tube and the bundle of insulated wires, and this probably explains the large differences in insertion times. The experimental time of 7.4 s for inserting the first inch (25 mm) of tube is suggested, and an additional time of 2.4 s per inch (25 mm) is added for a tube longer than 1 in. (25 mm).

Figure 4.31 shows two results for securing operations. The first involves heat shrinking the tube using a heat gun. The experimental time of 5.3 s per inch (25 mm) lies between those obtained from [9] and observations in Company 2. The wide variation in the heat-shrinking times could be attributed to the diameter of the tube to be shrunk. Also, a larger clearance between the inner diameter of

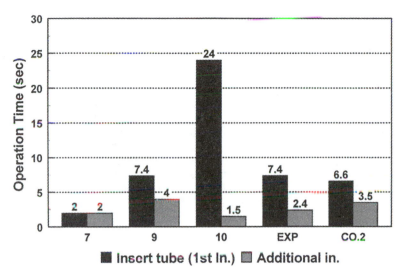

FIG. 4.30 Results for inserting a precut tube or sleeve. (From Ref. 6.)

the tube and the insulated wires will prolong the shrinking time. The experimental time of 5.3 s per inch (25 mm) appears reasonable.

The second operation in Fig. 4.31 involves acquiring an adhesive cable clamp, peeling off the protective layer, and pressing the clamp onto the equipment chassis. The experimental time of 9.4 s agreed fairly closely with observations in Company 2. For cable clamps that must be screwed onto the equipment chassis, a screwing-down operation into a tapped hole is necessary, and in Company 2, this operation took 27 s.

Figure 4.32 shows the times for labeling a wire. The operation involves peeling off a label and wrapping the label onto a wire. With the exception of [7], the times obtained from the various sources agreed fairly closely with the experimental time of 11.4 s. The time obtained from [7] (14.4 s) also involved labeling a cable with a marker.

4.5.4 Attachment

Figure 4.33 gives the times for the attachment of a bare wire to its mating part. The wire is first grasped with two hands (due to its length and flexibility), moved, positioned, and bent with a plier around its mating part. The mating part can be mounted in either a terminal block or a separate fastener such as a screw inserted into a tapped hole or a screw to be held in place with a nut. The wire is then secured. The attachment times for the various mating parts are shown in the figure, and the experimental times agreed closely with the times obtained from

FIG. 4.31   Results for (i) shrinking an inserted tube and (ii) installing an adhesive cable clamp. (From Ref. 6.)

FIG. 4.32   Results for labeling a wire. (From Ref. 6.)

FIG. 4.33   Results for attachment of bare wire to its mating part. (From Ref. 6.)

DFA (Chapter 3) with the exception of the time obtained for attaching a bare wire to a terminal block. The assembly time of 17.1 s can be assumed for attaching a bare wire to a terminal block, 23.3 s for a bare wire/screw attachment and 30.6 for a bare wire/screw and nut attachment.

Figure 4.34 presents the times for soldering a bare wire to its mating part. The bare wire is first grasped, moved, and positioned to its mating part. The wire is then soldered with a soldering iron. Before soldering, the wire may be bent using pliers, in which case the bending time must be included. From preliminary observations, it can be deduced that the time for soldering depends very much on whether the wire needs to be bent and the area to be soldered. The experimental times agreed fairly closely with DFA (Chapter 3) in the case where the wire is not bent before soldering and with observations in Company 2 in the case where the wire is bent before soldering. An attachment time of 21.1 s can be assumed for soldering a bare wire without bending and a time of 26.6 s can be assumed if the wire is first bent.

Figure 4.35 presents the times for wire-wrapping a bare end of an insulated solid wire around a terminal post. The operation involves acquiring a prestripped wire and inserting it into the tip of the wire-wrapping tool. The tool is then positioned over the terminal and the trigger is squeezed. The tip of the tool spins the wire around the terminal to form the desired attachment. The average time to wire-wrap a terminal post is 13 s.

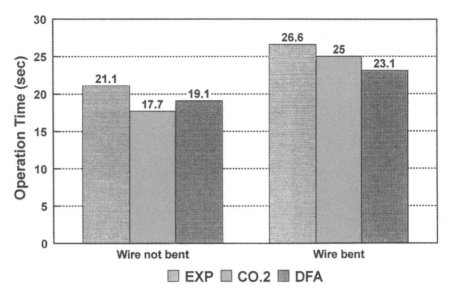

FIG. 4.34   Results for soldering a bare wire to its mating part. (From Ref. 6.)

FIG. 4.35   Results for wire wrapping around a terminal post. (From Ref. 6.)

Figures 4.36 and 4.37 give the times for attaching a wire terminal to its mating part. A wire terminal is first grasped, moved, and positioned to its mating part. It can be a push-on using a quick disconnect terminal or it can be fastened on a terminal block using a ring or fork terminal. The wire terminal (fork or ring) can also be fastened using a screw or a screw and nut combination. From the figure, it can be seen that the times taken from the various sources are fairly close, except for the difference in the experimental and DFA times for attaching a fork terminal to a terminal block. Based on the experimental values, the following can be assumed:

Attach a quick disconnect terminal—5.4 s.
Attach a fork and ring terminals, respectively, to a terminal block—12.5 s and 22.8 s.
Attach a terminated wire and secure with a screw—17.1 s.
Attach a terminated wire and secure with a screw and a nut—24.7 s.

Experiments were also performed on the various types of coupling devices found on connectors. The operation involved grasping a male cable connector (circular or rectangular) and inserting it into a female connector. Depending on the types of connector used, additional operations may be required to ensure that the connectors remain firmly coupled. No comparisons were made since there

FIG. 4.36   Results for attachment of a wire terminal to its mating part. (From Ref. 6.)

FIG. 4.37 Results for attachment of a wire terminal to its mating part. (From Ref. 6.)

was no information available from other sources. The average experimental times for attaching the various types of connectors were

Circular connector
    install only   5.2 s
    bayonet type   5.2 s
    friction type   6.7 s
    screw thread type   11.3 s
Rectangular type
    install only   6.5 s
    latch/snap-on type   8.1 s
    spring clip type   9.8 s
    screw (2×) type   24.0 s

4.6 ANALYSIS METHOD

A methodology has been developed [1] designed to allow estimates to be made of the labor involved in making electrical interconnections and in wire harness assembly. For a wire harness with normal complexity, there are three distinct steps:

1. *Preparation of the wires and cables:* This step includes cutting the wires or cables to length and terminating them. Connectors are sometimes attached to multiple wires at this stage. Also, some additional operations may be carried out such as marking, sealing, molding, sleeving, or labeling of the wires or connectors.
2. *Assembly of the harness on a board or jig:* This includes manually laying-in the wires and cables, inserting the wire ends into connectors if necessary, tying, taping, or lacing the bundles or wires, and/or sleeving the bundles. Testing is usually done at this stage and some labeling may also be carried out.
3. *Installation of the interconnections or the cable harness into the chassis or product:* This final step includes routing the wires or cables, insertion of the connectors into their mating parts, and the dressing and connection of any wire ends to appropriate terminals. Also included is the securing of the wires and cables to the chassis.

Sometimes, it is known exactly what steps are involved in the preparation, assembly, and installation of a particular wire harness assembly, including whether the preparation of the wires and cables will be carried out manually, with the aid of semiautomatic equipment, or automatically. For the designer who wishes to obtain assembly cost estimates at the earliest stages of design, some of these details will be unknown. To satisfy this need, it can be assumed, for example, that semiautomatic equipment will always be used for wire and cable preparation. This results in considerable simplification of the databases needed for the estimation of assembly time.

Ultimately, there will be a need for two distinct procedures.

1. The procedure introduced here, which is intended for designers who require approximate estimates of preparation, assembly, and installation costs during the early stages of product design.
2. A detailed procedure for those wishing to study various methods for the preparation and assembly of wire harness assemblies.

In the DFA method for manual assembly (Chapter 3), one worksheet is sufficient to record both the handling and insertion times for each item in a subassembly. As we have seen, three distinct steps are necessary for electrical connections and cable harness assembly. In addition, to analyze some operations, such as preparing wire or cable ends, it is convenient to combine each group of wire ends with similar terminations. For other operations, such as laying-in the wire on the board or jig, it is convenient to treat the complete lengths of wires or cables separately. For yet other operations, such as applying cable ties, it is convenient to account for these as one group in the later stages of analysis. This means that the overall analysis procedure is considerably more complicated than the DFA method for the manual assembly of mechanical parts.

4.6.1   Procedure

The analysis method involves the completion of worksheets using the data provided in the corresponding tables. Many of the activities to be recorded on these worksheets may be performed at various stages of wire harness assembly, depending upon the specific application and complexity of design. For example, insertion of lugged wire ends into connectors may be performed during preparation, harness assembly, or even final installation of the wire harness.

Therefore, to simplify the analysis procedure, separate worksheets have been developed for each activity. This system of analysis has proved to be the most versatile and efficient method considering the wide range of applications that exist. It relieves the designer from having to consider the order in which individual activities may be performed. Additionally, specific activities that do not apply to a particular application may be omitted from the analysis by simply skipping the irrelevant worksheets.

The following is a brief description of the activities associated with each of the worksheets and corresponding data charts. These are presented in a sequence that represents their most common order of occurrence in wire harness assembly and installation.

1. *Wire/cable preparation:* This includes activities such as cutting, stripping, crimping lugs, and tinting of wire ends. Additional operations such as insulation displacement and soldering of wire ends into connectors may be handled here.
2. *Assembly—wire/cable handling:* This involves the laying of individual wires or cables on an assembly board or jig.
3. *Assembly—insertion:* This is the insertion of lugged wire ends into connectors or the temporary securing of wire ends on an assembly board.
4. *Wire dressing:* This includes the selection, positioning, and individual dressing of wire or cable ends during either harness assembly or final installation.
5. *Installation—connector fastening:* This is the insertion and securing of previously wired and assembled connectors. These operations usually occur during final installation into the product.
6. *Installation—wire fastening:* This is the insertion and securing of individual exposed wire or cable ends. Specific operations include wire wrapping, screw fastening, and soldering. These operations usually occur during final installation into the product.
7. *Installation—lug fastening:* This is the insertion and securing of lugged wires (e.g., the screw fastening of a ring lug onto a terminal block). These operations usually occur during final installation in the product.
8. *Installation—routing:* This is the positioning of the wire harness or individual wires in the chassis (or parent assembly) during final installation. These

activities may include feeding wires or cables through channels, separating and routing individual legs of the harness, or coiling up excess lengths of wires or cables.

9. *Additional Operations:* These are additional operations that may be performed at any stage of wire harness assembly (e.g. preparation, assembly, and/or installation). Several of these operations are estimated per number of occurrences. Examples include labeling, tie wrapping, and taping single wraps or breakouts. Other operations are estimated for a given length of part or section to be covered. Examples including lacing, taping, and installing tubing over a continuous section of the harness.

Figure 4.38 shows, by way of example, worksheet No. 7 for lug fastening and a portion of the corresponding data chart. It can be seen that the leg of the harness is identified in the first column of the worksheet followed by the two-digit lug fastening code in the second column. This two-digit fastening code is obtained from the data chart and, in the example, the code 20 corresponds to the securing of one fork lug where the screw has previously been inserted into the terminal block and where there are no restrictions to the insertion and fastening operations. This code corresponds to an operation time of 10.4 s, which is entered into column 4 of the worksheet. The number of items to be fastened (eight in the example) is placed in column 3 and the resulting total time of 83.2 s is entered into column 5.

It should be noted that the times in chart No. 7 (Fig. 4.38) do not include the time for routing the wire or for dressing the wire ends prior to fastening.

4.6.2 Case Study

Figure 4.39 shows an installation consisting of 50 wires with 100 connections to be made. This example comes from a machine control unit manufactured in fairly small quantities. The various legs of the harness are lettered A through T.

Table 4.3 presents the wire run list showing, for each end of the wire, the harness leg, termination type, and connecting points. With the aid of such a chart and the drawing showing the location of wire ties, it is possible to complete the series of worksheets and obtain an estimate of the total wire preparation, harness assembly and installation times. Table 4.4 shows the results of such an analysis where it can be seen that the three principal activities each contribute significantly to the total time of almost one hour. Figure 4.40 shows a possible redesign employing ribbon cable and using mass termination techniques for the assembly of connectors onto the cable. In addition, the terminal block in the current design (Fig. 4.39) where legs A and B were connected has been eliminated. Finally, the connections between the two printed circuit boards have been replaced with one ribbon cable. It was assumed in the redesign that all the standard items (switches,

				no restrictions		restricted access	
				attachment of each lug	attachment of each additional lug at same location	attachment of each lug	attachment of each additional lug at same location
				0	1	2	3
quick connect	connect		0	2.0		6.0	
	connect & solder		1	14.0		18.0	
screw or nut previously inserted	fork lug		2	10.4	11.0	14.4	15.0
	ring lug	w/ screw	3	19.8	18.0	27.8	26.0
		w/ nut	4	14.6	5.0	22.6	13.0
screw or nut not previously inserted	nut fasten		5	13.0	5.0	21.0	13.0
	screw fasten		6	17.2	18.0	25.2	26.0
	screw & nut fasten		7	25.3	18.0	37.3	30.0

I.D. of leg	lug fastening code	number of items to be fastened	time per lug (sec)	total fastening time (sec)
1	2	3	4	(3x4)
A	20	8	10.4	83.2

times are in seconds

FIG. 4.38 Worksheet No. 7 and portion of data chart No. 7 for lug fastening. (Data partially based on Ref. 6.)

FIG. 4.39 Current wiring design for a control unit (length in feet).

TABLE 4.3 Wire Run List for Current Design of Control Unit

		From			To		
Wire ID	Length (ft)	Con. ID	No.	Termination type	Con. ID	No.	Termination type
1–6	1	A	1–6	Fork lug	C	18–23	Tin (solder and sleeve)
7	1	A	4	Fork lug	D	26	Quick con. lug
8	1	A	5	Fork lug	I	37	Tin (solder)
9	1	A	6	Fork lug	D	24	Ring lug
10–11	1	B	7–8	Fork lug	J	38–39	Lug
12	1	B	3	Fork lug	J	40	Lug
13–15	1	B	10–12	Fork lug	J	41–43	Lug
16	2	B	10	Fork lug	S	88	Lug
17	2	B	11	Fork lug	S	90	Lug
18	2	B	12	Fork lug	S	83	Lug
19	3	B	13	Fork lug	T	91	Ring lug
20	1	B	13	Fork lug	E	31	Tin (solder)
21	2.5	C	15	Tin (solder and sleeve)	P	79	Quick con. lug
22–24	2.5	C	20–22	Tin (solder and sleeve)	P	80–82	Quick con. lug
25	1	D	25	Quick con. lug	I	36	Tin (solder)
26–27	2.5	E	27–28	Tin (solder and sleeve)	Q	73–74	Lug
28	1	E	29	Tin (solder and sleeve)	F	32	Tin (solder)
29	1	E	30	Tin (solder and sleeve)	G	33	Tin (solder)
30	3	F	32	Tin (solder)	Q	75	Lug
31	3	G	33	Tin (solder)	Q	76	Lug
32	3	H	34	Tin (solder)	Q	73	Lug
33	3	H	35	Tin (solder)	Q	78	Lug
34–36	1.5	J	44–46	Lug	N	70–72	Lug
37–40	1.5	K	47–50	Lugs	L	56–59	Quick con. lugs
41–45	2	K	51–55	Lugs	M	60–64	Quick con. lugs
46–50	3	O	65–69	Lugs	R	83–87	Lugs

TABLE 4.4 Analysis Results for Current Design of Control Unit

	Time (s)	Time (min)
Preparation	800.1	13.3
Assembly	1236.7	21.6
Installation	1350	22.5
Totals	3446.8	57.4

FIG. 4.40 Proposed wiring design for a control unit.

TABLE 4.5 Analysis Results for
Proposed Design of Control Unit

	Time (s)	Time (min)
Preparation	576.8	9.6
Assembly	0	0
Installation	875.6	14.6
Totals	1452.4	24.2

transformers, etc.) and their connection methods could not be changed. Also, it was assumed that the two printed circuit boards could not be combined.

Analysis of this new design gives the results presented in Table 4.5 where it can be seen that the total assembly time has been reduced to 24.2 mm and that harness assembly has been eliminated.

Clearly, further significant improvements could be made through a complete redesign including the elimination of soldered connections and combination of the printed circuit hoards, but this case study serves to illustrate how the effects of design changes can be quantified in order to guide the designer to less costly and more easily manufactured products.

REFERENCES

1. Boothroyd, G., and Raucent, B., Factoring in the Labor Cost of Electrical Connections and Wire Harness Assembly, Connection Technol., June 1992, pp. 22–25.
2. Bilotta, A.J., Connections in Electronic Assemblies, Marcel Dekker, Inc., New York, 1985.
3. Power and Motion Control Reference Volume, Interconnections, Machine Des., Vol. 61, No. 12, 1989.
4. Markstein, H.W. (ed.), Flexible Circuits Show Design Versatility, Electronic Packaging and Production, Vol. 29, No. 4, 1989.
5. AMP, Flexible Film Products, Catalog 73–151, 1986.
6. Ong, N.S., and Boothroyd, G., Assembly Times for Electrical Connections and Wire Harnesses, Int. J. Adv. Manuf. Tech., 1991, Vol. 6, pp. 155–179.
7. Ostwald, P.F., AM Cost Estimator, McGraw-Hill, New York, 1985/86.
8. Matisoft, B.S., Handbook of Electronics Manufacturing Engineering, Van Nostrand Reinhold Co., New York, 1986.
9. Taylor, T., Handbook of Electronics Industry Cost Estimating Data, John Wiley & Sons, New York, 1985.
10. Funk, J.L. et al., Programmable Automation and Design for Manufacturing Economic Analysis, National Science Foundation, 1989.

5

Design for High-Speed Automatic Assembly and Robot Assembly

5.1 INTRODUCTION

Although design for assembly is an important consideration for manually assembled products and can reap enormous benefits, it is vital when a product is to be assembled automatically. The simple example shown in Fig. 5.1 illustrates this. The slightly asymmetrical threaded stud would not present significant problems in manual handling and insertion, whereas for automatic handling an expensive vision system would be needed to recognize its orientation. If the part were made symmetrical, automatic handling would be simple. For economic automatic assembly therefore, careful consideration of product structure and component part design is essential. In fact, it can be said that one of the advantages of introducing automation in the assembly of a product is that it forces a reconsideration of its design—thus offering not only the benefits of automation but also those of improved product design. Not surprisingly, the savings resulting from product redesign often outweigh those resulting from automation.

The example of the part in Fig. 5.1 illustrates a further point. The principal problems in applying automation usually involve the automatic handling of the parts rather than their insertion into the assembly. To quote an individual experienced in the subject of automatic assembly "if a part can be handled automatically, then it can usually be assembled automatically." This means that, when we consider design for automation, we will be paying close attention to the design of the parts for ease of automatic feeding and orienting.

asymmetrical -   **symmetrical -**
difficult to orient   **easy to orient**

FIG. 5.1   Design change to simplify automatic feeding and orienting.

In considering manual assembly we were concerned with prediction of the time taken to accomplish the various tasks such as grasp, orient, insert, and fasten. From a knowledge of the assembly worker's labor rate we could then estimate the cost of assembly. In automatic assembly, the time taken to complete an assembly does not control the assembly cost. Rather it is the rate at which the assembly machine or system cycles, because, if everything works properly, a complete assembly is produced at the end of each cycle. Then, if the total rate (cost per unit time) for the machine or system and all the operators is known, the assembly cost can be calculated after allowances are made for down-time. Thus, we shall be mainly concerned with the cost of all the equipment, the number of operators and technicians, and the assembly rate at which the system is designed to operate. However, so that we can identify problems associated with particular parts, we shall need to apportion the cost of product assembly between the individual parts and, for each part, we shall need to know the cost of feeding and orienting and the cost of automatic insertion.

In the following discussion we first look at product design for high-speed automatic assembly using special-purpose equipment and then we consider product design for robot assembly (i.e., using general-purpose equipment).

5.2   DESIGN OF PARTS FOR HIGH-SPEED FEEDING AND ORIENTING

The cost of feeding and orienting parts will depend on the cost of the equipment required and on the time interval between delivery of successive parts. The time between delivery of parts is the reciprocal of the delivery rate and is nominally equal to the cycle time of the machine or system. If we denote the required delivery or feed rate F_r (parts/min), then the cost of feeding each part C_f is given by

$$C_f = (60/F_r)R_f \text{ cents} \tag{5.1}$$

where R_f is the cost (cents/s) of using the feeding equipment.

Using a simple payback method for estimation of the feeding equipment rate R_f, this is given by

$$R_f = C_F E_o/(5760 P_b S_n) \text{ cents/s} \tag{5.2}$$

where C_F is the feeder cost ($), E_o is the equipment factory overhead ratio, P_b is the payback period in months and S_n is the number of shifts worked per day. The constant 5760 is the number of available seconds in one shift working for one month divided by 100 to convert dollars to cents.

For example, if we assume that a standard vibratory bowl feeder costs $5000 after installation and debugging, that the payback period is 30 months with 2 shifts working, and that the factory equipment overheads are 100% ($E_o = 2$), we get

$$R_f = 5000 \times 2/(5760 \times 30 \times 2)$$
$$= 0.03 \text{ cent/s}$$

In other words, it would cost 0.03 cents to use the equipment for 1 second. Supposing that we take this figure as the rate for a "standard" feeder and we assign a relative cost factor C_r to any feeder under consideration, then Eq. (5.1) becomes

$$C_f = 0.03(60/F_r)C_r \tag{5.3}$$

Thus, we see that the feeding cost per part is inversely proportional to the required feed rate and proportional to the feeder cost.

To describe these results in simple terms we can say that, for otherwise identical conditions, it would cost twice as much to feed each part to a machine with a 6 s cycle compared with the cost for a machine with a 3 s cycle. This illustrates why it is difficult to justify feeding equipment for assembly systems with long cycle times.

For the second result we can simply state that, for otherwise constant conditions, it would cost twice as much to feed a part using a feeder costing $10,000 compared with a feeder costing $5000.

If the feeding cost for a particular feeder is plotted against the required feed rate F_r on logarithmic scales, a linear relationship results, as shown in Fig. 5.2. It appears that the faster the parts are required, the lower the feeding cost. This is true only as long as there is no limit on the speed at which a feeder can operate. Of course, there is always an upper limit to the feed rate obtainable from a

FIG. 5.2   Effect of required feed rate on feeding cost.

particular feeder. We shall denote this maximum feed rate by F_m and we will consider the factors that affect its magnitude. However, before doing so let us look at its effect through an example.

Suppose the maximum feed rate from our feeder is 10 parts/min. Then if parts are required at a rate of 5 parts/min, the feeder can simply be operated more slowly involving an increased feeding cost as given by Eq. (5.3) and illustrated in Fig. 5.2. Suppose parts are required at a rate of 20 parts/min. In this case two feeders could be used, each delivering parts at a rate of 10 parts/min. However, the feeding cost per part using two feeders to give twice the maximum feed rate will be the same as one feeder delivering parts at its maximum feed rate. In other words, if the required feed rate is greater than the maximum feed rate obtainable from one feeder, the feeding cost becomes constant and equal to the cost of feeding when the feeder is operating at its maximum rate. This is shown in Fig. 5.2 by the horizontal line. If multiple feeders are used for increased feed rates, then the line will be saw-toothed as shown. However, in practice, the line can be smoothed by spending more on feeders to improve their performance when necessary.

From this discussion we can say that Eq. (5.3) holds true only when the required feed rate F_r is less than the maximum feed rate F_m, and when this is not the case the feeding cost is given by

$$C_f = 0.03(60/F_m)C_r \tag{5.4}$$

Now the maximum feed rate F_m is given by

$$F_m = 1500\, E/l \text{ parts/min} \tag{5.5}$$

where E is the orienting efficiency for the part and l (mm) is its overall dimension in the direction of feeding and where it is assumed that the feed speed is 25 mm/s.

To illustrate the meaning of the orienting efficiency E we can consider the feeding of dies (cubes with faces numbered 1 to 6). Suppose that if no orientation is needed, the dies can be delivered at a rate of 1 per second from a vibratory bowl feeder. However, if only those dies with the 6 side uppermost were of interest, a vision system could be employed to detect all other orientations and a solenoid-operated pusher could be used to reject them. In this case the delivery rate would fall to an average of 1 die every 6 seconds or a feed rate of 1/6 per second. The factor 1/6 is defined as the orienting efficiency E and it can be seen that the maximum feed rate is proportional to the orienting efficiency (Eq. (5.5)).

Now let us suppose our dies were doubled in size and that the feed speed or conveying velocity on the feeder track were unaffected. It would then take twice as long to deliver each die. In other words the maximum feed rate is inversely proportional to the length of the part in the feeding direction [Eq. (5.5)]

Equation (5.4) shows that when $F_r > F_m$, the feeding cost per part is inversely proportional to F_m. It follows that under these circumstances, the cost of feeding is inversely proportional to the orienting efficiency and proportional to the length of the part in the feeding direction.

This latter relationship illustrates why automatic feeding and orienting methods are only applicable to "small" parts. In practice this means that parts larger than about 8 in. in their major dimension cannot usually be fed economically.

When considering the design of a part and its feeding cost, the designer will know the required feed rate and the dimensions of the part. Thus F_r and l will be known. The remaining two parameters that affect feeding cost, namely, the orienting efficiency E and the relative feeder cost C_r, will depend on the part symmetry and the types of features that define its orientation. A classification system for part symmetry and features has been developed [1] and for each part classification the average magnitudes of E and C_r have been determined [2]. A portion of this system is presented in Figs. 5.3 to 5.5. Figure 5.3 shows how parts are categorized into basic types, either rotational or nonrotational. For rotational parts, their cylindrical envelopes are classified as discs, short cylinders, or long cylinders. For nonrotational parts, the subcategories are flat, long, or cubic depending on the dimensions of the sides of the rectangular envelope.

Figure 5.3 give the first digit of a three-digit shape code. Figure 5.4 shows how the second and third digits are determined for a selection of rotational parts (first digit 0, 1, or 2) and gives the corresponding values of the orienting efficiency E and the relative feeder cost C_r. Similarly, Fig. 5.5 shows how the second and third digits are determined for a selection of nonrotational parts (first digit 6, 7, or 8). The geometrical classification system was originally devised by Boothroyd and Ho [1] as a means of cataloging solutions to feeding problems.

FIG. 5.3 First digit of geometrical classification of parts for automatic handling.

FIG. 5.4 Second and third digits of geometrical classification for some rotational parts.

5.3 EXAMPLE

Suppose the part shown in Fig. 5.6 is to be delivered to an automatic assembly station working at a 5 s cycle. We now use the classification system and database to determine the feeding cost and we assume that the cost of delivering simple parts at 1 per second using our "standard" feeder is 0.03 cents per part.

FIG. 5.5 Second and third digits of geometrical classification for some nonrotational parts.

First we must determine the classification code for our part. Figure 5.6 shows that the rectangular envelope for the part has dimensions $A = 30$, $B = 20$, and $C = 15$ mm.

Thus $A/B = 1.5$ and $A/C = 2$. Referring to Fig. 5.3, since A/B is less than 3 and A/C is less than 4, the part is categorized as cubic nonrotational and is assigned a first digit of 8. Turning to Fig. 5.5, which provides a selection of data for nonrotational parts, we first determine that our example part has no rotational symmetry about any of its axes. Also, we must decide whether the part's orientation can be determined by one main feature. Looking at the silhouette of the part in the X direction, we see a step or projection in the basic rectangular shape and we realize that this feature alone can always be used to determine the part's orientation. This means that if the silhouette in the X direction is oriented as shown in Fig. 5.6, the part can be in only one orientation and, therefore, the second digit of the classification is 4. However, either the groove apparent in the view in the Y direction and the step seen in the view in the Z direction could also be used to determine the part's orientation. The procedure now is to select the feature giving the smallest third classification digit; in this case it is the step seen in the X direction. Thus the appropriate column number in Fig. 5.5 is 0. The

FIG. 5.6 Sample part.

three-digit code is thus 840 and corresponding values of orienting efficiency $E = 0.15$ and relative feeder cost $C_r = 1$.

Using the fact that the longest part dimension l is 30 mm and that the orienting efficiency E is 0.15, Eq. (5.5) gives the maximum feed rate obtainable from one feeder; thus

$$
\begin{aligned}
F_m &= 1500 \, E/l \\
&= 1500 \times 0.15/30 \\
&= 7.5 \text{ parts/min}
\end{aligned}
$$

Now, from the cycle time of 5 s the required feed rate F_r is 12 parts/min, which is higher than F_m. Therefore, since $F_r > F_m$ we use Eq. (5.4) and since $C_r = 1$ we get a feeding cost of

$$C_f = 0.03(60/F_m)C_r$$
$$= 0.03(60/7.5)1$$
$$= 0.24 \text{ cents}$$

5.4 ADDITIONAL FEEDING DIFFICULTIES

In addition to the problems of using the part's geometric features to orient it automatically, other part characteristics can make feeding particularly difficult. For example, if the edges of the parts are thin, shingling or overlapping can occur during feeding, which leads to problems with the use of orienting devices on the feeder track (Fig. 5.7).

Many other features can affect the difficulty of feeding the part automatically and can lead to considerable increases in the cost of developing the automatic feeding device. These features can also be classified as shown in Fig. 5.8, where, for each combination of features, an approximate additional relative feeder cost is given that should be taken into account in estimating the cost of automatic feeding.

5.5 HIGH-SPEED AUTOMATIC INSERTION

If a part can be sorted from bulk and delivered to a convenient location correctly oriented, a special-purpose mechanism or workhead can usually be designed that will place it in the assembly. Such workheads can generally be built to operate on

difficult to feed - parts overlap

easy to feed

FIG. 5.7 Parts that shingle or overlap on the feeder track.

				parts will not tangle or nest			
				not light		light	
				not sticky	sticky	not sticky	sticky
				0	1	2	3
parts do not tend to overlap during feeding	not delicate	non-flexible	0	0	1	2	3
		flexible	1	2	3	4	5
	delicate	non-flexible	2	1	2	3	4
		flexible	3	3	4	5	6

FIG. 5.8 Additional relative feeder costs for a selection of feeding difficulties.

a cycle as short as 1 second. Thus, for assembly machines operating on cycles greater than 1 second, the automatic insertion cost C_i will be given by

$$C_i = (60/F_r)R_i \tag{5.6}$$

where F_r is the required assembly rate (or feed rate of parts) and R_i is the cost (cents/s) of using the automatic workhead.

Again, using a simple payback method for estimation of the equipment rate R_i, this is given by

$$R_i = W_c E_o/(5760 P_b S_n) \text{ cents/s} \tag{5.7}$$

where W_c is the workhead cost (\$), E_o is the equipment factory overhead ratio, P_b is the payback period in months, and S_n is the number of shifts worked per day.

If we assume that a standard workhead costs \$10,000 after installation and debugging, that the payback period is 30 months with two shifts working, and the factory equipment overheads are 100% ($E_o = 2$), we get

$$R_i = 10,000 \times 2/(5760 \times 30 \times 2)$$
$$= 0.06 \text{ cents/s}$$

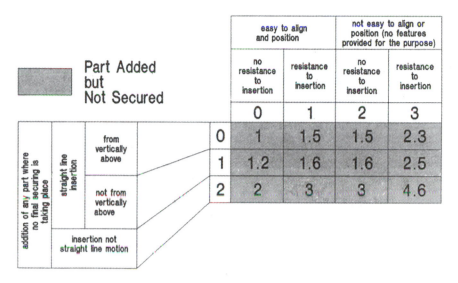

FIG. 5.9   Relative workhead costs W_r for a selection of automatic insertion situations.

In other words, it would cost 0.06 cents to use the equipment for 1 second. If we take this figure as the rate for a "standard" workhead and we assign a relative cost factor W_r to any workhead under consideration, then Eq. (5.6) becomes

$$C_i = 0.06(60/F_r)W_r \tag{5.8}$$

Thus, the insertion cost is inversely proportional to the required assembly rate and proportional to the workhead cost.

When considering the design of a part, the designer knows the required assembly rate F_r. For presentation of relative workhead costs, a classification system for automatic insertion similar to that for manual insertion was devised [2]. A portion of this system is shown in Fig. 5.9. It can be seen that this classification system is similar to that for manual insertion of parts except that the first digit is determined by the insertion direction rather than whether obstructed access or restricted vision occurs.

5.6 EXAMPLE

If the part shown in Fig. 5.6 is inserted horizontally into the assembly in the direction of arrow Y and it is not easy to align and position and not secured on insertion, then the appropriate classification is row 1, column 2 in Fig. 5.9. The automatic insertion code is thus 12, giving a relative workhead cost of 1.6.

For a cycle time of 5 s, the assembly rate F_r is 12 parts/min and Eq. (5.8) gives an insertion cost of

$$
\begin{aligned}
C_i &= 0.06(60/F_r)\ W_r \\
&= 0.06(60/12)1.6 \\
&= 0.48 \text{ cents}
\end{aligned}
$$

Thus the total handling and insertion cost C_t for this part is

$$
\begin{aligned}
C_t &= C_f + C_i \\
&= 0.24 + 0.48 \\
&= 0.72 \text{ cents}
\end{aligned}
$$

5.7   ANALYSIS OF AN ASSEMBLY

To facilitate the analysis of a complete assembly, a worksheet similar to that used for manual assembly analysis can be employed. Figure 5.10 shows the exploded view of a simple assembly before and after redesign which is to be assembled at a rate of 9.6 per minute and Fig. 5.11 presents the completed worksheets for automatic assembly analysis.

FIG. 5.10   Simple assembly.

No. of Repeats Part or Sub or Oper'n No.		Orientation Efficiency Handling Code	Maximum Feed Rate (parts/min.) Relative Feeder Cost	Handling Cost (cents) Handling Difficulty	Relative Workhead Cost Insertion Code		Insertion/ Operation Cost (cents) Insertion Difficulty		Figure for min. parts Total Cost (cents)		HIGH-SPEED AUTOMATIC ASSEMBLY ------------------ Name of Assembly- VALVE ------------------ Name of Part, Sub-
ID	RP	HC	OE	CR	FM	DF	CF	IC	WC	DI	CI	CA	NM	assembly or Operation
1	1	83100	0.20	1	4.8	12.4	0.40	00	1.0	6.3	0.38	0.69	1	housing
2	1	02000	0.40	1	21.4	6.3	0.20	02	1.5	9.4	0.56	0.63	1	plunger
3	1	00840	.*	*	***.*	**.*	*.**	-manual ass'y required-				7.13	0	gasket
4	1	00800	.*	*	***.*	**.*	*.**	-manual ass'y required-				6.67	1	cover
5	2	21000	0.90	1	122.7	6.3	0.20	39	1.8	11.3	0.68	1.44	0	screw

(a) original design

No. of Repeats Part or Sub or Oper'n No.		Orientation Efficiency Handling Code	Maximum Feed Rate (parts/min.) Relative Feeder Cost	Handling Cost (cents) Handling Difficulty	Relative Workhead Cost Insertion Code		Insertion/ Operation Cost (cents) Insertion Difficulty		Figure for min. parts Total Cost (cents)		HIGH-SPEED AUTOMATIC ASSEMBLY ------------------ Name of Assembly- NEWVALVE ------------------ Name of Part, Sub-
ID	RP	HC	OE	CR	FM	DF	CF	IC	WC	DI	CI	CA	NM	assembly or Operation
1	1	83100	0.20	1	4.8	12.4	0.40	00	1.0	6.3	0.29	0.69	1	housing
2	1	02000	0.40	1	21.4	6.3	0.20	02	1.5	9.4	0.43	0.63	1	plunger
3	1	00040	0.70	3	26.3	18.8	0.61	00	1.0	6.3	0.29	0.90	0	gasket
4	1	02000	0.40	1	15.0	6.3	0.20	38	0.8	5.0	0.23	0.43	1	cover

(b) redesign

FIG. 5.11 Completed worksheets for high-speed automatic assembly analysis of the assemblies in Fig. 5.10.

5.8 GENERAL RULES FOR PRODUCT DESIGN FOR AUTOMATION

The most obvious way in which the assembly process can be facilitated at the design stage is by reducing the number of different parts to a minimum. This subject was covered in the previous chapter dealing with manual assembly, where it was emphasized that simplification of the product structure can lead to substantial savings in assembly cost and parts cost. When considering product design for automation, it is even more important to consider reduction in the number of separate parts. For example, the elimination of a part would eliminate a

Old design   New design

FIG. 5.12   Redesign of part for ease of assembly. (From Ref. 3.)

complete station on an assembly machine–including the parts feeder, the special workhead, and the associated portion of the transfer device. Hence, the reduction in necessary investment can be substantial when product structure is simplified.

Apart from product simplification, automation can be facilitated by the introduction of guides and chamfers that directly facilitate assembly. Examples of this are given by Baldwin [3] and Tipping [4] in Figs. 5.12 and 5.13. In both examples, sharp corners are removed so that the part to be assembled can be

FIG. 5.13   Redesign to assist assembly. (From Ref. 4.)

guided into its correct position during assembly, which requires less control by the placement device or can even eliminate the need for a placement device.

Further examples in this category can be found in the types of screws used in automatic assembly. Screws that tend to centralize themselves in the hole give the best results in automatic assembly; Tipping [4] summarizes and grades the designs of screw points available as follows (Fig. 5.14):

1. Rolled thread point: very poor location; will not centralize without positive control on the outside diameter of the screws.
2. Header point: only slightly better than (1) if of correct shape.
3. Chamfer point: reasonable to locate.
4. Dog point: reasonable to locate
5. Cone point: very good to locate.
6. Oval point: very good to locate.

Tipping recommends that only cone and oval point screws be used in automatic assembly. However, industrial practice now favors dog point screws since these tend to be self-aligning once inserted.

Another factor to be considered in design is the difficulty of assembly from directions other than directly above. The designer should aim to allow for assembly in sandwich or layer fashion, each part being placed on top of the previous one. The biggest advantage of this method is that gravity can be used to assist in the feeding and placing of parts. It is also desirable to have workheads and feeding devices above the assembly station, where they will be accessible in the event of a fault due to the feeding of a defective part. Assembly from above

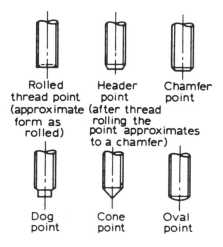

Rolled thread point (approximate form as rolled)

Header point (after thread rolling the point approximates to a chamfer)

Chamfer point

Dog point

Cone point

Oval point

FIG. 5.14 Various forms of screw points. (From Ref. 4.)

may also assist in the problem of keeping parts in their correct positions during the machine index period, when dynamic forces in the horizontal plane might tend to displace them. In this case, with proper product design using self-locating parts, the force due to gravity should be sufficient to hold the part until it is fastened or secured.

If assembly from above is not possible, it is probably wise to divide the assembly into subassemblies. For example, see the exploded view of a British power plug shown in Fig. 5.15; in the assembly of this product it would be relatively difficult to position and drive the two cord grip screws from below. The remainder of the assembly (apart from the main holding screw) can be conveniently built into the base from above. In this example the two screws, the cord grip, and the plug base could be treated as a subassembly dealt with prior to the main machine assembly.

It is always necessary in automatic assembly to have a base part on which the assembly can be built. This base part must have features that make it suitable for quick and accurate location on the work carrier. Figure 5.16a shows a base part for which it would be difficult to design a suitable work carrier. In this case, if a force were applied at *A*, the part would rotate unless adequate clamping were

FIG. 5.15 Assembly of three-pin power plug.

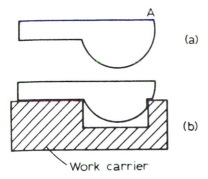

(a)

(b)

Work carrier

FIG. 5.16 Design of base part for mounting on work carrier.

provided. One method of ensuring that a base part is stable is to arrange that its center of gravity be contained within flat horizontal surfaces. For example, a small ledge machined into the part will allow a simple and efficient work carrier to be designed (Fig. 5.16b).

Location of the base part in the horizontal plane is often achieved by dowel pins mounted in the work carrier. To simplify the assembly of the base part onto the work carrier, the dowel pins can be tapered to provide guidance, as in the example shown in Fig. 5.17.

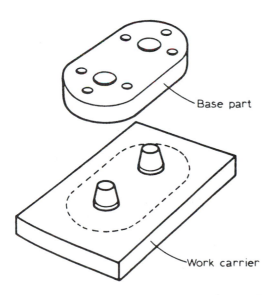

Base part

Work carrier

FIG. 5.17 The use of tapered pegs to facilitate assembly.

5.9 DESIGN OF PARTS FOR FEEDING AND ORIENTING

Many types of parts feeders are used in automatic assembly; but most feeders are suitable for feeding only a very limited range of part shapes and are not generally relevant when discussing the design of parts for feeding and orienting. The most versatile parts feeder is the vibratory bowl feeder, and the following section deals mainly with the aspects of the design of parts that facilitate feeding and orienting in these devices. Many of the points made, however, apply equally to other feeding devices. Three basic design principles can be enumerated:

1. Avoid designing parts that will tangle, nest, or shingle.
2. Make the parts symmetrical.
3. If parts cannot be made symmetrical, avoid slight asymmetry or asymmetry resulting from small or nongeometrical features.

   If parts tend to tangle or nest when stored in bulk, it can be almost impossible to separate, orient, and feed them automatically. Often a small, nonfunctional change in design will prevent this occurrence, and some simple examples of this are illustrated in Fig. 5.18.
   While the asymmetrical feature of a part can be exaggerated to facilitate orientation, an alternative approach is to deliberately add asymmetrical features

FIG. 5.18 Examples of redesign to prevent nesting or tangling. (From Ref. 5.)

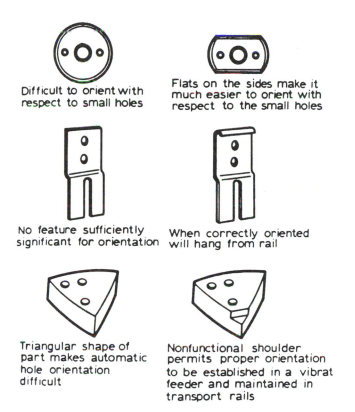

Difficult to orient with respect to small holes

Flats on the sides make it much easier to orient with respect to the small holes

No feature sufficiently significant for orientation

When correctly oriented will hang from rail

Triangular shape of part makes automatic hole orientation difficult

Nonfunctional shoulder permits proper orientation to be established in a vibrat feeder and maintained in transport rails

FIG. 5.19   Provision of asymmetrical features to assist in orientation. (From Ref. 5.)

for the purpose of orienting. The latter approach is more common, and some examples, given by Iredale [5], are reproduced in Fig. 5.19. In each case, the features that require alignment are difficult to utilize in an orienting device, so corresponding external features are deliberately added.

In the portion of the coding system shown in Fig. 5.4, those parts with a high degree of symmetry have codes representing parts easy to handle. There are, however, a wide range of codes representing parts that will probably be difficult to handle automatically; these parts create problems for designers unless assistance is provided.

Figure 5.20a shows a part that would be difficult to handle and Fig. 5.20b shows the redesigned part, which could be fed and oriented in a vibratory bowl feeder at a high rate. The subtle change in design would not be obvious to the designer without the use of the coding system. In fact, it might not have occurred to the designer that the original part was difficult to handle automatically.

(a)   (b)

Very difficult to orient   Possible to orient

FIG. 5.20   Less obvious example of a design change to simplify feeding and orienting.

It should be pointed out that, although the discussion above dealt specifically with automatic handling, parts that are easy to handle automatically will also be easy to handle manually. A reduction in the time taken for an assembly worker to recognize the orientation of a part and then reorient it results in considerable cost savings.

Clearly, with some parts, it will not be possible to make design changes that will enable them to be handled automatically: for example, very small parts or complicated shapes formed from thin strips are difficult to handle in an automatic environment. In these cases it is sometimes possible to manufacture the parts on the assembly machine or to separate them from the strip at the moment of assembly. Operations such as spring winding or blanking out thin sections have been successfully introduced on assembly machines in the past.

5.10   SUMMARY OF DESIGN RULES FOR HIGH-SPEED AUTOMATIC ASSEMBLY

The various points made in this discussion of parts and product design for automatic assembly are summarized below in the form of simple rules for the designer.

5.10.1   Rules for Product Design

1. Minimize the number of parts.
2. Ensure that the product has a suitable base part on which to build the assembly.

3. Ensure that the base part has features that enable it to be readily located in a stable position in the horizontal plane.
4. If possible, design the product so that it can be built up in layers, each part being assembled from above and positively located so that there is no tendency for it to move under the action of horizontal forces during the machine index period.
5. Try to facilitate assembly by providing chamfers or tapers that help to guide and position the parts in the correct position.
6. Avoid expensive and time-consuming fastening operations, such as screw fastening, soldering, and so on.

5.10.2   Rules for the Design of Parts

1. Avoid projections, holes, or slots that cause tangling with identical parts when placed in bulk in the feeder. This may be achieved by arranging the holes or slots to be smaller than the projections.
2. Attempt to make the parts symmetrical to avoid the need for extra orienting devices and the corresponding loss in feeder efficiency.
3. If symmetry cannot be achieved, exaggerate asymmetrical features to facilitate orienting or, alternatively, provide corresponding asymmetrical features that can be used to orient the parts.

5.11   PRODUCT DESIGN FOR ROBOT ASSEMBLY

As with product design for high-speed automatic assembly, one objective with robot assembly is to provide the designer with a means of estimating the cost of assembling the product—but in this case using robots. However, several important design aspects will be affected by the choice of robot assembly system; a choice which, in turn, is affected by various production parameters such as production volume and the number of parts in the assembly. Three representative types of robot assembly systems can be considered, namely:

1. Single-station with one robot arm
2. Single-station with two robot arms
3. Multistation with robots, special-purpose workheads, and manual assembly stations as appropriate.

For a single-station system, parts that required manual handling and assembly, and that must be inserted during the assembly cycle, present special problems. For reasons of safety it would usually be necessary to transfer the assembly to a location or fixture outside the working environment of the robot. This can be accomplished by having the robot place the assembly on a transfer device that carries the assembly to the manual station. After the manual operation has been completed, the assembly can be returned in a similar manner to within reach of the robot.

The use of special-purpose workheads for insertion or securing operations presents similar problems to those for manual assembly operations. Two different situations can be encountered. The first involves the insertion or placement of the part by the robot without it being secured immediately. This operation is then followed by transfer of the assembly to an external workstation to carry out the securing operation; a heavy press fit would be an example. The second situation is where a special-purpose workhead is engineered to interact directly at the robot workfixture. This might take the form of equipment activated from the sides of or underneath the workfixture to carry out soldering, tab bending, or twisting operations, spin riveting, etc., while the robot has to place and, if necessary, manipulate the part.

These major problems with single-station systems do not occur with a multistation system, where manual operations or special-purpose workheads can be assigned to individual stations as necessary. This illustrates why it is important to know the type of assembly system likely to be employed when the product is being designed.

In order to determine assembly costs, it is necessary to obtain estimates of the following:

1. The total cost of all the general-purpose equipment used in the system, including the cost of robots and any transfer devices and versatile grippers—all of which can be employed in the assembly of other products if necessary.
2. The total cost of all the special-purpose equipment and tooling, including special-purpose workheads, special fixtures, special robot tools or grippers, special-purpose feeders; and special magazines, pallets, or part trays.
3. The average assembly cycle time—that is the average time to produce a complete product or assembly.
4. The cost per assembly of the manual labor involved in machine supervision, loading feeders, magazines, pallets, or part trays and performing any manual assembly tasks.

Classification systems and databases have been developed for the purpose of cost estimating [2].

There is one classification and one data chart for each of the three basic robot assembly systems. In these charts, insertion or other required operations are classified according to difficulty. For each classification, and depending on the difficulty of the operation, relative cost and time factors are given that can be used to estimate equipment costs and assembly times. These cost and time estimate are obtained by entering data, from the appropriate chart, onto a worksheet for each part insertion or separate operation.

Figure 5.21 shows a portion of the classification system and database for a single-station one-arm robot assembly system. This portion of the system is for the situation where a part is being added to the assembly, but is not being secured immediately. The selection of the appropriate row (first digit) depends on the direction of insertion—an important factor influencing the choice of robot because the 4 degree-of-freedom Selective Assembly Compliance Robot Arm (SCARA) robot can only perform insertions along the vertical axis. The selection of the appropriate column (second digit) depends on whether the part needs a special gripper, clamping temporarily after insertion, and whether it tends to align itself during insertion. All of these factors affect either the cost of the tooling required or the time for the insertion operation or both.

When the row and columns have been selected for a particular operation, the figures in the box allow estimates to be made of the robot cost, the gripper tool cost, and the total time for the operation.

Let us suppose that a part is to be inserted along a horizontal axis, does not require a special gripper, requires temporary clamping, and is easy to align. For this operation, the code would be 12. In this case, the relative robot cost AR is 1.5. This means that if the basic capital cost of an installed standard 4 degree-of-freedom robot (including all controls, sensors, etc., and capable of only vertical insertions) is $60,000, a cost of $90,000 is assumed. This figure allows for a more

AR TP		Single-Station	part can be gripped & inserted using standard gripper or gripper used for previous part			
AG TG		One-Arm System				
TP – relative affective basic operation time			no holding down		part requires temporary holding or clamping	
AR – relative robot cost						
AG – relative additional gripper or tool cost			self-aligning	not easy to align	self-aligning	not easy to align
TG – relative time penalty for gripper or tool change			**0**	**1**	**2**	**3**
part added but not finally secured	using motion along or about the vertical axis	**0**	1.0 1.0	1.0 1.07	1.0 1.0	1.0 1.07
			0 0	0 0	1.0 0	1.0 0
	using motion along or about a non-vertical axis	**1**	1.5 1.0	1.5 1.07	1.5 1.0	1.5 1.07
			0 0	0 0	1.0 0	1.0 0
	involving motion along or about more than one axis	**2**	1.5 1.8	1.5 1.9	1.5 1.8	1.5 1.9
			0 0	0 0	1.0 0	1.0 0

FIG. 5.21 Portion of classification system and database for a single-station one-arm robot assembly system. (From Ref. 2.)

sophisticated robot, able to perform operations from directions other than above. In other words, there is a cost penalty of $30,000 for the basic equipment in the system because the "standard robot" cannot perform the operation required.

The value of the relative additional gripper or tool cost is 1.0. Since the part needs temporary clamping, special tooling mounted on the workfixture would be required. Thus, if the standard tooling or gripper cost $5000, the additional tooling needed would represent a cost penalty of $5000 in the form of special-purpose equipment.

The value of the relative basic operation time TP is 1.0. In this analysis method the basis for time estimates is the average time taken by the robot to move approximately 0.5 m, grasp the part, return, and insert the part when the motion is simple and no insertion problems exist. For a typical present-generation robot, this process might take 3 s. If this figure is used in the present example, then this is the basic time for the robot to complete the operation.

Finally, since the relative time penalty for gripper or tool change is zero, no additional time penalty is incurred and the total operation time is 3 s. A further time penalty must be added when the part to be inserted is not completely oriented by the part presentation device. In this case the robot arm must perform final orientation with the aid of a simple vision system and an additional 2 to 3 s must be added to the operation time.

In addition to the cost of the robot and the special tools or grippers, the costs of the part presentation must be estimated. Before this can be accomplished, it must be decided which part presentation method will be used for each part. In practice there are usually only two choices: the special-purpose feeder or the manually loaded magazine, pallet, or part tray.

The costs associated with part presentation can be divided into:

1. *Labor costs*, which include material handling (loading parts feeders or magazines), system tending (freeing jams in feeders, handling parts trays, etc.), and system changeover costs (changing of workfixture, feeders, and magazines and robot reprogramming);
2. *Equipment costs*, including feeder depreciation and the depreciation of special fixtures, special tooling, magazines, pallets, or part trays.

It can be assumed that the bulk material handling costs (i.e., dumping parts in bulk into feeder hoppers) are negligible compared with the cost of manually loading individual parts one-by-one into magazines, pallets, or part trays.

There are thus only three significant factors needed to estimate the cost of part presentation:

1. *Special-purpose feeders:* The cost of a special-purpose feeder, fully tooled and operating on the robot system, is assumed to be a minimum of $5000. The actual cost of a feeder, for a particular part, can be obtained from the

data presented earlier in this chapter where feeding and orienting costs were considered in detail.

2. *Manually loaded magazines:* The cost of one set of special magazines, pallets, or part trays for one part type is assumed to be $1000. For large parts this figure may considerably underestimate the actual cost and extra allowance should be made.

3. *Loading of magazines:* The time needed to hand-load one part into a magazine can be estimated to be the part-handling time, obtained from the data in Chapter 3, plus 1 s. Alternatively, a typical value of 4 s may be used.

It can seen that use of the classification systems and database allows the total cost of equipment and the cost of any manual assembly work to be estimated together with the assembly time for each part. These results provide the data necessary to predict assembly costs using each of the three robot assembly systems.

5.11.1 Summary of Design Rules for Robot Assembly

Many of the rules for product design for manual assembly and high-speed automatic assembly also apply to product design for robot assembly. However, when considering the suitability of a proposed design for robot assembly, careful consideration should be given to the need for any special-purpose equipment such as special grippers or special feeders. The cost of this equipment must be amortized over the total life volume of the product and for the mid-range volumes where robot assembly might be applied, this can add considerably to the cost of assembly.

The following are some specific rules to follow during product design [2]:

1. Reduce part count—this is a major strategy for reducing assembly, manufacture, and overhead costs irrespective of the assembly system to be used.

2. Include features such as leads, lips, chamfers, etc., to make parts self-aligning in assembly. Because of the relatively poor repeatability of many robot manipulators—when compared to dedicated workhead mechanisms—this is a vitally important measure to ensure consistent fault-free part insertions.

3. Ensure that parts which are not secured immediately on insertion are self-locating in the assembly. For multistation robot assembly systems, or one-arm single-station systems, this is an essential design rule. Holding down of unsecured parts cannot be carried out by a single robot arm, and so special fixturing is required which must be activated by the robot controller. This adds significantly to special-purpose tooling and, hence, assembly costs. With a two-arm single-station system, one arm can, in principle, hold down an unsecured part while the other continues the assembly and fastening

processes. In practice, this requires one arm to change end-of-arm tooling to a hold-down device; the system then proceeds with 50% efficiency while one arm remains immobile.

4. Design parts so that they can all be gripped and inserted using the same robot gripper. One major cause of inefficiency in robot assembly systems is the need for gripper or tool changes. Even with rapid gripper or tool change systems, each change to a special gripper and then back to the standard gripper is approximately equal to two assembly operations. Note that the use of screw fasteners always results in the need for tool changes since robot wrists can seldom rotate more than one revolution.

5. Design products so that they can be assembled in layer fashion from directly above (z-axis assembly). This ensures that the simplest, least costly, and most reliable 4 degree-of-freedom robot arms can accomplish the assembly tasks. It also simplifies the design of the special-purpose workfixture.

6. Avoid the need for reorienting the partial assembly or for manipulating previously assembled parts. These operations increase the robot assembly cycle time without adding value to the assembly. Moreover, if the partial assembly has to be turned to a different resting aspect during the assembly process, then this will usually result in increased workfixture cost and the need to use a more expensive 6 degree-of-freedom robot arm.

7. Design parts that can be easily handled from bulk. To achieve this goal avoid parts that

   Nest or tangle in bulk
   Are flexible
   Have thin or tapered edges that can overlap or "shingle" as they move along a conveyor or feed track
   Are so delicate or fragile that recirculation in a feeder would cause damage
   Are sticky or magnetic so that a force comparable to the weight of the part is required for separation
   Are abrasive and will wear the surfaces of automatic handling systems
   Are light so that air resistance will create conveying problems (less than 1.5 N/m^3 or 0.01 lb/in.^3)

8. If parts are to be presented using automatic feeders, then ensure that they can be oriented using simple tooling. Follow the rules for ease of part orientation discussed earlier. Note, however, that feeding and orienting at high speed is seldom necessary in robot assembly, and the main concern is that the features that define part orientation can be easily detected.

9. If parts are to be presented using automatic feeders, then ensure that they can be delivered in an orientation from which they can be gripped and inserted without any manipulation. For example, avoid the situation where a part can only be fed in one orientation from which it must be turned over for

insertion. This will require a 6 degree-of-freedom robot and special gripper, or a special 180° turn delivery track—both solutions leading to unnecessary cost increases.

10. If parts are to be presented in magazines or part trays, then ensure that they have a stable resting aspect from which they can be gripped and inserted without any manipulation by the robot. If the production conditions are appropriate, the use of robots holds advantages over the use of special-purpose workheads and some design rules can be relaxed. For example, a robot can be programmed to acquire parts presented in an array—such as in a pallet or part tray which has been loaded manually, thus avoiding many of the problems arising with automatic feeding from bulk. However, when making economic comparisons, the cost of manual loading of the magazines must be taken into account.

REFERENCES

1. Boothroyd, G., and Ho, C., Coding System for Small Parts for Automatic Handling, SME paper ADR76-13, Assemblex III Conference, Chicago, October 1976.
2. Boothroyd, G., and Dewhurst, P., Product Design for Assembly, Boothroyd Dewhurst, Inc., Wakefield, RI, 1986.
3. Baldwin, S.P., How to Make Sure of Easy Assembly, Tool Manuf. Eng., May 1966, p. 67.
4. Tipping, W.V., Component and Product Design for Mechanized Assembly, Conference on Assembly, Fastening and Joining Techniques and Equipment, PERA, 1965.
5. Iredale, R., Automatic Assembly—Components and Products, Metalwork Prod., April 8, 1964.

6

Printed Circuit Board Design for Manufacture and Assembly

6.1 INTRODUCTION

There has been a dramatic growth in the output of printed circuit boards (PCBs) in recent decades. This trend is symptomatic of the increased replacement of previously mechanical functions in products by electronics. A typical example of this change is the domestic washing machine, which a few years ago would have been controlled by an electromechanical cam timer, but which is now controlled electronically from a PCB. Similarly, the control of the spark timing and fuel intake of an automobile engine is now carried out through PCBs. It may be surprising to note that the biggest manufacturer of PCBs is reputed to be General Motors.

It is equally necessary to consider manufacturability at the early design stages for PCBs as it is for mechanical products. The use of quantitative tools to assess manufacturing difficulties and costs early in the design process is of great importance.

PCB manufacture is a rapidly developing field. New PCB designs, new component packaging, and new assembly techniques are continually being introduced. Manufacturers are constantly striving to achieve higher component densities, which make assembly increasingly difficult. Boards using through-hole mounted components are being replaced by boards that utilize surface-mounted devices (SMDs). In addition new techniques such as tape-automated bonding (TAB) and chip-on-board are increasingly being used. Tools used for assessing manufacturability issues for PCBs must be capable of expansion to account for these new developments. For the time being, discussion in this chapter is restricted to through-hole and surface-mount components, as these make up the majority of devices currently used.

6.2 DESIGN SEQUENCE FOR PRINTED CIRCUIT BOARDS

Design procedures for PCBs differ considerably from those used for mechanical devices. In addition the use of computer-aided design techniques has in general been much further developed and integrated into the design process than for mechanical products. The sequence for design of a PCB is as follows:

1. Development of a functional circuit schematic diagram to meet the design specification and performance of the circuit.
2. Circuit layout design—this leads to the artwork for the circuit, including component layout, routing of conductors, and component selection.

Layout design is a complex task involving many interrelated considerations, including

Component area
Number of sides
Number of boards
Volume computing (volumetric space taken up by the board)
Actual layout design, including component placement and conductor routing

A number of computer aids have been developed to assist with these design tasks, including autoplacement and routing, artwork preparation, and so on. However, many of these tasks are carried out without any assessment of costs until much later in the overall design process.

6.3 TYPES OF PRINTED CIRCUIT BOARDS

Printed circuit boards are manufactured in a variety of types and configurations. The choice of board type depends upon a number of factors, including:

1. The function of the board
2. Space available, related to component density
3. Component availability
4. Cost
5. Working environment
6. Any standards applied to the board.

The following variations of board types can be identified.

6.3.1 Number of Layers

Printed circuit boards may be single or multilayer. Single-layer boards have the circuit connections applied to one layer of insulation material only. Multilayer boards are built up from several layers of insulation materials, with parts of the circuit laid down between the layers. For two- or three-layer boards, which are relatively common, the intermediate layers serve only as ground and power planes. In some extreme cases the number of layers may be as high as 20 or more, but this is exceptional.

6.3.2 Number of Sides

Printed circuit boards may be referred to as single sided or double sided. In the first case the circuit is applied to only one side of the board and in the second case the circuit is laid down on both sides of the board. A further variation is that holes in the board may be either plated through to connect circuits on both sides or not plated through.

6.3.3 Board Materials

Various materials are used for the insulating layers of the board. The majority of boards are made from a reinforcing material and a thermosetting resin, but some ceramic boards are also used, particularly in military applications. The common combinations of resin and reinforcing materials are as follows:

Resin	Reinforcing material
Phenolics	Paper
	Cotton fabric
	Glass cloth
	Nylon
Epoxy	Paper
	Glass cloth
	Aramid cloth
Polymide	Glass cloth
	Aramid cloth
Alkyds	Glass material
Silicones	Glass cloth

6.3.4 Device Types

The discrete electronic components may be attached to the board in a variety of ways. The two main device types used are through-hole mounted and surface mounted. For through-hole devices attachment is by means of separate leads that pass through holes in the boards and are soldered to the circuit on the opposite side of the board. For surface-mounted devices, attachment is to pads on the same side of the board as the device is placed. Until recently, the use of through-hole mounted devices predominated, but more and more surface-mounted devices are being used, mainly because of the higher component densities that can be achieved. In general, it is easier from a manufacturing viewpoint if only through-hole or only surface-mounted devices are used, but many boards are made with a mixture of both types of device.

6.3.5 Bare Board Costs

The cost of manufacture of the bare boards is strongly influenced by the number of layers and size of the board, together with the materials of the board. Drilling and plating of holes for through-hole boards may also be a significant cost. Figure 6.1 [1] gives an approximate cost for different types of board per unit area as a guide to selection.

6.4 TERMINOLOGY

For those not familiar with the terms used in PCB manufacture, a glossary is included at the end of this chapter.

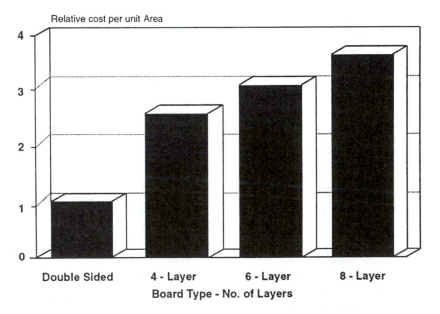

FIG. 6.1 Approximate cost per unit area of bare boards. (From Ref. 1.)

The term *insertion* is used to describe the process of placing a through-hole electrical component onto a printed circuit board so that its leads pass through the correct holes in the board or placing a surface-mount component onto the board in the required position. There are three methods of insertion: (1) dedicated automatic insertion machines, (2) manual assembly workers and semiautomatic insertion, and (3) robots.

Automatic insertion of axial (variable center distance, or VCD) components involves preforming the leads, inserting the component, and cutting and clinching the leads. Preforming and cut and clinch are done automatically as part of the insertion cycle and do not add to the cycle time or decrease the rate of insertion. Automatic insertion of dual in-line package components (DIPs) does not involve lead forming or cutting.

With manual insertion and semiautomatic insertion, all of the operations are performed manually in sequence. Thus, preforming before insertion and cut and clinch after insertion add to the total time for the insertion operation.

Robot insertion involves moving the robot arm to the component, grasping the component, realigning it if necessary, moving it to the correct board location, and insertion. If the component feeders or presenters cannot completely orient the component, then robot insertion will include final realignment and will increase the cycle time. As with manual assembly, preforming and cut and clinch are usually done separately.

Robots are generally used only to insert nonstandard components that otherwise would have been inserted manually, except that multistation robot assembly lines are now used for surface-mounted component placement. Nonstandard components or odd-form components are large or small or odd-shaped components that cannot be inserted by special-purpose machinery.

Before introducing the cost analysis for printed circuit board assembly, the assembly procedure will be described. The following section explains the various steps that can be included in a PCB assembly process; these steps are automatic, manual, and robotic insertion of components.

6.5 ASSEMBLY OF PRINTED CIRCUIT BOARDS

Printed circuit board (PCB) assembly involves mainly the insertion and soldering of electrical components into printed circuit boards. Component insertion is carried out manually or by high-speed dedicated machinery. Additionally, some manufacturers are employing robots to perform insertions—mainly those that otherwise would have to be performed manually because of the nonstandard shapes of the components. For high production volumes, most manufacturers use a combination of both automatic and manual insertion because odd-shaped or nonstandard components cannot be handled by the automatic insertion machines. However, it is desirable to use automatic insertion machines wherever possible, since they can operate much faster and with greater reliability than manual workers. For those PCBs manufactured in small batches and where the applications involve severe working environments, such as military applications, assembly is often entirely by hand.

Later in this chapter, data and equations are presented that can be used to estimate the cost of component insertion and soldering by dedicated automatic insertion machines, manual assembly workers, or robots.

6.5.1 Assembly Process for Through-Hole Printed Circuit Boards

Figure 6.2 shows all the possible steps in an assembly process for printed circuit boards [2]. The figure includes that portion of PCB manufacture where component insertions are carried out. Component presentation, repair for faulty insertions, and touch-up for faulty soldering are also included. However, steps that do not directly involve the addition of components to the board such as board preparation, where boards are given identification codes; board cut, where a series, of identical boards is cut from a larger panel; and final test, where functional testing of the board is done, are not included. Boards move through the steps as indicated by the horizontal lines on the figure. The vertical lines indicate the flow of components from inventory to the insertion stations.

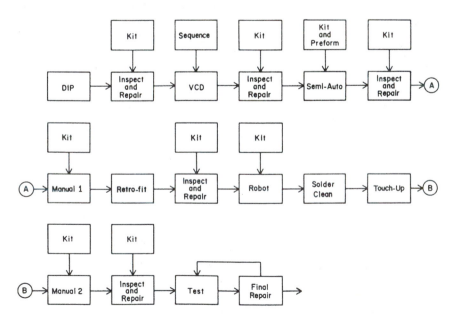

FIG. 6.2 Typical assembly process for printed circuit boards. (From Ref. 2.)

In order to minimize handling and processing time small boards are sometimes processed in the form of larger panels, which, after the components are inserted and soldered, are separated into the individual boards. Alternatively several separate boards can be mounted in one fixture to be processed together.

The first block in the assembly process shown in Fig. 6.2 indicates DIP. This refers to the automatic insertion of dual in-line package components (DIPs) and includes all integrated-circuit chips and chip sockets. The term "dual in-line package" refers to the two parallel rows of leads projecting from the sides of the package (Fig. 6.3). Typically, DIPs have between 4 and 40 leads; DIPs with more than 40 leads are infrequently used. The lead span, which is the distance between the two rows of leads, is standardized at 0.3, 0.4, or 0.6 in.

FIG. 6.3 Various electronic components (not to scale).

SOT

Pin-grid array

SOIC

PLCC

Chip

SOT

LCCC

Flat pack

Surface Mount Devices (SMD's)

Quad flat pack

FIG. 6.3 Continued

At the DIP station, components are inserted with an automatic DIP inserter (Fig. 6.4). Automatic insertion is carried out at high rates—approximately 2800 to 4500 per h or one component every 0.80 to 1.29 s. The DIP leads are inserted through predrilled holes in the board and are cut and clinched below the board (Fig. 6.5). The automatic insertion head moves only in the vertical direction, while the board is positioned below it on an x-y table that can also rotate. High-performance DIP insertion machines can insert DIPs having any of the three standard lead spans with no tooling change, but base models can only handle 0.3 in lead spans and 6 to 20 lead DIPs. To accommodate 2 and 4 lead DIPs a special insertion head must be employed at additional cost. To check for electrical

MAGAZINE   CHANNELS

CONTROL
PANEL

PRINTED
CIRCUIT
BOARD

X-Y
TABLE

FIG. 6.4 Automatic dual in-line package (DIP) insertion machine.

faults a component verifier may be employed that will stop the machine if certain preprogrammed electrical characteristics are not met.

Dual in-line package components are purchased from component manufacturers in long tubes called channels in which they are preoriented and stacked end-to-end. The channels are loaded onto the DIP inserting machine. Usually one channel is used for each DIP type on the board, but if large numbers of one type are used then more channels can be assigned to the same component. A magazine refers to a group of channels, usually about 15. If a high component mix is required, then additional magazines can be added to the machine. The machine's size and speed restrict the number of magazines. The insertion cycle time is longer when the channel is farther from the insertion head. Channels can be changed by an operator as the machine is running, which eliminates down-time caused by empty channels.

After DIP insertion, the next block in the assembly process shown in Fig. 6.2 indicates inspect and repair. In this inspection of the partially assembled board, the inspector is looking for faults that can be detected visually such as broken or bent leads or components inserted into the wrong holes. Components are either repaired and reinserted or discarded and replaced. Workers have available at each repair station all of the components inserted at the previous insertion station so

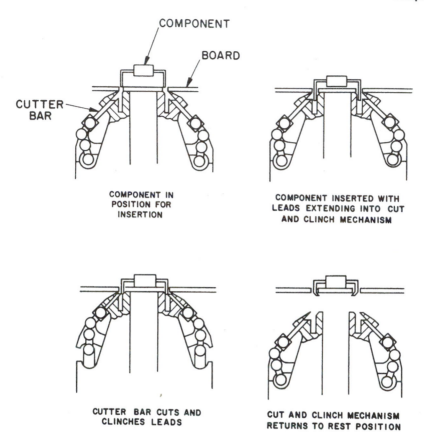

FIG. 6.5 Cut and clinch sequence.

that any component can be replaced. Inspect and repair stations can follow each insertion station. However, workers can never detect all of the faults, and these will inevitably have to be detected and corrected later in the manufacturing process.

The second insertion station in the assembly process is VCD insertion. This refers to the automatic insertion of axial-lead components (Fig. 6.3), also called VCDs (variable center distance components). These include resistors, capacitors, and diodes within the size limitations of the insertion head. Axial-lead components must usually have their leads bent at right angles prior to insertion, and the final lead span (called center distance) is variable.

At this second station axial-lead components are inserted with an automatic axial-lead inserter (Fig. 6.6). Axial-lead components are inserted at rates of

FIG. 6.6 Automatic axial-lead insertion machine.

approximately 9500 to 32,000 components per hour or one component every 0.11 to 0.38 s. Rates of 32,000 per hour can only be achieved with a dual-head insertion machine—a machine that inserts components into two identical boards simultaneously. Obviously single-head axial-lead insertion machines work at half the rate of dual-head machines. The rates are higher than those for automatic DIP insertion because components are fed on a spool in the correct sequence for insertion (Fig. 6.7). Movements of the insertion head are much less. Spools can hold large numbers of components and only have to be changed infrequently.

Automatic axial-lead insertion proceeds as follows: (i) components are stored on a spool in the correct order for insertion (this is accomplished by the automatic component sequencer described below); (ii) the spool is loaded manually onto the axial-lead inserter; (iii) during the automatic insertion cycle the leads are automatically cut to remove the component from the tape, then they are bent at right angles to the correct center distance and the component is positioned with the leads passing through the board; (iv) finally, the leads are cut and clinched below the board (Fig. 6.5).

FIG. 6.7   Axial-lead components on tape at insertion head.

As indicated in Fig. 6.2, axial-lead components are first mounted on a spool using a component sequencer. This machine (Fig. 6.8) arranges the components on a tape spool in the correct sequence for insertion. Axial-lead components are purchased on spools with one component type on each spool. Spools are loaded manually onto the sequencer for every type of component on the board and from these a master spool is created.

A component sequencer can handle components at a rate of approximately 10,000 to 25,000 per hour or one component every 0.14 to 0.36 s. Component leads are automatically cut to remove the components from their individual tapes in the correct order for insertion. Components are then retaped and rolled onto the master spool. Master spools are made off-line and in advance of the component insertion process.

A module for a sequencer is a group of dispensing spools, usually about 20 in number. If a larger component mix is needed, then additional modules can be added to the sequencer up to a maximum of about 240 dispensing spools, with the limit being imposed by the physical size of the machine. To check for faulty components or a component out of order a component verifier can be added to the sequencer.

MODULE

INPUT SPOOLS

OUTPUT SPOOL

SEQUENCED COMPONENTS

CONTROL PANEL

FIG. 6.8 Automatic axial-lead sequencing machine.

These last two processes, automatic axial-lead component insertion and sequencing, may be combined and carried out on one machine. Such a machine, about twice as large as a conventional automatic insertion machine, provides the advantage of eliminating the need for kitting and reduces setup time by bringing inventory to the assembly line. Since components are stored at the assembly line, different batches can be run—both large and small—by simply writing a new program for each batch.

In addition to these automatic insertion processes, PCB manufacturers may include stations for the automatic insertion of radial-lead components and single in-line package (SIP) components (Fig. 6.3). Automatic insertion machines for radial-lead components are similar to automatic axial-lead component insertion machine, and automatic SIP inserters are similar to automatic DIP inserters. Some DIP insertion machines have, as an option, the ability to insert SIP components using an additional insertion head. Also, the first three stations in Fig. 6.2 may each include duplicate machines if the number of components per board is high.

The next block in the PCB assembly process shown in Fig. 6.2 indicates semiauto. This refers to semiautomatic component insertion. i.e., machine-

assisted manual insertion. Inserted at this station are all DIP and axial-lead
components that cannot be machine inserted because of either their size or their
location on the board. Also inserted are radial-lead components, SIPs, and some
connectors. Wherever possible in high-volume assembly, semiautomatic insertion
is used instead of manual insertion, since it can reduce insertion times by 80%.

A semiautomatic insertion machine (Fig. 6.9) automatically presents the correct
component to the operator and indicates, by use of a light beam, the correct
location and orientation for the component on the board. The component is then
inserted by hand, with the lead sometimes being automatically cut and clinched.
Components are inserted in this way at rates of approximately one component
every 5 s. All components must have their leads preformed to the correct
dimensions for insertion prior to presentation to the operator. Typically, components are stored in a rotating tray. After receiving a signal from the operator, the
tray rotates so that only the section containing correct components is presented.
The light beam (which can be located either above or below the board) illuminates
the holes for the insertion and uses a symbol to indicate component polarity.

There are two manual assembly stations, manual 1 and manual 2, shown in the
PCB assembly process in Fig. 6.2. One station is before and the other after wave
soldering. The two stations are sometimes needed because some components

FIG. 6.9 Semiautomatic insertion machine.

cannot withstand the high temperatures of wave soldering. They are both manual insertion stations but may involve the use of special hand tools to facilitate the handling and insertion of certain components. Typically, manual assembly accounts for a high proportion of total assembly time, even when a relatively small number of manually assembled components are involved.

At the first manual assembly station large, nonstandard components are inserted. They are inserted manually because the part trays used in semiautomatic machines cannot accommodate many of these parts due to their large size. If a particular manufacturer does not use a semiautomatic machine, then all of the components that would have been inserted using this machine will be inserted manually. Also, components assembled mechanically (i.e., secured with screws or bolts) that are to be wave soldered are assembled here.

After the first manual assembly station is a block indicating retrofit (Fig. 6.2). Assembly here is also manual but involves only engineering change order (ECO) wires or jumper wires. These wires are cut to the required length from spools, and the ends are stripped and tinned. One end of the wire is soldered to a component lead on top of the board or is inserted through the board, the wire is routed around the board and cemented down at various points, and finally the other end is soldered to a component lead or inserted into the board. These ECOs are needed to satisfy certain customer options, to correct design problems, or sometimes because it is difficult to fit the entire etched circuit on a board without having to cross paths at some point. An alternative, where the circuit would have cross paths on one surface, is to use a multilayer circuit board (one with two or more circuits printed on it with insulating material separating the circuits).

An assembly station for robotic insertion is included in the assembly process (Fig. 6.2) and occurs after retrofit in the flow diagram. At this station robotic insertion of nonstandard components using a one-arm robot may be performed. Robot insertion is used mainly to reduce the amount of manual labor involved in PCB assembly.

Solder-Clean, the next station in the assembly process (Fig. 6.2), refers to wave soldering of the component leads on the underside or solderside of the board and the removal of excess flux applied to the solderside of the board prior to soldering. A conveyor carries the PCB assembly over a rounded crest or wave of solder so that the board impinges against the wave in passing. The purpose of applying a flux, usually an organic rosin type, is to remove oxides from the board that inhibit solderability. Cleaning is subsequently done to remove excess flux, which could cause corrosion and/or contamination.

The time taken for wave soldering and cleaning one board, or in the case of small boards, one panel or fixture, depends on the conveyor speed, with conveyor speed adjustment up to 20 fpm. Typically conveyor speeds are near 10 fpm, which yields a time of approximately 2 min for a board, panel, or fixture to pass through the wave solder and cleaning station. However, conveyor speeds are selected to yield specific solder contact times, which are usually about 3 s.

Immediately following solder and clean is touch-up. Touch-up is cleaning the solderside of the board to remove excess solder. This is necessary because of the tendency for icicling and bridging of the solder, which can cause short circuits in the electrical layout. Icicling and bridging are more prevalent on closely packed boards.

At the second manual assembly station (often referred to as final assembly) all the remaining components are inserted (Fig. 6.2). These include components that cannot be wave soldered because of their sensitivity to heat. Also, components that are secured mechanically are installed here. These include handles, some large electrolytic capacitors, connectors, and power transistors which are secured with screws or bolts and nuts. Lastly, some components, such as diodes and resistors, usually with axial leads, are soldered on the top of the board to the leads of other components. These operations, if necessary, are a type of engineering change order and there is a need to reduce the number of components inserted at this station because of the greater time involved with hand soldering compared to wave soldering. It is important to note that any through-hole components on the underside of the board must be manually inserted and manually soldered with consequent increases in manufacturing costs.

6.5.2 Assembly of Surface-Mounted Devices

The preceding discussion has dealt only with PCBs having through-hole components where the component leads pass through the board. However, surface-mount devices (SMDs) are increasingly being employed. These components have pads or leads that are soldered to corresponding areas on the surface of the board. Surface-mount devices include simple resistors in the form of a small rectangular prism and a variety of larger components such as flat packs, SOTs, PLCCs, SOICs and LCCCs (Fig. 6.3; see glossary for acronyms). Some surface-mount devices are the direct equivalent of through-hole devices, but several specific SMD packages have been developed. These devices can be mounted on either side of the board and are sometimes interspersed with through-hole components. Figure 6.10 shows the assembly sequence for a board with surface-mounted devices on the topside only, known as a type 1 board. Assembly of an SMD involves positioning the component on the board, which has previously had solder paste applied, usually by screen printing. Placement of SMDs is often done by specifically designed pick-and-place machines (Fig. 6.11). When all the SMDs have been positioned, the board is reflow soldered. This involves heating the solder paste until it flows into a uniform solder layer that permanently affixes the SMD pads or leads to the board. Passive SMD components such as resistors and capacitors can also be added to the underside of the board using adhesive. Soldering them takes place during wave solder, but reflow solder for both sides of the board is sometimes used. The components on

FIG. 6.10 Assembly sequence for topside-only SMD board (type 1 board).

FIG. 6.11 Pick-and-place machine for surface-mounted devices.

the underside of the board must be able to pass through the solder wave without suffering damage.

Many boards are now being manufactured with a mixture of through-hole and surface-mounted devices. This combination of components complicates the insertion and assembly sequence. The through-hole components can be mounted on the upper side of the board and the surface-mounted devices on the bottom (known as a type 2 board); then the assembly sequence is as shown in Fig. 6.12.

FIG. 6.12 Assembly sequence for type 2 board—through-hole devices topside and surface-mounted devices bottomside.

For this to be possible, the surface-mounted devices must be restricted to those able to pass through the solder wave. When surface-mounted devices are placed on both sides of the board (type 3 board), a more complex assembly sequence is required. Firstly, the topside SMDs must be placed (Fig. 6.13) and reflow soldered. The through-hole devices are then inserted and the board inverted for

FIG. 6.13 Assembly sequence for type 3 boards—through hole and surface-mounted topside and surface-mounted devices lowerside.

FIG. 6.14 Robot assembly of printed circuit boards.

the underside SMDs to be placed. Wave soldering is then used to attach these components.

There are many variations on the PCB assembly sequences described here. In particular, the use of robots is increasing, and for high-volume production the robots can be arranged in assembly line fashion as illustrated in Fig. 6.14. With such an arrangement, the components are usually presented in standardized pallets roughly oriented. The robot must then provide the final orientation prior to insertion and, for this purpose, vision systems are sometimes employed. Alternatively, some components or mechanical parts are delivered to a station using standard feeding and orienting techniques. The robot then acquires the preoriented component or part at the station.

6.6 ESTIMATION OF PCB ASSEMBLY COSTS

The materials necessary for an estimation of PCB assembly costs are presented at the end of this section. Databases giving the times for manual insertion and the costs for automatic and robot insertion are included and a worksheet is provided to assist in tabulating the results. Components and operations are entered on the worksheet in assembly order, one line for each basic type of component or operation.

The time for manual insertion, obtained from the database, is entered on the worksheet and then multiplied by the operator rate to give the insertion cost. After per component allowances for rework costs are added, the total operation cost is entered.

For automatic or robot insertion the cost is obtained directly from the database, and then adjusted for programming, setup, and rework.

For mechanical parts, the manual assembly times and costs can be obtained using the *Product Design for Assembly* handbook [3]. When all the operations have been entered the total cost is obtained by summing the figures in the cost column.

Figure 6.15 (pp. 240–243) shows, by way of example, a completed worksheet for a PCB assembly taken from a microcomputer. Such an assembly is commonly referred to as a logic board, and this particular board contains 69 DIPs, 1 DIP socket, 16 axial components, and 32 radial components. In addition, 2 parts are attached mechanically requiring 11 screws, nuts and washers. It is assumed that the DIPs, radials, and axials are autoinserted and the remaining components or parts manually inserted or assembled, except for one DIP, which is assembled into the corresponding DIP socket after wave solder. The completed worksheet for this example board gives a total estimated assembly cost of $3.48. Consideration of avoidable costs indicates that elimination of the 11 fasteners would save 48.0 cents—a surprisingly high figure for a board with only one nonelectrical component, namely, the end plate.

The operation costs for the automatic insertion processes include the cost of rework, which amounts to a total of 46.1 cents—clearly a significant item.

In estimating the cost of wave solder it was assumed that two boards would be processed together. However, the resulting cost of 75 cents is a significant item and in practice should be examined closely for accuracy.

Finally, it was assumed in this analysis that the manufacturer would have an automatic machine available for the insertion of the 32 radial components. It is interesting that if these components were to be inserted manually, an additional expense of $2.19 would be incurred—increasing the total cost of assembly by 63%.

6.6.1 Worksheet and Database for PCB Assembly Cost Analysis

Instructions

1. For soldered components, assemblies, or operations, use the database provided here. For all other parts, components, or operations, use the manual assembly data from the *Product Design for Assembly* handbook [3].

Completed PCB Assembly Worksheet

Name of PCB COLOR CARD	Operator rate Wa=0.6 cents/s		Batch size, Bs=1,000	No. of setups Nset= 10		No. of boards/ panel Nb= 2
Name of part, sub-assembly operation or soldered component	No. of pins or leads N1	No. of parts ops. comps. Rp	Total manual op.time (sec) Ta	Total op.cost (cents) Ca	Figs. for min. parts Nm	Description
Circuit board	–	1	4.0	2.4	1	place in fixture
DIP	20	11	–	19.9	–	auto. insert
DIP	14	56	–	82.9	–	auto. insert
axial	2	16	–	24.3	–	auto. insert
radial	2	32	–	47.0	–	auto. insert
DIP socket	24	1	18.0	13.0	–	man. insert
DIP	40	1	26.0	19.2	–	man. insert
screw	–	2	15.6	9.4	0	add
coax. connector	3	1	8.0	5.1	–	man. insert
display connector	9	1	8.0	5.6	–	man. insert
star washer	–	2	17.4	10.4	0	add & hold
nut	–	2	8.8	5.3	0	add & screw
end plate	–	1	4.5	2.7	1	add
lock washer	–	2	9.4	5.6	0	add
hex. nut	–	2	19.8	11.9	0	add & screw
screw	–	1	9.8	5.9	0	add & screw
wave solder	–	1	–	75.0	–	wave solder
DIP	24	1	4.0	2.4	1	add & snap fit
			Total	348.0		

Notes:
1. ta is inappropriate for auto or robot insertion unless hand soldering is carried out
2. Nmin is inappropriate for soldered electronic components

FIG. 6.15 Completed worksheet for sample PCB assembly.

2. Record the data on the worksheet in the following order of assembly:

a. Load PCB into fixture.

b. Insert all wave-soldered components.

   (i) auto inserted components

   (ii) robot inserted components

   (iii) manually inserted components

PCB Assembly Worksheet

Name of PCB	Operator rate	Batch size,	No.of setups	No. of boards/ panel
_____	Wa=_.__cents/s	Bs=_____	Nset=___	Nb=___

Name of part, sub-assembly operation or soldered component	No. of pins or leads N1	No. of parts ops. comps. Rp	Total manual op.time (sec) ta	Total op.cost (cents) Cop	Figs. for min. parts Nmin	Description
				Total	_____		

Note: 1. ta is inappropriate for auto or robot insertion unless hand soldering is carried out
2. Nmin is inappropriate for soldered electronic components

FIG. 6.15 Continued

    c. Wave solder.

    d. Insert and solder all manually soldered components.

        (i) autoinserted components

        (ii) robot inserted components

        (iii) manually inserted components

    e. Insert and secure all remaining nonsoldered components.

PCB Assembly Database - Manual Operations

Insertion of Components			
Components Types	Operations		Time(s)
Axials (VCDs)	bend leads, insert, cut and clinch leads		19.0
Radials or can-type ICs	insert component cut and clinch leads	basic time	10.0
		additional time per lead	1.8
SIP/SIP sockets or connectors	insert component	<= 80 leads	8.0
		> 80 leads	10.0
Posts, DIP/DIP sockets, pin-grid arrays or odd-form components	insert component	basic time	6.0
		additional time per lead or post	0.5
SMDs	add and solder using special fixture or tool		10.0

Other Manual Operations		
Part	Operation	Time(s)
Sleeve	cut one sleeve and add to a lead	15.0
Jumper wire (ECO)	cut, strip and tin ends, insert and solder	60.0
Heat sink	add to transistor	25.0
Lead or post	hand-solder one lead or post	6.0

Note: Much of this data has been adapted from
      Boothroyd & Shinohara [4].

FIG. 6.15 Continued

PCB Assembly Database - Automatic Operations

Insertion Costs per Component (cents)		
Component Type	Auto, Cai	Robot, Cri
Axial (VCD)	1.2*	5.0
Radial	1.2*	5.0
SIP/SIP socket	0.8	5.0
DIP/DIP socket	0.8	5.0
Connector	1.0	5.0
Small SMD (2 connections)	0.2	5.0
Large SMD (>2 connections)	1.0	5.0

* Note: includes cost of sequencing

Associated Costs (dollars)	
Wave or reflow-solder**	$1.50

**Note: If several boards are contained in one panel or secured in a
single fixture divide this figure by Nb the number of boards per panel or
fixture in order to obtain the cost of wave or reflow solder per board.
This figure also includes the time to place the boards or panel in the
soldering fixture.

FIG. 6.15 Continued

If parts and mechanical fasteners are associated with any soldered components, then list them with the appropriate component.

PCB Assembly: Equations for Total Operation Cost, C_{op}

Manual:

$$C_{op} = t_a W_a + R_p C_{rw} M_f \text{ cents}$$

Autoinsertion machine:

$$C_{op} = R_p(C_{ai} + C_{ap}/B_s + C_{rw}A_f) + N_{set}C_{as}/B_s \text{ cents}$$

Robot insertion machine:

$$C_{op} = R_p(C_{ri} + C_{rp}/B_s + C_{rw}R_f) + N_{set}C_{rs}/B_s \text{ cents}$$

where

A_f = average number of faults requiring rework for each autoinsertion (0.002)

B_s = total batch size

C_{ai} = cost of autoinsertion (obtained from database)

C_{ap} = programming cost per component for autoinsertion machine (150 cents)

C_{as} = setup cost per component type for autoinsertion machine (150 cents)

C_c = cost of replacement component when rework is needed

C_{ri} = cost of robot insertion (obtained from database)

C_{rp} = programming cost per component for robot system (150 cents)

C_{rs} = setup cost per component type for robot system (150 cents)

$C_{rw} = T_{rf}N_1 W_a + C_c$ and is the average rework cost per faulty component (cents)

M_f = average number of faults requiring rework for each manual insertion (0.005)

N_1 = number of leads or posts on one component

N_{set} = estimated number of setups per batch

R_f = average number of faults requiring rework for each robot insertion (0.002)

R_p = number of components

t_a = total time for all manual operations (calculated from database figures)

R_{rf} = average estimated time to rework one component fault per lead or post (30 s)

W_a = rate for manual operations, cents/s

6.7 CASE STUDIES IN PCB ASSEMBLY

The analyses of two case studies are now presented.

6.7.1 Measuring Instrument Connector Board

Figures 6.16 shows the layout of components on the upper and lower side of a small PCB. Figures 6.17 shows the cost analysis for the assembly of this board assuming the following:

Labor rate	$36/h
Production quantity	1000
Number of batches	2
Boards per panel	4

It is assumed that all DIP sockets, axials, and radials are autoinserted and that all posts and connectors on the upper side of the board are semiauto-inserted

FIG. 6.16 Layout of components for a small PCB. (a) Upperside; (b) lowerside.

before wave solder. The single connector on the underside of the board must be hand inserted and hand soldered after wave soldering has occurred. The total cost of assembly of this board is $5.46, and of this total, $2.17 is associated with the single hand-inserted and hand-soldered connector on the underside of the board. This highlights the importance of avoiding manually inserted and soldered parts as much as possible. The cost of this particular board could almost be halved by a simple redesign to place this single connector on the upper side of the board.

ASSEMBLY NAME - SAMPLE   FILENAME - SAMPLE.MOO

	Description	No. of items	Figs. min. parts	Operation Time, sec.	Operation cost $	Part(s) Cost $	Tooling Cost k$
Total Assembly Time 410							
Time for Main Assembly 410							
Circuit board	place in fixture	1	1	4.0	0.04	0.00	0.0
DIP/DIP socket	auto. insert - wave	4	-	-	0.06	0.00	0.0
DIP/DIP socket	auto. insert - wave	5	-	-	0.10	0.00	0.0
DIP/DIP socket	auto. insert - wave	2	-	-	0.05	0.00	0.0
Axial (VCD)	auto. insert - wave	4	-	-	0.07	0.00	0.0
Radial	auto. insert - wave	27	-	-	0.45	0.00	0.0
Post	semi-auto insert - w	8	-	20.8	0.28	0.00	0.0
Connector	semi-auto insert - w	4	-	12.8	0.18	0.00	0.0
Connector	semi-auto insert - w	1	-	3.2	0.07	0.00	0.0
Connector	semi-auto insert - w	1	-	3.2	0.05	0.00	0.0
Connector	semi-auto insert - w	1	-	3.2	0.06	0.00	0.0
wave/reflow solder	wave/reflow solder	1	-	-	0.38	-	-
Connector	man. prep & insert-h	1	-	8.0	0.13	0.00	0.0
Hand solder	hand solder leads	34	-	204.0	2.04	0.00	0.0
screws connector	add & hold down	4	0	30.2	0.30	0.00	0.0
screw standoff	add & hold down	2	0	14.6	0.15	0.00	0.0
8 pin dip'	add & snap fit	4	4	20.2	0.20	0.00	0.0
16 pin dips	add & snap fit	5	5	23.7	0.24	0.00	0.0
24 pin dip	add & snap fit	2	2	9.5	0.09	0.00	0.0
reorientation	reorient & adjust	1	-	4.5	0.05	0.00	0.0
nut	add & screw fasten	4	0	31.7	0.32	0.00	0.0
stand off	add & screw fasten	2	0	16.0	0.16	0.00	0.0

Boothroyd Dewhurst, Inc.

Current Date: 03-26-1991
Date of Analysis: 03-26-1991
Time of Analysis: 01:09:34

FIG. 6.17 Assembly cost analysis of PCB in Fig. 6.16.

```
ASSEMBLY NAME - SAMPLE    FILENAME - SAMPLE.MOO

CUMULATIVE DATA FOR FINAL ASSEMBLY AND SUB-ASSEMBLIES
```

```
!--------------------------------------------------!
!                   Assembly data                  !
!--------------------------------------------------!
! Total assembly cost (dollars)            5.46    !
!                                                  !
! Total manual assembly time (seconds)     410     !
!                                                  !
! Total number of operations (inc. repeats)  118   !
!                                                  !
! Total soldered electronic components      58     !
!                                                  !
! Total parts and non-soldered                     !
!   electronic components *                 24     !
!                                                  !
! Theoretical minimum number of                    !
!   parts or pre-assembled items *          12     !
!                                                  !
! Average assembly cost for soldered               !
!   electronic components (cents/component)  3.24  !
!                                                  !
! Assembly efficiency (percent)**           23     !
!                                                  !
! Labor rate                                36.00  !
!--------------------------------------------------!
```

```
!--------------------------------------------------!
! Total tooling cost (k$)                   0.0    !
!                                                  !
! Total part/component cost (dollars)       0.00   !
!                                                  !
! Number of components/parts/subs for              !
!             which cost data is available   0     !
!                                                  !
! Number of components/parts/subs for              !
!             which no cost data is available 117  !
!--------------------------------------------------!
```

* These are cumulative totals and do not include analyzed
  sub-assemblies as individual items - only the parts in
  those sub-assemblies. Thus, these figures will not
  necessarily agree with the totals of the parts for each
  sub-assembly.
** Assembly efficiency calculation does not
   include soldered electronic components

FIG. 6.17 Continued

It is interesting to compare the assembly cost with the cost of complete manual assembly of the board. For a labor cost of $36/h, the cost of complete manual assembly is $29.60. Figure 6.18 shows the effect of varying the quantity of boards required on the assembly cost per board. For the autoinserted board it is necessary to reduce the quantity to only six boards before the cost is more than for manual assembly with a $36/h labor rate. This illustrates why, if automatic insertion equipment is available, it is usually economical to use this even for small board quantities or for a few autoinsertable components. In addition, for manual assembly in this case it is necessary to reduce the labor cost to around $5 per hour before the assembly cost is less than for the autoinserted board. It is obvious why the assembly of PCBs is often carried out in countries where manual labor costs are considerably lower than those in the United States.

6.7.2 Power Supply

Many electronic products, in particular power supplies, contain a number of large, odd-shaped components, and many mechanical parts (screws, washers, heat sinks, brackets, etc.) are needed to secure them. As a result, power supplies present particular assembly difficulties. Part of this can be attributed to a lack of consideration by components suppliers to features that facilitate ease of assembly. Figure 6.19 [3] shows an exploded view of a typical small power supply. The large can capacitor at the top right requires some 20 items, mainly mechanical, to mount it to the board. Similarly, a large number of mechanical parts are required

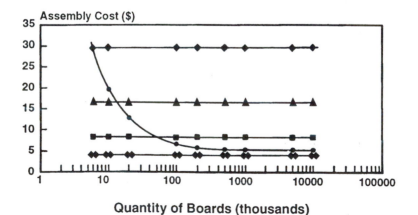

FIG. 6.18   Effect of production volume on costs for PCB in Fig. 6.16; ●, auto assembly; ◆, manual ($36/h); ▲, manual ($20/h); ■, manual ($10/h); ◆◆, manual ($5/h).

FIG. 6.19 Components in a small power supply board. (From Ref. 3.)

for the two power transistors at the left-hand side. Figure 6.20 shows a breakdown of the costs for assembly of this power supply determined by the method described in Section 6.6. The total manual assembly cost is $29.08. Considerable savings could be made by eliminating the jumper wires (saving $5.81), inserting the radial components automatically (saving $5.16), and simplifying the assembly of the large can capacitor ($5.08). Obviously the availability of this type of cost information early in the design process guides designers in directions that will have the greatest influence on assembly cost reduction.

6.8 PCB MANUFACTURABILITY

Manufacturability guidelines and specifications are available throughout the electronics industry to facilitate the design of PCBs. A survey of these resources has been undertaken [5], with data collected from the Department of Defense

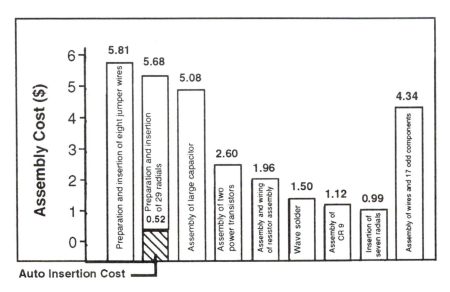

FIG. 6.20 Breakdown of assembly costs for small power supply.

(DOD), military contractors, electronics manufacturers, and PCB consulting groups, including some generally available design guides [5–9].

The Department of Defense has created and imposed on industry several specifications to direct the design and manufacture of PCBs for military applications. A typical example is MIL-STD-275E, Military Standard Printed Wiring for Electronic Equipment [10], which is widely invoked on military contracts and impacts the production of PCBs.

With military products, the requirements imposed on the design often differ from those imposed on commercial PCBs, particularly in terms of quality and reliability needs. This is attributable to the environment in which the products are used and the potentially catastrophic results of failure. For example, one widely used military requirement is that of conformal coating, which is seldom found on commercial PCBs. The function of the coating, which is a uniform clear covering, usually of urethane applied by a spray, is to protect the PCB from dust particles and the possibility of electrical shorting This typifies the high reliability requirements characteristic of military products. Compared with commercial electronics DFM documents, military electronics DFM documents usually require more stringent manufacturing processes. Increased reliability needs often have associated increased production cost in order to ensure high product dependability in harsh environments.

Documents available for PCB manufacturability guidelines can be classified into three groups.

1. General Specifications. General specifications are purely descriptive items of information relating to the design of PCBs. The information normally includes data on both the PCB geometry and the functional requirements for the design. However, this information can easily affect the manufacture of the product. For example, the design specification: "Component hole diameter must be increased at least an additional 0.25 mm (above the lead diameter) to allow for lead position variation" [6] is clearly concerned with component insertion efficiency. Failure to comply with this requirement will likely have its impact on production and not on product functionality.

Military specifications, by their name and nature, clearly fall under this heading of PCB DFM documents. When products are designed in accordance with military specifications, the design engineer is obliged to ensure that the design follows all of the applicable items that are identified. The requirements identified in general specifications have varying degrees of manufacturability impact. General specifications provide no quantitative information on how selected design features will impact manufacturing cost.

2. Manufacturability Guidelines. Manufacturability guidelines suggest how to improve product manufacturability by identifying process capabilities, sound processing rules, and process restrictions to designers. Normally, these manufacturability guidelines have been compiled by manufacturing engineers who have a working knowledge of PCB production. For example, a typical PCB manufacturer's manufacturability guideline states: "Avoid the use of heat sensitive components which cannot withstand wave solder." This requirement is intended to remind designers to avoid the requirement, by design selection, of an expensive hand-soldering operation.

Although there are certain similarities between manufacturability guidelines, most differ in length, focus and format. For example, a design group at one company uses a manufacturability guideline that employs a "checklist" format and addresses 160 items associated with the overall development of the product. Efficient manufacture is only one of several areas addressed. In contrast, sometimes manufacturability guidelines are solely intended to assist designers' decision making where such decisions impact manufacturing cost performance. These manufacturability guidelines address several different areas of manufacturing, including automatic insertion, manual assembly, and wave soldering. References are often made to the preferred design features and to the associated reasoning. For example, a typical guideline might be: "Board should have a less than or equal to 1.5 length to width ratio." Reason: "Prevent board warpage through wave solder."

3. Manufacturability Rating Systems. Manufacturability rating systems are used by PCB designers as a means of measuring how proposed PCB designs compare to an ideally producible design. In this way, various design alternatives may be measured for relative manufacturing effectiveness by comparison with a single standard. These systems normally use manufacturability guideline information in a format that allows the design engineer to estimate a rating value or metric. Many companies have developed manufacturability rating systems for internal use only and maintain their confidentiality. This is understandable because of the cost associated with developing such a system and the competitive advantages its use may bring about.

6.9 DESIGN CONSIDERATIONS

The matrix shown in Figure 6.21 can be used to summarize some of the PCB DFM information available [5]. The matrix identifies Companies A through J, together with the referenced sources of publicly available documents. The type of

	Reference	Company A	Company B	Company C	Company D	Company E	Company F	Company G	Company H	Company I	Company J	(9)	(11)	(13)	(12)	(16)	(18)	(14)	(2)
	Document Type*	G	G	G	G	G	G	G	R	R	R	S	S	S	G	R	G	R	R
1	Component Geometry	X		X		X	X	X	X	X	X	X	X	X	X			X	
2	Standard Parts Listing				X				X		X								X
3	Component Requirement		X	X	X			X	X	X	X				X	X		X	
4	Board Geometry	X		X	X		X	X	X		X	X		X			X		X
5	Directional Preference	X	X	X	X	X	X	X	X	X		X							
6	Marking Requirement	X	X	X	X				X	X			X						
7	Board Size	X	X	X	X	X	X	X	X		X	X		X					X
8	Grand Plane Requirement		X	X					X			X		X	X				
9	Solder Resist Requirement		X	X			X	X				X			X	X			X
10	To Be Avoided Features	X		X	X			X	X	X	X			X		X	X	X	X
11	Hole to Head Clearance	X		X	X	X						X		X	X				
12	Tooling Requirement	X		X	X		X	X	X		X			X		X			X
13	Testing Requirement	X							X		X								X
14	Generally Preferred Features	X	X	X	X	X		X	X	X	X		X		X	X		X	X

S = Specification
G = Guideline
R = Manufacturability Rating System

FIG. 6.21 Summary of PCB manufacturability information. (From Ref. 5.)

document is indicated (i.e., general specification (S), manufacturability guideline (G), or manufacturability rating system (R)). Fourteen types of design requirements are identified in the rows of the matrix, and these are described separately in the following sections.

6.9.1 Component Geometry and Spacing

These aspects of the design are often specified for compatibility with existing assembly equipment. For example, Company F gives: "Axial Component Specification: A) Lead Diameter: .020 in Min to .032 in Max ... D) Component Body Length: 0.60 in Max. ..." Compliance with these component dimensions will ensure that the design is within the automatic insertion range for a particular piece of equipment.

Not all geometrical component specifications in the systems relate to manufacturability. For example, bend radius of an axial-lead component is often specified to ensure stress relief for component leads subjected to continual thermal cycling; see Fig. 6.22. Component heights are also specified so that the final product will physically fit into the assembly in which it is to be used. Also, spacing between component leads is frequently determined by electrical functionality.

Figures 6.23 and 6.24 show typical recommendations for component spacing on through-hole and surface-mount boards respectively.

Lead diameter in inch	Minimum radius (R) inch
Up to 0.69 mm (0.027 inches)	1 diameter
From 0.71 mm to 1.19 mm (0.028 inches to 0.047 inches)	1.5 diameters
1.22 mm (0.048 inches) and larger	2 diameters

FIG. 6.22 Typical PCB bend radius specification. (From Ref. 6.)

Clearances required for various component locations and sequences of assembly; bl = body length; bd = body diameter; wd = wire diameter; cd = center distance. Dimensions indicated are minimum tolerance unless otherwise indicated.

Y = Machine limitation (e.g., 0.100")
X = Machine limitation (e.g., 0.100")

FIG. 6.23 Typical spacing specification chart for automatic insertion. (From Ref. 8.)

FIG. 6.24 Typical example of PCB geometry specification. (From Ref. 8.)

6.9.2 Standard Parts Listing

This item identifies whether the PCB DFM document includes a standard parts list that indicates the preferred components with regard to performance, purchase cost, and manufacturing compatibility. Such parts may be referred to as "preferred," "off-the-shelf (OTS)" or "catalog" components. All of the documents reviewed emphasize the importance of standard component use, but not all include the standard parts list. Where such lists exist, component specification has been based upon manufacturing equipment compatibility, cost, availability, and successful history of use on past designs.

6.9.3 Component Compatibility

This item identifies whether the PCB DFM document specifies that components should have certain characteristics associated with the method of assembly and processes to be used. For example, Company C specifies that designers should: "Avoid components that cannot withstand 6 second exposure to soldering temperatures of 525°F." This requirement is intended to help avoid the necessity for manually assembling and soldering a component, which would result in added cost. Figure 6.21 shows that such component compatibility requirements are widespread among the industries surveyed. However, compliance with such requirements can be ensured if the designer is provided with a standard list of preferred components. The compilation of such a list needs an investment of considerable time by the company, and only 3 of the 10 industries surveyed have preferred parts lists.

6.9.4   Board Geometry

This item identifies whether the PCB DFM document specifies physical board parameters such as hole size and grid size. The hole size is the diameter of the PCB hole (after plating) into which the component lead is inserted. The grid size defines the spacing between the holes. Spacing, hole size, and grid size can determine the effectiveness of automatic insertion equipment use.

6.9.5   Directional Preference

This item is concerned with preferred component layout on the board for ease of automatic insertion. Layout in a uniform direction can best meet this need, as well as minimizing certain problems associated with wave soldering. It is common for DFM documents to specify that a component should be placed on a PCB in a uniform direction with respect to other components in order to improve the use of automatic equipment. An example of how several companies specify this feature is shown in Fig. 6.25.

With regard to the wave solder process, the orientation of the component can also potentially affect the failure rates of the product. The quality of soldering is associated with component orientation relative to the motion of the PCB through the solder wave. One manufacturer has suggested that the preferred component orientation is perpendicular to the solder wave direction, as illustrated in Fig. 6.26. Component clinching is used in automatic, semiautomatic, and manual insertion to secure components in their proper plated-through holes during transport and soldering.

6.9.6   Marking Requirements

This item identifies whether the PCB DFM document specifies how PCB identification marks or component marks should appear after the assembly is completed. Lettering heights, marking materials, board-marking positions, and bar-coding requirements are often specified in PCB DFM documents.

6.9.7   Board Size

This item covers the specification of acceptable board sizes that are compatible with all existing processing equipment. An example of how one general specification depicts this board size is illustrated in Fig. 6.27.

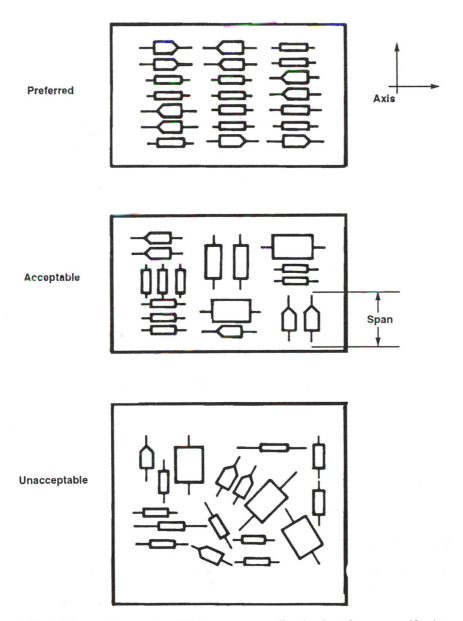

FIG. 6.25 Typical example of PCB component directional preference specification. (From Ref. 6.)

Improper Design Feature

Direction of clinch
is perpendicular to wave
resulting in poor solder
flow in plated through hole

Solder Wave

Proper Design Feature

Direction of clinch
is parallel with wave

Solder wave

FIG. 6.26   Proper component clinch orientation for wave solder.

6.9.8   Ground Plane Requirements

In a PCB design, a ground plane is often included. The ground plane is normally a continuous sheet of metal used as a common reference point for circuit returns, electrical shielding or, heat dissipation [11]. If not properly designed, the ground plane will draw heat away from the plated-through hole, thereby adversely affecting the soldering process. Company D's manufacturability guidelines, for example, state that designers should: "thermally relieve (or provide a large diameter clearance of approximately 0.250") in the area of the plated-though hole locations connected to large ground and power planes."

FIG. 6.27 Typical example of PCB size specification. (From Ref. 8.)

6.9.9 Solder Resist Requirements

This concerns the application of solder resist material to the PCB. Solder resist is a coating material used to mask or to protect selected areas of a printed wiring board circuit pattern from the action of soldering. Preferred practices are often specified such that the resist does not adversely affect the areas that must be soldered. For example, Company C manufacturability guidelines state that designers should "avoid solder mask (resist) in the area of the plated-through holes."

6.9.10 Features to Be Avoided

This item is concerned with the specification of features that would detract from PCB manufacturing efficiency. These are features that would generally result in the addition of nonstandard manufacturing operations. Such operations normally result in added cost. Examples of features that would introduce potentially avoidable costs include but are not limited to: mechanical hardware, adhesives, jumper wires, and components that cannot be inserted or placed automatically. Such features, which can be regarded as the "do-nots" of PCB design, are summarized in Table 6.1.

TABLE 6.1 Primary PCB Manufacturability Reduction Items

Feature to avoid	Reason
1. Adhesives for securing components to PCBs or thermal compound to improve heat conduction	Must be applied manually under special handling conditions
2. Components that are:	
Nonmachine insertable	Require manual assembly
Chemically sensitive to cleaning process	Require manual assembly and special cleaning provisions
Heat sensitive to soldering process	Require manual assembly and hand soldering
Mounted on the solder side of the PCB	Require manual assembly and special mounting provisions
Axial leaded and mounted vertically	Require manual assembly and special preforming
Off the board (piggyback)	Require manual assembly and hand soldering
Component sockets	Require manual assembly
3. Manual component soldering	Very cost ineffective in comparison to automatic soldering; also error prone and threatening to product reliability
4. Jumper wires	Relatively cost-intensive assembly process normally used to correct faulty designs; parts must be formed and installed manually
5. Mechanical parts	Parts such as washers, nuts, and screws must be installed manually
6. Sleeved leads/or leads that require special trimming	Must be applied and trimmed manually; entrap solder, flux, and cleaning chemicals
7. Spacers under components	Must be applied manually

6.9.11 Plated-Through Hole to Lead Clearance

This consideration is critical for insertion as well as soldering. The clearance controls the capillary action of solder up through the plated hole. It must also be sufficient for repeatedly successful automatic lead insertion. Table 6.2 lists the plated-through hole to lead clearances as specified by several PCB DFM sources. Based on the results of a study at the Soldering Technology International Laboratory [7], the optimal clearance for a majority of PCB assemblies has been indicated to be 0.017 in. This value can be seen to be consistent with several of the other clearance recommendations identified in Table 6.2.

6.9.12 Tooling

This item identifies whether the PCB DFM document includes reference to tooling such as tool hole sizes or geometry to fit existing fixtures. The objective of conforming to this requirement is to avoid the costly need for new tools. Many documents pictorially define locations and ranges of values for these requirements. One such example is illustrated in Fig. 6.28.

TABLE 6.2 Hole-to-Lead Clearances (HLC)

Reference	HLC (in.)
Company A	0.015
Company B	Not specified
Company C	0.020
Company D	0.015 +/− 0.033
Company E	0.010 +/− ht[a]
Company F	0.015
Company G	Not specified
Company H	Not specified
Company I	Not specified
Company J	Not specified
(8)	0.020 max.
	0.006 min.
(10)	0.010 to 0.020
(11)	0.017
(12)	0.10 min.
	0.027 max.

[a] ht, hole tolerance.

FIG. 6.28 Typical example of PCB tooling requirement specification. (From Company D.)

6.9.13 Testing

This item addresses the design compatibility with existing methods of testing. The requirement may address the interconnects between the PCB and test equipment or merely some preferred design feature that will simplify product testing. As an example of the latter, Company A manufacturability guidelines state that: "It is always preferable to design the board such that testing can be accomplished from one side."

6.9.14 Generally Preferred Features

This item refers to the list of features that are typically provided as recommendations to the designers from engineers who have production experience with past designs. The features that are common to many PCB documents are intended to increase the efficiency of the required processes. They include such items as:

"Panelize boards to minimize board handling through all phases of production" (Company E). "Identify that the board has static sensitive components by screening the Electrostatic Discharge (ESD) symbol on the board" [13]. "Denote the direction that the PCB will travel through the wave solder by use of an arrow on the board" [13].

6.10 GLOSSARY OF TERMS

The following is a brief description of some of the terminology used in this chapter. The physical configurations of some electrical components, the manual assembly operations and some terms pertaining to automatic insertion equipment are given.

Axial-lead component: Electrical component of cylindrical shape with two leads exiting from opposite ends of the component in line with its axis (Fig. 6.3). The component is sometimes called a VCD (variable center distance). The most common axial-lead components are resistors, capacitors, and diodes.

Can-type IC: A cylindrical integrated circuit, packaged so that the leads form a circular pattern (Fig. 6.3). This multileaded radial component can have from 3 to 12 leads.

Cement: Due to the requirements for ruggedness placed on some PCBs, components can be cemented to the board to reduce the effects of vibrations. This sometimes requires that the component leads be bent prior to cementing.

Center distance: Distance between leads when formed for insertion. This term applies to two-leaded components and DIPs; also termed *lead span.*

Channel: Plastic container in the form of a long tube, in which a number of DIPs are placed in single file and oriented for dispensing to an insertion machine; also called *stick.*

Chip resistor, chip capacitor: Small passive SMD packages with pads at both ends for mounting to the board.

Clinch: The bending of a component lead end after insertion through the PCB. This temporarily secures the component prior to soldering. In a full clinch, the lead is bent to contact the terminal area. In a modified clinch, the lead is partially bent to a predetermined intermediate angle.

Component: Any electrical device to be attached to the PCB.

DIP (dual in-line package (Fig. 6.3): A rectangular integrated circuit, packaged so as to terminate in two straight rows of pins or leads.

ECO (engineering change order): A component or insulated jumper wire, installed manually, which is needed when the electrical circuit cannot be etched onto the board without crossing paths at some point. Often the leads are not inserted into the board but are manually soldered to the leads of components already assembled to the board. ECO wires are also referred to as jumper wires. (See Retrofit.)

Fault: Any error that causes the assembled PCB to fail testing or inspection procedures and requires rework.

Hybrid: PCB populated by both surface-mount and through-hole components also referred to as *mixed-mounting technology.*

Insertion: Process whereby the component is grasped, prepared if necessary, and placed on the board, temporarily secured if necessary.

LCCC (leadless ceramic chip carrier): An SMD package made of a ceramic material that can withstand high temperatures and can be hermetically sealed. It does not have leads, but has pads around all four sides of its perimeter.

Kitting: Preparing a package of parts, usually with instructions for assembly, to facilitate manual assembly.

Magazine: A unit containing a group of dispensing channels, usually about 15, used for an automatic DIP insertion machine.

Module: A unit containing a group of dispensing spools, usually 20 or less, used with an automatic axial-lead component sequencer.

Nonstandard component: Any component that cannot be inserted by dedicated, automatic machinery because of its physical characteristics (i.e., size, shape, lead span, etc.) These are also called "odd-form" components.

Pallet: A tray where components are arranged in a known position and orientation.

PCB (printed circuit board): An insulating board onto which an electrical circuit has been printed in the form of conductive paths. Contains drilled holes into which the leads of components are inserted. It is also known as a *printed wiring board* (PWB).

PLCC (plastic leaded chip carrier): A package in which an integrated circuit chip is mounted to form an SMD. It is made of a plastic material that can withstand high temperatures and has rolled-under leads around all four sides of its perimeter.

Preform: Forming the leads of a component to the correct dimensions prior to insertion. Axial-lead components usually have their leads bent at right angles for insertion, and DIPs sometimes require lead or pin straightening. Radial-lead components. may have their leads notched or a stand-off or spacer installed, which maintains the required clearance between the component and the board. Can-type ICs and transistors often need a type of lead forming called "form a," which refers to the profile of the leads after forming.

Radial-lead component: Electrical components with leads at right angles to the body (Fig. 6.3). Examples are disc capacitors, 'kidney" or 'jellybean" capacitors, cermet resistors, etc.

Reflow solder: Process by which surface-mount devices become secured to the PCB. Components are placed onto solder paste that has been added to pads on the board, often by screen printing. Heat is applied to the board and the solder paste melts to attach the devices to the board.

Retrofit: A type of ECO that involves only the assembly of wires (jumper wires) to the PCB. Can refer to an assembly station in the PCB assembly process where only wires are assembled.

Rework: Repair a fault. This usually means severing the leads of the component and removing it, removing the individual leads from the PCB holes, cleaning the holes, inserting a new component, and soldering its leads. The operations are performed manually and are time-consuming and expensive.

SIP (single inline package): An integrated circuit—usually a resistor network or a connector packaged so as to terminate in one straight row of pins or leads (Fig. 6.3).

Sleeve: An insulating plastic tube slid manually onto the lead of a component to insertion to guard against electrical short circuits (Fig. 6.3).

SMD (surface-mount device): A component (often leadless) that is secured to the surface of he board.

SOIC (small-outline integrated circuit): An SMP package made of a plastic material that can sustain high temperatures and has leads in a gull-wing shape along its two longer sides (Fig. 6.3).

SOT (small-outline transistor): An SMD package for discrete components with about four leads in a gull-wing shape along two sides (Fig. 6.3).

Spacer: This can be a small plastic ring (Fig. 6.3) used to keep a minimum clearance between the component and the board. It is usually cemented to the board before the component is inserted. Some components use temporary spacers that are removed after the component is secured. Some spacers are provided with holes corresponding to each lead (Fig. 6.3).

Spool: The package for holding taped axial-lead components.

Standard component: Any component that can be inserted by an automatic insertion machine.

Stick: A plastic container in which a number of DIPs are aligned in single file and are oriented for dispensing to an automatic insertion machine.

Tin: Providing a layer of solder on the surface of leads prior to insertion.

Touch-up: Cleaning the underside or solderside of a PCB after wave soldering to remove any excess solder, which can cause short circuits.

Transistor: A small component whose body has a cylindrical envelope except for one flat face, with three leads at right angles to the body (Fig. 6.3.).

VCD (variable center distance): The capability of an axial-lead component insertion head to vary the distance between leads when forming and inserting an axial-lead component (Fig. 6.3). The term is also used to refer to an axial-lead component. The terms *adjustable span* and *variable span* can also be used.

Wave solder: To automatically solder all the leads on an assembled PCB by conveying it, at a slight incline, over a wave of solder.

REFERENCES

1. Grezesik, A., Layer Reduction Techniques, Circuit Des., August 1990, p. 21.
2. John, J., and Boothroyd, G., Economics of Printed Circuit Board Assembly, Report No. 6, Economic Application of Assembly Robots Project, University of Massachusetts, Amherst April 1985.
3. Boothroyd, G., and Dewhurst, P., Product Design for Assembly, Boothroyd Dewhurst, Inc., Wakefield, RI, 1987.
4. Boothroyd, G., and Shinohara, T., Component Insertion Times for Electronics Assembly, Int. J. Adv. Manuf. Technol., 1 (5), 1986, p. 3.
5. Timmins, J.M., and Dewhurst, P., Printed Circuit Board Assembly—A compilation of Generally Accepted Guidelines, Report No. 39, Product Design for Manufacture Series, University of Rhode Island, Kingston, 1992.
6. IPC-CM-770 C., Component Mounting, Institute for Interconnecting and Packaging Electronic Circuits, Lincolnwood, IL, 1987.
7. Raby, J., Proceedings from 1986 Seminars—Soldering Technology International, San Dimas, CA, 1987.
8. Henson, H.L., Design Guidelines for Printed Circuit Board Assemblies, HS Enterprises, 1985.
9. Skaggs, C., MS85-893 Design for Electronic Assembly, Proceedings of SME Conference, Dearborn, MI, 1985.
10. MIL-STD-275E, Military Standard Printed Wiring for Electronic Equipment, Department of Defense, Washington, DC 1984.
11. IPC-T-50, Terms and Definitions for Electronics Interconnecting and Packaging, Institute for Interconnecting and Packaging Electronic Circuits, Lincolnwood, IL; 1986.
12. Edington, M., Tools of Simultaneous Engineering, PC Assembly, 1988, p. 32.
13. Marty Christensen, Artwork and Board Producibility, NEPCON—National Electronics Packaging and Production Conference, Des Plaines, IL, 1979.
14. Melander, W., and Mast, K. Design for Manufacturability: It's Not Just Design Rules Anymore, Electronics Test Magazine—Proceedings from ATE (Automatic Test Engineering) Conference East, Boston, MA, 1986.
15. Duck, T., Design for Manufacturing Integration, Production Eng. (Lond) Vol. 65, September 1986.
16. Bello, D., Design Impacts on the Wave Solder Process, Printed Circuit Des. June 1987.

7

Design for Machining

7.1 INTRODUCTION

In machining, material is removed from the workpiece until the desired shape is achieved. Clearly, this a wasteful process, and many engineers feel that a main objective should be to design components that do not require machining. Since most manufacturing machines are designed to remove metal by machining, the view that machining should be avoided must be considered impracticable for the immediate future. However, the trend toward the use of "near net shape" processes that conserve material is clearly increasing, and when large-volume production is involved, this approach should be foremost in the designer's mind.

In this chapter we first introduce the common machining processes. Then we consider the ways in which the work material can be readily changed to the desired form by machining, and the ways in which the surfaces of the component are finished. Finally, an introduction to early cost estimating for designers is presented.

All machine tools provide a means of (i) holding a cutting tool or abrasive wheel, (ii) holding a workpiece, and (iii) providing relative motion between them in order to generate the required surfaces.

7.2 MACHINING USING SINGLE-POINT CUTTING TOOLS

Lathes are designed to rotate the workpiece and feed the cutting tool in the direction necessary to generate the required machined surface.

The workpiece is gripped in a chuck or collet or is mounted on a faceplate mounted on the end of the main spindle of the machine. The rotation of the workpiece is provided by an electric motor driving the main spindle through a series of gears. Cutting tools can be driven or fed parallel to or normal to the axis of rotation of the workpiece.

Modern lathes are provided with computer control of all workpiece and tool motions. These are known as computer numerical control (CNC) lathes, and the tools can be fed in any direction in the horizontal plane to generate a required contour on the workpiece. Figure 7.1 shows a cylindrical surface being generated in cylindrical turning.

The feed motion setting on the lathe is the distance moved by the tool during each revolution of the workpiece. The feed f is defined as the displacement of the tool relative to the workpiece, in the direction of feed motion, per stroke or per revolution of the workpiece or tool. Thus, to turn a cylindrical surface of length l_w, the number of revolutions of the workpiece is l_w/f, and the machining time t_m is given by

$$t_m = l_w/(fn_w) \tag{7.1}$$

where n_w is the rotational speed of the workpiece.

It should be emphasized that t_m is the time for one pass of the tool (one cut) along the workpiece. This single pass does not necessarily mean, however, that the machining operation is completed. If the first cut is designed to remove a large amount of material at high feed (roughing cut), the forces generated during the operation will probably have caused significant deflections in the machine structure. The resulting loss of accuracy may necessitate a further machining operation at low feed (finish cut) to bring the workpiece diameter within the limits specified and to provide a smooth machined surface. For these reasons, the

FIG. 7.1 Cylindrical turning.

workpiece is often machined oversize during the roughing cut, leaving a small amount of material that will subsequently be removed during the finishing cut. For very large depths it may be necessary to provide several rough cuts or passes.

Figure 7.2 illustrates five typical lathe operations: cylindrical turning, facing, boring, external threading, and cut-off. In each case, the primary motion and the feed motion, together with certain other terms and dimensions, are indicated. In any machining operation the workpiece has three important surfaces:

1. The work surface, the surface on the workpiece to be removed by machining
2. The machined surface, the desired surface produced by the action of the cutting tool
3. The transient surface, the part of the surface formed on the workpiece by the cutting edge and removed during the following cutting stroke, during the following revolution of the tool or workpiece, or, in other cases (e.g., a thread-turning operation), (Fig. 7.2d) during the following pass of the tool.

In Fig. 7.2a, which shows the geometry of a cylindrical-turning operation, the cutting speed at the tool corner is given by $\pi d_{\mathrm{m}} n_{\mathrm{w}}$, where n_{w} is the rotational speed of the workpiece, and d_{m} is the diameter of the machined surface. The maximum value of the cutting speed is given by $\pi d_{\mathrm{w}} n_{\mathrm{w}}$, where d_{w} is the diameter of the work surface. Thus, the average, or mean, cutting speed v_{av} is given by

$$v_{\mathrm{av}} = \pi n_{\mathrm{w}}(d_{\mathrm{w}} + d_{\mathrm{m}})/2 \qquad\qquad (7.2)$$

FIG. 7.2 Lathe operations: (a) Cylindrical turning, (b) facing, (c) boring, (d) external threading, (e) parting or cut-off. (From Ref. 7.)

FIG. 7.2 Continued

The metal removal rate Z_w is the product of the mean cutting speed and the cross-sectional area of the material being removed A_c. Thus

$$\begin{aligned} Z_w &= A_c\, v_{av} \\ &= \pi f a_p\, n_w (d_w + d_m)/2 \\ &= \pi f a_p\, n_w (d_m + a_p) \end{aligned} \qquad (7.3)$$

This same result could have been obtained by dividing the total volume of metal removed by the machining time t_m.

For a given work material machined under given conditions, the unit power or the energy required to remove a unit volume of material, p_s, can be measured. This factor is mainly dependent on the work material, and if its value is known, the power P_m required to perform any machining operation can be obtained from

$$P_m = p_s Z_w \qquad (7.4)$$

Finally, if the overall efficiency of the machine tool motor and drive systems is denoted by E_m, the electrical power P_e consumed by the machine tool is given by

$$P_e = P_m/E_m \qquad (7.5)$$

Approximate values of the unit power p_s for various work materials are presented at the end of this chapter.

An operation in which a flat surface is generated by a lathe is shown in Fig. 7.2b and can be performed by feeding the tool in a direction at right angles to the axis of workpiece rotation. This operation is known as facing, and when the rotational speed of the workpiece is constant, the cutting speed at the tool corner varies from a maximum at the beginning of the cut to zero when the tool reaches the center of the workpiece.

The machining time t_m is given by

$$t_m = d_m/(2 f n_w) \qquad (7.6)$$

The maximum cutting speed v_{max} and the maximum metal removal rate $Z_{w,max}$ are given by

$$v_{max} = \pi n_w\, d_m \qquad (7.7)$$
$$Z_{w,max} = \pi f a_p\, n_w\, d_m \qquad (7.8)$$

With modern CNC lathes, the rotational speed of the workpiece can be gradually increased during a facing operation as the tool moves toward the center of the workpiece. In this case the machining time is reduced. However, as the tool approaches the center of the workpiece, the maximum rotational speed of the spindle will be encountered, and machining will then proceed at this maximum speed and, consequently, with diminishing cutting speed.

Figure 7.2c shows an internal cylindrical surface being generated. This operation is termed boring and can only be used to enlarge an existing hole in the workpiece. If the diameter of the work surface is d_w and the diameter of the machined surface is d_m, the mean cutting speed is given by Eq. (7.2) and the metal removal rate by

$$Z_w = \pi f a_p \, n_w (d_m - a_p) \tag{7.9}$$

Finally, the machining time t_m is given by Eq. (7.1) if l_w is taken as the length of the hole to be bored.

The lathe operation illustrated in Fig. 7.2d is known as external threading, or single-point screw cutting. The combined motions of the tool and workpiece generate a helix on the workpiece and are obtained by setting the relationship between rotational speed and tool feed to give the required pitch of the machined threads. The machining of a thread necessitates several passes of the tool along the workpiece, each pass removing a thin layer of metal from one side of the thread. The feed is applied in increments, after each pass of the tool, in a direction parallel to the machined surface. In calculating the production time, allowance must be included for the time taken to return the tool to the beginning of the cut, to increment the feed, and to engage the feed drive.

The last lathe operation to be illustrated (Fig. 7.2e) is used when the finished workpiece is to be separated from the bar of material. It is known as a cut-off operation and produces two machined surfaces simultaneously. As with a facing operation at constant rotational speed, the cutting speed and hence the metal removal rate varies from a maximum at the beginning of the cut to zero at the center of the workpiece. The machining time is given by Eq. (7.6) and the maximum metal removal rate by Eq. (7.8). Again, with modern CNC lathes, the rotational speed of the workpiece can be controlled to give constant cutting speed until the limiting rotational speed is reached.

Multispindle automatic lathes are used for high volume or mass production of small components machined from work material in bar form. The various motions of these lathes are controlled by specially machined cams, and the operations are completely automatic, including the feeding of the workpiece through the hollow spindles.

The vertical-boring machine operates on the same principle as a lathe, but it has a vertical axis and is used for large components. Like the lathe, this machine rotates the workpiece and applies continuous, linear feed motion to the tool.

Another machine that uses single-point tools and has a rotary primary motion is a horizontal-boring machine. This machine is needed mostly for heavy, noncylindrical workpieces in which an internal cylindrical surface is to be machined. In general, the words "horizontal" or "vertical" used when describing a machine tool refer to the orientation of the machine spindle that provides primary motion (main spindle). Thus, in the horizontal borer, the main spindle is horizontal.

The principal feature of the horizontal-boring machine is that the workpiece remains stationary during machining, and all the generating motions are applied to the tool. The most common machining process is boring and is shown in Fig. 7.3. Boring is achieved by rotating the tool, which is mounted on a boring bar connected to the spindle, and then feeding the spindle, boring bar, and tool along the axis of rotation. A facing operation can be carried out by using a special toolholder (Fig. 7.4) that feeds the tool radially as it rotates. The equations developed earlier for the machining time, and the metal removal rate in boring and facing also apply to this machine.

The planer is suitable for generating flat surfaces on very large parts. With this machine a linear primary motion is applied to the workpiece and the tool is fed at right angles to this motion (Fig. 7.5). The primary motion is normally accomplished by a rack-and-pinion drive using a variable-speed motor, and the feed motion is intermittent. The machining time t_m and metal removal rate Z_w can be estimated as follows:

$$t_m = b_w/(fn_r) \tag{7.10}$$

where b_w is the width of the surface to be machined, n_r is the frequency of cutting strokes, and f is the feed. The metal removal rate Z_w during cutting is given by

$$Z_w = fa_p v \tag{7.11}$$

where v is the cutting speed and a_p is the depth of cut (the depth of the layer of material to be removed).

FIG. 7.3 Boring on a horizontal-boring machine. (From Ref. 7.)

FIG. 7.4 Facing on a horizontal-boring machine. (From Ref. 7.)

FIG. 7.5 Production of a flat surface on a planer. (From Ref. 7.)

7.3 MACHINING USING MULTIPOINT TOOLS

A drilling machine (or drill press) can perform only those operations in which the tool is rotated and fed along its axis of rotation (Fig. 7.6). The workpiece always remains stationary during the machining process. On small drill presses, the tool is fed by the manual operation of a lever (known as sensitive drilling). The most common operation performed on this machine is drilling with a twist drill to generate an internal cylindrical surface. A twist drill has two cutting edges, each of which removes its share of the work material.

The machining time t_m is given by

$$t_m = l_w/(fn_t) \tag{7.12}$$

where l_w is the length of the hole produced, f is the feed (per revolution), and n_t is the rotational speed of the tool.

The metal removal rate Z_w may be obtained by dividing the volume of material removed during one revolution of the drill by the time for one revolution. Thus

$$Z_w = (\pi/4)fd_m^2 n_t \tag{7.13}$$

FIG. 7.6 Drilling on a drill press. (From Ref. 7.)

where d_m is the diameter of the machined hole. If an existing hole of diameter d_w is being enlarged, then

$$Z_w = (\pi/4)f(d_m^2 - d_w^2)n_t \tag{7.14}$$

Twist drills are usually considered suitable for machining holes having a length of no more than five times their diameter. Special drills requiring special drilling machines are available for drilling deeper holes.

The workpiece is often held in a vise bolted to the machine worktable. The drilling of a concentric hole in a cylindrical workpiece, however, is often carried out on a lathe.

Several other machining operations can be performed on a drill press, and some of the more common ones are illustrated in Fig. 7.7. The center-drilling operation produces a shallow, conical hole with clearance at the bottom. This center hole can provide a guide for a subsequent drilling operation to prevent the drill point from "wandering" as the hole is started. The reaming operation is intended for finishing a previously drilled hole. The reamer is similar to a drill but has several cutting edges and straight flutes. It is intended to remove a small amount of work material only, but it considerably improves the accuracy and surface finish of a hole. The spot-facing (or counterboring) operation is designed to provide a flat surface around the entrance to a hole; this flat surface can provide a suitable seating for a washer and nut, for example.

FIG. 7.7 Some drill-press operations. (a) Center drilling, (b) reaming, (c) spot-facing. (From Ref. 7.)

There are two main types of milling machines: horizontal and vertical. In the horizontal-milling machine the milling cutter is mounted on a horizontal arbor (or shaft) driven by the main spindle.

The simplest operation, slab milling, is used to generate a horizontal surface on the workpiece, as shown in Fig. 7.8.

When estimating the machine time t_m in a milling operation, it should be remembered that the distance traveled by the cutter will be larger than the length of the workpiece. This extended distance is illustrated in Fig. 7.9 in which it can be seen that the cutter travel distance is given by $l_w + \sqrt{a_e(d_t - a_e)}$, where l_w is the length of the workpiece, a_e the depth of cut, and d_t the diameter of the cutter. Thus the machining time is given by

$$t_m = [l_w + \sqrt{a_e(d_t - a_e)}]/v_f \tag{7.15}$$

where v_f is the feed speed of the workpiece.

The metal removal rate Z_w will be equal to the product of the feed speed and the cross-sectional area of the metal removed, measured in the direction of feed motion. Thus, if a_p is equal to the workpiece width,

$$Z_w = a_e\, a_p\, v_f \tag{7.16}$$

Figure 7.10 shows some further horizontal-milling operations. In form cutting, the special cutter has cutting edges shaped to form the cross section required on the workpiece. These cutters are generally expensive to manufacture, and form milling is used only when the quantity to be produced is sufficiently large. In slotting, a standard cutter is used to produce a rectangular slot in a workpiece. Similarly in angular milling, a standard cutter machines a triangular slot. The straddle-milling operation shown in the figure is only one of an infinite variety of

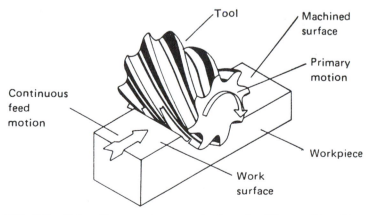

FIG. 7.8 Slab milling on a knee-type horizontal milling machine. (From Ref. 7.)

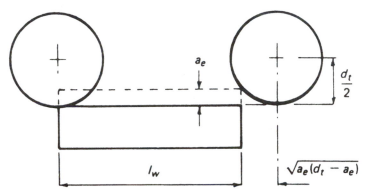

FIG. 7.9 Relative motion between a slab-milling cutter and the workpiece during machining time. (From Ref. 7.)

operations that can be carried out by mounting more than one cutter on the machine arbor. In this way, combinations of cutters can machine a wide variety of cross-sectional shapes. When cutters are used in combination, the operation is often called gang milling.

A wide variety of operations involving the machining of horizontal, vertical, and inclined surfaces can be performed on a vertical-milling machine. As the name of the machine implies, the spindle is vertical.

A typical face-milling operation, where a horizontal flat surface is being machined, is shown in Fig. 7.11. The cutter employed is known as a face-milling cutter.

In estimating the machine time t_m allowance should again be made for the additional relative motion between the cutting tool and workpiece. As can be seen in Fig. 7.12, the total motion when the path of the tool axis passes over the workpiece is given by $(l_w + d_t)$, and, therefore, the machining time is given by

$$t_m = (l_w + d_t)/v_f \tag{7.17}$$

where l_w is the length of the workpiece, d_t is the diameter of the cutter, and v_f is the feed speed of the workpiece.

When the path of the tool axis does not pass over the workpiece,

$$t_m = [l_w + 2\sqrt{a_e(d_t - a_e)}]/v_f \tag{7.18}$$

where a_e is the width of the cut in vertical milling.

The metal removal rate Z_w in both cases is given by Eq. (7.16).

Various vertical-milling machine operations are illustrated in Fig. 7.13.

FIG. 7.10 Some horizontal-milling operations. (a) Form cutting, (b) slotting, (c) straddle milling, (d) angular milling. (From Ref. 7.)

Another machine using multipoint tools is the broaching machine. In broaching, the machine provides the primary motion (usually hydraulically powered) between the tool and workpiece, and the feed is provided by the staggering of the teeth on the broach, each tooth removing a thin layer of material (Fig. 7.14). Since the machined surface is usually produced during one pass of the tool, the machining time t_m is given by

$$t_m = l_t/v \tag{7.19}$$

FIG. 7.11 Face milling on a vertical-milling machine. (From Ref. 7.)

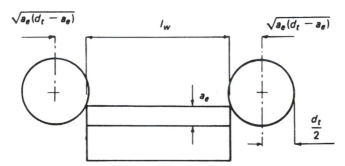

FIG. 7.12 Relative motion between the face-milling cutter and the workpiece during machining time. (From Ref. 7.)

FIG. 7.13 Some vertical-milling machine operations. (a) Horizontal surface, (b) slot, (c) dovetail, (d) T slot. (From Ref. 7.)

where l_t is the length of the broach and v is the cutting speed. The average metal removal rate Z_w can be estimated by dividing the total volume of metal removed by the machining time.

Broaching is widely used to produce noncircular holes. In these cases the broach can be either pulled or pushed through a circular hole to enlarge the hole to the shape required or to machine a keyway, for example (Fig. 7.15). Broaches must be designed individually for the particular job and are expensive to manufacture. This high cost must be taken into account when comparing broaching to slower, alternative machining methods.

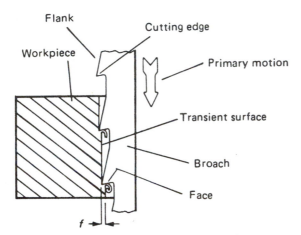

FIG. 7.14   Broaching on a vertical-broaching machine. (From Ref. 7.)

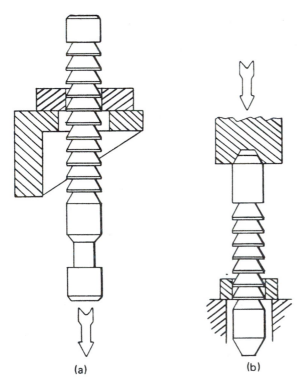

FIG. 7.15   Methods of broaching a hole. (a) Pull broach; (b) Push broach. (From Ref. 7.)

The production of internal and external screw threads can be accomplished by the use of taps and dies. These multipoint tools can be thought of as helical broaches.

In Fig. 7.16, a tap is fed into a prepared hole and rotated at low speed. The relative motion between a selected point on a cutting edge and the workpiece is, therefore, helical; this motion is the primary motion. All the machining is done by the lower end of the tap, where each cutting edge removes a small layer of metal (Fig. 7.16) to form the thread shape; the fully shaped thread on the tap serves to clear away fragments of chips that may collect. A die has the same cutting action as a tap, but is designed to produce an external thread.

Internal threading using taps can be carried out on turret lathes and drill presses. External threading using dies can be carried out on turret lathes and special screw-cutting machines.

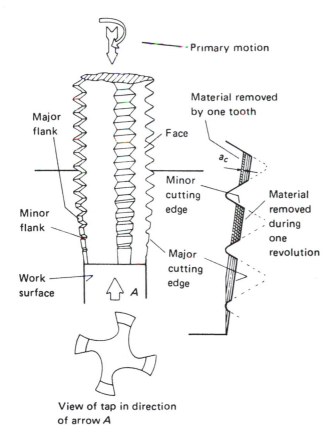

FIG. 7.16   Tapping. (From Ref. 7.)

7.4 MACHINING USING ABRASIVE WHEELS

Abrasive wheels (or grinding wheels) are generally cylindrical, disc-shaped, or cup-shaped (Fig. 7.17). The machines on which they are used are called grinding machines, or grinders; they all have a spindle that can be rotated at high speed and on which the grinding wheel is mounted. The spindle is supported by bearings and mounted in a housing; this assembly is called the wheel head. A belt drive from an electric motor provides power for the spindle. The abrasive wheel consists of individual grains of very hard material (usually silicon carbide or aluminum oxide) bonded in the form required.

Abrasive wheels are sometimes used in rough grinding, where material removal is the important factor; more commonly abrasive wheels are used in finishing operations, where the resulting smooth surface finish is the objective.

In the metal-cutting machine tools described earlier, generation of a surface is usually obtained by applying a primary motion to either the tool or workpiece and a feed motion to either the tool or the workpiece. In grinding machines, however, the primary motion is always the rotation of the abrasive wheels, but often two or more generating (feed) motions are applied to the workpiece to produce the desired surface shape.

In horizontal-spindle surface grinding (Fig. 7.18) the principal feed motion is the reciprocation of the worktable on which the work is mounted; this motion is known as the traverse. Further feed motions may be applied either to the wheel head, by moving it down (known as infeed), or to the table, by moving it parallel to the machine spindle (known as cross-feed). In Fig. 7.18 a horizontal surface is being generated on a workpiece by a cross-feed motion. This feed motion, which is intermittent, is usually applied after each stroke or pass of the table. The amount of cross-feed f may, therefore, be defined as the distance the tool advances across the workpiece between each cutting stroke. The operation is known as traverse grinding.

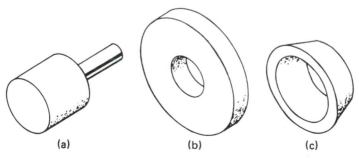

(a)   (b)   (c)

FIG. 7.17 Common shapes of abrasive wheels. (a) Cylindrical, (b) disc, (c) cup. (From Ref. 7.)

Work
surface

Primary motion

Grinding wheel

Machined surface

Intermittent
feed motion
(cross-feed)

Workpiece
motion
(traverse)

Workpiece

FIG. 7.18 Surface grinding on a horizontal-spindle surface grinder. (From Ref. 7.)

Figure 7.19 shows the geometries of both traverse grinding and plunge grinding on a horizontal-surface grinder. From Fig. 7.19a, the metal removal rate in traverse grinding is given by

$$Z_w = fa_p v_{trav} \tag{7.20}$$

where f is the cross-feed per cutting stroke, a_p is the depth of cut, and v_{trav} is the traverse speed.

The machining time t_m is given by

$$t_m = b_w/(2fn_r) \tag{7.21}$$

where n_r is the frequency of reciprocation and b_w is the width of the workpiece.

In a similar way, for the plunge-grinding operation (Fig. 7.19b), the metal removal rate is given by Eq. (7.20).

Before estimating the machining time in the plunge-grinding operation, it is necessary to describe a phenomenon known as "sparking-out." In any grinding operation where the wheel is fed in a direction normal to the work surface (infeed), the feed f, which is the depth of the layer of material removed during one cutting stroke, will initially be less than the nominal feed setting on the machine. This feed differential results from the deflection of the machine tool elements and workpiece under the forces generated during the operation. Thus, on completion of the theoretical number of cutting strokes required, some work material will still have to be removed. The operation of removing this material, called sparking-out, is achieved by continuing the cutting strokes with no further application of feed until metal removal becomes insignificant (no further sparks

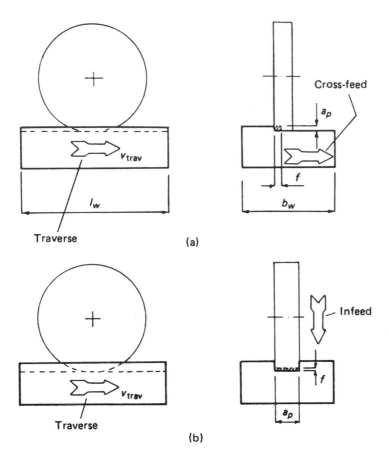

FIG. 7.19 Horizontal-spindle surface-grinding operations. (a) Traverse grinding; (b) plunge grinding. (From Ref. 7.)

appear). If the time for sparking-out is denoted by t_s, the machining time in plunge grinding is given by

$$t_m = (a_t/2fn_r) + t_s \tag{7.22}$$

where a_t is the total depth of work material to be removed.

In vertical-spindle surface grinding (Fig. 7.20) a cup-shaped abrasive wheel performs an operation similar to face milling. The worktable is reciprocated and the tool fed intermittently downward; these motions are known as traverse and infeed, respectively. A horizontal surface is generated on the workpiece, and because of the deflection of the machine structure, the feed f will initially be less

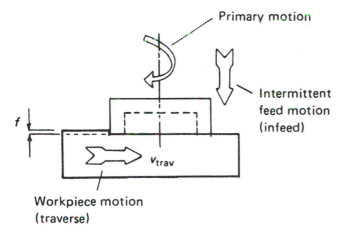

FIG. 7.20   Surface grinding on a vertical-spindle surface grinder. (From Ref. 7.)

than the feed setting on the machine tool. This means that sparking-out is necessary, as in plunge grinding on a horizontal-spindle machine.

The metal removal rate is given by

$$Z_w = f a_p v_{trav} \tag{7.23}$$

where a_p is equal to the width of the workpiece and v_{trav} is the traverse speed.

The machining time is given, as in plunge grinding on a horizontal-spindle machine, by Eq. (7.22).

Larger vertical-spindle surface grinders are available with rotary worktables on which several workpieces can be mounted. The machining time per part for this type of grinder is given by

$$t_m = \left(\frac{a_t}{f n_w} + t_s \right) / n \tag{7.24}$$

where n_w is the rotational speed of the worktable and n is the number of workpieces mounted on the machine.

In cylindrical grinding (Fig. 7.21) the workpiece is supported and rotated between centers. The headstock provides the low-speed rotational drive to the workpiece and is mounted, together with the tailstock, on a worktable that is reciprocated horizontally using a hydraulic drive. The grinding-wheel spindle is horizontal and parallel to the axis of workpiece rotation, and horizontal, hydraulic feed can be applied to the wheel head in a direction normal to the axis of workpiece rotation; this motion is known as infeed.

Figure 7.21 shows a cylindrical surface being generated using the traverse motion—an operation that can be likened to cylindrical turning, where the single-

FIG. 7.21 Cylindrical grinding. (From Ref. 7.)

point cutting tool is replaced by a grinding wheel. In fact, grinding attachments are available that allow this operation to be performed on a lathe.

The geometries of traverse and plunge grinding on a cylindrical grinder are shown in Fig. 7.22. In traverse grinding, the maximum metal removal rate is closely given by.

$$Z_{w,max} = \pi f d_w\, v_{trav} \tag{7.25}$$

where d_w is the diameter of work surface, v_{trav} is the traverse speed, and f is the feed per stroke of the machine table (usually extremely small compared to d_w). The machining time will be given by Eq. (7.22).

FIG. 7.22 Cylindrical-grinding operations. (a) Traverse grinding; (b) plunge grinding. (From Ref. 7.)

In the plunge-grinding operation shown in Fig. 7.22b, the wheel is fed into the workpiece, without traverse motion applied, to form a groove. If v_f is the feed speed of the grinding wheel, d_w the diameter of the work surface, and a_p the width of the grinding wheel, the maximum metal removal rate is given by

$$Z_{w,max} = \pi a_p d_w v_f \qquad (7.26)$$

and the machining time is

$$t_m = (a_t/v_f) + t_s \qquad (7.27)$$

where a_t is the total depth of material to be removed and t_s is the sparking-out time.

In internal grinding (Fig. 7.23), the wheel head supports a horizontal spindle and can be reciprocated (traversed) in a direction parallel to the spindle axis. A small cylindrical grinding wheel is used and is rotated at very high speed. The workpiece is mounted in a chuck or on a magnetic faceplate and rotated. Horizontal feed is applied to the wheel head in a direction normal to the wheel spindle; this motion is known as infeed. Again, traverse and plunge grinding can be performed, the geometries of which are shown in Fig. 7.24.

Traverse grinding is shown in Fig. 7.24a, and the maximum removal rate, which occurs at the end of the operation, is given by

$$Z_{w,max} = \pi f d_m v_{trav} \qquad (7.28)$$

FIG. 7.23 Internal grinding. (From Ref. 7.)

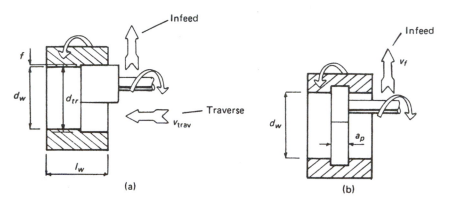

FIG. 7.24 Internal-grinding operations. (a) Traverse grinding; (b) plunge grinding. (From Ref. 7.)

where f is the feed, v_{trav} is the traverse speed, and d_m is the diameter of the machined surface. The machining time is again given by Eq. (7.22).

Finally, in plunge grinding (Fig. 7.24b) the maximum removal rate is given by

$$Z_{w,max} = \pi a_p d_m v_f \qquad\qquad (7.29)$$

and the machining time by Eq. (7.27).

Now that the various machine tools, machining operations, and basic equations for metal removal rate and machining time have been introduced, we can turn our attention to those design factors that affect the cost of machining and to cost estimating for designers.

7.5   STANDARDIZATION

Perhaps the first rule in designing for machining is to design using standard components as much as possible. Many small components, such as nuts, washers, bolts, screws, seals, bearings, gears, and sprockets, are manufactured in large quantities and should be employed wherever possible. The cost of these components will be much less than the cost of similar, nonstandard components. Clearly, the designer will need catalogues of the standard items available; these can be obtained from suppliers. Supplier information is provided in standard trade indexes, where companies are listed under products. However, there is a danger in overemphasizing standardization. Many of the impressive successes brought about by the application of DFMA procedures were only made possible by breaking away from standardization. For example, the IBM "proprinter" was successful mainly because the designers departed from the customary approach to the design of dot-matrix printers. They included a novel new mechanism for driving the print head; they also introduced new plastic materials for the base and departed from the use of standard components for securing important items such as the power transformer and drive motors. Taken to extremes, a slavish adherence to company "standards" will prevent innovation in design.

A second rule is, if possible, minimize the amount of machining by preshaping the workpiece. Workpieces can sometimes be preshaped by using castings or welded assemblies or by metal deformation processes, such as extrusion, deep drawing, blanking, or forging. Obviously, the justification for preforming of workpieces will depend on the required production quantity. Again standardization can play an important part when workpieces are to be preformed. The designer may be able to use preformed workpieces designed for a previous similar job; because the necessary patterns for castings or the tools and dies for metal-forming processes are already available.

Finally, even if standard components or standard preformed workpieces are not available, the designer should attempt to standardize on the machined features to be incorporated in the design. Standardizing on machined features means that the appropriate tools, jigs, and fixtures will be available, which can reduce manufacturing costs considerably. Examples of standardized machined features might include drilled holes, screw threads, keyways, seatings for bearings, splines, etc. Information on standard features can be found in various reference books.

7.6 CHOICE OF WORK MATERIAL

When choosing the material for a component, the designer must consider applicability, cost, availability, machinability, and the amount of machining required. Each of these factors influences the others, and the final optimum choice will generally be a compromise between conflicting requirements. The applicability of various materials depends on the component's eventual function and is decided by such factors as strength, resistance to wear, appearance, corrosion resistance, etc. These features of the design process are outside the scope of this chapter, but once the choice of material for a component has been narrowed, the designer must then consider factors that help to minimize the final cost of the component. It should not be assumed, for example, that the least expensive work material will automatically result in minimum cost for the component. For example, it might be more economical to choose a material that is less expensive to machine (more machinable) but has a higher purchase cost. In a constant cutting-speed, rough-machining operation, the production cost C_{pr} per component is given by

$$C_{pr} = Mt_l + Mt_m + (Mt_{ct} + C_t)t_m/t \tag{7.30}$$

where M is the total machine and operator rate, t_1 is the nonproductive time, t_m is the machining time (time the machine tool is operating), t is the tool life (machining time between tool changes), t_{ct} is the tool changing time, C_t is the cost of providing a sharp tool, including the cost of regrinding and/or the depreciation of the insert holder and insert where applicable.

The machining time is given by

$$t_m = K/v \tag{7.31}$$

where K is a constant for the particular operation and v is the cutting speed.

Also, the tool life t is given by Taylor's tool life equation:

$$vt^n = v_r t_r^n \tag{7.32}$$

where v_r and t_r are the reference cutting speed and tool life, respectively, and n is the Taylor tool life index, which is mainly dependent on the tool material. Usually, for high-speed steel tools n is assumed to be 0.125 and for carbide tools it is assumed to be 0.25.

If Eqs. (7.31) and (7.32) are substituted into Eq. (7.30), and the resulting expression differentiated, it can be shown that the cutting speed v_c for minimum cost is given by

$$v_c = v_r(t_r/t_c)^n \tag{7.33}$$

where t_c is the tool life for minimum cost and is given by

$$t_c = [(1/n) - 1](t_{ct} + C_t/M) \tag{7.34}$$

Thus, if Eqs. (7.30) through (7.34) are combined, the minimum cost of production C_{min} is given by

$$C_{min} = Mt_1 + \frac{MK}{(1-n)v_r}\left(\frac{t_c}{t_r}\right)^n \tag{7.35}$$

The first term in this expression is the cost of the nonproductive time on the machine tool and will not be affected by the work material chosen or by the amount of machining carried out on the workpiece. The second term is the cost of the actual machining operation, and for a given machine and tool design it depends on the values of n, v_r, t_r, and K. The factor n depends mainly on the tool material; $v_r t_r^n$ is a measure of the machinability of the material; K is proportional to the amount of machining to be carried out on the workpiece and can be regarded as the distance traveled by the tool cutting edge corner relative to the workpiece during the machining operation. For a given operation on a given machine tool and with a given tool material it is shown in Eq. (7.34) that the tool life for minimum cost would be constant and hence [from Eq. (7.35)] that the machining costs would be inversely proportional to the value of $v_r t_r^n$. Since v_r is the cutting speed giving a tool life of t_r, more readily machined materials have a higher value of $v_r t_r^n$ and hence give a lower machining cost.

Taking, for example, a machining operation using high-speed steel tools ($n = 0.125$) and a low carbon steel workpiece and typical figures of $M = \$0.00833/\text{s}$, $t_1 = 300$ s, $t_c = 3000$ s, $K = 183$ m (600 ft), and $v_r = 0.76$ m/s (150 ft/min) when $t_r = 60$ s, then from Eq. (7.35) the minimum production cost per component C_{min} is \$6.22. If, however, an aluminum workpiece for which a typical value of v_r is 3.05 m/s (600 ft/mm) when t_r is 60 s could be used, the use of aluminum would reduce the production cost to \$3.43. In other words, an additional amount equal to the difference between these two costs could be spent for each workpiece in order to employ the more machinable material, i.e., as much as \$2.79 additional per workpiece.

Clearly, the designer should try to select work materials that will result in minimum total component cost.

7.7 SHAPE OF WORK MATERIAL

With the exception of workpieces that are to be partially formed before machining, such as forgings, castings, and welded structures, the choice of the shape of the work material depends mainly on availability. Metals are generally sold in plate, sheet, bar, or tube form (Table 7.1) in a wide range of standard sizes.

TABLE 7.1 Standard Material Shapes and Ranges of Sizes

Name	Size	Shape
Plate	6–75 mm (0.25–3 in.)	
Sheet	0.1–5 mm (0.004–0.2 in.)	
Round bar or rod	3–200 mm dia. (0.125–8 in. dia.)	
Hexagonal bar	6–75 mm (0.25–3 in.)	
Square bar	9–100 mm (0.375–4 in.)	
Rectangular bar	3 × 12–100 × 150mm (0.125 × 0.5–4 × 6 in.)	
Tubing	5 mm dia., 1 mm wall–100 mm dia., 3 mm wall (0.1875 in. dia., 0.035 in. wall–4 in. dia., 0.125 in. wall)	

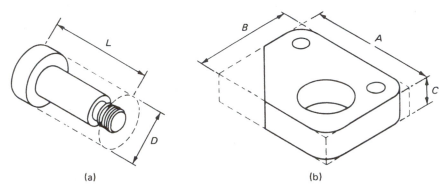

(a)   (b)

FIG. 7.25   Basic component shapes. (a) Rotational; (b) nonrotational. (From Ref. 7.)

The designer should check on the standard shapes and sizes from the supplier of raw material and then design components that require minimal machining.

Components manufactured from a circular or hexagonal bar or tube are generally machined on those machine tools that apply a rotary primary motion to the workpiece; these types of components are called rotational components (Fig. 7.25a). The remaining components are manufactured from square or rectangular bar, plate, or sheet and are called nonrotational components (Fig. 7.25b). Components partially formed before machining can also be classified as either rotational or non-rotational components.

Some of the machining techniques used to alter the initial workpiece shape will now be described and will help to illustrate some further design rules for machined components.

7.8   MACHINING BASIC COMPONENT SHAPES

7.8.1   Disc-Shaped Rotational Components ($L/D \leqslant 0.5$)

Rotational components where the length-to-diameter ratio is less than or equal to 0.5 may be classified as discs. For diameters to approximately 300 mm (12 in.) the workpiece would generally be gripped in a lathe chuck; for larger diameters it would be necessary to clamp the workpiece on the table of a vertical borer. The simplest operations that could be performed would be machining of the exposed face and drilling, boring, and threading a concentric hole. All these operations could be performed on one machine without regripping the workpiece (Fig. 7.26). The realization that neither the unexposed face nor a portion of the outer cylindrical surface can be machined leads to some general guidelines for design: if possible, design the component so that machining is not necessary on the unexposed surfaces of the workpiece when it is gripped in the work-

Stepped surface (facing)	Bore (drilling)	Stepped bore (counterboring)	Threaded bore (threading or tapping)	Typical component

FIG. 7.26 Some ways of machining a disc-shaped workpiece. (From Ref. 7.)

holding device. Also, the diameters of the external features should gradually increase, and the diameters of the internal features should gradually decrease from the exposed face.

Of course, with the examples shown in Fig. 7.26 it would probably be necessary to reverse the workpiece in the chuck to machine the opposite face. However, if its diameter were less than about 50 mm (2 in.), the desired surfaces could probably be machined on the end of a piece of bar stock and the component then separated from the bar by a parting or cut-off operation (Fig. 7.27). It should be remembered that when a workpiece must be reversed in the chuck, the concentricity of features will be difficult to maintain (Fig. 7.28).

When machined surfaces intersect to form an edge, the edge is square; when surfaces intersect to form an internal corner, however, the edge is rounded to the

FIG. 7.27 Parting finished components from bar stock. (From Ref. 7.)

FIG. 7.28 Machining of components stepped to both ends. (From Ref. 7.)

shape of the tool corner. Thus the designer should always specify radii for internal corners. When the two intersecting faces are to form seatings for another component in the final assembly, the matching corner on the second component should be chamfered to provide clearance (Fig. 7.29). Chamfering ensures proper seating of the two parts and presents little difficulty in the machining operations.

On rotational components some features may be necessary that can only be produced by machine tools other than those that rotate the workpiece. Consequently, the batch of workpieces requiring these features will have to be stacked temporarily and then transferred to another machine tool that may be in another part of the factory. This storage and transfer of workpieces around a factory presents a major organizational problem and adds considerably to the manufacturing costs. Thus, if possible, the components should be designed to be machined on one machine tool only.

FIG. 7.29 Rounded corners and chamfers. (From Ref. 7.)

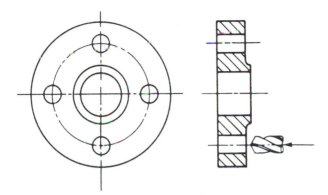

FIG. 7.30   Drilling a pattern of auxiliary holes.

Plane-machining operations may also be required on a rotational component. Such operations might be carried out on a milling machine. Finally, auxiliary holes (those not concentric with the component axis) and gear teeth may be required. Auxiliary holes would be machined on a drill press and would generally form a pattern as shown in Fig. 7.30. Axial auxiliary holes are usually the easiest to machine, because one of the flat surfaces on the workpiece can be used to orient it on the work-holding surface. Thus, the designer should avoid auxiliary holes inclined to the workpiece axis. Gear teeth would be generated on a special gear-cutting machine, and this process is generally slow and expensive.

7.8.2   Short, Cylindrical Components ($0.5 < L/D < 3$)

The workpieces from which short, cylindrical components are produced would often be in the form of bar stock, and the machined component would be separated from the workpiece by parting or cut-off as was shown in Fig. 7.27. The whole of the outer surface of this type of component can be machined without interference from the jaws of the chuck. However, it is important for the designer to ensure (if possible) that the diameters of a stepped internal bore are gradually decreasing from the exposed end of the workpiece and that no recesses or grooves are required on the surface produced in the parting or cut-off operation (Fig. 7.31).

7.8.3   Long, Cylindrical Rotational Components ($L/D \geqslant 3$)

Long, cylindrical rotational components would often be supported between centers or gripped at the headstock end by a chuck and supported by a center at the other end. If the L/D ratio is too large, the flexibility of the workpiece creates a problem because of the forces generated during machining. Thus, the

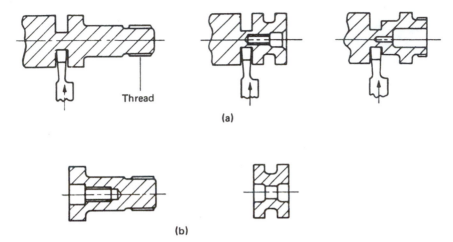

FIG. 7.31   Machining components from bar stock. (a) Components that can be parted off complete; (b) components that cannot be parted off complete. (From Ref. 7.)

designer should ensure that the workpiece, when supported by the work-holding devices, is sufficiently rigid to withstand the machining forces.

When a rotational component must be supported at both ends for machining of the external surfaces, internal surfaces of any kind cannot be machined at the same time. In any case, with slender components, concentric bores would necessarily have large length-to-diameter ratios and would be difficult to produce. Thus, the designer should try to avoid specifying internal surfaces for rotational components having large L/D ratios.

A common requirement on a long, cylindrical component is a keyway, or slot. A keyway is usually milled on a vertical-milling machine using an end-milling cutter (Fig. 7.32a) or on a horizontal-milling machine using a side- and face-milling cutter (Fig. 7.32b). The shape of the end of the keyway is determined by the shape of the milling cutter used, and the designer, in specifying this shape, is specifying the machining process.

Before the ways of changing basic nonrotational shapes by machining operations are discussed, some general points should be noted regarding undesirable features on rotational components. These undesirable design features can be categorized as follows:

1.   Features impossible to machine.
2.   Features extremely difficult to machine that require the use of special tools or fixtures.
3.   Features expensive to machine even though standard tools can be used.

FIG. 7.32 Machining of a keyway. (a) Vertical milling, (b) horizontal milling. (From Ref. 7.)

In considering the features of a particular design it should be realized that

1. Surfaces to be machined must be accessible when the workpiece is gripped in the work-holding device.
2. When the surface of workpiece is being machined, the tool and tool-holding device must not interfere with the remaining surfaces on the workpiece.

Figure 7.33a shows an example of a component with external surfaces impossible to machine. This is because during the machining of one of the cylindrical surfaces, the tool would interfere with the other cylindrical surface. Figure 7.33b shows a component that would be extremely difficult to machine on a lathe because when the hole is drilled, the workpiece would have to be supported in a special holding device. Even if the workpiece were transferred to a drill press for the purpose of drilling the hole (in itself an added expense), a milled preparation would be required (Fig. 7.33c) to prevent the drill from deflecting sideways at the beginning of the drilling operation.

Figure 7.34 shows two examples where the tool or toolholder would interfere with other surfaces on the workpiece. The small radial hole shown in Fig. 7.34a would be difficult to machine because a special long drill would be required. The internal recess shown in the component in Fig. 7.34b could not be machined because it would be impossible to design a cutting tool that would reach through the opening of the bore.

Figure 7.35a shows a screw thread extending to a shoulder. Extending a screw thread to a shoulder would be impossible because when the lathe carriage is

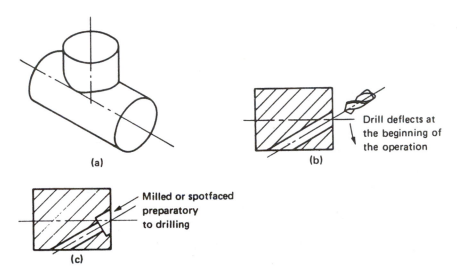

FIG. 7.33 Difficulties arising when nonconcentric cylindrical surfaces are specified. (a) Impossible to machine, (b) difficult to machine, (c) can be machined on a drill press. (From Ref. 7.)

FIG. 7.34 Design features to avoid in rotational parts. (a) Special drill required to machine radial hole, (b) impossible to machine internal recess. (From Ref. 7.)

FIG. 7.35 Machined screw threads on stepped components. (a) Impossible to machine, (b) good design with run-out groove. (From Ref. 7.)

disengaged from the lead screw at the end of each pass, the threading tool generates a circular groove in the workpiece. Thus it is necessary to provide a run-out groove (Fig. 7.35b) so that the threading tool will have clearance and not interfere with the remaining machined surfaces.

7.8.4 Nonrotational Components ($A/B \leqslant 3$, $A/C \geqslant 4$)

Extremely thin, flat components should be avoided because of the difficulty of work holding while machining external surfaces. Many flat components would be machined from plate or sheet stock and would initially require machining of the outer edges. Outer edges would generally be machined on either a vertical- or horizontal-milling machine. Figure 7.36 shows the simplest shapes that can be

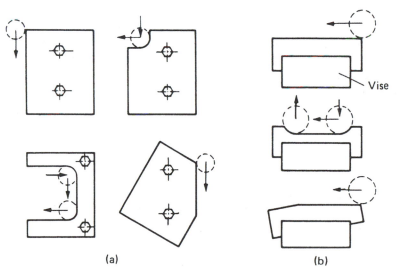

FIG. 7.36 Milling external shape of flat components. (a) Vertical milling (plan view), (b) horizontal milling (front view). (From Ref. 7.)

generated on the edge of a flat component. It can be seen that internal corners must have radii no smaller than the radius of the milling cutter used.

In general, the minimum diameter of cutters for horizontal milling (about 50 mm [2 in.] for an average machine) is much larger than the diameter of a cutter for vertical milling (about 12 mm [0.5 in.]). Thus, small internal radii would necessitate vertical milling. However, as can be seen from Fig. 7.36, a flat workpiece that must be machined around the whole periphery is generally clamped to the machine worktable with a spacer beneath the workpiece smaller than the finished component. This means of work holding requires at least two bolt holes to be provided in the workpiece. In horizontal milling the workpiece can be gripped in a vise.

With flat components required in reasonably large batch sizes, manufacturing costs can often be considerably reduced by simultaneous machining of a stack of workpieces.

Sometimes large holes (principal bores) are required in nonrotational components. These principal bores are generally normal to the two large surfaces on the component and require machining by boring. This operation could be performed on a lathe (Fig. 7.37a), where the workpiece would be bolted to a faceplate, or on a vertical borer (Fig. 7.37b), where the workpiece would be bolted to the rotary worktable. For small parts, however, where high accuracy is required, the bores would be machined on a jig borer. A jig borer is similar to a vertical-milling machine, but the spindle is fed vertically and can hold a boring tool (Fig. 7.37c). From these examples it can be seen that where possible, principal bores should be cylindrical and normal to the base of the component. It can also be seen that a spacer is required between the workpiece and the work-holding surface.

The next type of secondary machining operation to be considered is the provision of a series of plane surfaces such as the machining of steps, slots, etc., in one of the large surfaces on the workpiece. If possible, plane-surface machining should be restricted to one surface of the component only, thus avoiding the need for reclamping the workpiece. Plane surfaces might be machined on milling machines, or, in very large workpieces (such as machine beds), on planing machines. Figure 7.38 shows a variety of plane-surface machining operations, and it can be seen that plane-machined surfaces should, if possible, be either parallel or normal to the base of the component. Also, internal radii need not be specified for the milling operations, because the corners of the teeth on milling cutters are usually sharp.

Finally, auxiliary holes might be required in flat components; these would generally be machined on a drill press. Similar requirements to those discussed for the machining of auxiliary holes in disc-shaped rotational components apply. Thus, auxiliary holes should, if possible, be cylindrical and normal to the base of the component and preferably related by a pattern to simplify positioning of the workpiece for drilling.

FIG. 7.37 Machining of principal bores in nonrotational workpieces. (a) Lathe, (b) vertical borer, (c) jig borer. (From Ref. 7.)

7.8.5 Long, Nonrotational Components ($A/B > 3$)

Long, nonrotational components would often be machined from rectangular- or square-section bar stock. Extremely long components should be avoided because of work-holding difficulties. The most common machining operations would be drilling and milling. Machined surfaces parallel to the principal axis of the component should be avoided because of the difficulties of holding down the entire length of the workpiece. Instead, the designer should, if possible, utilize work material preformed to the cross section required.

7.8.6 Cubic, Nonrotational Components ($A/B < 3$, $A/C < 4$)

Cubic components should be provided with at least one plane surface that can initially be surface-ground or milled to provide a base for work-holding purposes and a datum for further machining operations.

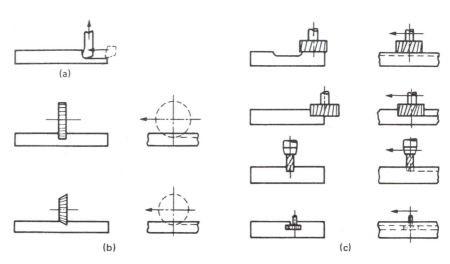

FIG. 7.38 Plane-surface machining of flat components. (a) Shaping or planing, (b) horizontal milling, (c) vertical milling. (From Ref. 7.)

If possible, the outer machined surfaces of the component should consist of a series of mutually perpendicular plane surfaces parallel to and normal to its base. In this way, after the base has been machined, further machining operations can be carried out on external surfaces with minimal reclamping of the workpiece. Figure 7.39, for example, shows a cubic workpiece where the external exposed surfaces can all be machined on a vertical-milling machine without reclamping. From this figure it can be seen that sharp internal corners parallel to the base can be machined readily but that sharp internal corners normal to the base should be avoided.

The workpiece shown in Fig. 7.39 is blocklike, but others may be hollow or boxlike. Main bores in cubic components will often be machined on a horizontal-boring machine. For ease of machining, internal cylindrical surfaces should be concentric and decrease in diameter from the exposed surface of the workpiece. Also, where possible, blind bores should be avoided because in horizontal boring the boring bar must usually be passed through the workpiece. Internal machined surfaces in a boxlike cubic component should be avoided unless the designer is certain that they will be accessible.

With small cubic components it is possible to machine pockets or internal surfaces using an end-milling cutter as shown in Fig. 7.40. Again it can be seen that internal corners normal to the workpiece base must have a radius no smaller than that of the cutter. Usually, the same cutter will be used to clear out the pocket after machining the outer shape, and the smaller the cutter diameter the longer it

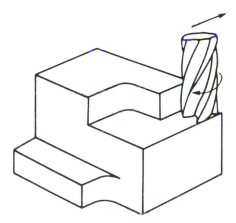

FIG. 7.39 Milling outer surface of a cubic workpiece. (From Ref. 7.)

will take to perform this operation. Consequently, the cost of the operation will be related to the radii of the vertical internal corners. Thus, internal corners normal to the workpiece base should have as large a radius as possible.

Finally, cubic components will often have a series of auxiliary holes. Auxiliary holes should be cylindrical and either normal to or parallel to the base of the

Cleared after
machining outline

FIG. 7.40 Milling a pocket in a blocklike cubic workpiece. (From Ref. 7.)

component; they should also be in accessible positions and have L/D ratios that make it possible to machine them with standard drills. In general, standard drills can produce holes having L/D ratios as large as 5.

Figure 7.41a shows examples of features that would be difficult and expensive to produce in nonrotational components. In the first case the internal vertical corners are shown sharp; these features cannot be produced with standard tools. In the second case the through hole has an extremely large L/D ratio and would be difficult to produce even with special deep-hole drilling techniques. Figure 7.41b shows examples of machined features virtually impossible to produce because a suitable tool cannot be designed that would reach all the internal surfaces. Figure 7.42 shows the design of some blind holes. A standard drill produces a hole with a conical blind end, and therefore the machining of a hole with a square blind end requires a special tool. Thus, the end of a blind hole should be conical. If the blind hole is to be provided with a screw thread, the screw thread will be tapped, and the designer should not specify a fully formed thread to the bottom of the blind hole since this type of screw thread is impossible to produce.

Holes that have a dogleg, or bend, should be avoided if possible. A curved hole (Fig. 7.42) is clearly impossible to machine; however, drilling a series of through holes and plugging unwanted outlets can often achieve the desired effect although this operation is expensive.

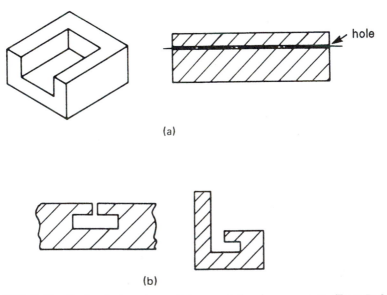

(a)

(b)

FIG. 7.41 Design features to avoid in nonrotational components. (From Ref. 7.)

Difficult   Easy   Impossible   Easy

Impossible   Impossible   Plug   Possible

FIG. 7.42   Design of blind holes. (From Ref. 7.)

7.9 ASSEMBLY OF COMPONENTS

Most machined components must eventually be assembled, and the designer should give consideration to the assembly process. Design for ease of assembly is treated in Chapter 3; however, one or two aspects of this subject that affect machining are mentioned here. The first requirement is, of course, that it should be physically possible to assemble the components. Obviously, the screw thread on a bolt or screw should be the same as the mating thread on the screwed hole into which the bolt or screw is to be inserted. Some assembly problems, however, are not quite so obvious. Figure 7.43 shows some impossible assembly situations, and it is left to the reader to decide why the components cannot be assembled properly.

A further requirement is that each operating machined surface on a component should have a corresponding machined surface on the mating component. For example, where flanges on castings are to be bolted together, the area around the bolt holes should be machined perpendicular to the hole (spot-faced, for example) to provide a proper seating for the bolt heads, nuts, or washers. Also, internal corners should not interfere with the external corner on the mating component. Figures 7.29 and 7.35 give examples of how this interference can be avoided. Finally, incorrect specification of tolerances can make assembly difficult or even impossible.

FIG. 7.43   Components that cannot be assembled. (From Ref. 7.)

7.10   ACCURACY AND SURFACE FINISH

A designer will not generally want to specify an accurate surface with a rough finish or an inaccurate surface with a smooth finish. When determining the accuracy and finish of machined surfaces, it is necessary to take into account the function intended for the machined surface. The specification of too-close tolerances or too-smooth surfaces is one of the major ways a designer can add unnecessarily to manufacturing costs. Such specifications could, for example, necessitate a finishing process, such as cylindrical grinding after rough turning, where an adequate accuracy and finish might have been possible using the lathe that performed the rough-turning operation. Thus, the designer should specify the widest tolerances and roughest surface that will give acceptable performance for operating surfaces.

As a guide to the difficulty of machining to within required tolerances it can be stated that

1.   Tolerances from 0.127 to 0.25 mm (0.005 to 0.01 in.) are readily obtained.
2.   Tolerances from 0.025 to 0.05 mm (0.001 to 0.002 in.) are slightly more difficult to obtain and will increase production costs.
3.   Tolerances 0.0127 mm (0.0005 in.) or smaller require good equipment and skilled operators and add significantly to production costs.

Figure 7.44 illustrates the general range of surface finish that can be obtained in different operations. It can be seen that any surface with a specified surface finish of 1 μm (40 μin.) arithmetical mean or better will generally require separate finishing operations, which substantially increases costs. Even when the surface can be finished on the one machine, a smoother surface requirement will mean increased costs.

To illustrate the cost increase as the surface finish is improved, a simple turning operation can be considered. If a tool having a rounded corner is used under ideal cutting conditions, the arithmetical mean surface roughness R_a is related to the feed by

$$R_a = 0.0321f^2/r_\varepsilon \tag{7.36}$$

where f is the feed and r_ε is the tool corner radius.

The machining time t_m is inversely proportional to the feed f and related by the equation

$$t_m = l_w/fn_w \tag{7.37}$$

where l_w is the length of the workpiece and n_w is the rotational speed of the workpiece.

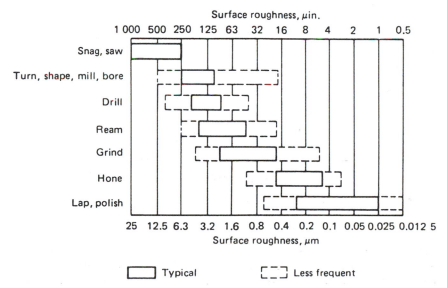

FIG. 7.44 General range of surface roughness obtainable by various machining operations. (From Ref. 7.)

Substitution of f from Eq. (7.36) in (7.37) gives the machining time in terms of the specified surface finish:

$$t_{\mathrm{m}} = 0.18 l_{\mathrm{w}} / [n_{\mathrm{w}} (R_{\mathrm{a}} r_{\varepsilon})^{0.5}] \qquad (7.38)$$

Thus, the machining time (and hence the machining cost) is inversely proportional to the square root of the surface finish. Figure 7.45 shows the relationship between production cost and surface finish for a typical turning operation. It can be seen that the costs rise rapidly when low values of surface finish are specified.

For many applications, a smooth, accurate surface is essential. This smooth, accurate surface can most frequently be provided by finish grinding. When specifying finish grinding, the designer should take into account the accessibility of the surfaces to be ground. In general, surfaces to be finish-ground should be

FIG. 7.45 Effect of specified surface roughness on production costs in a turning operation, where the corner radius $r_{\varepsilon} = 0.03$ in. (0762 mm), the rotational frequency of the workpiece $n_{\mathrm{w}} = 200$ rpm (3.33 s^{-1}), and the length of the workpiece $l_{\mathrm{w}} = 34$ in. (864 mm). (From Ref. 7.)

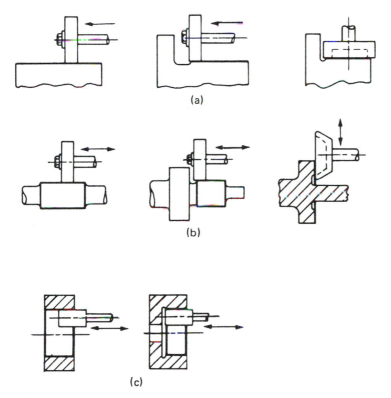

FIG. 7.46 Surfaces that can readily be finish-ground. (a) Surface grinding, (b) cylindrical grinding, (c) internal grinding. (From Ref. 7.)

raised and should never intersect to form internal corners. Figure 7.46 shows the types of surfaces that are most readily finish-ground using standard-shaped abrasive wheels.

7.11 SUMMARY OF DESIGN GUIDELINES

This section lists the various design guidelines that have been introduced, providing a summary of the main points a designer should keep in mind when considering the design of machined components.

Standardization

1.   Utilize standard components as much as possible.
2.   Preshape the workpiece, if appropriate, by casting, forging, welding, etc.

3. Utilize standard pre-shaped workpieces, if possible.
4. Employ standard machined features wherever possible.

Raw Materials

5. Choose raw materials that will result in minimum component cost (including cost of production and cost of raw material).
6. Utilize raw materials in the standard forms supplied.

Component Design
  a. General

7. Try to design the component so that it can be machined on one machine tool only.
8. Try to design the component so that machining is not needed on the unexposed surfaces of the workpiece when the component is gripped in the work-holding device.
9. Avoid machined features the company is not equipped to handle.
10. Design the component so that the workpiece, when gripped in the work-holding device, is sufficiently rigid to withstand the machining forces.
11. Verify that when features are to be machined, the tool, toolholder, work, and work-holding device will not interfere with one another.
12. Ensure that auxiliary holes or main bores are cylindrical and have L/D ratios that make it possible to machine them with standard drills or boring tools.
13. Ensure that auxiliary holes are parallel or normal to the workpiece axis or reference surface and related by a drilling pattern.
14. Ensure that the ends of blind holes are conical and that in a tapped blind hole the thread does not continue to the bottom of the hole.
15. Avoid bent holes or dogleg holes.

  b. Rotational Components

16. Try to ensure that cylindrical surfaces are concentric, and plane surfaces are normal to the component axis.
17. Try to ensure that the diameters of external features increase from the exposed face of the workpiece.
18. Try to ensure that the diameters of internal features decrease from the exposed face of the workpiece.
19. For internal corners on the component, specify radii equal to the radius of a standard rounded tool corner.
20. Avoid internal features for long components.
21. Avoid components with very large or very small L/D ratios.

c. Nonrotational Components

22. Provide a base for work holding and reference.
23. If possible, ensure that the exposed surfaces of the component consist of a series of mutually perpendicular plane surfaces parallel to and normal to the base.
24. Ensure that internal corners normal to the base have a radius equal to a standard tool radius. Also ensure that for machined pockets, the internal corners normal to the base have as large a radius as possible.
25. If possible, restrict plane-surface machining (slots, grooves, etc.) to one surface of the component
26. Avoid cylindrical bores in long components.
27. Avoid machined surfaces on long components by using work material preformed to the cross section required.
28. Avoid extremely long or extremely thin components.
29. Ensure that in flat or cubic components, main bores are normal to the base and consist of cylindrical surfaces decreasing in diameter from the exposed face of the workpiece.
30. Avoid blind bores in large cubic components.
31. Avoid internal machined features in cubic boxlike components.

Assembly

32. Ensure that assembly is possible.
33. Ensure that each operating machined surface on a component has a corresponding machined surface on the mating component.
34. Ensure that internal corners do not interfere with a corresponding external corner on the mating component.

Accuracy and Surface Finish

35. Specify the widest tolerances and roughest surface that will give the required performance for operating surfaces.
36. Ensure that surfaces to be finish-ground are raised and never intersect to form internal corners.

7.12 COST ESTIMATING FOR MACHINED COMPONENTS

Designers normally have a reasonable knowledge of the factors to bear in mind when attempting to minimize manufacturing costs for machined components and the previous section listed some design rules that might be followed. Ultimately, however, the designer will need to know the magnitude of the effects of design decisions on manufacturing costs. The need for a method of estimating these costs is highlighted when considering the design of a product for ease of assembly. Techniques have been available for some time for analyzing an assembly for handling and insertion costs incurred as each part is added, and they are described in Chapter 3. As a result of such analyses, many suggestions for design simplifications arise—often involving the elimination of individual parts or components. However, it is also necessary to have methods for quickly estimating the cost of these parts and the cost of tooling so that the total savings in products costs can be quantified.

Before embarking on a discussion of how an estimating method for designers of machined components might be developed, we should consider how the requirements for such a method differ from conventional cost-estimating procedures. These conventional procedures are meant to be applied after the component has been designed and its production planned. Thus, every step in production is known and can be estimated with a high degree of accuracy. During the early stages of design, however, the designer will not wish to specify, for example, all the work-holding devices and tools that might be needed; most likely, detailed design will not yet have taken place. Indeed, a final decision on the specific work material might not have been made at this stage. Thus, what is wanted is an approximate method requiring the minimum of information from the designer and assuming that the ultimate design will avoid unnecessary manufacturing expense and the component will be manufactured under reasonably economic conditions.

Perhaps the simplest approach would be to have the designer specify the shape and size of the original workpiece and the quantity of material to be removed by machining. Then, with data on typical material costs per unit weight, an estimate can be made of the cost of the material needed to manufacture the component. If an approximate figure is available for the average cost of removal of each cubic inch of the material by machining, an estimate can also be made of the machining cost.

Unfortunately, this very simple approach will not take adequate account of the nonproductive costs involved in a series of machining operations. For example, if 1 in^3 of material were to be removes in one pass by a simple turning operation, the nonproductive costs would be quite small—the component would only need be loaded into the lathe and unloaded once and the cutting tool would only be set and the feed engaged once. Compare this with 1 in^3 of the same material removed by a combination of turning, screw cutting, milling, and drilling. In this case the nonproductive costs accumulate and become a highly significant factor in the ultimate cost of the machined component; especially when the machined component is relatively small.

What is needed is a method that forms a compromise between this over-simplified approach and the traditional detailed cost-estimating methods used by manufacturing and industrial engineers.

TABLE 7.2 Approximate Costs in Dollars per Pound for Various Metals (to convert to dollars per kilogram multiply by 2.2)

	Density				Sheet	Plate	
	lb/in^3	Mg/m^3	Bar	Rod	< 0.5 in.	> 0.5 in.	Tube
Ferrous							
Carbon steel	0.283	7.83	0.51	0.51	0.36	0.42	0.92
Alloy steel	0.31	8.58	0.75	0.75	1.20	—	—
Stainless steel	0.283	7.83	1.50	1.50	2.50	2.50	—
Tool steel	0.283	7.83	6.44	6.44	—	6.44	—
Nonferrous							
Aluminum alloys	0.10	2.77	1.93	1.93	1.95	2.50	4.60
Brass	0.31	8.58	0.90	1.22	1.90	1.90	1.90
Nickel alloys	0.30	8.30	5.70	5.70	5.70	5.70	—
Magnesium alloys	0.066	1.83	3.35	3.35	6.06	6.06	3.35
Zinc alloys	0.23	6.37	1.50	1.50	1.50	1.50	—
Titanium alloys	0.163	4.51	15.40	15.40	25.00	25.00	—

7.12.1 Material Cost

Often, the most important factor in the total cost of a machined component is the cost of the original workpiece. This material cost will frequently form more than 50% of the total cost and, therefore, should be estimated with reasonable care. Table 7.2 gives densities and approximate costs in dollars per pound for a variety of materials in the basic shapes normally available. Provided the designer can specify the volume of material required for the original workpiece, the material cost can easily be estimated. Although the figures in Table 7.2 can be used as a rough guide, the designer would be able to obtain more accurate figures from material suppliers.

7.12.2 Machine Loading and Unloading

Nonproductive costs are incurred every time the workpiece is loaded into (and subsequently unloaded from) a machine tool. An exhaustive study of loading and unloading times has been made by Fridriksson [1]; he found that these times can be estimated quite accurately for a particular machine tool and work-holding device if the weight of the workpiece is known. Some of Fridriksson's results are presented in Table 7.3, which can be used to estimate machine loading and unloading times. To these figures must be added the times for turning coolant on and off, cleaning the work-holding or clamping device, etc.

TABLE 7.3 Loading and Unloading Times (s) Versus Workpiece Weight

	Workpiece weight				
Holding device	0–0.2 0–0.4	0.2–4.5 0.4–10	4.5–14 10–30	14–27 (kg) 30–60 (lb)	Crane
Angle plate (2 U-clamps)	27.6	34.9	43.5	71.2	276.5
Between centers, no dog	13.5	18.6	24.1	35.3	73.1
Between centers, with dog	25.6	40.2	57.4	97.8	247.8
Chuck, universal	16.0	23.3	31.9	52.9	—
Chuck, independent (4 jaws)	34.0	41.3	49.9	70.9	—
Clamp on table (3 clamps)	28.8	33.9	39.4	58.7	264.6
Collet	10.3	15.4	20.9	—	—
Faceplate (3 clamps)	31.9	43.3	58.0	82.1	196.2
Fixture, horizontal (3 screws)	25.8	33.1	41.7	69.4	274.7
Fixture, vertical (3 screws)	27.2	38.6	53.3	—	—
Hand-held	1.4	6.5	12.0	—	—
Jig	25.8	33.1	41.7	—	—
Magnet table	2.6	5.2	8.4	—	—
Parallels	14.2	19.3	24.8	67.0	354.3
Rotary table or index plate (3 clamps)	28.8	36.1	44.7	72.4	277.7
"V" Blocks	25.0	30.1	35.6	77.8	365.1
Vise	13.5	18.6	24.1	39.6	174.2

Source: After Ref. 1.

7.12.3 Other Nonproductive Costs

For every pass, cut, or operation carried out on one machine tool, further nonproductive costs are incurred. In each case the tool must be positioned, perhaps the feed and speed settings changed, the feed engaged, and then, when the operation is completed, the tool must be withdrawn. If different tools are employed, then the times for tool engagement or indexing must also be taken into

TABLE 7.4 Some Nonproductive Times for Common Machine Tools

Machine tool	Time to engage tool; etc.[a] (s)	Basic setup time (h)	Additional setup per tool (h)
Horizontal band saw	—	0.17	—
Manual turret lathe	9	1.2	0.2
CNC turret lathe	1.5	0.5	0.15
Milling machine	30	1.5	—
Drilling machine	9	1.0	—
Horizontal-boring machine	30	1.3	—
Broaching machine	13	0.6	—
Gear hobbing machine	39	0.9	—
Grinding machine	19	0.6	—
Internal grinding machine	24	0.6	—
Machining center	8	0.7	0.05

[a] Average times to engage tool, engage and disengage feed, change speed or feed. (Includes change tool for machining center.)

account. Some time elements for these tasks for different types of machine tools are presented in Table 7.4.

Also included In Table 7.4 are estimates of the basic setup time and additional setup time per cutting tool. The total set-up time must be divided by the size of the batch in order to obtain the setup time per component.

7.12.4 Handling Between Machines

One of the costs to be considered is that incurred in moving batches of partially machined workpieces between machines. Fridriksson [1] made a study of this by assuming that stacks of pallets of workpieces were moved around the factory using forklift trucks. He developed the following expression for t_f, the transportation time for a round trip by a forklift truck

$$t_f = 25.53 + 0.29(l_p + l_{rd}) \text{ s} \tag{7.39}$$

where l_p is the length of the pathway between machines and l_{rd} is the distance the truck must travel to respond to a request—both lengths are measured in feet.

Assuming that $(l_p + l_{rd})$ is 450 ft (137 m) on average and that for every trip with a load of full pallets, a trip must be made with empty pallets, the total time is

$$t_f = 315 \text{ s} \tag{7.40}$$

If a full load of pallets and workpieces is 2000 lb, the number of workpieces of weight W transported will be 2000/W and the time per workpiece t_{tr} will be given by

$$t_{tr} = 315/(2000/W) = 0.156W \text{ s} \tag{7.41}$$

Thus, for a workpiece weighing 10 lb, the effective transportation time is only 1.6 s, which is small compared with typical loading and unloading times for that size workpiece (Table 7.3). However, allowances for transportation time can be added to the loading and unloading times, and these will become significant for large workpieces.

7.12.5 Material Type

The so-called machinability of a work material has been one of the most difficult factors to define and quantify. In fact, it is impossible to predict the difficulty of machining a material from a knowledge of its composition or its mechanical properties, without performing a machining test. Nevertheless, it is necessary for the purposes of cost estimating to employ published data on machinability. Perhaps the best source of such data, presented in the form of recommended cutting conditions, is the *Machining Data Handbook* [2].

7.12.6 Machining Costs

The machining cost for each cut, pass, or operation is incurred during the period between when the feed is engaged and, finally, disengaged. It should be noted that the tool would not be cutting for the whole of this time because allowances for tool engagement and disengagement must be made—particularly for milling operations. However, typical values for these allowances can be found and are presented for various operations in Table 7.5 as correction factors to be applied to the actual machining time.

For an accurate estimation of actual machining time, it is necessary to know the cutting conditions, namely, cutting speed, feed, and depth of cut in single-point tool operations, and the feed speed, depth of cut, and width of cut in multipoint tool operations. Tables giving recommended values for these parameters for different work materials can fill large volumes, such as the Metcut *Machining Data Handbook* [2].

Analysis of the selection of optimum machining conditions shows that the optimum feed (or feed per tooth) is the largest that the machine tool and cutting tool can withstand. Then, selection of the optimum cutting speed can be made by minimizing machining costs [see Eq. (7.33)]. The product of cutting speed and feed in a single-point operation gives a rate for the generation of the machined surface that can be measured in in^2/min, for example. The inverse of such rates is presented by Ostwald [3] for a variety of workpiece and tool materials and for

TABLE 7.5 Allowances for Tool Approach

Operations	Allowances	
Turn, face, cut-off bore,	$t'_m = t_m + 5.4$	$d_m > 2$
groove, thread	$t'_m = t_m + (1.35 d_m^2)$	$d_m \leqslant 2$
Drill (twist) (approach)	$t'_m = t_m(1 + 0.5 d_m/l_w)$	
Drill (twist) (start)	$t'_m = t_m + (88.5/vf))d_m^{1.67}$	
Helical, side, saw, and key	$l'_w = l_w + 2(a_e(d_t - a_e)^{0.5} + 0.066 + 0.011 d_t)$	
slot milling		
Face and end milling	$l'_w = l_w + d_t + 0.066 + 0.011 d_t$	
Surface grinding	$l'_w = l_w + d_t/4$	
Cyl. and int. grinding	$l'_w = l_w + w_t$	
All grinding operations	$a'_r = a_r + 0.004$	$a_r \leqslant 0.01$
	$a'_r = a_r + 0.29(a_r - 0.01) \quad \begin{cases} a_r > 0.01 \\ a_r \leqslant 0.024 \end{cases}$	
	$u'_r = a_r + 0.008$	$a_r > 0.024$
Spline broaching	$l_t = -5 + 15 d_m + 8 l_w$	
Internal keyway broaching	$l_t = 20 + 40 w_k + 85 d_k$	
Hole broaching	$l_t = 6 + 6 d_m + 6 l_w$	

t_m = machining time, s; d_m = diameter of machined surface, in.; l_w = length of machined surface in direction of cutting, in.; v_f = speed $\times$ feed, in^2/min (Table 7.6); a_e = depth of cut or depth of groove in milling, in.; d_t = diameter of cutting tool, in.; w_t = width of grinding wheel, in.; a_r = depth of material removed in rough grinding, in.; l_t = length of tool, in.; w_k = width of machined keyway, in.; d_k = depth of machine keyway, in.
Source: Adapted from Ref. 3.

different roughing and finishing operations. A problem arises, however, when applying the figures for roughing operations. For example, Ostwald recommends a cutting speed of 500 ft/mm (2.54 m/s) and a feed of 0.02 in. (0.51 mm) for the rough machining of low carbon steel (170 Bhn) with a carbide tool. For a depth of cut of 0.3 in. (7.6 mm) this would mean a metal removal rate of 36 in^3/min (9.82 μm^3/s). The *Machining Data Handbook* [2] quotes a figure of 1.35 hp min/in^3 (3.69 GJ/m^3) (unit power) for this work material. Thus, the removal rate obtained in this example would require almost 50 hp (36 kW). Since a typical medium-sized machine tool would have a 5 to 10 hp motor (3.7 to 7.5 kW) and an efficiency of around 70%, it can be seen that the recommended conditions could not be achieved except for small depths of cut. Under normal rough-machining circumstances, therefore, a better estimate of machining time would be obtained from the unit horsepower (specific cutting energy) for the material, the volume of material to be removed, and the typical power available for machining, as described earlier in this chapter.

For multipoint tools such as milling cutters, the chip load (feed per tooth) and the cutting speed are usually recommended for given tool materials. However, in these cases the machining time is not directly affected by the cutting speed but by the feed speed, which is controlled independently of the cutter speed. Thus, assuming that the optimum cutting speed is being employed, the feed speed that will give the recommended feed per tooth can be used to estimate the machining time. Again, a check must be made that the power requirements for the machine tool are not excessive.

7.12.7 Tool Replacement Costs

Every time a tool needs replacement because of wear, two costs are incurred: (1) the cost of machine idle time while the operator replaces the worn tool, and (2) the cost of providing a new cutting edge or tool. The choice of the best cutting speed for particular conditions is usually made by minimizing the sum of the tool replacement costs and the machining costs, since both of these are affected by changes in the cutting speed.

The minimum cost of machining a feature in one component on one machine tool is given by Eq. (7.35). If the expressions for machining time t_m [Eq. (7.31)] and cutting speed v_c [Eq. (7.33)] are substituted, the minimum cost of production can be expressed by

$$C_{min} = Mt_l + Mt_{mc}/(1 - n) \tag{7.42}$$

where t_{mc} is the machining time when the optimum cutting speed for minimum cost is used.

It can be seen that the factor $1/(1 - n)$ applied to the machining time will allow for tool replacement costs provided that the cutting speed for minimum cost is always employed. The factor would be 1.14 for high-speed steel tooling and 1.33 for carbides.

Under those circumstances where use of optimum cutting conditions would not be possible because of power limitations, it is usually recommended that the cutting speed be reduced. This is because greater savings in tool costs will result than if the feed were reduced. When the cutting speed has been reduced, with a corresponding increase in the machining time, the correction factor given by Eq. (7.42) will overestimate tool costs. If t_{mp} is the machining time where the cutting speed v_{po} giving maximum power is used, then the production cost C_{po} for maximum power will be given by

$$C_{po} = Mt_l + Mt_{mp} + (Mt_{ct} + C_t)t_{mp}/t_{po} \tag{7.43}$$

where t_{po} is the tool life obtained under maximum power conditions, which, from Taylor's tool life equation, is

$$t_{po} = t_c(v_c/v_{po})^{1/n} \tag{7.44}$$

The tool life t_c under minimum cost conditions is given by Eq. (7.34), and substituting Eqs. (7.44) and (7.34) in Eq. (7.43) and using the relation in Eq. (7.31) gives

$$C_{po} = Mt_l + Mt_{mp}\{1 + [n/(1-n)](t_{mc}/t_{mp})^{1/n}\} \tag{7.45}$$

Thus, Eq. (7.45) can be used instead of Eq. (7.42) when the cutting speed is limited by the power available on the machine tool and, therefore, when $t_{mp} > t_{mc}$.

7.12.8 Machining Data

In order to employ the approach described above, it is necessary to be able to estimate, for each operation, the machining time t_{mc} for minimum cost conditions and the machining time t_{mp} where the cutting speed is limited by power availability. It was shown earlier that machining data for minimum cost for single-point tools, presented in handbooks, can be expressed as speed × feed (vf), or the rate at which the machined surface can be generated. Table 7.6 gives typical values of vf for several material classifications selected and for lathe operations using high-speed tools or brazed carbide tools. These values were adapted from the data in the *Machining Data Handbook* [2]. Analysis of the handbook data shows that if disposable insert carbide tools are to be used, then the data for brazed carbide tools can be multiplied by an average factor of 1.17.

When turning a surface of diameter d_m for a length l_w, the figures for vf given in Table 7.6 would be divided into the surface area ($A_m = \pi l_w d_m$) to give the machining time t_{mc}.

Thus

$$t_{mc} = 60A_m/(vf) \text{ s} \tag{7.46}$$

For an estimate of the machining time t_{mp} for maximum power it is necessary to know the power available for machining and the unit power p_s (specific cutting energy) for the work material. Table 7.6 gives average values of p_s for the selection of work materials employed here.

When estimating the power available for machining P_m, it should be realized that small components will generally be machined on small machines with lower power available, whereas larger components will be machined on large higher-powered machines. For example, a small lathe may have less than 2 hp available for machining, whereas an average-sized lathe may have 5 to 10 hp available. A larger vertical lathe will perhaps have 10 to 30 hp available. Typical power values for a selection of machines are presented in Fig. 7.47, where the horsepower

TABLE 7.6 Machining Data

Material	Hardness	Turning, facing and boring			Drilling and reaming (1 in.)	
		v_f (in^2/min)[a]		p_s (hp/in^3/min)[b]	v_f (in^2/min)[a]	p_s (hp/in^3/min)[b]
		HSS	Brazed carbide		HSS	
Low carbon steel (free machining)	150–200	25.6	100	1.1	33.0	0.95
Low carbon steel	150–200	22.4	92	1.35	13.4	1.2
Medium and high carbon steel	200–250	18.2	78	1.45	15.1	1.4
Alloy steel (free machining)	150–200	23.7	96	1.3	16.4	1.15
Stainless, ferritic (annealed)	135–185	12.6	48	1.55	9.4	1.35
Tool steels	200–250	12.8	54	1.45	6.2	1.4
Nickel alloys	80–360	9	42	2.25	14.3	2.0
Titanium alloys	200–275	12.6	24	1.35	7.9	1.25
Copper alloys (soft) (free machining)	40–150	76.8	196	0.72	38.4	0.54
Zinc alloys (die cast)	80–100	58.5	113	0.3	51.1	0.2
Magnesium and alloys	49–90	162	360	0.18	75.2	0.18
Aluminum and alloys	30–80	176	352	0.28	79.8	0.18

Factor	For Finishing Disposable insert		Turning, facing, boring	Milling
k_f			0.60	0.89
k_i			1.17	1.13

Factor	For Drilling and reaming	Tool diameter (in.) (mm)							
		1/16	1/8	1/4	1/2	3/4	1	1.5	2 in.
		1.59	3.18	6.35	12.7	19.7	25.4	38.1	50.8 mm
k_h		0.08	0.19	0.35	0.60	0.83	1.00	1.23	1.47

Factor	For Deep holes	Length/diameter ratio					
		<2	3	4	5	6	8
k_d		1.00	0.81	0.72	0.56	0.52	0.48

[a] To convert in²/min to m²/min, multiply by 6.45×10^{-4}.
[b] To convert hp min/in³ to GJ/m³ multiply by 2.73.

All data are for rough machining. For finish machining multiply by k_f.
For cut-off or form tool operations, multiply by 0.2.
The term *carbide* refers to tools with brazed carbide inserts. For tools with disposable carbide inserts, multiple by k_i.
Data for drilling are for 1.0 in. diameter tools with hole depth/diameter less than 2.
For sawing, multiply the data for turning with HSS tools by 0.33.
For tap or die threading, multiply data for turning with HSS tools by 10 and divide by TPI (threads per inch); for standard threads TPI $= 2.66 + 4.28/d_m$.
For single-point threading, multiply result for die threading by number of passes, approximately 100/TPI, and add tool engagement time for each pass.
Source: Adapted from Ref. 2.

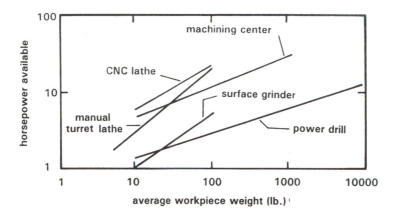

FIG. 7.47 Relation between horsepower and workpiece weight for some machine tools.

available for machining P_m is plotted against the typical weight capacity of the machine.

The machining time for maximum power is given by

$$t_{mp} = 60 V_m p_s / P_m \text{ s} \tag{7.47}$$

where V_m is the volume of material to be removed in the machining operation. If a_p is the depth of cut, then V_m is given approximately by $\pi d_m l_w a_p$. However, for a facing or cut-off operation carried out at constant rotational speed, the power limitations apply only at the beginning of the cut and the machining time will be longer than that given by Eq. (7.47).

It was pointed out earlier that for milling operations, it is convenient to estimate machining time from a knowledge of the feed speed v_f that will give the recommended feed per tooth. Data for milling selected materials are presented in Table 7.7.

The machining time t_m for recommended conditions is thus given by

$$t_{mc} = 60 l_w / v_f \text{ s} \tag{7.48}$$

where l_w is the length of the feature to be milled. However, it is important to note that this result must be corrected for the approach and overtravel distances, which will often be as large as the cutter diameter.

The machining time for maximum power is given by Eq. (7.47), but, again, corrections must be made for cutter approach and overtravel.

TABLE 7.7 Machining Data for Milling Operations

Material	Hardness (Bhn)	Milling v_f (in/min)[a]				p_s (hp/in^3/min)[b]
		Side and face		End (1.5 in.)		
		HSS	Brazed carb.	HSS	Brazed carb.	
Low carbon steel (free machining)	150–200	19.2	52.9	4.5	15.7	1.1
Low carbon steel	150–200	13.5	43.3	2.2	9.9	1.4
Medium and high carbon steel	200–250	10.8	37.3	1.8	8.9	1.6
Alloy steel (free machining)	150–200	13.7	40.2	2.7	10.5	1.3
Stainless, ferretic (annealed)	135–185	14.0	41.0	2.4	6.0	1.7
Tool steels	200–250	6.7	23.7	0.9	4.5	1.5
Nickel alloys	80–360	4.1	7.7	1.0	—	2.15
Titanium alloys	200–275	3.9	13.2	1.5	7.1	1.25
Copper alloys (soft) (free machining)	40–150	50.5	108.3	9.9	20.7	0.72
Zinc alloys (die cast)	80–100	28.0	60.1	9.8	16.0	0.4
Magnesium and alloys	40–90	77.0	240.6	27.5	55.0	0.18
Aluminum and alloys	30–80	96.2	216.5	20.4	36.7	0.36

[a] To convert in/min to mm/s multiple by 0.423.
[b] To convert hp/min/in^3 to GJ/m^3 multiply by 2.73.
Source: Adapted from Ref. 2.

7.12.9 Rough Grinding

Limitations on the rate at which grinding operations can be carried out depend on many interrelated factors, including the work material, the wheel grain type and size, the wheel bond and hardness, the wheel and work speeds, downfeed, infeed, the type of operation, the rigidity of the machine tool, and power available. It appears that, assuming adequate power, these limitations can be summarized in terms of the maximum metal removal rate per unit width of the grinding wheel Z_w/w_t. For example, the *Machining Data Handbook* [2] gives the following recommendations for the rough grinding of annealed free-machining low carbon steel on a horizontal-spindle reciprocating surface grinder:

Wheel speed: 5500 to 6500 ft/min
Table speed: 50 to 150 ft/min
Downfeed: 0.003 in/pass
Crossfeed: 0.05–0.5 in/pass (1/4 wheel width maximum)
Wheel: A46JV (aluminum oxide grain, size 46, grade J, vitrified bond)

If the wheel width w_t were 1 in. and an average table speed (work speed) of 75 ft/min were employed, then a downfeed of 0.003 in. and a maximum crossfeed of 0.25 in. would give a metal removal rate Z_w of 0.68 in^3/min. In a plunge-grinding operation, the wheel width would be equal to the width of the groove to be machined and the rough grinding time t_{gc} for recommended conditions would be given by

$$t_{gc} = 60 V_m / Z_w \tag{7.49}$$

where V_m is the volume of metal to be removed and Z_w is the metal removal rate (in^3/min). If the groove depth a_d were 0.25 in. and the groove length l_w were 4 in., the grinding time would be

$$t_{gc} = 60 a_d w_t l_w / Z_w = 60(1)(0.25)(4)/0.68 = 88.2 \text{ s}$$

The *Machining Data Handbook* [2] also gives values of the unit power (specific cutting energy) for the surface grinding of various materials. The unit power p_s depends to a large extent on the downfeed, and for a downfeed of 0.003 in., 13 hp min/in^3 would be required for carbon steel. In our example, the removal rate for a 2 in. wide groove would be

$$Z_w = 60(2)(0.25)4/88.2 = 1.36 \text{ in}^3/\text{min}$$

and the power required P_m would then be given by

$$P_m = p_s Z_w = 13(1.36) = 17.7 \text{ hp}$$

Clearly, for a particular rough-grinding operation, it will be necessary to check the grinding time t_{gp} when maximum power is used, and this will be given by

$$t_{gp} = 60 V_m p_s / P_m \tag{7.50}$$

The estimated rough-grinding time t_{gr} would be given by the grinding time t_{gc} for recommended conditions or the grinding time t_{gp} for maximum power, whichever is the largest. Table 7.8 gives recommendations for typical conditions for the horizontal-spindle surface grinding of selected materials. These recommendations are expressed in terms of Z_w/w_t, the metal removal rate per unit width of wheel in rough grinding, and the corresponding unit power p_s.

If the operation is one of plunge grinding, the width of the grinding wheel will be known. In a traverse operation, the width of the wheel will depend mainly on the grinding machine.

TABLE 7.8 Machining Data for Horizontal-Spindle Surface Grinding

Material	Hardness (Bhn)	Z_w/w_t (in^2/min)	p_s (hp/in^3/min)
Low carbon steel (free machining)	150–200	0.68	13
Low carbon steel	150–200	0.68	13
Medium and high carbon steel	200–250	0.68	13
Alloy steel (free machining)	150–200	0.68	14
Stainless, ferretic (annealed)	135–185	0.45	14
Tool steels	200–250	0.68	14
Nickel alloys	80–360	0.15	22
Titanium alloys	200–275	0.9	16
Copper alloys (soft) (free machining)	40–150	0.89	11
Zinc alloys (die cast)	80–100	0.89[a]	6.5[a]
Magnesium and alloys	40–90	0.89	6.5
Aluminum and alloys	30–80	0.89	6.5

[a] Estimated values.
For external cylindrical grinding, multiply Z/w_t by 1.24 and multiply p_s by 0.81.
For internal grinding, multiply Z_w/w_t by 1.15 and p_s by 0.87.
Source: Adapted from Ref. 2.

In a plunge-grinding operation. the depth of material to be removed will be specified by the geometry of the finished workpiece. In a traverse-grinding operation, it is necessary to remove the rough-grinding stock left by the previous machining operation.

7.12.10 Finish Grinding

The time for a finish-grinding operation is usually determined by the desired surface finish. This means that the metal removal rate must be slow enough to generate an acceptable surface finish, and it therefore becomes independent of the parameters affecting the removal rate in rough grinding. From the *Machining Data Handbook* [2], typical average values of the removal rate per inch of wheel width are 0.16 in^3/min for horizontal-spindle surface grinding, 0.08 in^3/min for cylindrical grinding, and 0.06 in^3/min for internal grinding. Recommended stock allowances for finish grinding range from 0.002 to 0.003 in. for horizontal grinding and 0.005 to 0.01 in. for cylindrical grinding.

7.12.11 Allowance for Grinding Wheel Wear

In his analysis of the economics of internal grinding, Lindsay [4] shows that the costs per component associated with wheel wear and wheel changing are proportional to the metal removal rate during rough grinding and that the wheel costs due to dressing and finish grinding are negligible in comparison. Thus, the total cost C_g of a grinding operation will be given by

$$C_g = Mt_c + Mt_{gr} + C_w \tag{7.51}$$

where M is the total rate for the machine (including direct labor, depreciation, and overhead), t_c is a constant time that includes the wheel-dressing time (assumed to occur once per component), the loading and unloading time, the wheel advance and withdrawal time, and the finish-grinding time, t_{gr} is the rough-grinding time, and C_w represents the wheel wear and wheel-changing costs. If we substitute

$$C_w = k_1 Z_w \tag{7.52}$$

where k_1 is a constant and Z_w is the metal removal rate during rough grinding, and

$$t_{gr} = k_2/Z_w \tag{7.53}$$

where k_2 is a constant, into Eq. (7.51) we get

$$C_g = Mt_c + Mk_2/Z_w + k_1 Z_w \tag{7.54}$$

Differentiating with respect to Z_w and equating to zero for minimum cost, we find that the optimum condition arises when the wheel wear and wheel changing costs [represented by the third term on the right of Eq. (7.54)] are equal to the rough-grinding costs (represented by the second term). This means that if optimum conditions are used in a grinding operation, wheel wear and wheel-changing costs can be allowed for by multiplying the rough-grinding time by a factor of 2.

However, it was pointed out earlier that the recommended conditions may exceed the power P_m available for grinding. In this case, the metal removal rate must be reduced—resulting in a reduction in wheel wear and wheel-changing costs and an increase in rough-grinding costs with a consequent increase in the total operation costs.

If Z_{wc} and Z_{wp} are the metal removal rates for optimum (recommended) and maximum power conditions respectively, the corresponding costs C_c and C_p are given by

$$C_c = Mt_c + 2Mk_2/Z_{wc} \tag{7.55}$$
$$C_p = Mt_c + Mk_2/Z_{wp} + k_1 Z_{wp} \tag{7.56}$$

Also, since for optimum conditions

$$k_1 Z_{wc} = Mk_2/Z_{wc} \tag{7.57}$$

we can obtain, after substitution and rearrangement, the following expression for the cost C_p, under maximum power conditions.

$$
\begin{aligned}
C_p &= Mt_c + \frac{Mk_2}{Z_{wc}}\left(\frac{Z_{wc}}{Z_{wp}} + \frac{Z_{wp}}{Z_{wc}}\right) \\
&= Mt_c + Mt_{gp}\left[1 + \left(\frac{t_{gc}}{t_{gp}}\right)^2\right]
\end{aligned}
\tag{7.58}
$$

where t_{gc} and t_{gp} are the rough-grinding times for recommended and maximum power conditions given by Eqs. (7.49) and (7.50), respectively, and where $t_{gp} > t_{gc}$.

This means that a multiplying factor equal to the term in square brackets in Eq. (7.58) can be used to adjust the rough grinding time and thereby allow for wheel wear and wheel-changing costs. Under circumstances where the recommended grinding conditions can be used (i.e., when $t_{gp} < t_{gc}$) the multiplying factor is equal to 2. If, for example, because of power limitations the metal removal rate were 0.5 of the recommended rate, then t_{gp} would be equal to $2t_{gc}$ and the correction factor would be 1.25. Under these circumstances, the rough-grinding costs would be double those for recommended conditions and the wheel costs would be one-half those for recommended conditions.

Example

Suppose the diameter d_w of a stainless steel bar is 1 in. and it is to be traverse ground for a length l_w of 12 in. If the wheel width w_t is 0.5 in., the power available P_m is 3 hp and the rough-grinding stock a_r left on the radius is 0.005 in., we get
Volume of metal to be removed

$$
\begin{aligned}
V_m &= \pi d_w a_r l_w \\
&= \pi(1)(0.005)(12) = 0.189 \text{ in}^3
\end{aligned}
$$

From Table 7.8 the recommended metal removal rate per unit width of wheel Z_w/w_t is 0.45 in^2/min for horizontal-spindle surface grinding. Using a correction factor of 1.24 for cylindrical grinding, the rough-grinding time for recommended conditions is given by

$$
\begin{aligned}
t_{gc} &= 60 V_m/Z_w \\
&= 60(0.189)/(1.24)(0.45)(0.5) = 40.65 \text{ s}
\end{aligned}
$$

However, Table 7.8 gives a unit power value of $p_s = 14$ hp in^3/min for stainless steels and, therefore, with a correction factor of 0.81, the rough-grinding time for maximum power would be

$$t_{gp} = 60 V_m p_s / P_m$$
$$= 60(0.189)(14)(0.81)/(3) = 42.9 \text{ s}$$

Thus, insufficient power is available for optimum grinding conditions and the condition for maximum power must be used. Finally, using the multiplying factor to allow for wheel costs, we get a corrected value t'_{gp} for the rough-grinding time of

$$t'_{gp} = t_{gp}[1 + (t_{gc}/t_{gp})^2]$$
$$= 42.9[1 + (40.6/42.9)^2]$$
$$= 42.9(1.9) = 81.3 \text{ s}$$

As explained earlier, the metal removal rate for finish grinding is basically independent of the material and is approximately 0.08 in^3/min per inch of wheel width in cylindrical grinding. For the present example where the wheel width is 0.5 in., this would give a removal rate Z_w of 0.05 in^3/min with a correction factor of 1.24 applied. Assuming a finish-grinding radial stock allowance of 0.001 in., the volume to be removed is

$$V_m = \pi(1)(0.001)(12) = 0.038 \text{ in}^3$$

and the finish-grinding time is

$$t_{gf} = 60(0.038)/(0.05) = 45.6 \text{ s}$$

7.12.12 Allowance for Spark-Out

In spark-out, the feed is disengaged and several additional passes of the wheel or revolutions of the workpiece are made in order to remove the material remaining because of machine and workpiece deflections. Since the number of passes is usually given, this is equivalent to removing a certain finish stock. For estimating purposes, the finish grinding time can be multiplied by a constant factor of 2 to allow for spark-out.

7.12.13 Examples

We shall first estimate the machining cost for a facing operation on a free-machining steel bar 3 in. (76.2 mm) diameter and 10 in. (254 mm) long, where

0.2 in. (5.1 mm) is to be removed from the end of the bar using a brazed-type carbide tool. The surface area to be generated is

$$A_m = [\pi/4](3)^2 = 7.07 \text{ in}^2 \text{ (4.5 m.m}^2\text{)}$$

and the volume of metal to be removed is

$$V_m = [\pi/4](3)^2(0.2) = 1.41 \text{ in}^3 \text{ (23.1 m}^3\text{)}$$

For this work material–tool material combination, Table 7.6 gives a value of vf (speed × feed) of 100 in^2/min. (0.065 m^2/min), and from Eq. (7.46) the machining time is

$$t_{mc} = 60(2)(7.07/100) = 8.5 \text{ s}$$

where the factor of 2 allows for the gradually decreasing cutting speed when constant rotational speed is used in a facing operation.

The weight of the workpiece is estimated to be

$$W = [\pi/4](3)^2(10)(0.28) = 20 \text{ lb (9.07 kg)}$$

From Fig. 7.47 the power available for machining on a CNC chucking lathe would be approximately given by

$$P_m = 10 \text{ hp (7.76 kW)}$$

Table 7.6 gives a value of specific cutting energy (unit power) of 1.1 hp min/in^3 (3 GJ/m^3) and so, from Eq. (7.47), the machining time at maximum power is

$$t_{mp} = 60(2)(1.1)(1.41/10) = 18.6 \text{ s}$$

Again, the factor of 2 has been applied for facing a solid bar.

It can be seen that, in this case, the conditions for minimum cost cannot be used because of power limitations and that a machining time of 18.6 s will be required. Now we can apply the factor given by Eq. (7.45) to allow for tool costs.

For carbide tools the Taylor tool life index is approximately 0.25, and since the ratio t_{mc}/t_{mp} is 8.5/18.6, or 0.46, the correction factor is

$$\{1 + [0.25/(1 - 0.25)](0.46)^{(1/0.25)}\} = 1.01$$

and the corrected machining time is 18.6 (1.01), or approximately 18.9 s.

In this example the correction factor for tool costs is quite small because cutting speeds below those giving minimum costs are being employed. If optimum speeds could be used, the correction factor would be 1.33 and the corrected machining time would be 11.2 s.

Finally, in Fig. 7.48, data are presented on the typical cost of various machine tools, where it can be seen that, for the present example of a CNC lathe, a cost of about $80,000 would be appropriate. Assuming that the total rate for the operator

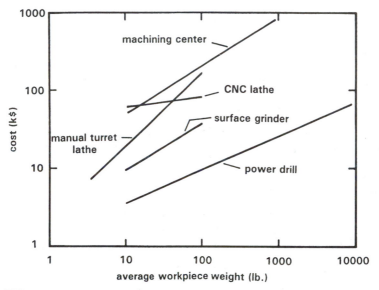

FIG. 7.48 Relation between cost and workpiece weight for some machine tools.

and the machine would be $30 per hour, or $0.0083 per second, the machining cost for the facing operation would be $0.157.

Thus, using the approach described in this chapter, it is possible to estimate the cost of each machined feature on a component. For example, Fig. 7.49 shows a

FIG. 7.49 Turned component. Batch size, 1500; workpiece, 3.25 in. dia. × 10.25 in. long; material, low carbon free-machining steel.

turned component with the machining cost for each feature indicated. The small, nonconcentric hole and the keyway are relatively expensive features. This is because in order to machine them, the component had to be loaded on separate machine tools—significantly increasing the nonproductive costs. The designer who is able to make these estimates would clearly be encouraged to reconsider the securing operations that necessitated these features and thereby reduce the overall manufacturing costs of the product.

7.12.14 Approximate Cost Models for Machined Components

During the initial conceptual stages, the designer or design team will be considering a variety of solutions to the design problem. Selection of the most promising design may involve trade-offs between the cost of machined components and components manufactured by other methods. However, the designer or design team will not be in a position to specify all the details necessary to complete the type of analysis presented in the previous section. In fact, the information in the early stages of design may consist only of the approximate dimensions of the component, the material, and a knowledge of its main features. Surprisingly, it is possible to obtain fairly accurate estimates of the cost of a component based on a limited amount of information. These estimates depend on historical data regarding the types of features usually found in machined components and the amount of machining typically carried out.

As an example we can consider a rotational component machined on a CNC turret lathe. In a study of the turning requirements for British industry [5] it was found that the average ratio of the weight of metal removed to initial workpiece weight was 0.62 for light engineering. Also, in light engineering, only 2% of workpieces weighed over 60 lb (27 kg) and therefore required lifting facilities, and 75% of the workpieces were turned from bar. Usually, the proportion of initial volume of material removed by internal machining is relatively small for geometrical reasons.

The British survey [5] also showed a direct correlation between the length-to-diameter ratio and the diameter of turned components.

Using this type of data, Fig. 7.50 shows, for a low-carbon-steel turned component, the effect of the finished size of the component on the rough machining, finish machining, and nonproductive times per unit volume. It can be seen that, as the size of the component is reduced below about 5 in^3 (82 μm^3), the time per unit volume and hence the cost per unit volume increases dramatically—particularly for the nonproductive time. This increase is to be expected for the nonproductive time because it does not reduce in proportion to the weight of the component. For example, even if the component size were reduced to almost zero, it would still take a finite time to place it in the machine, to make speed and feed settings, and to start the cutting operations. For the rough-

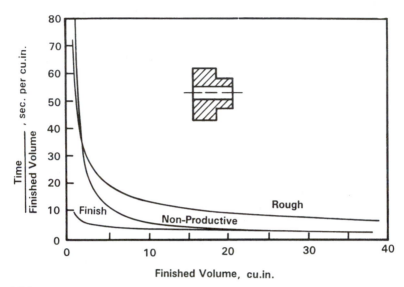

FIG. 7.50 Effect of component size on rough machining, finish machining, and nonproductive times per unit volume.

machining time, the higher times per unit volume for small components are a result of the reduced power available with the smaller machines used. The finish-machining time is proportional to the area machined. It can be shown that the surface area per unit volume (or weight) is higher for smaller components—thus leading to higher finish-machining times per unit volume.

These results have not taken into account the cost of the work material, and Fig. 7.51 shows how the total cost of a finished steel component varies with component size. This total cost is broken down into material cost and machining cost, and it can be seen that material cost is the most important factor contributing to the total cost even though 62% of the original workpiece was removed by machining—a figure that results in relatively high rough-machining costs. In fact, for the larger parts, about 80% of the total cost is attributable to material costs.

From the results of applying the approximate cost models it is possible to make the following observations.

1. For medium-sized and large workpieces, the cost of the original workpiece mainly determines the total manufactured cost of the finished component.
2. The cost per unit volume or per unit weight of small components (less than about 5 in^3 or 82 μm^3) increases rapidly as size is reduced because:

    a. The nonproductive times do not reduce in proportion to the smaller component size.

FIG. 7.51 Effect of component size on total cost for machining a free-machining steel workpiece (cost 50 cents/lb) with an inserted carbide tool.

    b.   The power available and hence the metal removal rate is lower for components.

    c.   The surface area per unit volume to be finish-machined is higher for smaller components.

This is illustrated in Fig. 7.52, where a series of turned components are shown, each being one-tenth the volume of the previous component. Although the cost per unit volume for the material is the same for all the components, the machining cost per unit volume increases rapidly as the components become smaller. For example, the total cost of the smallest component is $4.00 per in^3 whereas for the largest it is $0.55 per in^3. Stated another way, it would clearly be much less expensive to machine one of the largest components rather than 1000 of the smallest components when using the same types of machines.

3.   The choice of tool materials and optimum machining conditions only affects the finish-machining time. Since finish machining represents only about 25% of the total manufacturing cost, which in turn represents only about 20% of the total component cost for larger components, the effects of changes in tool materials or recommended conditions can be quite small under many conditions.

4.   The factors to be taken into account in making early estimates of machining costs are:

	A	B	C	D
Volume of finished part (in³)	40.0	4.0	0.4	0.04
Material cost/in³ of finished part	0.44	0.44	0.44	0.44
Manufacturing cost/in³	0.11	0.35	1.32	3.56
Total cost/in³ of finished part	0.55	0.79	1.76	4.00

FIG. 7.52 Costs (dollars) for a series of turned components.

a.   The amount of material to be removed. This factor directly affects the material costs per unit volume of the finished product and, less importantly, the rough-machining time.

b.   The rate for the machine tool and operator.

c.   The power available for machining and the specific cutting energy of the work material.

d.   The nonproductive times—especially for smaller components.

e.   The surface area to be finish machined.

f.   The recommended finish-machining conditions, which are in turn affected by the work material and tool material used.

5.   The factors that affect the nonproductive times are:

a.   The number of times the component must be clamped in a machine tool—each clamping involves transportation, loading and unloading, and setup.

b. The number of separate tool operations required—each operation requires tool indexing, and other associated activities and increases setup costs.

It was found in previous studies [6] that common workpieces can be classified into seven basic categories. Knowledge of the workpiece classification and the production data not only allows the cost of the workpiece to be estimated, but also allows predictions to be made of the probable magnitudes of the remaining items necessary for estimates of nonproductive costs and machining costs.

For example, for the workpiece shown in Fig. 7.49, the total cost of the finished component was estimated to be $24.32, a figure obtained from knowledge of the work material, its general shape classification and size, and its cost per unit volume. A cost estimate for this component, based on its actual machined features and using approximate equations based on the type of data listed above gave a total cost of $22.83, which is within 6%. A more detailed estimate obtained using the traditional cost-estimating methods presented in this chapter gave a total cost of $22.95. Thus, approximate methods, using the minimum of information, can give estimates surprisingly close to the results of analysis carried out after detailed design has taken place.

REFERENCES

1. Fridriksson, L. Non-productive Time in Conventional Metal Cutting. Report No. 3, Design for Manufacturability Program, University of Massachusetts, Amherst, February 1979.
2. Machining Data Handbook. Vols. 1 and 2, 3rd ed. Metcut Research Association Inc., 1980.
3. Ostwald, P.F. AM Cost Estimator, McGraw-Hill, New York 1985/1986.
4. Lindsay, R.P. Economics of Internal Abrasive Grinding. SME Paper MR 70-552, 1970.
5. Production Engineering Research Association. Survey of Turning Requirements in Industry, 1963.
6. Production Engineereing Research Associatioin. Survey of Machining Requirements in Industry, 1963.
7. G. Boothroyd and W.A. Knight. Fundamentals of Machining and Machine Tools. 2nd ed. Marcel Dekker, New York, 1989.

8

Design for Injection Molding

8.1 INTRODUCTION

Injection molding technology is a method of processing predominantly used for thermoplastic polymers. It consists of heating thermoplastic material until it melts, then forcing this melted plastic into a steel mold, where it cools and solidifies. The increasingly sophisticated use of injection molding is one of the principal tools in the battle to produce elegant product structures with reduced part counts. Perhaps the most widely recognized, innovative product development in this context was the Proprinter developed by IBM as the domestic competitor to the Japanese personal printers. Plastic components in the Proprinter incorporated the functions of cantilever springs, bearings, support brackets, and fasteners into single snap-fit components. The result of this integration of features into single complex parts was a reduction of part count from 152 to 32, with a corresponding reduction of assembly time from 30 to 3 min [1], when compared with the Epson printer which IBM had previously been remarketing from Japan.

In order to exploit the versatility of injection molding technology for economical manufacture, it is necessary to understand the basic mechanisms of the process and related aspects of the molding equipment and materials used. Also, since injection molding is a process that utilizes expensive tooling and equipment, it is vital to be able to obtain part and tooling cost estimates at the earliest stages of design. Only in this way can the design team be sure that the choice of the process is correct and that maximum economic advantage will be obtained from the process. For these reasons, this chapter will first present a review of injection molding materials and the injection molding process. This will be followed by the description of a procedure for estimating the cost of injection molded parts, which is applicable to the early phase of product design.

8.2 INJECTION MOLDING MATERIALS

It is not possible to injection-mold all polymers. Some polymers, like PTFE (poly tetrafluoro ethylene), cannot be made to flow freely enough to make them suitable for injection molding. Other polymers, such as a mixture of resin and glass fiber in woven or mat form, are unsuitable by their physical nature for use in the process. In general, polymers that are capable of being brought to a state of fluidity can be injection-molded.

The vast majority of injection molding is applied to thermoplastic polymers. This class of materials consists of polymers that always remain capable of being softened by heat and of hardening on cooling, even after repeated cycling. This is because the long-chain molecules always remain separate entities and do not form chemical bonds to one another. An analogy can be made to a block of ice that can be softened (i.e., turned back to liquid), poured into any shape cavity, then cooled to become a solid again. This property differentiates thermoplastic materials from thermosetting ones. In the latter type of polymer, chemical bonds are formed between the separate molecule chains during processing. This chemical bonding, referred to as cross-linking, is the hardening mechanism. Thermosetting polymers are generally more expensive to mold than thermoplastics and represent only about 5% of plastics processing. In this chapter we concentrate exclusively on injection molding of thermoplastics.

In general, most thermoplastic materials offer high impact strength, good corrosion resistance, and easy processing with good flow characteristics for molding complex designs. Thermoplastics are generally divided into two classes: crystalline and amorphous. Crystalline polymers have an ordered molecular arrangement, with a sharp melting point. Because of the ordered arrangement of molecules, the crystalline polymers reflect most incident light and generally appear opaque. They also undergo a high shrinkage or reduction in volume during solidification. Crystalline polymers usually are more resistant to organic solvents and have good fatigue and wear-resistance properties. Crystalline polymers also generally are denser and have better mechanical properties than amorphous polymers. The main exception to this rule is polycarbonate, which is the amorphous polymer of choice for high-quality transparent moldings and has excellent mechanical properties.

The mechanical properties of thermoplastics, while substantially lower than those of metals, can be enhanced for some applications by adding glass fiber reinforcement. This takes the form of short-chopped fibers, a few millimeters in length, which are randomly mixed with the thermoplastic resin. The fibers can occupy up to one-third of the material volume to considerably improve the material strength and stiffness. The negative effect of this reinforcement is usually a decrease in impact strength and an increase in abrasiveness. The latter also has an effect on processing, since the life of the mold cavity is typically reduced from about 1,000,000 for plain resin parts to about 300,000 for glass-filled parts.

Perhaps the main weakness of injection-molded parts is the relatively low service temperatures to which they can be subjected. Thermoplastic components can only rarely be operated continuously above 250°C, with an absolute upper service temperature of about 400°C. The temperature at which a thermoplastic can be operated under load can be defined qualitatively by the heat deflection temperature. This is the temperature at which a simply supported beam specimen of the material, with a centrally applied load, reaches a predefined deflection. The

TABLE 8.1 Commonly Used Polymers in Injection Molding

Thermoplastic	Yield strength (MN/m²)	Elastic modulus (MN/m²)	Heat deflection temperature (°C)	Cost ($/kg)
High-density polyethylene	23	925	42	0.90
High-impact polystyrene	20	1900	77	1.12
Acrylonitrile-butadiene-styrene (ABS)	41	2100	99	2.93
Acetal (homopolymer)	66	2800	115	3.01
Polyamide (6/6 nylon)	70	2800	93	4.00
Polycarbonate	64	2300	130	4.36
Polycarbonate (30% glass)	90	5500	143	5.54
Modified polyphenylene oxide (PPO)	58	2200	123	2.75
Modified PPO (30% glass)	58	3800	134	4.84
Polypropylene (40% talc)	32	3300	88	1.17
Polyester terephthalate (30% glass)	158	11,000	227	3.74

temperature value obviously depends upon the conditions of the test and the allowed deflection, and for this reason, the test values are only really useful for comparing different polymers.

Table 8.1 lists the more commonly molded thermoplastics together with typical mechanical property values.

8.3   THE MOLDING CYCLE

The injection molding process cycle for thermoplastics consists of three major stages as shown in Fig. 8.1: (1) injection or filling, (2) cooling, and (3) ejection and resetting. During the first stage of the process cycle, the material in the molten state is a highly nonlinear viscous fluid. It flows through the complex mold passages and is subject to rapid cooling from the mold wall, on the one hand, and internal shear heating, on the other. The polymer melt then undergoes solidification under the high packing and holding pressure of the injection system. Finally the mold is opened, the part is ejected, and the machine is reset for the next cycle to begin.

8.3.1   Injection or Filling Stage

The injection stage consists of the forward stroke of the plunger or screw injection unit to facilitate flow of molten material from the heating cylinder through the nozzle and into the mold. The amount of material to be transferred into the mold is referred to as the shot. The injection stage is accompanied by a gradual increase in pressure. As soon as the cavity is filled, the pressure increases

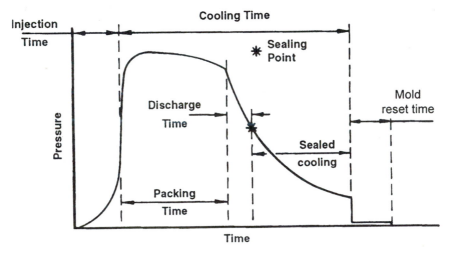

FIG. 8.1   Injection molding cycle.

rapidly, and packing occurs. During the packing part of the injection stage, flow of material continues, at a slower rate, to account for any loss in volume of the material due to partial solidification and associated shrinkage. The packing time depends on the properties of the materials being molded. After packing, the injection plunger is withdrawn or the screw is retracted and the pressure in the mold cavity begins to drop. At this stage, the next charge of material is fed into the heating cylinder in preparation for the next shot.

8.3.2 Cooling or Freezing Stage

Cooling starts from the first rapid filling of the cavity and continues during packing and then following the withdrawal of the plunger or screw, with the resulting removal of pressure from the mold and nozzle area. At the point of pressure removal, the restriction between the mold cavity and the channel conveying material to the cavity, referred to as the gate of the mold, may still be relatively fluid, especially on thick parts with large gates. Because of the pressure drop, there is a chance for reverse flow of the material from the mold until the material adjacent to the gate solidifies and the sealing point is reached. Reverse flow is minimized by proper design of the gates such that quicker sealing action takes place upon plunger withdrawal [2,3].

Following the sealing point, there is a continuous drop in pressure as the material in the cavity continues to cool and solidifies in readiness for ejection. The length of the sealed cooling stage is a function of the wall thickness of the part, the material used, and the mold temperature. Because of the low thermal conductivity of polymers, the cooling time is usually the longest period in the molding cycle.

8.3.3 Ejection and Resetting Stage

During this stage, the mold is opened, the part is ejected, and the mold is then closed again in readiness for the next cycle to begin. Considerable amounts of power are required to move the often massively built molds, and mold opening and part ejection are usually executed by hydraulic or mechanical devices. Although it is economical to have quick opening and closing of the mold, rapid movements may cause undue strain on the equipment, and if the mold faces come into contact at speed, this can damage the edges of the cavities. Also, adequate time must be allowed for the mold ejection. This time depends on the part dimensions, which determine the time taken for the part to fall free of moving parts between the machine platens. For parts to be molded with metal inserts, resetting involves the reloading of inserts into the mold. After resetting, the mold is closed and locked, thus completing one cycle.

FIG. 8.2   Injection molding system.

8.4   INJECTION MOLDING SYSTEMS

An injection molding system consists of the machine and mold for converting, processing and forming raw thermoplastic material, usually in the form of pellets, into a part of desired shape and configuration. Figure 8.2 shows a schematic view of a typical injection molding system. The major components of an injection molding system are the injection unit, the clamp unit, and the mold. These are described briefly in the following sections.

8.4.1   Injection Unit

The injection unit has two functions: to melt the pellets or powder, and to inject the melt into a mold. The most widely used types of injection units are (1) conventional units, consisting of a cylinder and a plunger that forces the molten plastic into the mold cavity, and (2) reciprocating screw units, consisting of a barrel or cylinder and a screw that rotates to melt and pump the plastic mix from the hopper to the end of the screw and then moves forward to push the melt into the mold.

Of the two types, reciprocating screw injection units are considered to be better designed because of their improved mixing action. The motion of the polymer melt along the screw flights helps to maintain a uniform melt temperature. It also facilitates better blending of the materials and any coloring agents, resulting in the delivery of more uniform melt to the mold. Because of these advantages, reciprocating screw units are found on the majority of modern injection molding machines.

The injection units are usually rated with two numbers: the first is the shot capacity, defined as the maximum volume of polymer that can be displaced by one forward stroke of the injection plunger or screw. The shot capacity for other materials can be calculated by using the ratio of specific gravities. It is usually recommended that an injection unit be selected so that the required shot sizes fall within 20 to 80% of the rated capacity.

The second rating number is the plasticating rate, which is the amount of material that can be plasticized or softened into a molten form by heating in the cylinder of the machine in a given time. This number is usually expressed as the number of pounds of polystyrene material that the equipment can heat to molding temperature in one hour. For further information on polymer processing requirements, consult Refs. 2–6.

8.4.2 Clamp Unit

The clamp unit has three functions: to open and close the mold halves, to eject the part, and to hold the mold closed with sufficient force to resist the melt pressure inside the mold as it is filled. The required holding force typically varies between 30 and 70 MN/m^2 of projected area of the part (approximately 2 to 5 $ton/in.^2$). The pressure developed in the mold during filling and packing, and the shrinkage of the part onto its cores, may cause the part to stick, thus making the separation of the two mold halves difficult. The magnitude of the initial opening force required depends on packing pressure, material, and part geometry (depth and draft) and is approximately equal to 10 to 20% of the nominal clamp force [3,7].

There are two common types of clamp designs:

1. *Linkage or toggle clamp*: This design utilizes the mechanical advantage of a linkage to develop the force required to hold the mold during the injection of the material. Mechanical toggle clamps have very fast closing and opening actions and are lower in cost than alternative systems. The major disadvantage is that the clamp force is not precisely controlled, and for this reason the clamps are found only on smaller machines.
2. *Hydraulic clamp units*: These use hydraulic pressure to open and close the clamp and to develop the force required to hold the mold closed during the injection phase of the cycle. The advantages of this type of design are long-term reliability and precise control of clamp force. The disadvantage is that hydraulic systems are relatively slow and expensive compared to toggle clamp systems.

After the mold halves have opened, the part that has a tendency to shrink and stick to the core of the mold (usually the half of the mold furthest from the injection unit) has to be ejected by means of an ejector system provided in the clamping unit. The force required to eject the part is a function of material, part geometry, and packing pressure and is usually less than 1% of the nominal clamp force [8].

8.5 INJECTION MOLDS

Molds for injection molding are as varied in design, degree of complexity, and size as are the parts produced from them. The functions of a mold for thermoplastics are basically to impart the desired shape to the plasticized polymer and then to cool the molded part.

A mold is made up of two sets of components: (1) the cavities and cores, and (2) the base in which the cavities and cores are mounted. The size and weight of the molded parts limit the number of cavities in the mold and also determine the equipment capacity required. From consideration of the molding process, a mold has to be designed to safely absorb the forces of clamping, injection, and ejection. Also, the design of the gates and runners must allow for efficient flow and uniform filling of the mold cavities.

8.5.1 Mold Construction and Operation

Figure 8.3 illustrates the parts in a typical injection mold. The mold basically consists of two parts: a stationary half (cavity plate), on the side where molten polymer is injected, and a moving half (core plate) on the closing or ejector side of the injection molding equipment. The separating line between the two mold halves is called the parting line. The injected material is transferred through a central feed channel, called the sprue. The sprue is located on the sprue bushing and is tapered to facilitate release of the sprue material from the mold during mold opening. In multicavity molds, the sprue feeds the polymer melt to a runner system, which leads into each mold cavity through a gate.

The core plate holds the main core. The purpose of the main core is to establish the inside configuration of the part. The core plate has a backup or support plate. The support plate in turn is supported by pillars against the U-shaped structure known as the ejector housing, which consists of the rear clamping plate and spacer blocks. This U-shaped structure, which is bolted to the core plate, provides the space for the ejection stroke also known as the stripper stroke. During solidification the part shrinks around the main core so that when the mold opens, part and sprue are carried along with the moving mold half. Subsequently, the central ejector is activated, causing the ejector plates to move forward so that the ejector pins can push the part off the core.

Both mold halves are provided with cooling channels through which cooled water is circulated to absorb the heat delivered to the mold by the hot thermoplastic polymer melt. The mold cavities also incorporate fine vents (0.02 to 0.08 mm by 5 mm) to ensure that no air is trapped during filling.

FIG. 8.3 Mold.

8.5.2 Mold Types

The most common types of molds used in industry today are (1) two-plate molds, (2) three-plate molds, (3) side-action molds, and (4) unscrewing molds.

A two-plate mold consists of two active plates (cavity and core plates in Fig. 8.3) into which the cavity and core inserts are mounted, as shown in Fig. 8.4. In this mold type, the runner system, sprue, runners, and gates solidify with the part being molded and are ejected as a single connected item. Thus the operation of a two-plate mold usually requires continuous machine attendance. The machine operator must spend time separating the runner system from parts (which break easily at the narrow gates) and periodically cutting the runner systems into small pieces that can be reintroduced into the machine hopper. This task is accomplished in an ancillary piece of equipment that operates like a small wood chipper, and the chips are somewhat confusingly referred to as regrind.

The three-plate mold consists of: (1) the stationary or runner plate, which contains the sprue and half of the runner; (2) the middle or cavity plate, which contains the other half of the runner, the gates, and cavities and is allowed to float when the mold is open; and (3) the movable or core plate, which contains the cores and the ejector system. This type of mold design facilitates separation of the runner system and the part when the mold opens; the two items usually fall into separate bins below the mold.

For very high rates of production, full automation can be achieved with a hot runner system, sometimes called a runnerless molding system. In this system, which also uses three main plates, the runner is contained completely in the fixed plate, which is heated and insulated from the rest of the cooled mold. The runner section of the mold is not opened during the molding cycle. The advantages of

FIG. 8.4 Two-plate injection mold.

this design are that there are no side products (gates, runner, or sprues) to be disposed of or reused, and there is no need for separation of the gate from the part. Also, it is possible to maintain a more uniform melt temperature.

Side-acting molds are used in molding components with external depressions or holes parallel to the parting plane. These features are sometimes referred to as undercuts or cross-features. These undercuts prevent molded parts from being removed from the cavity in the axial direction and are said to create a die-locked situation. The usual way of providing the side action needed to release the part is with side cores mounted on slides. These are activated by angle pins, or by air or hydraulic cylinders that pull the side cores outward during opening of the mold. Because of this action, the side core mechanisms are often referred to as side-pulls.

The mechanism of an angle-pin side action mold for a part with an undercut formed by a hole is illustrated in Fig. 8.5. The slide, which carries the secondary side core pin, is moved by the angle pin mounted in the stationary half of the

FIG. 8.5 Angle-pin-activated side-pull.

mold. As the two halves of the mold move apart during mold opening, the slide, which is mounted on the moving plate, is forced to move sideways by the angle of the pin. This allows the undercut to become free of the core pin, and the part can then be ejected. Note that side-pull is needed for each cross-feature or group of cross-features that lie on a particular axis. Molds have been built with as many as nine separate side-pulls to release particularly complex parts.

The mechanism for unscrewing molds is shown in Fig. 8.6. The rack and pinion gear mechanism shown in the illustration is the most common method used to free the undercuts formed by internal or external threads. With this method, a gear rack moved by a hydraulic cylinder engages with a spur gear attached to the threaded core pin. The rotating action imparted to the core pin through the gear transmission thus frees the undercuts formed by the threads. This additional unscrewing mechanism increases the mold cost and mold maintenance cost to a great extent, but eliminates the need for a separate thread-cutting operation. Note that external thread forms with axes that lie on the molding plane can be separated from the mold without the need for an unscrewing device.

A final category of mold mechanisms is required to mold depressions or undercuts on the inside of plastic parts. The design of a part with internal die-locking features of this type requires the mold maker to build the core pin retraction device within the main core. This is clearly much more difficult and expensive to manufacture than a corresponding side-pull on the outside of the cavity. In the latter case, adequate plate area can readily be provided for the machining of slideways. For this reason, the need for internal core retraction mechanisms, called lifters by mold makers, should be avoided whenever possible. The obvious way to do this is to replace internal depressions with either external ones or through holes.

FIG. 8.6 Unscrewing mold.

The injection molding process can handle unusual shapes because many different types of gates, ejection systems, and mold movement mechanisms can be combined into one mold. The resulting mold may be highly complex and often extremely expensive to both build and maintain. However, the general rule for efficient manufacture is to incorporate as many features as possible into a single molded part, provided, of course, that the required number of parts to be produced is sufficiently large to justify the tool costs.

8.5.3 Sprue, Runner, and Gates

A complete runner system with sprue and gates is shown in Fig. 8.4. The sprue bushing acts as an inlet channel for molten material from the heating chamber into the mold or runner system. The solid material in the form of a carrot called the "sprue" acts as a transition from the hot molten thermoplastic in the chamber to the relatively cooler mold. Runner systems are used in multiple cavity molds and act as channels to connect the sprue bush to the cavity gates. The gate, a constriction between the feed system and the mold cavity, serves several purposes. It freezes rapidly and prevents material from flowing out of the cavity when the injection pressure is removed. It provides an easy way of separating moldings from the runner system. It also suddenly increases the rate at which the polymer is sheared, which helps to align the polymer chains for more effective cavity filling.

In multiple-cavity molds, great care is taken to balance the runner system in order to produce identical parts. Different-length runners of the same cross-sectional area would result in different cavity pressures with resulting size and density variations among the multicavity parts.

8.6 MOLDING MACHINE SIZE

Determination of the appropriate size of an injection molding machine is based primarily on the required clamp force. This in turn depends upon the projected area of the cavities in the mold and the maximum pressure in the mold during mold filling. The former parameter is the projected area of the part, or parts if a multicavity mold is used, and runner system, when viewed in the direction of the mold opening, i.e., the area projected onto the surface of the mold cavity plate. The value for this parameter should not include any through holes molded in the direction of the mold opening. Thus, for a 15 cm diameter plain disk, the projected area is 176.7 cm^2. However, if the disk has a single 10 cm diameter through hole in any position, the projected area is 98.2 cm^2 since this is the area over which the polymer pressure will act during filling.

The size of the runner system depends upon the size of the part. Typical runner volumes as a percentage of part volume are shown in Table 8.2. As a first

TABLE 8.2 Runner Volumes (Du Pont)

Part volume (cm^3)	Shot size (cm^3)	Runner %
16	22	37
32	41	28
64	76	19
128	146	14
256	282	10
512	548	7
1024	1075	5

approximation, these figures will also be applied to give the projected area of the runner system as a percentage of the projected area of the part. Note, however, this is strictly correct only if a part is flat and if the runner system is the same thickness as the part.

Estimation of polymer pressure in the cavity during mold filling is a much more difficult problem. The flow characteristics of polymers are highly nonlinear, and mathematical models for mold filling can only be analyzed for individual runner and cavity geometries through the use of computer-intensive numerical procedures. However, it appears that as a general rule, approximately 50% of the pressure generated in the machine injection unit is lost because of the flow resistance in the sprue, runner systems, and gates [9]. This rule will be applied extensively in the costing analyses later in this chapter.

Example

A batch of 15 cm diameter disks with a thickness of 4 mm are to be molded from acrylonitrile-butadiene-styrene (ABS) in a six-cavity mold. Determine the appropriate machine size:

1. The projected area of each part equals 177 cm^2. From Table 8.2 the percentage increase in area due to the runner system is approximately 15%. Thus the total projected shot area will be

$$6 \times 1.15 \times 177 = 1221.3 \text{ cm}^2$$

2. The recommended injection pressure for ABS from Table 8.5 is 1000 bars. Thus the maximum cavity pressure is likely to be 500 bars, or $500 \times 10^5 \text{ N/m}^2$.

3. The estimate of maximum separating force F is thus given by

$$F = (1221.3 \times 10^{-4}) \times 500 \times 10^5 \text{ N} = 6106.5 \text{ kN}$$

Thus, if the available machines are those listed in Table 8.4, then the appropriate machine would be the one with a maximum clamp force of 8500 kN.

This machine must be checked to ensure that it has large enough shot size and clamp stroke. The required shot size is the volume of the six disks plus the volume of the runner system. This equals

$$6 \times 1.15 \times (177 \times 0.4) = 489 \text{ cm}^3$$

which is easily within the maximum machine shot size of 3636 cm^3.

The final check on machine suitability is the available machine clamp stroke. For the 8500 kN machine this is shown to be 85 cm. This stroke is sufficient to mold a hollow part up to a depth of approximately 40 cm. For such a part, the 85 cm stroke would separate the molded part from both the cavity and the core with a clearance of approximately 5 cm for the part to fall between the end of the core and the cavity plate. This stroke is thus excessive for the molding of 4 mm thick flat disks. The stroke is, however, adjustable, and to speed up the machine cycle it would be reduced in this case to just a few centimeters.

8.7 MOLDING CYCLE TIME

After the appropriate machine size for a particular molded part has been established, the molding cycle time can next be estimated. This estimation is essential in any consideration of the merits of alternative part designs or the choice of alternative polymers. As described earlier in this chapter, the molding cycle can be effectively divided into three separate segments: injection or filling time, cooling time, and mold-resetting time. Time estimates for these three separate segments will be established in this section.

8.7.1 Injection Time

A precise estimate of injection time requires an extremely difficult analysis of the polymer flow as it travels through the runners, gates, and cavity passages. This type of analysis would clearly not be justified as a basis for initial comparisons of alternative part design concepts. At this stage of design the position and number of gates and the size of the runner system would not be known. To circumvent this problem some major simplifying assumptions will be made about the machine performance and the polymer flow. First, modern injection molding machines are equipped with powerful injection units specifically to achieve the required flow rates for effective mold filling. It will be assumed that at the commencement of filling, the full power of the injection unit is utilized, and that the polymer pressure at the nozzle of the injector is that recommended by the polymer supplier. Under these circumstances, which may not be realizable for a particular mold design, the flow rate, using elementary mechanics, would be given by

$$Q = P_j/p_j \, \mathrm{m}^2/\mathrm{s} \tag{8.1}$$

where

P_j = injection power, W

and

p_j = recommended injection pressure, $\mathrm{N/m}^2$

In practice, the initial flow rate will gradually decrease as the mold is filled, due to both flow resistance in the mold channels and a constriction of the channels as the polymer solidifies against the walls. It will further be assumed that the flow rate suffers a constant deceleration to reach an insignificantly low value at the point at which the mold is nominally filled. Under these circumstances, the average flow rate would be given by

$$Q_{av} = 0.5 P_j/p_j \, \mathrm{m}^3/\mathrm{s} \tag{8.2}$$

and the fill time would be estimated as

$$t_f = 2V_s p_j/P_j \, \mathrm{s} \tag{8.3}$$

where

V_s = required shot size, m^3

Example

For the 15 cm diameter disks molded in a six-cavity mold, described in Sec. 8.6, the required shot size is 489 cm³. The recommended injection pressure for ABS is 1000 bars, or 100 MN/m². The available power at the injection unit of the 8500 kN machine is 90 kW.

Thus the estimated fill time is

$$t_f = 2 \times (489 \times 10^{-6}) \times (100 \times 10^6)/(90 \times 10^3)$$
$$= 1.09 \mathrm{\ s}$$

8.7.2 Cooling Time

In the calculation of cooling time, it is assumed that cooling in the mold takes place almost entirely by heat conduction. Negligible heat is transferred by convection since the melt is highly viscous and it is clear that radiation cannot contribute to the heat loss in a totally enclosed mold.

An estimation of cooling time can be made by considering the cooling of a polymer melt of initial uniform temperature T_i, between two metal plates, distance h apart and held at constant temperature T_m. This situation is analogous to cooling of the wall of an injection-molded component between the mold cavity and core. The variation of temperature across the wall thickness and with changing time is described by the one-dimensional heat conduction equation:

$$\frac{\partial T}{\partial t} = \alpha \frac{\partial^2 T}{\partial x^2} \tag{8.4}$$

where

$x =$ coordinate distance from center plane of wall normal to the plate surface, mm
$T =$ temperature, °C
$t =$ time, s
$\alpha =$ thermal diffusivity coefficient, mm^2/s

The thermal conductivity of thermoplastic materials is about three orders of magnitude lower than that of the steel mold. Under this situation it is reasonable to neglect the thermal resistance of the mold, which then merely becomes a heat sink at assumed constant temperature T_m. A classical series solution to this boundary condition applied to Eq. (8.4) has been given by Carslaw and Jaeger [10].

Ballman and Shusman [11] suggested an estimate of the cooling time based on truncating the Carslaw and Jaeger general series solution to just the first term. Mold opening and ejection are assumed to be permissible when the injected polymer has cooled to the point where the highest temperature in the mold (at the thickest wall center plane) equals T_x, the recommended ejection temperature. With this assumption the first-term solution for the cooling time is given by

$$t_c = \frac{h_{max}^2}{\pi^2 \alpha} \log_e \frac{4(T_i - T_m)}{\pi(T_x - T_m)} \text{ s} \tag{8.5}$$

where

$h_{max} =$ maximum wall thickness, mm
$T_x =$ recommended part ejection temperature, °C
$T_m =$ recommended mold tempeature, °C
$T_i =$ polymer injection temperature, °C
$\alpha =$ thermal diffusivity coefficient, mm^2/s

The data needed for making cooling-time predictions are given in Table 8.5 (see p. 358), which contains a list of the most widely injection-molded thermoplastics. It should be noted that Eq. (8.5) tends to underestimate the cooling time for very thin wall moldings. One reason is that for such parts the thickness of the runner system is often greater than the parts themselves and the greater delay is needed to ensure that the runners can be ejected cleanly from the mold. It is suggested that 3 s be taken as the minimum cooling time even if Eq. (8.5) predicts a smaller value.

The most important observation to make about Eq. (8.5) is that for a given polymer, with given molding temperatures, the cooling time varies with the square of the wall thickness of the molded part. This is the principal reason why injection molding is often uneconomical for thick-walled parts. It should also be noted that Eq. (8.5) applies only to a rectangular slab which is representative of the main wall of an injection-molded part. For a solid cylindrical section a correction factor of $2/3$ should be used on the diameter, since cooling takes place more rapidly for this boundary condition. Thus, a 3 mm thick flat part with a 6 mm diameter cylindrical projection would have an equivalent maximum thickness of $2/3 \times 6 = 4$ mm.

8.7.3 Mold Resetting

Mold opening, part ejection, and mold closing times depend upon the amount of movement required for part separation from the cavity and core and on the time required for part clearance from the mold plates during free fall. The summation of these three machine operation times is referred to as the resetting time. Approximate times for these machine operations have been suggested by Ostwald [12] for three general part shape categories: flat, box-shaped, and deep cylindrical parts. These are given in Table 8.3.

It is clear that these estimates can only be viewed as very rough approximations since they do not include the effect of part size. Part size influences resetting time in two ways. First, the projected area of the part together with the number of cavities determines the machine size and hence the power available for mold opening and closing. Second, the depth of the part, of course, determines the amount of mold opening required for part ejection.

In order take account of the preceding factors, use will be made of injection molding machine data where typical maximum clamp strokes and dry cycle times for various sizes of molding machines are given. Dry cycle time is defined as the time required to operate the injection unit and then to open and close an appropriately sized mold by an amount equal to the maximum clamp stroke. Table 8.4 gives values of these parameters for a wide range of currently available injection molding machines.

It should be realized that the dry cycle time given by a machine supplier bears little relationship to the actual cycle time when molding parts. This is because the

TABLE 8.3 Machine Clamp Operation Times (s)

	Flat	Box	Cylindrical
Mold open	2	2.5	3
Part eject	0	1.5	3
Mold close	1	1	1

TABLE 8.4 Injection Molding Machines

Clamping force (kN)	Shot size (cc)	Operating cost ($/h)	Dry cycle times (s)	Maximum clamp stroke (cm)	Driving power (kW)
300	34	28	1.7	20	5.5
500	85	30	1.9	23	7.5
800	201	33	3.3	32	18.5
1100	286	36	3.9	37	22.0
1600	286	41	3.6	42	22.0
5000	2290	74	6.1	70	63.0
8500	3636	108	8.6	85	90.0

dry cycle time is based on an empty injection unit, and it takes only milliseconds to inject air through the mold. Moreover, there is obviously no required delay for cooling, and the machine clamp is operated during both opening and closing at maximum stroke and at maximum safe speed.

In practice, the clamp stroke is adjusted to the amount required for the molding of any given part. If the depth of the part is given by D cm, then for the present time estimation purposes it will be assumed that the clamp stroke is adjusted to a value of $2D + 5$ cm. This will give the mold opening required for complete separation of the part from the cavity and matching core with a clearance of 5 cm for the part to fall away.

It can be noted from Table 8.3 that mold opening usually takes place more slowly than mold closing. This is because during mold opening, ejection of the part takes place usually with a significant level of force to separate the part from the core onto which it will have shrunk. Rapid mold opening may thus result in warping or fracture of the molded part. For present time estimation purposes it will be assumed that opening takes place at 40% of the closing speed; this corresponds to the average of Ostwald's data in Table 8.3.

The precise motion of a clamp unit depends upon the clamp design and its adjustment. To obtain a simple estimate of resetting time it will be assumed that for a given clamp unit the velocity profile during a clamp movement (opening or closing) will have identical shape irrespective of the adjusted stroke length. Under these conditions, the time for a given movement will be proportional to the square-root of the stroke length.

Thus, if the maximum clamp stroke is L_s for a given machine and the dry cycle time is t_d, then the time for clamp closing at full stroke will be assumed equal to $t_d/2$. However, if a part of depth D is to be molded, then the adjusted clamp stroke

TABLE 8.5 Processing Data for Selected Polymers

Thermoplastic	Specific gravity	Thermal diffusivity (mm²/s)	Injection temp. (°C)	Mold temp. (°C)	Ejection temp. (°C)	Injection pressure (bars)
High-density polyethylene	0.95	0.11	232	27	52	965
High-impact polystyrene	1.59	0.09	218	27	77	965
Acrylonitrile-butadiene-styrene (ABS)	1.05	0.13	260	54	82	1000
Acetal (homopolymer)	1.42	0.09	216	93	129	1172
Polyamide (6/6 nylon)	1.13	0.10	291	91	129	1103
Polycarbonate	1.20	0.13	302	91	127	1172
Polycarbonate (30% glass)	1.43	0.13	329	102	141	1310
Modified polyphenylene oxide (PPO)	1.06	0.12	232	82	102	1034
Modified PPO (30% glass)	1.27	0.14	232	91	121	1034
Polypropylene (40% talc)	1.22	0.08	218	38	88	965
Polyester teraphthalate (30% glass)	1.56	0.17	293	104	143	1172

will be $2D + 5$ cm and the time for mold closing will be

$$t_{close} = 0.5t_d[(2D + 5)/L_s]^{1/2} \tag{8.6}$$

If we now use the assumption of 40% opening speed and a dwell of 1 s for the molded part to fall between the plates, then this gives an estimate for mold resetting as

$$t_r = 1 + 1.75t_d[(2D + 5)/L_s]^{1/2} \tag{8.7}$$

Examples of Resetting Time

Assume that plain 15 cm diameter cylindrical cups, with a depth of 20 cm, are to be manufactured from ABS in a six-cavity mold. From the example in Sec. 8.6 we know that the appropriate machine size is 8500 kN and from Table 8.4 the corresponding values of dry cycle time, t_d, and maximum clamp stroke, L_s are

8.6 s and 85 cm, respectively. Substituting the values of $D = 20$, $L_s = 85$, and $t_d = 8.6$ into Eq. (8.7) gives an estimated resetting time of 12.0 s.

If the depth of the cylindrical cups is 10 cm, then the estimate of resetting time changes to 9.2 s, while if 15 cm diameter disks with a thickness of only 3 mm are to be molded, then Eq. (8.7) predicts the resetting time to be 4.9 s. These estimates can be seen to be at some variance with the Ostwald data, which were only reasonable averages for typical small parts.

8.8   MOLD COST ESTIMATION

The skills needed for mold design and construction differ substantially from those required for all the other steps in the injection molding process. As a consequence, mold design usually takes place in isolation from the various other functions involved. This presents a serious hurdle to the exchange of information and ideas between the tool maker and molder on one side, and the part designer on the other. Desirable changes in part design often become evident only after major investments in tooling and testing have already been made. The consequences of such a belated recognition can be very significant in terms of final cost and part quality. On the other hand, mold cost estimations made during the concept design stage itself will help in identifying acceptable part and mold configurations before actual investment in the mold is made.

The mold cost can be broken down into two major categories: (1) the cost of the prefabricated mold base consisting of the required plates, pillars, guide bushings, etc., and (2) cavity and core fabrication costs. These will be discussed separately in the following sections.

8.8.1   Mold Base Costs

From a survey of currently available prefabricated mold bases, it has been shown by Dewhurst and Kuppurajan [13] that mold base cost is a function of the surface area of the selected mold base plates and the combined thickness of the cavity and core plates. Figure 8.7 shows mold base costs plotted against a single parameter based on area and thickness values. The data in Fig. 8.7 can be represented by

$$C_b = 1000 + 0.45 \, A_c h_p^{0.4} \tag{8.8}$$

where
C_b = cost of mold base, $
A_c = area of mold base cavity plate, cm^2
h_p = combined thickness of cavity and core plates in mold base, cm

The selection of an appropriate mold base is based on the depth of the part, its projected area, and the number of cavities required in the mold. In addition to the cavity size, extra allowance has to be given for molds with mechanical action

FIG. 8.7   Principal mold base cost driver.

side-pulls and other complicated mechanisms, such as unscrewing devices for the molding of screw threads.

To determine the appropriate mold base size for a particular part, it is necessary to imagine the molded part (or parts for a multicavity operation) embedded within the mold base plates. The part(s) must have adequate clearance from the plate surfaces (and from each other) to provide the necessary rigidity against distortion from the cavity pressure during molding and to allow space for cooling channels and any moving core devices. Typically, the minimum clearance between adjacent cavities and between cavity surface and the edges and rear surfaces of cavity plates should be 7.5 cm. The extra plate size required to accommodate side-pulls and unscrewing devices will depend upon the actual mechanisms used. However, for the purpose of the present cost-estimating procedure, it will be assumed that side-pulls or side unscrewing devices require twice the minimum clearance from the edges, while rear unscrewing devices require a doubling of the material at the rear of the cavity. Thus one side-pull will increase the plate width or length by an additional 7.5 cm. Additional side-pulls will result in further plate size increases, so that four or more pulls, one or more on each side of a part, will require a plate that is 15 cm larger in both length and width. It should also be noted that use of two side-pulls restricts the mold design to a single row of cavities, while use of three or more usually implies single-cavity operation.

Example

As an application of the preceding rules, assume that 10 cm diameter plain cylindrical cups with a depth, D_d, of 15 cm are to be molded in a six-cavity mold. A 3 × 2 array of cavities with clearances as specified above gives the required plate area A_c is 2550 cm^2. The combined cavity and core plate thickness h_p is $h_d + 15 = 30$ cm. Hence the mold base cost parameter $A_c h_p^{0.4}$ is 9940 cm$^{2.4}$ and applying this value to the graph in Fig. 8.7 leads to an estimated mold base cost of $5500.

If a more complex cylindrical part of the same size is imagined, with two diametrically opposed holes in the side surfaces and an internal thread, the estimated plate size increases will be as follows. The cavity plate will now hold a single row of six cavities in order to accommodate the diametrically opposed side-pulls. Using 15 cm clearance along each side of the cavities to house the side core mechanisms, the plate area is 112.5 × 40 or 4500 cm^2. To support the unscrewing device, the combined plate thickness increases to an assumed value of 37.5 cm, which results in a new value of $A_c h_p^{0.4}$ equal to 19,179 cm$^{2.4}$. Referring to Fig. 8.7, this corresponds to a new mold base cost of approximately $9500.

Mold makers will often increase the clearances between cavities as the cavity area increases. From assessment of a large number of molds, it seems the typical clearance may increase by about 0.5 cm for every 100 cm^2 of cavity area. This rule of thumb would have a marginal effect on the estimated mold base costs in the preceding example. However, it should be applied for larger parts to obtain a better estimate of mold base size and cost. The costs above are only for the mold base with square flat plates. The costs to manufacture the necessary cavities and moving cores are discussed in the next section.

8.8.2   Cavity and Core Manufacturing Costs

Initial cost estimates in the present work are based on the use of a standard two-plate mold. The decisions regarding the use of three-plate molds, hot runner systems, etc., can only be made by comparing the increased cost of the mold system with the reduced machine supervision associated with semiautomatic or fully automatic operation.

As discussed in the last section, mold making starts with the purchase of a preassembled mold base from a specialist supplier. The mold base includes the main plates, pillars, bushings, etc. However, in addition to the manufacture of the cavities and cores, a substantial amount of work has to be performed on the mold base in order to transform it into a working mold. The main tasks are the deep hole drilling of the cooling channels and the milling of pockets in the plates to receive the cavity and core inserts. Additional tasks are associated with custom work on the ejector plate and housing to receive the ejection system, the insertion of extra support pillars where necessary, and the fitting of electrical and coolant systems. A rule of thumb [14] in mold manufacture is that the purchase price of the mold base should be doubled to account for the custom work that has to be performed on it.

Determination of the cost of an injection mold involves a knowledge of the number of ejector pins to be used. This information would not usually be available at the early stages in part design. Discussions with experienced mold makers have indicated that the number of ejector pins is governed by such factors as the size of the part, the depth of the main core, the depth and closeness of ribs, and other features contributing to part complexity. However, analysis of a range of parts for which the corresponding number of ejector pins could be determined yielded no strong relationships between number of pins and part depth, part size, or part complexity. The closest relationship was found to be based simply on the projected cross-sectional area of the parts at right angles to the direction of molding. With some considerable scatter, the number of ejector pins used was found to be approximately equal to the square root of the cross-sectional area when measured in square centimeters, i.e.,

$$N_e = A_p^{0.5} \tag{8.9}$$

where
N_e = number of ejector pins required
A_p = projected part area, cm^2

Equation (8.9) will be used in the estimation of the cost of the ejection system for a molded part. An investigation of mold-making costs by Sors et al. [15] suggests an approximate value of 2.5 manufacturing hours for each ejector pin, and this will be used in the present work. From Eq. (8.9) this gives the additional number of manufacturing hours for the ejection system of a part as

$$M_e = 2.5 \times A_p^{0.5} \text{ h} \tag{8.10}$$

It is recognized that part ejection is not always accomplished through the use of ejector pins. Nevertheless, Eq. (8.10) represents a reasonable basis for estimating the cost of an ejection system at the concept design stage.

The geometric complexity of a part to be molded is handled in the present mold cost estimation scheme by assigning a complexity score on the range 0 to 10 for both the inner and outer surface of the part. The number of mold-manufacturing hours, associated with the geometrical features of the part, for one cavity and matching core(s) is then estimated from

$$M_x = 5.83(X_i + X_o)^{1.27} \text{ h} \tag{8.11}$$

where X_i and X_o are the inner and outer complexity of the part, respectively.

This empirical relationship was obtained by Archer [16] from analysis of a wide range of injection-molded parts ranging from small brackets to large cabinets and items of furniture.

It is expected that for rapid cost estimating a quick judgment can be made as to the appropriate complexity numbers. However, in order to gain confidence in the assignment of the different levels of geometrical complexity, a simple complexity counting procedure has been established as described below.

Geometrical Complexity Counting Procedure

Count all separate surface segments on the part inner surface. The inner surface is the surface that is in contact, during molding, with the main core and other projections or depressions in the core plate. Surface segments are either planar or have a constant or smoothly changing curvature. The junction of different surface segments can be a sudden change (discontinuity) in either slope or curvature. The complexity of the inner surface is given by

$$X_i = 0.1 N_{sp} \qquad (8.12)$$

where N_{sp} is the number of surface patches.

This procedure is repeated for the part outer surface to obtain the outer surface complexity level. When counting surface patches, small connecting blend surfaces should not be included.

When counting multiple identical features on the surface of a part, a power index of 0.7 may be used to account for the savings that come from machining identical features in the mold. For example, if the surface of a part is covered by 100 spherical dimples, then the equivalent number of surface patches to be counted is $100^{0.7} = 25$.

Let us take as an example, a plane conical component with recessed base that is to be injection-molded; see Fig. 8.8. The inner and outer surface complexity levels are established as follows. The inner surface comprises the following surface segments:

1. Main conical surface
2. Flat base

Thus

$$X_i = 0.1 \times 2 = 0.2$$

The outer surface comprises:

1. Main conical surface
2. Flat annular base
3. Cylindrical recess in the base
4. Flat recessed base

FIG. 8.8 Surface segments of plain conical components (all dimensions in milli-meters).

Thus

$$X_o = 0.1 \times 4 = 0.4$$

In addition to geometrical complexity, the size of the part to be molded clearly also affects the cost of the cavity and core inserts. Building on a part area relationship given by Sors et al. [15], Archer [16] has shown, from analysis of a wide range of injection molds, that for parts with very simple geometry the number of manufacturing hours for one cavity and core can be represented by

$$M_{po} = 5 + 0.085 \times A_p^{1.2} \text{ h} \tag{8.13}$$

where

$$A_p = \text{part projected area, cm}^2$$

The sum of the point scores from Eqs. (8.10), (8.11), and (8.13) provides a base estimate of the number of manufacturing hours to make one cavity and core and the ejection system for a part of given size with a known degree of geometrical complexity. However, in order to complete a mold cost-estimating system six additional important factors need to be considered. These are:

1. The need for retractable side-pulls or internal core lifters
2. The requirement for one or more unscrewing cores to produce molded screw threads
3. The surface finish and appearance specified for the part
4. The average tolerance level applied to the part dimensions

5. The requirement for one or more surfaces to be textured (e.g., checkered, leathergrain finish, etc.)
6. The shape of the surface across which the cavity and core separate: referred to in die design as the parting line

From discussion with a number of mold makers, it appears that the slideways and associated angle pins or withdrawal mechanisms for a side-pull, excluding manufacture of the core, will have an associated manufacturing time of 50 to 80 h. However, constructing an internal mechanism in the main core (sometimes called a lifter) to retract an internal core pin is substantially more difficult and may take between 100 and 200 h. More difficult still is the building of an unscrewing mechanism for the molding of screw threads, which may require 200 to 300 tool-making hours. For the present early costing procedure, manufacturing hours for side-pulls, internal lifters, or unscrewing devices will be assumed to correspond to the average of these estimates.

The incorporation of texture into mold cavity surfaces is usually carried out by specialist companies that offer a wide range of standard texture patterns. It appears that the cost of texturing is proportional to both the complexity and size of the part and that a fairly good estimate is obtained by allowing 5% of the basic cavity manufacturing cost. Shallow lettering that can be etched or engraved into the mold can be considered equivalent to texture and costed in the same way.

The hand finishing of cavities required to produce high-quality surfaces on molded parts is extremely costly and time-consuming. The time involved is clearly dependent on the size of the cavity, its geometrical complexity, and the required appearance of the molded part. In this context it is necessary to differentiate between opaque and transparent parts. For opaque parts, the required appearance can be separated into four categories: not critical, standard (Society of Plastics Engineers No. 3 finish), high gloss (SPE No. 2), and highest gloss (SPE No. 1). On the other hand, transparent parts are generally produced according to only two categories: standard finish and with some internal flaws permissible, or highest quality with internal blemishes unacceptable. These two categories are more difficult to achieve than SPE No. 3 and No. 1 finishes respectively, for opaque parts. From discussions with mold makers it appears that the time taken to finish a cavity and core to achieve the required appearance levels can be represented as a percentage increase applied to the basic time for cavity and core manufacture. For the present estimating system, this translates into a percentage increase to the sum of the manufacturing hours predicted by Eqs. (8.11) and (8.13). Reasonable percentage values for the different part appearance categories are given in Table 8.6.

The tolerances given to the dimensions of an injection-molded part must clearly be within the capabilities of the process. These capabilities will be addressed later in this chapter. However, the part tolerances also indirectly

TABLE 8.6 Percentage Increases for Different Appearance Levels

Appearance	Percentage increase
Not critical	10
Opaque, standard (SPE #3)	15
Transparent, standard internal flaws or waviness permissible	20
Opaque, high gloss	25
Transparent, high quality	30
Transparent, optical quality	40

affect the cost of mold manufacture. The reason for this is that the mold maker will be required to work within a small portion of the part tolerances in order to leave the remainder of the tolerance bands to cover variations in the molding process. Tighter tolerances will thus result in more careful cavity and core manufacture. Evidence from mold makers suggests that this effect, while less significant than surface finish requirements, depends on the number of features and dimensions rather than on the part size. In terms of the present cost-estimating procedure, the part tolerance affects the time estimate for geometrical complexity given by Eq. (8.11). Acceptable percentage increases, which should be applied to the result of Eq. (8.11), for the six different tolerance levels are given in Table 8.7.

The final consideration is the shape of the plane separating the cavity and core inserts. Whenever possible, the cavity and core inserts should be mounted in flat opposing mold plates. This results in a flat parting plane (straight parting line) which only requires surface grinding to produce a well-fitting mold. Flat bent parts, or hollow parts whose edge, separating the inner and outer surface, does not lie on a plane, cannot be molded with a flat parting plane. For these cases, the parting surface should be chosen from the six classifications given in Table 8.8.

TABLE 8.7 Percentage Increases for Tolerance

Tolerance level	Description of tolerances	Percentage increase
0	All greater than ±0.5 mm	0
1	Most approx. ±0.35 mm	2
2	Several approx. ±0.25 mm	5
3	Most approx. ±0.25 mm	10
4	Several approx. ±0.05 mm	20
5	Most approx. ±0.05 mm	30

TABLE 8.8 Parting Surface Classification

Parting surface type	Factor (f_p)
Flat parting plane	0
Canted parting surface or one containing a single step	1.25
Two to four simple steps or a simple curved surface	2
Greater than four simple steps	2.5
Complex curved surface	3
Complex curved surface with steps	4

For each of these separate categories, industrial data suggest that the additional number of manufacturing hours required to manufacture the mold is approximately proportional to the square root of the cavity area as given by the following relationship.

$$M_s = f_p A_p^{1/2} \text{ h} \tag{8.14}$$

where
A_p = projected area of cavity, cm^2
f_p = parting plane factor given in Table 8.8
M_s = additional mold manufacturing hours for nonflat parting surface

8.9 MOLD COST POINT SYSTEM

Following the preceding discussions, a point system for mold cavity and core cost estimating can now be established. The main cost drivers will simply be listed in order, and associated graphs or tables will be referred to for determination of the appropriate number of points. The mold manufacturing cost is determined by equating each point to one hour of mold manufacture.

(i)   Projected Area of Part (cm^2)
—refer to Eqs. (8.10) and (8.13), which include points for the size effect on manufacturing cost plus points for an appropriate ejection system.

(ii)   Geometric Complexity
—identify complexity ratings for inner and outer surfaces according to the procedure described earlier.
—Apply Eq. (8.11) to determine the appropriate point score

(iii)   Side-Pulls
—identify number of holes or apertures requiring separate side-pulls (side cores) in the molding operation.

—Allow 65 points for each side-pull.

(iv) Internal Lifters

—Identify number of internal depressions or undercuts requiring separate internal core lifters.

—Allow 150 points for each lifter.

(v) Unscrewing Devices

—Identify number of screw threads that would require an unscrewing device.

—Allow 250 points for each unscrewing device.

(vi) Surface Finish/Appearance

—Refer to Table 8.6 to identify the appropriate percentage value for the required appearance category.

—Apply the percentage value to the sum of the points determined for (i) and (ii) to obtain the appropriate point score related to part finish and appearance.

(vii) Tolerance Level

—Refer to Table 8.7 to identify the appropriate percentage value for the required tolerance category.

—Apply the percentage value to the geometrical complexity points determined for (ii) to obtain the appropriate point score related to part tolerance.

(viii) Texture

—If portions of the molded part surface require standard texture patterns, such as checkered, leather grain, etc., then add 5% of the point scores from (i) and (ii).

(ix) Parting Plane

—Determine the category of parting plane from Table 8.8 and note the value of the parting plane factor, f_p.

—Use f_p to obtain the point score from Eq. (8.14).

To determine the cost to manufacture a single cavity and matching core(s) the total point score is multiplied by the appropriate average hourly rate for tool manufacture.

Example

It is anticipated that 2,000,000 plain hollow conical components are to be molded in acetal homopolymer. The component, illustrated in Fig. 8.8, has a material volume of 78 cm^3 and a projected area in the direction of molding of 78.5 cm^2. The mold manufacturing points are first established as follows:

		Points
(i)	Projected Area of Part [substitute $A_p = 78.5$ cm^2 into Eqs. (8.10) and (8.13)]	43
(ii)	Geometrical Complexity (established earlier as $X_i = 0.2$ and $X_o = 0.4$ for this part—apply Eq. (8.11) for points)	3
(iii)	Number of Side-Pulls	0
(iv)	Number of Internal Lifters	0
(v)	Number of Unscrewing Devices	0
(vi)	Surface Finish/Appearance (Opaque high gloss; see Table 8.6—add 25% of 43 + 3)	11.5
(vii)	Tolerance Level (category 1; see Table 8.7— insignificant effect for low complexity)	0
(viii)	Texture	0
(ix)	Parting Plane (category 0)	0

Total point score = 57.5

Assuming an average rate of $40 per hour for mold manufacturing, the estimated cost for one activity and core is found to be 57.5 × $40 or $2,300.

8.10 ESTIMATION OF THE OPTIMUM NUMBER OF CAVITIES

A major economic advantage of injection molding is its ability to make multiple parts in one machine cycle through the use of multicavity molds. Sometimes the cavities in a mold may be for different parts that are to be used together in the same product. This type of mold is referred to as a family mold. It is used infrequently since it requires the existence of a family of parts, made of the same material and having similar thicknesses. It also has the obvious disadvantage of individual reject parts always requiring the remanufacture of the entire family.

The common practice is to use multicavity molds to make sets of identical parts with each molding cycle. The motivation is to reduce processing cost through an initially higher mold investment. The effect of the chosen number of cavities on part cost can be dramatic. This means that the cost of alternative designs of a particular part can only be estimated if the appropriate number of cavities is known. The identification of the appropriate number of cavities for a particular part will be explored in this section.

When a multicavity mold is used, three principal changes occur:

1. A larger machine with a greater hourly rate is needed than for a single-cavity mold.
2. The cost of the mold is clearly greater than for a single-cavity one.
3. The manufacturing time per part decreases in approximately inverse proportion to the number of cavities.

In order to identify the optimum number of cavities, the increase in hourly rate with machine size increase must be known. Also, an estimate must be available of the cost of a multicavity mold compared to the cost of a single-cavity mold for the same part. With regard to the first requirement, Fig. 8.9 shows a national survey of injection molding machine rates that was carried out by *Plastics Technology Magazine* [17]. It can be seen that the hourly rate can be represented almost

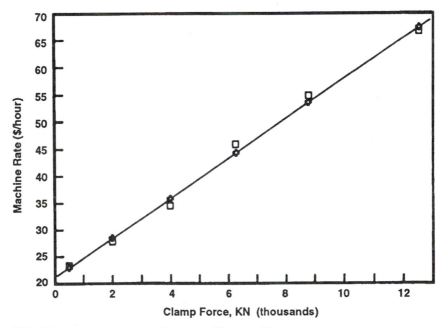

FIG. 8.9 National average injection molding machine rates.

precisely as a linear relationship based on the machine clamp force; i.e., machine hourly rate is

$$C_r = k_1 + m_1 F \ \$/h \tag{8.15}$$

where

$F =$ clamp force, kN
$k_1, m_1 =$ machine rate coefficients

For the latest machine rate data, obtained by the authors and given in Table 8.4, $k_1 = 25\$/h$ and $m_1 = 0.0091\$/h/kN$.

Turning now to the cost of multicavity mold manufacture, evidence in the literature [18] suggests that the cost of multiple cavity and core inserts, compared with the cost of one unique cavity/core set, follows an approximate power law relationship. That is, if the cost of one cavity and matching core is given by C_1, then the cost, C_n, of producing identical sets of the same cavity and core can be represented by

$$C_n = C_1 n^m \tag{8.16}$$

where m is a multicavity mold index and n is the number of identical cavities.

Testing of this relationship for a wide range of multicavity molds suggests that a reasonable value of m for most molding applications is 0.7. This value can be interpreted as an approximate rule that doubling the number of cavity and core inserts will always involve a cost increase of 62%, since $2^{0.7}$ is 1.62.

Savings also occur in the mold base cost per cavity when increasing the number of cavities. This can readily be established from the discussion of mold base costs in the previous section. However, the savings depend upon the cavity area, smaller cavities being associated with larger savings. Nevertheless, to allow a simple analysis to be performed, it will be assumed that a power law relationship similar to Eq. (8.16) applies equally to mold bases and with the same value for the power index. With this assumption the cost of a complete production mold satisfies the same relationship:

$$C_{cn} = C_{c1} n^m \tag{8.17}$$

where
$C_{c1} =$ cost of single-cavity mold
$C_{cn} =$ cost of n-cavity mold
$n =$ number of cavities
$m =$ multicavity mold index

Using the preceding relationship, the cost, C_t, of producing N_t molded components can be expressed as

$$C_t = \text{processing cost} + \text{mold cost} + \text{polymer cost}$$
$$= (N_t/n)(k_1 + m_1 F)t + C_{c1}n^m + N_t C_m \tag{8.18}$$

where

$t = $ machine cycle time, h
$C_m = $ cost of polymer material per part, \$

Assuming that an infinite variety of different clamp force machines were available, then one with just sufficient force would be chosen since hourly rate increases with clamp force (Fig. 8.3). Thus we can write

$$F = nf \tag{8.19}$$

where f is the separating force on one cavity. Substituting Eq. (8.19) into (8.18) gives

$$C_t = N_t(k_1 f/F + m_1 f)t + C_{c1}(F/f)^m + N_t C_m \tag{8.20}$$

For a given part, molded in a particular polymer, of which N_t are required, the only variable in Eq. (8.20) is clamp force F. Thus the minimum value of C_t will occur when dC_t/dF is equal to zero or when

$$-N_t k_1 ft/F^2 + mC_{c1}F^{(m-1)}/f^m = 0 \tag{8.21}$$

Multiplying Eq. (8.21) by F and substituting $n = F/f$ gives the expression for the optimum number of cavities as

$$n = (N_t k_1 t/(mC_{c1}))^{1/(m+1)} \tag{8.22}$$

A number of simplifying assumptions were used to derive Eq. (8.22), and for this reason the value predicted for the optimum number of cavities should be regarded as a first approximation. However, Eq. (8.22) is very easily applied and provides a reasonable basis for comparing alternative designs during the concept phase of a new injection-molded part.

8.11 DESIGN EXAMPLE

Figure 8.10 shows an injection-molded cover manufactured by a U.S. automobile company. The cover has a flange that has thickened pads at locations around the periphery, where bolts secure it to the assembly. The main body of the part is 2 mm thick and the bolt pads have a thickness of 4.6 mm.

The design can be considered to be produced from a 2 mm thick basic shape, referred to in injection molding jargon as the main wall. Features are then added to the main wall. In the present design the features are:

FIG. 8.10 Heater core cover.

1. The eight triangular stiffening ribs (called gussets) that support each pad to the side wall
2. The six through holes
3. The four pads of thickness 2.6 mm that rest on top of the main wall to give a combined thickness of 4.6 mm

Adding features to the main wall will always increase the mold cost. In addition, if they result in an increase in wall thickness, then the cycle time will also increase.

Using the point system in Sec. 8.8, it can readily be shown that for the cover design the increase in mold cost due to all eighteen features is approximately 23%, or $1150, and that the cost of one cavity and core set would be approximately $8000. Thus, for a fairly modest production volume of 50,000 parts the added mold cost is only 2.3 cents, or 0.13 cents per feature.

In contrast, the cooling equation in Sec. 8.6 shows that the addition of the thickened pads will increase the cooling time by factor $(4.6/2)^2$. This results in an additional processing cost per part of 12.3 cents.

Thus, if we wish to reduce the cost of the cover, the obvious way is to reduce the material thickness in the stiffened areas around the bolt clearance holes. The way in which this can be achieved is through the use of ribbed structures, as shown in Fig. 8.11. The projecting circular rib around each hole is known as a boss, and it is supported by a network of intersecting straight ribs as shown. The rib structure can be deep enough to give equivalent stiffness to the solid 4.6 mm thick pads. Also, the recommended rib thickness would be two-thirds of the main wall thickness, or 1.67 mm. For a production volume of 50,000 parts the new design would have an associated cavity and core cost of approximately $10,500, corresponding to an increase in mold cost per part of 5 cents. However, this would be more than offset by the 12.3 cents decrease in processing cost. Just as

FIG. 8.11 Proposed redesign of heater core cover.

important, the new design is likely to result in high-quality, distortion-free moldings. The existing design, in contrast, is difficult to mold because of the different wall thicknesses. The problems are due to the continued cooling of the pads after the main wall has fully solidified. This results in a buildup of residual or locked-in stresses as the pad material continues to shrink while constrained by the surrounding solidified wall.

8.12 INSERT MOLDING

Insert molding refers to the common practice of molding small metal items, such as pins or bushings, into injection-molded parts. Most typically, the inserts are hand-loaded into the mold cavity, or cavities, prior to closing of the mold and activation of the injection unit. For this reason, insert molding machines are of a vertical design so that inserts can be placed into a horizontal cavity plate when the mold is open. Machine rates can be assumed to be the same as for conventional molding machines. An approximate estimate of the cost of an insert-molded part can be obtained by adding 2 s per insert to the molding cycle time. Thus for a four-cavity mold, with two inserts per cavity, the cycle time would be increased by 16 s.

For high-volume manufacture of insert-molded parts, special-purpose machines can be obtained that employ multiple stations. The simplest type has a shuttle table that moves two separate molds alternately into the molding machine. With this system, inserts can be loaded into one mold while the other one is being filled and allowed to cool. Following this procedure, insert loading takes place within the machine cycle but with a higher-cost machine and a larger mold investment.

It should be noted that insert molding does not find universal favor. Many manufacturing engineers feel that the high risk of mold damage due to misplaced inserts offsets any advantage of the process. The alternative is simply to mold the depressions needed to accept inserts, which can then be secured later by ultrasonic welding. This process is described briefly later in the chapter.

8.13 DESIGN GUIDELINES

Suppliers of engineering thermoplastics have in general provided excellent support to the design community. Several have published design manuals or handbooks that are required reading for those designing injection molding components. Information can be obtained from them on the design of ribbed structures, gears, bearings, spring elements, etc.

The interested reader may wish to write, in particular, to Du Pont, G.E. Plastics Division, or Mobay Corporation for design information associated with their engineering thermoplastics.

Generally accepted design guidelines are listed below.

1. Design the main wall of uniform thickness with adequate tapers or draft for easy release from the mold. This will minimize part distortion by facilitating even cooling throughout the part.
2. Choose the material and the main wall thickness for minimum cost. Note that a more expensive material with greater strength or stiffness may often be the best choice. The thinner wall this choice allows will reduce material volume to offset the material cost increase. More important, the thinner wall will significantly reduce cycle time and hence processing cost.
3. Design the thickness of all projections from the main wall with a preferred value of one-half of the main wall thickness and do not exceed two-thirds of the main wall thickness. This will minimize cooling problems at the junction between the projection and main wall, where the section is necessarily thicker.
4. If possible, align projections in the direction of molding or at right angles to the molding direction lying on the parting plane. This will eliminate the need for mold mechanisms.
5. Avoid depressions on the inner surfaces of the part, which would require moving core pins to be built inside the main core. The mechanisms to produce these movements (referred to in mold making as lifters) are very expensive to build and maintain. Through holes on the side surfaces, instead of internal depressions, can always be produced with less expensive side-pulls.
6. If possible, design external screw threads so that they lie in the molding plane. Alternatively, use a rounded or rolled-type thread profile which can be stripped from the cavity or core without rotating. In the latter case, polymer suppliers should be consulted for material choice and appropriate thread profile and depth.

In addition to these general rules, design books should be consulted for design tips and innovative design ideas. Many of these are concerned with methods for producing undercuts and side features without using mold mechanisms. This will be explored a little further in the next section when snap fit elements are discussed.

One important cautionary note should be made with regard to design guidelines. Guidelines should never be followed when doing so may have a negative effect on the cost or quality of the assembly as a whole. This applies particularly to the guidelines aimed at avoiding mechanisms in the mold. The only valid rule in this regard is that the need for mold mechanisms should be recognized by the designer, and if they are unavoidable, then their cost should be justified in the early stages of design. This cost may be simply the cost of the mechanisms or it may also include the increase in processing cost if they restrict the number of cavities below the optimum value. The worst case is a high-volume component with side-pulls or unscrewing devices on all sides so that it must be made uneconomically in a single-cavity mold.

8.14 ASSEMBLY TECHNIQUES

One of the major advantages of injection molding is its ability to easily incorporate, in the molded parts, effective self-securing techniques. In the present context, self-securing refers to the ability to achieve a secure assembly without the use of separate fasteners or a separate bonding agent. Two of these self-securing techniques are also widely used with metal parts: press fitting and riveting. With press fitting much larger interferences are possible with injection-molded parts than with metal parts. This has the advantage of requiring less precise tolerance control for the press fit. The negative aspect of plastic press fit joints is that the material is constantly under stress and will invariably relax over a period of time to produce some degradation of the joint strength. Testing of press fits under the expected loading conditions is therefore essential. With regard to riveting, integral rivets are of course easily produced by adding inexpensive feature to the mold. On assembly the rivet heads are also easily formed by cold heading or by the use of heated forming tools; the latter operation is sometimes referred to as staking.

A third self-securing method, which is unique to plastic parts, is the use of ultrasonic welding. This is a method of joining two or more plastic molded parts through the generation of intermolecular frictional heat at the assembly interfaces. Ultrasonic welding equipment simply involves a special fixture to hold and clamp the parts and through which a high-frequency vibration of approximately 20 kHz is passed. The detail design of the butting or overlapping joints is critical for successful joining of the parts. Figure 8.12 shows typical recommended joint

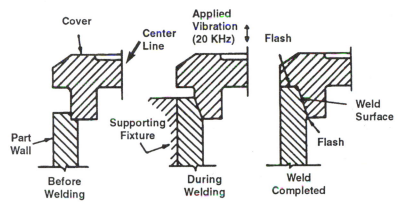

FIG. 8.12   Ultrasonic welding joint design by Du Pont.

designs. Ultrasonic welding is a good economic choice where sealed joining is required, since the equipment is relatively inexpensive and the process is fast. Welding is accomplished typically in about 2 s.

The final and most widely recognized self-securing method for molded parts is through the use of snap fit elements. These can be separated into two main types. The first type was developed for mating parts of circular cross-section and involves the use of a cylindrical undercut and mating lip, as shown in Fig. 8.13. The male partner of the mating pair may be molded with the parting plane along its axis with very little added cost, simply the cost of adding the groove feature to the cavity and core geometries. However, if the male cylinder cannot be designed at right angles to the direction of mold opening, then its separation from the mold will require the use of side-pulls. In contrast, the female or undercut part is almost

FIG. 8.13   Annular snap joint design.

always stripped off the core by the mold ejection system and is, therefore, inexpensively produced. Snap fits of this type are only truly satisfactory for circular parts. The further part shapes deviate from circular, the more difficult it becomes to eject parts from the mold that can be assembled satisfactorily.

The second type of snap fit design involves the use of one or more cantilever snap elements, as shown in Fig. 8.14. If possible, the cantilever snap element and its mating undercut should be designed for molding without mold mechanisms; one such design is shown in Fig. 8.14a. The alternative design, illustrated in Fig. 8. 14b, will require two side-pulls for the molding of the cantilever and the undercuts. However, even in the latter case, for a large enough production volume the extra mold costs can be easily outweighed by the subsequent savings in assembly cost.

Holes produced by core rods

(a)

(b)

FIG. 8.14   Cantilever snap fit elements. (a) Undercuts formed by core rods; (b) undercuts formed by side-pulls.

Polymer manufacturers should be contacted for detailed information on snap fit design.

REFERENCES

1. Dewhurst, P., and Boothroyd, G., Design for Assembly in Action, Assembly Eng., January 1987.
2. Rosato, D.V. (Ed.), Injection Molding Handbook, Van Nonstrand Reinhold, New York, 1986.
3. Bown, J., Injection Molding of Plastic Components, McGraw-Hill (UK), 1979.
4. MacDermott, C.P., Selecting Thermoplastics for Engineering Applications, Marcel Dekker, New York, 1984.
5. Bernhardt, E.C. (ed.), Computer-Aided Engineering for Injection Molding, Hanser Publishers, Munich, 1983.
6. Design Handbook for Dupont Engineering Polymers, E.I. du Pont de Nemours and Co. Inc., 1986.
7. Farrell, R.E., Injection Molding Thermoplastics, Modern Plastics Encyclopedia, 1985–86, pp. 252–270.
8. Khullar, P., A Computer-Aided Mold Design System for Injection Molding of Plastics, Ph.D. Dissertation, Cornell University, 1981.
9. Gordon Jr., B.E., Design and Development of a Computer Aided Processing System with Application to Injection Molding of Plastics, Ph.D. Thesis, Worcester Polytechnic Institute, Worcester, MA, November 1976.
10. Carslaw, H.S., and Jaeger, J.C., Conduction of Heat in Solids, Clarendon Press, Oxford, 1986.
11. Ballman, P., and Shusman, R., Easy Way to Calculate Injection Molding Set-Up Time, Modern Plastics, McGraw-Hill, New York, 1959.
12. Ostwald, P.F. (ed.), American Cost Estimator, American Machinist, McGraw-Hill, New York, 1985.
13. Dewhurst, P., and Kuppurajan, K., Optimum Processing Conditions for Injection Molding, Report No. 12, Product Design for Manufacture Series, University of Rhode Island, Kingston, February 1987.
14. Schuster, A., Injection Mold Tooling, Society of Plastic Engineers Seminar, New York, September 30–October 1, 1987.
15. Sors, L., Bardocz, L., and Radnoti, I., Plastic Molds and Dies, Van Nostrand Reinhold, New York, 1981.
16. Archer, D., Economic Model of Injection Molding, M.S. Thesis, University of Rhode Island, Kingston, 1988.
17. Plastics Technology Magazine, June 1987.
18. Reinbacker, W.R., A Computer Approach to Mold Quotations, PACTEC V, 5th Pacific Technical Conference, Los Angeles, February 1980.

9

Design for Sheet Metalworking

9.1 INTRODUCTION

Parts are made from sheet metal in two fundamentally different ways. The first way involves the manufacture of dedicated dies which are used to shear pieces of required external shape, called blanks, from metal stock that is in strip form. The strip stock may be in discrete lengths that have been cut from purchased sheets or may be purchased as long lengths supplied in coil form. With this method of manufacture, dies are also used to change the shape of the blanks, by stretching, compressing, or bending, and to add additional features through piercing operations. The dies are mounted on vertical presses into which the sheet metal stock may be manually loaded or automatically fed from coil.

The alternative method of manufacture involves the use of computer numerically controlled (CNC) punching machines which are used to make arrays of sheet metal parts directly from individual sheets. These machines usually have a range of punches available in rotating turrets and are referred to as turret presses. The method of operation is to first produce all of the internal part features in positions governed by the spacing of parts on the sheet. The external contours of the parts are then produced through punching with curved or rectangular punches or by profile cutting. The latter operation is usually performed by plasma or laser cutting attachments affixed to the turret press. Parts produced on a turret press are essentially flat, although internal features may protrude above the sheet surface. For this reason it is common practice to carry out secondary bending operations, if required, on separate presses. These are typically performed on wide, shallow bed presses, called press brakes, onto which standard bending tools are mounted.

TABLE 9.1   Standard U.S. Sheet Metal Thickness

Steels		Aluminum alloys (mm)	Copper alloys (mm)	Titanium alloys (mm)
Gage no.	(mm)			
28	0.38	0.41	0.13	0.51
26	0.46	0.51	0.28	0.63
24	0.61	0.63	0.41	0.81
22	0.76	0.81	0.56	1.02
20	0.91	1.02	0.69	1.27
19	1.07	1.27	0.81	1.60
18	1.22	1.60	1.09	1.80
16	1.52	1.80	1.24	2.03
14	1.91	2.03	1.37	2.29
13	2.29	2.29	2.06	2.54
12	2.67	2.54	2.18	3.17
11	3.05	3.17	2.74	3.56
10	3.43	4.06	3.17	3.81
8	4.17	4.83	4.75	4.06
6	5.08	5.64	6.35	4.75

Using either of these manufacturing methods, sheet metal parts can be produced with a high degree of geometrical complexity. However, the complex geometries are not free form, in the sense of molding or casting but are usually achieved through a combination of individual features that must conform to strict guidelines. These guidelines will be discussed in Sec. 9.6.

Sheet metal is available from metal suppliers in sheet or coil form, in a variety of sizes and thicknesses, for a wide range of different alloys. Table 9.1 shows the range of gage thicknesses available for the four alloy types that represent almost all of the materials used in sheet metalworking. For historical reasons, steels are ordered according to gage numbers whereas other material types have just a thickness designation. Steels are the most widely used sheet metal group. The reason for this is evident in Table 9.2, which gives typical properties and comparative costs of a sample of materials from the four alloy groups. The tensile strain values are for the materials in an annealed or lightly cold-worked condition suitable for forming. The tensile strain value of 0.22 for commercial-quality steel gives it excellent forming qualities and it has high strength and elastic modulus at very low cost. The combination of modulus and cost gives it unsurpassed stiffness per unit cost in sheet form, and this is the reason for its dominance in the manufacture of such items as automobile and major-appliance body components.

In this chapter we will concentrate on sheet metal components that can be made using either dedicated dies or alternatively on turret presses. This limits the

TABLE 9.2   Sheet Metal Properties and Typical Costs

Alloy	Cost ($/kg)	Scrap value ($/kg)	Specific gravity	UTS (MN/m²)	Elastic modulus (GN/m²)	Max. tensile strain
Steel, low-carbon commercial quality	0.80	0.09	7.90	330	207	0.22
Steel, low-carbon, drawing quality	0.90	0.09	7.90	310	207	0.24
Stainless steel T304	6.60	0.40	7.90	515	200	0.40
Aluminum, 1100, soft	3.00	0.80	2.70	90	69	0.32
Aluminum, 1100, half hard	3.00	0.80	2.70	110	69	0.27
Aluminum, 3003, hard	3.00	0.80	2.70	221	69	0.02
Copper, soft	9.90	1.90	8.90	234	129	0.45
Copper, 1/4 hard	9.90	1.90	8.90	276	129	0.20
Titanium, Grade 2	19.80	2.46	4.50	345	127	0.20
Titanium, Grade 4	19.80	2.46	4.50	552	127	0.15

discussion to flat, shallow formed or bent parts with a variety of feature types. Deep formed parts, which must be made by the process of deep drawing on special double-action presses, will not be considered here.

9.2   DEDICATED DIES AND PRESSWORKING

A typical sheet metal part is produced through a series of shearing and forming operations. These may be carried out by individual dies on separate presses or at different stations within a single die. The latter type of die is usually termed a progressive die, and in operation the strip is moved incrementally through the die while the press cycles. In this way the punches at different positions along the die produce successive features in the part. We will first consider the use of individual dies.

9.2.1   Individual Dies for Profile Shearing

Sheet metal dies are manufactured by mounting punches and die plates into standard die sets. The die sets, as shown in Fig. 9.1, consist of two steel or cast iron plates that are constrained to move parallel to one another by pillars and bushings mounted on the separate plates. Small die sets typically have two guide pillars, whereas larger ones have four. In operation the lower plate is mounted to the bed of a mechanical or hydraulic press and the upper plate is attached to the moving press platen. As the press cycles, the die set opens and closes so that the

FIG. 9.1   Die set.

punches and dies mounted on the two plates move in precise alignment. Figure
9.2 shows a typical mechanical press with a die set mounted in position.

When individual die sets are used, the first operation is typically shearing of
the external profile of the part. The way in which this is carried out can be divided
into three categories depending on the part design. The most efficient method is a
simple cut-off operation, which applies to parts that have two parallel edges and
that "jigsaw" together along the length of the strip. For the basic cut-off
operation, the trailing edge of the part must be the precise inverse of the leading
edge, as shown in Fig. 9.3.

Parts designed for cut-off operations may not have the aesthetically pleasing
shapes required for some applications. However, for purely functional parts, cut-
off type designs have the advantage of simple tooling and the minimization of
manufactured scrap. The term "manufactured" scrap refers to the scrap sheet
metal produced as a direct result of the manufacturing process, as opposed to the
scrap metal of defective parts. For cut-off-type designs the only manufactured
scrap is the sheet edges left over from the shearing of purchased sheets into part-
width strips. Some scrap also results from the ends of the strips as they are cut up
into parts. Shearing of sheets is normally carried out on special presses called
power shears, which are equipped with cutting blades and tables for sliding sheets
forward against adjustable stops.

For situations where a sheet metal part can be designed with two parallel
edges, but where the ends cannot jigsaw together, the most efficient process to
produce the outer contour is with a part-off die. This die employs two die blocks

FIG. 9.2 Mechanical press.

and a punch that passes between them to remove the material separating the ends of adjacent parts. The principal design rule for this process is that the sheared ends should not meet the strip edges at an angle less than about 15 degrees. This ensures that a good-quality sheared edge is produced with a minimum of tearing and edge distortion at the ends of the cut. Thus full semicircular ends or corner blend radii should be avoided. A simple part that could be produced with a part-

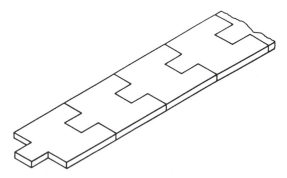

FIG. 9.3 Cut-off part design.

FIG. 9.4   Part-off part design.

off die is illustrated in Fig. 9.4. The part-off process offers the same advantage as cut-off in that the part edges are produced inexpensively with a minimum of scrap by power shear operations. The die, however, is a little more complex than a cut-off die, involving the machining and fitting of an extra die block. Scrap is also increased because adjacent parts must be separated by at least twice the sheet metal thickness to allow adequate punch strength. The main elements of cut-off and part-off dies are illustrated in Fig. 9.5.

For sheet metal parts that do not have two straight parallel edges, the die type used to shear the outer profile is called a blanking die. A typical blanking die is shown in Fig. 9.6. This illustrates the blanking of circular disks, but the shape of the "blank" can be almost any closed contour. The disadvantage of blanking as opposed to cut-off or part-off is mainly the increase in manufactured scrap. This arises because the edges of the part must be separated from the edges of the strip by approximately twice the sheet metal thickness to minimize edge distortion. Thus, extra scrap equal in area to four times the material thickness multiplied by part length is produced with each part. In addition, blanking dies are more expensive to produce than cut-off or part-off dies. The reason for this is that the blanking die has an additional plate, called a stripper plate, which is positioned above the die plate with separation sufficient to allow the sheet metal strip to pass between. The stripper plate aperture matches the contour of the punch so that it uniformly supports the strip while the punch is removed from it on the upward stroke of the press. Note that in comparison the cut-off die has a simple spring action hold-down block to keep the strip from lifting during the shearing operation.

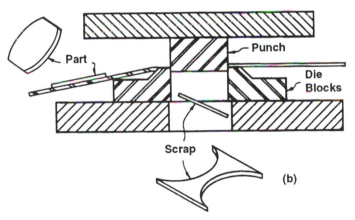

FIG. 9.5 Die elements of cut-off and part-off dies. (a) Cut-off die. (b) Part-off die.

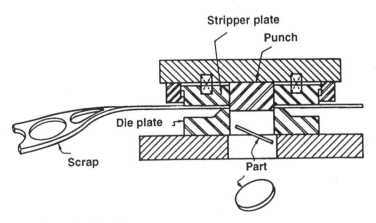

FIG. 9.6 Blanking die.

A less common design for the contour of a sheet metal part is shown in Fig. 9.7. This uses a part-off die to produce parts whose ends are 180 degrees symmetric. The opposite part ends shown in Fig. 9.7 have a similar appearance, but this need not be the case. If both ends are symmetric, then adjacent parts can be arranged on the strip at a 180 degree orientation to each other. With this design the portion that is normally removed as scrap in a part-off die is now an additional part. Each press stroke thus produces two parts, and the die is called a cut-off and drop-through die. The general symmetry rule for this type of part seems not to have been applied in practice, and the only examples appear to be simple trapezoid-shaped parts. A problem with cut-off and drop-through is associated with the nature of the shearing process, which tends to produce a rounded edge on the die side of the part from the initial deformation as the sheet is pressed downward against the die edge. However, final separation of the part from the strip is by brittle fracture, which leaves a sharp edge, or burr, on the punch side of the part. Thus parts made by cut-off and drop-through have the sharp edges on opposite sides of adjacent parts. This lack of edge consistency may be unacceptable for some applications.

Irrespective of the die type used, the sharp edges produced by punching must be removed. This deburring process is carried out, for small parts, by tumbling them in barrels with an abrasive slurry. For larger parts, the usual practice is to pass the flat parts, before forming, through abrasive belt machines. In either case, the added cost is small.

9.2.2 Cost of Individual Dies

Zenger and Dewhurst [1] have investigated the cost of individual dies. For each type of die the cost always includes a basic die set as shown in Fig. 9.1. Current

FIG. 9.7 Part design for cut-off and drop-through.

costs of die sets were found to be directly proportional to the usable area between the guide pillars and to satisfy the following empirical equation:

$$C_{ds} = 120 + 0.36\, A_u \qquad (9.1)$$

where

C_{ds} = die set purchase cost, \$
A_u = usable area, cm^2

A comparison of Eq. (9.1) with a range of commercially available die sets is shown in Fig. 9.8.

To estimate the cost of the tooling elements such as die plate, punch, punch retaining plate, stripper plate, etc., a manufacturing point system was developed. The system includes the time for manufacturing the die elements and for assembly and tryout of the die. Assembly includes custom work on the die set, such as the drilling and tapping of holes and the fitting of metal strips or dowel pins to guide the sheet metal stock in the die.

The basic manufacturing points were found to be determined by the size of the punch and by the complexity of the profile to be sheared. Profile complexity is measured by index X_p as

$$X_p = P^2/(LW) \qquad (9.2)$$

where

P = perimeter length to be sheared, cm
L, W = length and width of the smallest rectangle that surrounds the punch, cm

For a blanking die, or a cut-off and drop-through die, L and W are the length and width of the smallest rectangle that surrounds the entire part. For a part-off die, L

FIG. 9.8 Die set cost versus usable area.

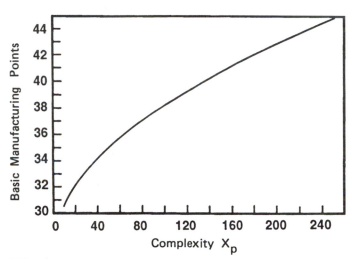

FIG. 9.9   Basic manufacturing points for blanking die.

is the distance across the strip while W is the width of the zone removed from between adjacent parts. For a cut-off die, L and W are the dimensions of a rectangle surrounding the end contour of the part. Note that for either cut-off or part-off, a minimum punch width W of about 6 mm should be allowed to ensure sufficient punch strength. Basic manufacturing points for blanking dies are shown in Fig. 9.9. This basic point score is then multiplied by a correction factor for the plan area of the punch; see Fig. 9.10. Zenger [2] has shown that the basic manufacturing points for a part-off die are about 9% less than for a blanking die, while those for a cut-off die are approximately 12% less than for blanking. Note that this does not represent the differences in die costs since the punch envelope area LW will be less for cut-off and part-off dies and X_p will also generally be smaller for these processes.

For die manufacturing, where computer-controlled wire electrodischarge machining is used to cut the necessary profiles in die blocks, punch blocks, punch holder plates, and stripper plates, each manufacturing point in Fig. 9.9 corresponds to one equivalent hour of die making. This also includes the time for cutting, squaring, and grinding the required tool steel blocks and plates. Note that, as for injection molding, the cost of the die materials is insignificant compared to the cost of die making.

The estimated point score from Figs. 9.9 and 9.10 does not include the effect of building more robust dies to work thicker-gage or higher-strength sheet metal, or to make very large production volumes of parts. To accommodate such requirements it is usual practice to use thicker die plates and correspondingly thicker punch holder plates, stripper plates, and larger punches. This allows the

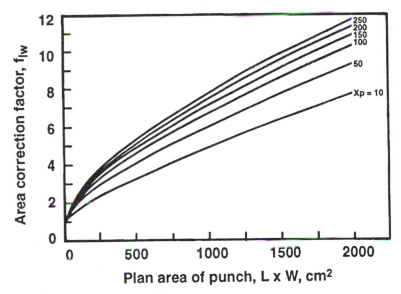

FIG. 9.10 Area correction factor.

die plate to handle longer-term abuse and also provides additional material for the greater number of times that the punch and die faces must be surface-ground to renew edge sharpness. Recommendations on die plate thickness h_d given by Nordquist [3] fit quite well with the relationship

$$h_d = 9 + 2.5 \times \log_e(U/U_{ms})Vh^2 \text{ mm} \tag{9.3}$$

where

U = the ultimate tensile stress of the sheet metal to be sheared
U_{ms} = the ultimate tensile stress of annealed mild steel
V = required production volume, thousands
h = sheet metal thickness, mm

In practice, in U.S. industry the value of h_d is usually rounded to the nearest one-eighth of an inch to correspond with standard tool steel stock sizes.

The manufacturing points in Fig. 9.9 were determined for the condition

$$(U/U_{ms})Vh^2 = 625 \tag{9.4}$$

or

$$h_d = 25 \text{ mm}$$

Zenger and Dewhurst [1] have shown that the cost of dies changes with die plate thickness approximately according to a thickness factor f_d given by

$$f_d = 0.5 + 0.02h_d \tag{9.5}$$

or

$$f_d = 0.75$$

whichever is the larger.

Thus the manufacturing points M_p for a blanking die are given by

$$M_p = f_d f_{1w} M_{po} \tag{9.6}$$

where

M_{po} = basic manufacturing points from Fig. 9.9
f_{1w} = plan area correction factor from Fig. 9.10
f_d = die plate thickness correction factor from Eq. (9.5)

Example

A sheet metal blank is 200 mm long by 150 mm wide and has plain semicircular ends with radius 75 mm; see Fig. 9.11a. It is proposed that 500,000 parts should be manufactured using 16 gage low carbon steel.

Estimate the cost of a blanking die to produce the part and the percentage of manufactured scrap that would result from the blanking operation.

If the part were redesigned with 80 mm radius ends as shown in Fig. 9.11b, it could then be produced with a part-off die. What would be the die cost and percentage of manufactured scrap for this case?

The required blank area is 200×150 mm^2. If 50 mm space is allowed around the part for securing of the die plate and installation of strip guides, then the required die set usable area A_u is

$$A_u = (20 + 2 \times 5) \times (15 + 2 \times 5) = 750 \text{ cm}^2$$

and so from Eq. (9.1) the cost of the die set will be given by

$$C_{ds} = 120 + (0.36 \times 750) = \$390$$

For the design shown in Fig. 9.11a the required blanking punch would have perimeter P equal to 571 mm and cross-sectional dimensions L, W equal to 150 and 200 mm, respectively. Thus the perimeter complexity index X_p is given by

$$X_p = 571^2/(150 \times 200) = 10.9$$

| 150 | 200 | 150 | 200 |

(a)   (b)

FIG. 9.11   Sheet metal part (dimensions in mm). (a) Blanking design. (b) Part-off design.

The basic manufacturing point score from Fig. 9.9 is thus $M_{po} = 30.5$. With plan area LW equal to 300 cm^2 the correction factor from Fig. 9.10 is approximately 2.5. For 500,000 parts of thickness 1.52 mm (equivalent to 16 gage), the die plate thickness from Eq (9.3) is $h_d = 26.6$ mm. The die plate thickness correction factor from Eq. (9.5) is thus $f_d = 1.03$.

Total die manufacturing points are therefore

$$M_p = 1.03 \times 2.5 \times 30.5 = 78.5$$

Assuming \$40/h for die making, the cost of a blanking die is estimated to be

Blanking die cost $= 390 + 78.5 \times 40 = \$3530$

The area of each part is

$$A_p = 251.7 \text{ cm}^2$$

Since the separation between each part on the strip and between the part and the strip edges should be 3.04 mm (equal to twice the material thickness), the area of sheet used for each part is

$$A_s = (200 + 3.04) \times (150 + 2 \times 3.04) \text{ mm}^2$$
$$= 316.9 \text{ cm}^2$$

Thus the amount of manufactured scrap is given by

Scrap percent $= (316.9 - 251.7)/316.9 \times 100$
$$= 20.6$$

For the alternative design shown in Fig. 9.11b, the perimeter to be sheared is the length of the two 80 mm arcs, which can be shown to be given by

$$P = 388.9 \text{ mm}$$

With 3.04 mm separating the parts end to end on the strip, the cross-sectional dimensions L, W of the part-off punch equal 106.5 and 150 mm, respectively.

Thus the complexity index X_p, is given by

$$X_p = 388.9^2/(106.5 \times 150) = 9.5$$

With the part plan area equal to 300 cm^2 as before, the manufacturing points are the same as for the blanking die.

Since part-off dies are typically 9% less expensive than blanking dies for the same C_{px} value, and the values of f_d and f_{lw} are unchanged, the total die manufacturing hours are

$$M_p = 0.91 \times 1.03 \times 2.5 \times 30.5 = 71.4 \text{ h}$$

Assuming \$40/h for die making as before, the cost of a part-off die is estimated to be

Part-off die cost $= 390 + 71.4 \times 40 = \$3,250$

The area of each part shown in Fig. 9.11b can be shown to be 257.9 cm^2. Since the edges of the strip now correspond to the edges of the part, the area of sheet used for each part is

$$A_s = (200 + 3.04) \times 150 \text{ mm}^2$$
$$= 304.6 \text{ cm}^2$$

Thus the amount of manufactured scrap for the part-off design is

$$\text{Scrap percent} = (304.6 - 257.9)/304.6 \times 100$$
$$= 15.3$$

The change in percent scrap between the two designs is somewhat artificial since the redesign has a slightly larger area than the original one. If the end profile curves of the new design were designed to cut the edges at approximately 20 degrees and enclose the same area of 251.7 cm^2, the percentage of manufactured scrap would equal 17.4.

9.2.3 Individual Dies for Piercing Operations

A piercing die is essentially the same as a blanking die except that the material is sheared by the punching action to produce internal holes or cut-outs in the blank. Thus the die illustrated in Fig. 9.6 could also be a piercing die for punching circular holes into the center of a previously sheared blank. However, piercing dies are typically manufactured with several punches to simultaneously shear all of the holes required in a particular part.

It has been shown by Zenger [2] that with piercing dies, the individual punch areas have only a minor effect on final die cost. The main cost drivers are the number of punches, the size of the part, and the perimeter length of the cutting edges of any nonstandard punches. For cost estimation purposes a nonstandard punch is one with cross-sectional shape other than circular, square, rectangular, or obround as illustrated in Fig. 9.12. These standard punch shapes are available at low cost in a very large number of sizes. Any punch shapes other than those in Fig. 9.12 will be referred to as nonstandard.

Following the procedure developed by Zenger, a manufacturing point score is determined for a piercing die from three main components. First, based only on the area of the part to be pierced, the base manufacturing score is given by

$$M_{\text{po}} = 23 + 0.03LW \text{ h} \tag{9.7}$$

FIG. 9.12   Standard punch shapes.

where $L, W =$ length and width of the rectangle that encloses all the holes to be punched, cm.

Equation (9.7) predicts the number of hours to manufacture the basic die block, punch retaining plate, stripper plate, and die backing plate. This must be added to the time to manufacture the punches and to produce the corresponding apertures in the die block. This time depends upon the number of required punches and the total perimeter of punches. From a study of the profile machining of punches from punch blocks and of apertures in die blocks, Zenger [2] has shown that the manufacturing time M_{pc} for custom punches can be represented approximately by

$$M_{pc} = 8 + 0.6P_p + 3N_p \text{ h} \tag{9.8}$$

where
$P_p =$ total perimeter of all punches, cm
$N_p =$ number of punches

Equation (9.8) is used for estimating the time to manufacture nonstandard or custom punches and for cutting the corresponding die apertures. For the standard punch shapes, shown in Fig. 9.12, typical supplier costs for punches and die plate inserts (called die buttons) can be divided by the appropriate tool manufacturing hourly rate to obtain the equivalent number of manufacturing hours. With this approach, Zenger has shown that manufacturing hours M_{ps} for standard punches and die inserts, and for the time to cut appropriate holes in the punch retaining plate and die plate, can be given by

$$M_{ps} = KN_p + 0.4N_d \text{ h} \tag{9.9}$$

where
$K = 2$ for round holes
$\quad = 3.5$ for square, rectangular, or obround holes
$N_p =$ number of punches
$N_d =$ number of different punch shapes and sizes

Example

Determine the cost of the piercing die to punch the three holes in the part shown in Fig. 9.13. The rectangle that surrounds the three holes has dimensions

FIG. 9.13 Part design with three punched holes.

120 × 90 mm, and the nonstandard "C"-shaped hole has a perimeter length equal to 260 mm.

The base manufacturing score from Eq. (9.7) is

$$M_{po} = 23 + 0.03(12 \times 9) = 26 \text{ h}$$

The number of hours required to manufacture the custom punching elements for the nonstandard aperture is, from Eq. (9.8),

$$M_{pc} = 8 + 0.6 \times 26 + 3 = 26.6 \text{ h}$$

The equivalent manufacturing time for the punches, die plate inserts, etc., for the two "standard" circular holes is, from Eq. (9.9),

$$M_{ps} = 2 \times 2 + 0.4 \times 1 = 4.4 \text{ h}$$

If 50 mm space is allowed around the part in the die set, then the required plate area is given by

$$A_u = (20 + 2 \times 5) \times (10 + 2 \times 5) = 600 \text{ cm}^2$$

which gives a die set cost of $336.

Thus the estimated piercing die cost, assuming $40/h for die making, is

$$336 + (26 + 26.6 + 4.4) \times 40 = \$2,616$$

9.2.4 Individual Dies for Bending Operations

Bends in sheet metal parts are typically produced by one of two die-forming methods. The simplest method is by using a v-die and punch combination as shown in Fig. 9.14a. This is the least expensive type of bending die, but it suffers because of the difficulty of precisely positioning the metal blank and a resulting lack of precision in the bent part. The alternative method, which allows greater control of bend location on the part, is the wiper die shown in Fig. 9.14b. This

FIG. 9.14 Basic bending tools (a) v-die. (b) Wiper die.

method is most commonly used for the high-volume production of parts [4]. With the use of dedicated bending dies it is common practice to produce multiple bends in a single press stroke. The basic die block configurations for doing this are the u-die (which is a double-wiper die) shown in Fig. 9.15a and the z-die (double v-die) illustrated in Fig. 9.15b. It can readily be visualized how a combination of die blocks and punches using v-forming and wiper techniques can form a combination of several bends in one die. With the use of die blocks that can move under heavy spring pressure, a combination of bends can be made that displace the material upward and downward. For example, the part shown in Fig. 9.16 can be formed in a single die. In this case a z-die first forms the "front step." The lower die block then proceeds to move downward against spring pressure so that stationary wiper blocks adjacent to the three other sides displace the material upward.

In order to determine the number of separate bending dies required for a particular part, the following rules may be applied.

1. Bends that lie in the same plane, such as the four bends surrounding the central area in Fig. 9.16, can usually be produced in one die.
2. Secondary reverse bends in displaced metal, such as the lower step in Fig. 9.16, can often be produced in the same die using a z-die action.

FIG. 9.15 Basic methods of producing multiple bends. (a) u-die. (b) z-die.

FIG. 9.16   Multiple bends produced in one die.

3.   Secondary bends in displaced metal that would lead to a die-locked condition
     will usually be produced in a separate die.

     For example, consider the part shown in Fig. 9.17. Bends a, c, and d or bends
a, b, and d could be formed in one die by a combination of a wiper die and a

FIG. 9.17   Part design requiring two bending dies.

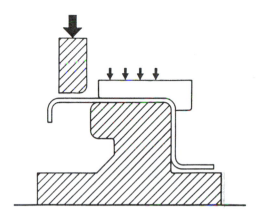

FIG. 9.18   Wiper-die arrangement to produce bend b in Fig. 9.17.

z-die. The remaining bend would then require a second wiper die and a separate press operation. For example, bend b could be produced in the second die using a tooling arrangement, as shown in Fig. 9.18.

Referring once more to the early cost-estimating work by Zenger [2], the following relationships were established from investigations of the cost of bending dies. The system is based on a point score that relates directly to tool manufacturing hours as before. First, based on the area of the flat part to be bent and the final depth of the bent part, the base die manufacturing score for bending is given by:

$$M_{po} = (18 + 0.023LW) \times (0.9 + 0.02D) \tag{9.10}$$

where

L, W = length and width of rectangle that surrounds the part, cm
D = final depth of bent part, cm, or 5.0, whichever is larger

An additional number of points is then added for the length of the bend lines to be formed and for the number of separate bends to be formed simultaneously. These are given by

$$M_{pn} = 0.68L_b + 5.8N_b \tag{9.11}$$

where

L_b = total length of bend lines, cm
N_b = number of different bends to be formed in the die

Finally, the cost of a die set must be added according to Eq. (9.1).

Example

The part shown in Fig. 9.16 is produced from a flat blank 44 cm long by 24 cm wide. There are five bends and the total length of the bend lines is 76 cm. The final height of the formed part from the top edge of the box to the bottom of the step is 12 cm.

Thus Eq. (9.10) gives

$$M_{po} = [18 + 0.023 \times (44 \times 24)] \times (0.88 + 0.02 \times 12)$$
$$= 42.3 \times 1.12 = 47.4 \, h$$

The additional points for bend length and multiple bends are

$$M_{pn} = 0.68 \times 76 + 5.8 \times 5 = 80.7 \, h$$

If 5.0 cm clearance is allowed around the part in the die set, then the cost of the die set is estimated from Eq. (9.1) as

$$C_{ds} = 120 + 0.36 \times (54 \times 34) = \$780$$

Finally, assuming \$40/h for tool making, the cost of the bending die is given by

$$C_d = 780 + (47.4 + 80.7) \times 40 = \$5900.$$

9.2.5 Miscellaneous Features

Other features commonly produced in sheet metal parts by regular punching operations are lances, depressions, hole flanges, and embossed areas.

A lance is a cut in a sheet metal part that is required for an internal forming operation. This may be for the bending of tabs or for the forming of bridges or louver openings. In producing a lance the cutting edges of the punch are pressed only partway through the material thickness, sufficient to produce the required shear fracture.

Depressions are localized shallow-formed regions produced by pressing the sheet downward into a depression in the die plate with a matching profile punch. The punch and die surfaces in this case are analogous to the cavity and core in injection molding, and the "cavity" is filled by localized stretching of the sheet metal. Patterns of long, narrow depressions, called beads, are often formed onto the open surfaces of sheet metal parts in order to increase bending stiffness. In a depression the sheet material reduces in thickness as a result of being stretched around the punch profile. For example, in the depression shown on the left side of the part in Fig. 9.19, assume the material is stretched by approximately 15% in every direction. Because the volume of metal stays constant after forming, the thickness will have been reduced by approximately 30%. In contrast, the embossed region shown on the right side of the part in Fig. 9.19 is reduced in thickness by direct compression between punch and die. In this case the required punch pressures are much larger than for the material stretching involved in depression forming. For this reason, embossed areas are usually small, with only modest reductions in thickness.

FIG. 9.19   Shallow formed and embossed regions of sheet metal part.

Finally, hole flanges are produced by pressing a taper or bullet-nosed cylindrical punch into a smaller punched hole. The material is thus stretched by entry of the larger punch and displaced in the direction of punch travel. Because of ductility limitations of sheet metals, typically hole flanges can only be formed to a height of two or three times the sheet metal thickness.

The cost of dies for these miscellaneous operations can be determined from the equations for the costs of piercing dies given in Sec. 9.2.3. Equation (9.7) is used to determine the base cost of the die plates, punch blocks, etc. The additional cost of punch and die machining is then obtained from Eq. (9.8). In this case parameter P_p is the perimeter of the forming or cutting punches to be used. With the appropriate rate for tool making and an appropriate die set cost, these equations can provide an approximate estimate of die cost for lancing, hole swaging, or the forming of simple depressions or embossed areas. However, if the required formed areas of a part have surface details or patterns, then the cost of machining appropriate die surfaces should be added. Empirical equations given in Chapter 8, for the machining of geometrical details in matching cavities and cores, can be used for this purpose. Combining Eqs. (8.10) and (8.11) from Chapter 8, the number M_{px} of additional hours of punch and die machining is given by

$$M_{px} = 0.13N_{sp}^{1.27} \text{ h} \qquad\qquad (9.12)$$

where
$N_{sp} =$ total number of separate surface patches to be machined on punch faces and matching die surfaces

9.2.6 Progressive Dies

For the manufacture of sheet metal parts in very large quantities, the hand-loading of parts into individual dies at different presses is unnecessarily inefficient. If the quantities to be produced can justify the additional tooling expense, then it is usual practice to use a multistation die on a single press. Stations within the die carry out the different piercing, forming, and shearing operations as the sheet metal is transported incrementally through the die. To carry out this incremental movement, the sheet metal is supplied from coil, which has to be purchased in the required width, and which is fed through the die automatically by coil feeding equipment mounted at the side of the press. An example of multistation operation is illustrated in Fig. 9.20. It can be seen that the technique is to produce features on the part at the different stations and then to separate the part from the strip at the last one. The illustration in Fig. 9.20 shows the last station as a blanking operation. However, as described in Sec. 9.2.1, if the part can be designed appropriately, then the separation from the strip can take place with a part-off or cut-off operation with reduced scrap and lower die cost. It should be noted that for complex-shaped parts, the perimeter will usually be sheared in increments at the different stations with only the final parts of the profile being sheared at the last station. This allows a more uniform distribution of shearing forces among the different stations, resulting in balanced loads on the die. It also enables bending

FIG. 9.20 Multistation die operation with strip feed.

operations to be performed with wiper dies when portions of the perimeter around the bend have been removed. The two additional holes in the strip skeleton shown in Fig. 9.20 are punched at the first station and then engaged with taper-nosed punches at the second station. This pulls the strip into more precise registration between the stations so that part accuracy does not depend on the accuracy of the strip feeding mechanism.

The design of progressive dies is an art form with only basic principles of alignment, required clearances, material ductility, and loading distribution to guide the tool engineer. Die casting or injection molding tools for the same part produced by different designers will be almost identical. In contrast, progressive dies for the same sheet metal part may be entirely different, with different numbers of stations and different combinations of shearing and forming punches. Under these circumstances early cost estimating of progressive dies can only be approximate. For such estimates, at the sketch stage of part design, a rule of thumb quoted in the literature [5] will be used. This can be stated as

$$C_{pd} = 2C_{id} \tag{9.13}$$

where

$C_{pd} =$ cost of single progressive die
$C_{id} =$ cost of individual dies for blanking, cut-off or part-off, piercing, and
forming operations for the same part

As reported by Zenger [2] the factor 2 seems to agree with tool cost quotes for other than very simple or very complex parts. For the latter the appropriate factor can be as high as 3, whereas for very simple parts a factor of 1.5 may be more appropriate.

9.3 PRESS SELECTION

A selection of typical mechanical presses used for sheet metal stamping operations is given in Table 9.3. In choosing the appropriate press for a given part, the main considerations will be the press bed size and the required press force. For shearing operations, the required force is simply determined from the shear length, gage thickness, and material shear strength. For metals, the material strength in shear, S, is approximately half of the strength in simple tension. Moreover, during shearing, the strains build up rapidly in the narrow shear zone extending between the punch and die edges; see Fig. 9.21. This means that strain hardening must be taken into account and that the tensile strength at failure, denoted by U (ultimate tensile strength), is more appropriate than the initial yield

TABLE 9.3 Mechanical Presses

Bed size					
Width (cm)	Depth (cm)	Press force (kN)	Operating cost ($/hr)	Maximum press stroke (cm)	Strokes (per min)
50	30	200	55	15	100
80	50	500	76	25	90
150	85	1750	105	36	35
180	120	3000	120	40	30
210	140	4500	130	46	15

strength. Accordingly, the required force f for such operations as blanking, piercing, lancing, etc., is given by

$$f = 0.5 U h l_s \text{ kN} \tag{9.14}$$

where
$h =$ gage thickness, m
$l_s =$ length to be sheared, m

As an example, circular disks 50 cm in diameter are to be blanked from No. 6 gage commercial-quality, low-carbon steel. From Tables 9.1 and 9.2 the thickness of 6 gage steel is 5.08×10^{-3} m and the ultimate tensile strength U is 330×10^3 kN/m². The required blanking force is thus given by

$$f = 0.5 \times (330 \times 10^3) \times (5.08 \times 10^{-3}) \times (\pi \times 50 \times 10^{-2})$$
$$= 1316.6 \text{ kN}$$

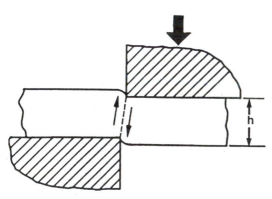

FIG. 9.21 Shearing operation.

Thus, the 1750 kN press in Table 9.3, which can be seen to have sufficient bed size, would be the appropriate choice.

For bending or shallow forming operations, the required forces are usually much less than for shearing. For example, for the bending operation illustrated in Fig. 9.22 assume that the inside bend radius, r, equals twice the gage thickness, h. Under these conditions, as the material is bent around the die profile, through increasing angle θ, the length of the outer surface increases to $3h\theta$. The length of the centerline of the material (the neutral axis in simple bending) remains approximately constant at $2.5h\theta$. The strain in the outer fibers of the material is thus

$$e = (3h\theta - 2.5h\theta)/2.5h\theta = 0.2 \qquad (9.15)$$

The strain decreases to zero from the outer fibers to the centerline, and then becomes compressive, increasing to approximately -0.2 on the inside surface. The average magnitude of strain in the bent material is thus $0.5e$ under these conditions. To obtain an approximate value of the required force we can consider the energy balance in the process. The work done per unit volume on the material as it forms around the die is the product of stress and strain. If we assume that the punch radius also equals twice the thickness, then the 90 bend will be completed when the punch moves down, while in contact with the part, through a distance of approximately $5h$. At this point the volume of material subjected to bending is

$$V = \pi((3h)^2 - (2h)^2)L_b/4 = 5\pi h^2 L_b/4 \qquad (9.16)$$

where L_b is the bend length.

The energy balance can thus be represented approximately by

$$0.5e \times U \times 5\pi h^2 L_b/4 = f \times 5h \qquad (9.17)$$

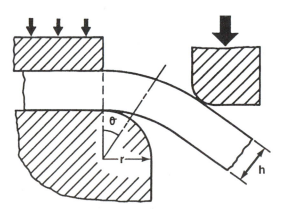

FIG. 9.22 Wiper die bending operation.

where f is the average press force that moves through distance $5h$. Thus, under these conditions, the press force f is given by

$$f = 0.08UhL_b \, kN \tag{9.18}$$

Comparing Eq. (9.18) with Eq. (9.14) shows that under typical bending conditions the required force is only about 15% of the force required for shearing of the same length and gage thickness. Eary and Reed [4] give an empirical relationship for wiper die bending as

$$f = 0.333UL_b h^2 / (r_1 + r_2) \, kN \tag{9.19}$$

where r_1 is the profile radius of the punch and r_2 is the profile radius of the die. This agrees almost precisely with Eq. (9.18) for the condition $r_1 = r_2 = 2h$.

For shallow forming, as illustrated on the left side of the part in Fig. 9.19, the material being displaced downward into the die form is subjected to stretching in every direction. During this process, the tensile stresses in the stretching material are transmitted from the perimeter of the depression in toward its center. If the walls of the depression make angle θ with the part surface (Fig. 9.19), and the stress is assumed to be approximately equal to the ultimate tensile stress, then the vertical resisting force from the walls can be given as

$$f = Uh \sin \theta L \, kN \tag{9.20}$$

where L is the perimeter of the depression. This is equal to the required punch force. Thus, for a depression with almost vertical walls ($\sin \theta = 1$) the required punch force can approach twice the force required to shear the material around the same perimeter.

Finally, the region on the right side of the part in Fig. 9.19, which has been reduced in thickness, is referred to as having been embossed or coined. This type of bulk forming of sheet metal between punch and die surfaces involves very high compressive stresses. Because the displaced material must flow sideways across the face of the die, this introduces constraints that make the process much less efficient than pure compression. The required force for an embossing operation is thus

$$f = \phi UA \, kN \tag{9.21}$$

where
A = area to be embossed
ϕ = constraint factor > 1

As the size of the embossed region increases, factor ϕ increases exponentially. Because of this, extensive embossing should be avoided whenever possible. The alternative is to produce required surface patterns through shallow forming, although much less precise pattern details are possible with this form-stretching process.

9.3.1 Cycle Times

When using individual dies for sheet metalworking, the presses must be manually operated. This involves hand loading of parts into the dies. In the case of shearing (blanking, part-off, cut-off, or piercing) the part can be automatically ejected from the die after the press operation. However, for a bending or forming operation, the press operator must also remove the parts from the die, which increases the cycle time further.

Ostwald [6] has shown that the time to load a blank or part into a mechanical press, operate the press, and then remove the part following the press operation is proportional to the perimeter of the rectangle that surrounds the part. This time can be given by

$$t = 3.8 + 0.11(L + W) \text{ s} \tag{9.22}$$

where

L, W = rectangular envelope length and width, cm

For shearing or piercing of flat parts, for which automatic press ejection would be appropriate, two-thirds of the time given by Eq. (9.22) should be used. Also, for the first press operation, the material may be supplied in strips that have been power-sheared from large sheets into the appropriate width. These strips are then loaded individually into the die and manually indexed forward after each press operation. The power shearing time per part is typically small, since the strip for several parts is produced by each shear. It will be assumed that this time is balanced by the reduced press time per part from strip loading of the first die.

For press working with progressive dies, the cycle time is governed by the size of the press and its reciprocating speed when operated continuously. Typical operation speeds for a variety of press sizes are given in Table 9.3.

Example

For the part shown in Fig. 9.20 compare the cycle times and processing costs for using individual dies to those for progressive die working. The part is to be made from No. 8 gage stainless steel for which the ultimate tensile stress is 515 MN/m^2.

The outer perimeter of the part equals 370 mm and the thickness of No. 8 gage steel is 4.17 mm. From Eq. (9.14) the required shear force for blanking the outer perimeter is

$$f_1 = 0.5 \times (515 \times 10^3) \times (4.17 \times 370 \times 10^{-6})$$
$$= 397 \text{ kN}$$

For piercing the obround cutout with perimeter 149 mm, the required force is

$$f_2 = 160 \text{ kN}$$

Finally, the force required for bending the tab across an approximate 25 mm bend line, with assumed 6 mm tool profile radii, is given from Eq. (9.19) as

$$f_3 = 0.333 \times 515 \times 10^3 \times (25 \times 10^{-3}) \times (4.172 \times 10^{-6})/((6+6) \times 10^{-3})$$
$$= 6.2 \text{ kN}$$

Referring to Table 9.3, it can now be seen that the blanking operation would require the 500 kN press, and the piercing and bending operations could be carried out on the smallest 200 kN press.

Individual Dies

For the blanking and piercing operations, we can assume automatic ejection of the blanks and scrap. The cycle time for these two operations will thus be approximated as two-thirds of the time for loading and unloading given by Eq. (9.22):

$$t_1 = 0.67 \times (3.8 + 0.11(10 + 11.5)) = 0.67 \times 5.4$$
$$= 3.6 \text{ s}$$

For the bending operation, part unloading is required and thus

$$t_2 = 5.4 \text{ s}$$

Finally, applying the press hourly rates from Table 9.3 gives the processing cost per part as

$$C_p = [(3.6/3600) \times 76 + (3.6/3600) \times 55 + (5.4/3600) \times 55] \times 100 \text{ cents}$$
$$= 21.4 \text{ cents}$$

Progressive Die

Using a progressive die the required press force will be approximately

$$f = f_1 + f_2 + f_3 = 563 \text{ kN}$$

The space required for the four die stations is

$$4 \times 100 + 3(2 \times 4.17) = 418.5 \text{ mm}$$

From Table 9.3 the appropriate press has 1750 kN press force, an operating cost of 105 \$/h, and a press speed of 35 strokes/min.

The estimated cycle time per part is thus

$$t = 60/35 = 1.7 \text{ s}$$

and the processing cost per part is

$$C_p = (1.7/3600) \times 105 \times 100$$
$$= 5.0 \text{ cents}$$

9.4 TURRET PRESSWORKING

An alternative to the use of dedicated dies for the manufacture of sheet metal parts is the numerically controlled turret press. This is a machine, as illustrated in Fig. 9.23, that contains punches and matching dies in two rotary magazines or turrets. Depending upon the machine size, turrets may contain as many as 72 different dies for a variety of punching operations.

The lower turret rests in the center of the press bed, the surface of which is covered with steel balls that freely rotate in spherical sockets and that project just above the bed surface. For this reason the press bed is sometimes referred to as a ball table. A large metal sheet placed onto the press bed can thus slide easily to different positions between the two turret faces. This sliding is accomplished by gripping an edge of the sheet in two clamps attached to linear (X, Y) slideways. The slideways are under numerical control, which allows precise positioning of the sheet under the active punch in the machine turret. The turret is also controlled numerically so that while the sheet is moving to the next punching position, the turrets can be rotated to bring the desired punch and die into play.

The advantage of the turret press is that it uses general-purpose punches and dies, which can be used to manufacture a wide range of different parts. Changing from the manufacture of one part to another usually involves changing only one or two of the punches and dies. This is accomplished in a matter of a few minutes. Moreover, the punches and dies fit into standard holders and can be purchased from tool suppliers in a large variety of standard profiles or custom-ordered. In either case, punches and dies typically range from less than $150 for simple cutting punches and dies up to about $750 for punches and dies for cutting complex shapes or doing localized forming operations. The disadvantages of

Clamps mounted on
X,Y slideways

Indexing rotary turrets
containing punches
and matching dies

FIG. 9.23 Turret press.

turret press working are that any forming operations must be shallow enough for the part to pass between the two turret faces. Moreover, any forming of depressions, or of lanced areas to produce projections or louvers, must be made in an upward direction. That is, the forming punch must be in the lower turret and the corresponding die in the upper one. This is necessary so that the sheet can continue to slide smoothly over the ball table. Also, formed areas must be limited in height so that they can still pass between the turret faces. The limitation depends upon the particular machine but may be of the order of 15 mm. This height limitation also applies to bends; if parts require bending, this must usually be carried out as a separate operation after the turret press operations are completed. Such bending operations are usually carried out on a special press called a press brake, which will be described later.

The other disadvantage of turret press working is that punching operations are carried out sequentially, and in consequence the cycle time per part may be much longer than with dedicated die sets.

Because of these disadvantages, turret presses are most often used when parts are required in relatively small quantities. In this case the insignificant cost of the tools easily outweighs the disadvantage of longer cycle times. However, there is no simple rule for choosing the appropriate process, since the changeover point from one process to the other depends on the number of features, the type of features, and the likely cost of dedicated tooling.

A typical turret press manufactured part is shown in Fig. 9.24. The part is laid out to cover the sheet, usually in a regular array pattern as shown. The turret press is programmed to punch one of the two hole types first in every row and column position on the sheet. The turret then rotates to the punch for the other hole and punches that one in the appropriate position for every part. Finally, cut-off

FIG. 9.24 Layout of parts produced by turret press. (————) Subsequent bend lines.

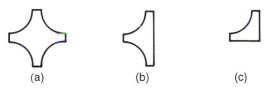

FIG. 9.25   Turret press radius tools.

punches are used to shear the outer part perimeters. For straight edges, narrow cutting punches typically about 6 mm thick are used to separate parts. As shown in Fig. 9.24, the punching positions are programmed to leave small regions, called microtabs, still connecting all adjacent parts. This allows the punching to continue without frequent stoppages for part removal. The sheet can then be removed at the end of the machine cycle with all of the parts attached. Parts can then be knocked out of the sheet, or with thin-gauge material separated by simply shaking the sheet. In industrial jargon the parts are referred to as shaker parts. External radii such as the two lower corners of the parts shown in Fig. 9.24 are produced with a radius tool, as shown in Fig. 9.25. For the example part, tool (a) could be used or tool (b) in a different orientation than shown. It should also be noted that the punching of the external profile requires a cut-off punch in two different orientations for the vertical and horizontal edges, respectively. For most turret presses this will be accomplished by having two different cut-off punches at different turret positions. However, a more recent innovation is the use of so-called indexers, or tool holders, which can be rotated under numerical control. These are not currently used on a widespread basis. However, they open up the possibility of efficient turret press manufacture of a much wider variety of part geometries.

Turret presses are less efficient when required to produce more complex curved edges. In this case one method is to use a circular punch to create the curved edge through successive closely spaced hits. This procedure, known as nibbling, produces a scalloped edge, as shown on the lower-right corner profile in Fig. 9.26. The height of the scallops can be reduced by using a larger punch or reducing the pitch between hits. For the internal curved cutout in Fig. 9.26 the size of the circular punch is determined by the width of the slot, and both slot edges are produced simultaneously during the nibbling operation. Most turret presses have a nibble mode in which the press cycles continuously at high speeds. This allows rapid incremental moves of the sheet to be used for efficient nibbling. The nibbling characteristics of a typical turret press are given in Table 9.4.

An alternative to nibbling for curved edges is the use of turret presses equipped with profile-cutting attachments. Machines are available with either plasma or laser cutting devices. These are typically affixed to the machine structure at a precise distance from the center of the active turret punch. This distance is simply placed in the numerical control file as a coordinate offset prior

FIG. 9.26 Part geometry that requires nibbling or profile cutting.

to the commencement of profile cutting. The sheet is moved to the appropriate start point, the cutting torch is switched on, and movement is then continued around the required profile path. The cutting speed for either laser or plasma cutting varies with the thickness of material and the curvature of the path being followed; tighter curves require a slower speed to maintain a given tolerance level. Table 9.5 gives typical average cutting speeds for 3 mm gage thickness of different metals. Plasma cutting is generally faster than laser cutting, but the former process produces a somewhat less precise and heat-affected edge. For thicker gages the speed difference becomes more pronounced, and from cutting speed data in the literature [7], it appears that the effect of thickness on cutting speed S can be given by

$$S = S_p \times (3/h)^{0.5} \text{ mm/s for plasma cutting} \tag{9.23}$$

and

$$S = S_e \times (3/h) \text{ mm/s for laser cutting} \tag{9.24}$$

TABLE 9.4 Typical Turret Press Manufacturing Characteristics

Machine setup time	20 min
Loading plus unloading time per sheet	
One 750×750 mm sheet	24 s
One 1200×3600 mm sheet	72 s
Average speed between punching	0.5 m/s
Nibbling speed	120 stroke/min
Maximum form height	6 mm
Machine rate, including programming costs	72 \$/h

TABLE 9.5 Plasma and Laser Cutting Speeds for 3 mm Thick Material

	Typical speed, mm/s	
Material class	Plasma, S_p	Laser, S_e
Carbon steel	60	40
Stainless steel	60	35
Aluminum alloy	75	15
Copper alloy	75	20
Titanium alloy	50	20

where

S_p, S_e = plasma and laser cutting speed values, respectively, from Table 9.5 for 3 mm thick material

h = material gage thickness, mm

9.5 PRESS BRAKE OPERATIONS

A press brake is a mechanical press with a bed several feet wide and only a few inches deep. General-purpose bending tools, usually v-blocks or wiper-die blocks and matching punches, are mounted at positions along the bed. At each punch position a back stop is mounted on the rear of the bed in order for the part to be positioned correctly with respect to the bending tool. The method of operation is for the press operator to pick up the flat sheet metal part, turn it to the correct orientation, push it against the back stop at the first tool position, and then operate the press by pressing a foot pedal. If more bends are required, then the part is turned and moved to the next punching position and the operation repeated. After the last bend the part is stacked on a pallet beside the press. The manufacturing characteristics of a typical press brake are given in Table 9.6.

Example

The part shown in Fig. 9.24 is to be manufactured from 16 gage (1.52 mm thick), commercial-quality, low-carbon steel in standard sheet size 48 in. (1219.2 mm) by 60 in. (1524 mm).

The maximum number of parts which can be made from each sheet is 135, from a pattern of 15 rows with 9 parts in each row and 6 mm separation between each part. The area of each part is 94.6 cm^2. The volume of each part is given by

$$94.6 \times 0.152 = 14.38 \text{ cm}^3$$

TABLE 9.6 Typical Press Brake Manufacturing Characteristics

Machine setup time	45 min
Time to load, position, brake, and stack	
One 200 mm × 300 mm part	8.50 s
One 400 mm × 600 mm part	13.00 s
Time to position and brake for each additional bend	
One 200 mm × 300 mm part	4.25 s
One 400 mm × 600 mm part	6.50 s
Machine rate	28 $/h

and the volume of material used for each part is

$$(121.92 \times 152.4) \times 0.152/135 = 20.92 \text{ cm}^3$$

Using data from Table 9.2, the cost of material per part is

$$20.92 \times 7.90 \times 10^{-3} \times 80 = 13.2 \text{ cents}$$

and the resale value of scrap per part is

$$(20.92 - 14.38) \times 7.90 \times 10^{-3} \times 6 = 0.3 \text{ cents}$$

The number of hits required to produce each part can be estimated as follows:

Internal holes	10 hits
Outside radii	8 hits
Outside edges	11 hits

In estimating the 11 hits for the outside straight edges, account is taken of the simultaneous generation of adjacent part edges around portions of the perimeter.

A typical sheet movement speed for a modern CNC turret press is 0.5 m/s, so the time for sheet movement between punching is of the order of 0.1 s. However, this time does not include the dwell time for punching, the acceleration and deceleration between stops, and the periodic delays for turret rotation to change tools. Time studies carried out on a variety of turret press parts [8] suggest that an average of 0.5 s per hit is appropriate for early cost estimating where hole spacing is of the order of 50 mm or less. For large parts with significantly greater distances between holes, extra sheet movement times of 0.1 s for each additional 50 mm can be added.

For the example part, the punching time per part is estimated to be

$$t_1 = N_h \times 0.5 \tag{9.25}$$

where

N_h = number of hits

Thus for the present example

$$t_1 = 29 \times 0.5 = 14.5 \text{ s}$$

From the data in Table 9.4, if the time for sheet loading plus unloading is assumed to be proportional to the sheet perimeter, it can be expressed as

$$t_2 = 2.0 + 0.15(L + W) \text{ s} \tag{9.26}$$

where

L, W = sheet length and width, cm

For the sheet size used for the example part, the loading plus unloading time is given by

$$t_2 = 2.0 + 0.15(121.92 + 152.4) = 43 \text{ s}$$

Note that separation of the 135 parts from the sheet can be carried out during the total machine cycle time of 135×14.5 s or 32.6 m. During this time previously punched parts may also be passed between an automatic belt deburring machine to remove sharp punched edges.

The turret press cycle time per part is thus

$$t = 14.5 + 43/135 = 14.8 \text{ s}$$

The cost of turret press operations per part, using the machine rate of $72/h from Table 9.4, is estimated to be

$$C_1 = 14.8 \times (72 \times 100)/3600 = 29.6 \text{ cents}$$

Applying linear interpolation to the press brake data in Table 9.6 gives the following empirical relationship for brake bending.

$$t = 2(1 + N_b) + 0.05(2 + N_b)(L + W) \tag{9.27}$$

where

N_b = number of required bending operations
L, W = length and width of part, cm

For the example part, $N_b = 3$, $L = 15$, and $W = 7.5$ and Eq. (9.27) predicts a press brake cycle time per part of

$$t = 8 + 0.05 \times 5 \times 22.5 = 13.6 \text{ s}$$

The cost for press brake operations per part is thus

$$C_2 = 13.6 \times (28 \times 100)/3600 = 10.6 \text{ cents}$$

Finally, the estimated processing part cost is given by

$$C_p = 13.2 - 0.3 + 29.6 + 10.6 = 53.1 \text{ cents}$$

9.6 DESIGN RULES

In the design of sheet metal stampings the first consideration is the shape of the external perimeter. As discussed in Sec. 9.2.1, for parts that are to be manufactured with dedicated dies, it is advantageous to design the outer profile with parallel straight edges defining the part width. To allow for satisfactory shearing in cut-off or part-off operations, the end profiles should meet the straight edges at angles no less than 15 degrees. Whether or not this is possible, the profile shape should not contain narrow projections or notches that will require narrow weak sections in either punches or die plates; see dimensions marked "a" in Fig. 9.27.

Similar considerations for the avoidance of weak tool sections apply to internal punched holes. That is, small holes or narrow cut-outs that will require fragile punches should be avoided. In addition, internal punched holes should be separated from each other, and from the outside edge, with sufficient clearance to avoid distortion of narrow sections of the workpiece material during punching. The accepted rule of thumb is that both feature dimensions and feature spacings should be at least twice the material thickness. With reference to the part shown in Fig. 9.27, satisfactory blanking and punching will require that dimensions labeled "a" through "d" should all be greater than or equal to twice the gage thickness. Note that all profile radii, such as dimension "e," are subjected to the same rule of thumb. In this case the concern is the associated corner radii in the die plate. Radii

FIG. 9.27 Critical dimensions in the design of a sheet metal blank.

equal to at least twice the gage thickness will minimize the corner stress concentrations in the die plate, which may lead to crack formation and failure. Finally, it is good practice to incorporate relief cut-outs dimensioned as "d," at the ends of proposed bend lines that terminate at internal corners in the outer profile. These circular relief cut-outs will be part of the die profile for blanking or will be punched before the adjacent outer profile in turret press working. However, if for any reason holes that intersect the outer profile must be punched later, then the diameter should be at least three times the gage thickness to accommodate the offset loading to which the punch will be subjected.

When formed features are being considered, the principal design constraint is the maximum tensile strain the material can withstand; this is usually called the material ductility. Typical ductility values are given in Table 9.2. Thus, if a lanced and formed bridge as shown in Fig. 9.28 is to be incorporated into a component made from low-carbon, commercial-quality steel, the ratio of L to H can be calculated as follows. Assume that the transition or ramp from the surface to the top of the bridge is 45 degrees. The length along the bridge from end to end is approximately

$$\text{Bridge length} = L - 2H/\tan(45) + 2H/\sin(45)$$
$$= L + 0.82H \tag{9.28}$$

Assuming uniform stretching of the bridge, the tensile strain along the bridge is thus

$$e = 0.82H/L \tag{9.29}$$

Thus if the maximum permissible strain in tension is 0.22 (as given in Table 9.2), then from Eq. (9.29) successful forming will be assured if

$$L > 3.7H \tag{9.30}$$

FIG. 9.28 Lanced and formed bridge.

This corresponds approximately to a rule of thumb quoted in the literature that the length of bridges should be greater than 4 times their height. However, such rules are frequently based on experience with press working of annealed low-carbon steel. For different materials or varying geometries, such as changing the ramp angles in the preceding example, the tensile strains must be estimated and compared to the permissible maximum value.

A common example of a lanced and formed feature in sheet metal parts is the louver. Louvers are often formed as groups of parallel slots in the sides of sheet metal enclosures for air circulation and cooling purposes. Figure 9.29 shows a section through a louver. The length of the front edge of the louver must be greater than a certain multiple of the louver opening height H, determined by the material ductility and the end ramp angles exactly as in the bridge calculation. However, stretching also occurs at right angles to the louver edge where the material is stretched upward into a circular arc as shown. This will not cause material failure, since the front edge of the louver will be pulled backward as the tensile stress develops in the surface. The consideration here, and the choice of radius R in Fig. 9.29, is more one of appearance and the amount of space taken up by a single louver.

Another type of feature, which involves stretching along a sheared edge, is the hole flange. Figure 9.30 shows a sectional view of this feature type. Hole flanging is often carried out to provide increased local thickness for tapping of screw threads or for assembly with self-tapping screws. The hole flange is formed by pressing a taper-nosed punch of diameter D into a smaller punched hole of diameter d. The tensile strain around the top edge of the formed flange is thus

$$e = (D - d)/D \tag{9.31}$$

FIG. 9.29 Lanced and formed louver.

FIG. 9.30   Formed hole flange.

and this value must be less than the permissible material ductility. The limit of the ratio D/d, due to limited ductility, limits the amount of material displaced and in turn the height of hole flanges that can be produced. Typical values of flange height in sheet steel components, for example, range between 2 and 3 times the material gage thickness.

In the design of beads or ribs used to stiffen open surfaces of sheet metal parts, the cross-sectional geometry as shown in Fig. 9.31 is important. Ribs may be circular in section as shown, or are sometimes V-shaped. In any case, for a required height, H, the width and shape of the rib must be chosen so that the required amount of stretching across the rib does not exceed the material ductility. The radius at the base of the rib, in Fig. 9.31, must also be greater than a certain

FIG. 9.31   Cross section of rib.

value to prevent overstraining the material on the underside of the part. This may result from the bending effect along the sides of the rib and will be considered next.

As discussed in Sec. 9.3, the maximum tensile strain in bending is in the outer fibers of the sheet on the outside of the bend and is governed by the ratio of inside bend radius, r, to sheet gage thickness, h. For a bend through any angle θ, the length of the outer surface is

$$L_s = (r + h)\theta \tag{9.32}$$

and the length of the surface in the center of the sheet, on which lies the neutral axis of bending, is

$$L_o = (r + h/2)\theta \tag{9.33}$$

Hence the strain on the outer surface is

$$e = (L_s - L_o)/L_o = 1/(1 + 2r/h) \tag{9.34}$$

Radius r is defined precisely by the profile radius of the bending tool: either the convex radius of the die block for a wiper die or the convex radius of the punch in a v-die.

In any case the minimum acceptable radius value can be obtained from Eq. (9.34) and the ductility of the material to be bent. For example, for low-carbon, commercial-quality steel with ductility 0.22, Eq. (9.34) gives

$$e = 0.22 = 1/(1 + 2r/h)$$

or

$$r = 1.77h$$

A rule of thumb often quoted in the literature is that the inside bend radius should be greater than or equal to twice the sheet thickness. This is, in fact, the limiting value for a material with 20% ductility.

An additional consideration with respect to bending is the placing of other features next to bend lines. For the part shown in Fig. 9.32 the slots would almost certainly have to be punched after the bending operation. This is because the small separation, l, of the edges of the slots from the bend line would result in distortion of the slots during bending if they were punched first. This would give rise to the need for a more expensive piercing die, since a die block would have to be machined with a matching step to support the nonflat part. Even worse, if the part contains other holes or slots that are now on nonparallel surfaces to the one shown, then two separate dies and operations are needed for punching where one would otherwise have been sufficient. The rule of thumb in stamping is that the edge of circular holes should preferably be 2 times the sheet thickness from the

FIG. 9.32   Punched slots adjacent to a bend.

beginning of a bend. For slots parallel to a bend this clearance should increase to 4 times sheet thickness.

The manufacture of small flat sheet metal parts can be performed with a high degree of precision. Blanked parts or punched holes with maximum dimensions up to 10 cm can be held to tolerances of approximately ±0.05 mm. However, as part size increases, precision is more difficult to control, and for a part with dimensions as large as 50 cm permissible tolerances are in the range of ±0.5 mm. The requirement for tolerances much tighter than these guideline values may call for features to be machined at greatly increased cost. For formed parts, or formed features, variation tends to be larger and minimum tolerances attainable are in the range of ±0.25 mm for small parts. This includes bending when dedicated bending dies are used. Thus a tight tolerance between punched holes, which are on parallel surfaces separated by bends, would require the holes to be punched after bending at greater expense. If the holes are on nonparallel surfaces, then machining may be necessary to obtain the required accuracy. Finally, in the design of turret press parts to be bent on press brakes, it should be noted that the inaccuracies of this bending process are substantially worse than with dedicated dies. Attainable tolerances between bent surfaces and other surfaces, or features on other surfaces, range from ±0.75 mm for small parts up to ±1.5 mm for large ones.

Finally, an important consideration in the design of any sheet metal part should be the minimization of manufactured scrap. This is accomplished by designing part profiles so that they can be nested together as closely as possible on the strip or sheet. Also, if individual dies are to be used, then the part should be designed if possible for cut-off or part-off operations. Figure 9.33 illustrates the type of design changes that should always be considered. The cut-off design lacks the elegance of the rounded end profiles. Nevertheless, the acute sharp corner will be

FIG. 9.33 Design changes of a three-hole bracket for minimization of manufactured scrap.

removed during deburring, and for many applications this type of design may be perfectly functional.

REFERENCES

1. Zenger, D., and Dewhurst, P., Early Assessment of Tooling Costs in the Design of Sheet Metal Parts, Report No. 29, Department of Industrial and Manufacturing Engineering, University of Rhode Island, Kingston, August 1988.
2. Zenger, D.C., Methodology for Early Material/Process Cost of Estimating, Ph.D. Thesis, University of Rhode Island, Kingston, 1989.
3. Nordquist, W.N., Die Designing and Estimating, 4th ed., Huebner Publishing, Cleveland, 1955.
4. Eary, D.F., and Reed, E.A., Techniques of Pressworking Sheet Metal, 2nd ed., Prentice-Hall, Englewood Cliffs, NJ, 1974.
5. Bralla J.G., Handbook of Product Design for Manufacturing, McGraw-Hill, New York, 1987.
6. Ostwald, P.F., AM Cost Estimator, McGraw-Hill, New York, 1986.
7. Wick, C., Benedict, J.T., and Veilleux, R.F., Tool and Manufacturing Engineers Handbook, Vol. 2: Forming, Society of Manufacturing Engineers, Dearborn, MI, 1984.
8. Donovan, J.R., Computer-Aided Design of Sheet Metal Parts, M.S. Thesis, University of Rhode Island, Kingston, 1992.

10

Design for Die Casting

10.1 INTRODUCTION

The die casting process, also called pressure die casting, is a molding process in which molten metal is injected under high pressure into cavities in reusable steel molds, called dies, and held under pressure during solidification. In principle, the process is identical to injection molding with a different class of materials. Die casting can, in fact, produce parts that have identical geometries to injection-molded ones. The reverse is also true, and much of the increase in the use of injection molding since the mid 1980s has been as a substitute for part types that were previously die cast. In many cases, this has been a wise substitution resulting in decreased parts costs. However, for structural parts, particularly those for which thick-wall injection moldings are required, die casting can often be the better selection. The analysis of die casting costs in this chapter closely parallels the early costing procedure for injection molding given in Chapter 8. This is intended to allow comparisons of the two processes to be made with a minimum of redundant effort.

10.2 DIE CASTING ALLOYS

The four major types of alloys that are die-cast are zinc, aluminum, magnesium, and copper-based alloys. The die casting process was developed in the 19th century for the manufacture of lead/tin alloy parts. However, lead and tin are now very rarely die-cast because of their poor mechanical properties. A tabulation of the specific gravity, mechanical properties, and cost of commonly used examples of the four principal alloy groups is given in Table 10.1.

TABLE 10.1   Commonly Used Die Casting Alloys

Alloy[a]	Specific gravity	Yield strength (MN/m^2)	Elastic modulus (GN/m^2)	Cost $(\$/kg)$
Zamak[1]	6.60	220	66	1.78
Zamak 5[1]	6.60	270	73	1.74
A13[2]	2.66	130	130	1.65
A360[2]	2.74	170	120	1.67
ZA8[3]	6.30	290	86	1.78
ZA27[3]	5.00	370	78	1.94
Silicon brass 879[4]	8.50	240	100	6.60
Manganese[4] bronze 865	8.30	190	100	6.60
AZ91B[5]	1.80	150	45	2.93

[a] Alloy types: (1) zinc, (2) aluminum, (3) zinc-aluminum, (4) copper, (5) magnesium.

The most common die casting alloys are the aluminum alloys. They have low density, good corrosion resistance, are relatively easy to cast, and have good mechanical properties and dimensional stability. Aluminum alloys have the disadvantage of requiring the use of cold-chamber machines, which usually have longer cycle times than hot-chamber machines owing to the need for a separate ladling operation. The distinction between hot- and cold-chamber machines will be discussed in some detail later in this chapter.

Zinc-based alloys are the easiest to cast. They also have high ductility and good impact strength, and therefore can be used for a wide range of products. Castings can be made with very thin walls, as well as with excellent surface smoothness, leading to ease of preparation for plating and painting. Zinc alloy castings, however, are very susceptible to corrosion and must usually be coated, adding significantly to the total cost of the component. Also, the high specific gravity of zinc alloys leads to a much higher cost per unit volume than for aluminum die casting alloys, as can be deduced from the data in Table 10.1.

Zinc-aluminum (ZA) alloys contain a higher aluminum content (8–27%) than the standard zinc alloys. Thin walls and long die lives can be obtained, similar to standard zinc alloys, but as with aluminum alloys, cold-chamber machines, which require pouring of the molten metal for each cycle, must usually be used. The single exception to this rule is ZA8 (8% Al), which has the lowest aluminum content of the zinc-aluminum family.

Magnesium alloys have very low density, a high strength-to-weight ratio, exceptional damping capacity, and excellent machinability properties.

Copper-based alloys, brass and bronze, provide the best mechanical properties of any of the die casting alloys, but are much more expensive. Brasses have high strength and toughness, good wear resistance, and excellent corrosion resistance.

TABLE 10.2 Typical Die Life
Values per Cavity

Alloy	Die life
Zinc	500,000
Zinc-aluminum	500,000
Aluminum	100,000
Magnesium	180,000
Copper	15,000

One major disadvantage of copper-based alloy casting is the short die life caused by thermal fatigue of the dies at the extremely high casting temperatures.

Die life is influenced most strongly by the casting temperature of the alloys, and for that reason is greatest for zinc and shortest for copper alloys. The typical number of castings per die cavity is given in Table 10.2. However, this is only an approximation since casting size, wall thickness, and geometrical complexity also influence the wear and eventual breakdown of the die surface.

10.3 THE DIE CASTING CYCLE

In the casting cycle, first the die is closed and locked. The molten metal, which is maintained by a furnace at a specified temperature, then enters the injection cylinder. Depending on the type of alloy, either a hot-chamber or cold-chamber metal-pumping system is used. These will be described later. During the injection stage of the die casting process, pressure is applied to the molten metal, which is then driven quickly through the feed system of the die while air escapes from the die through vents. The volume of metal must be large enough to overflow the die cavities and fill overflow wells. These overflow wells are designed to receive the lead portion of the molten metal, which tends to oxidize from contact with air in the cavity and also cools too rapidly from initial die contact to produce sound castings. Once the cavities are filled, pressure on the metal is increased and held for a specified dwell time during which solidification takes place. The dies are then separated, and the part extracted, often by means of automatic machine operation. The open dies are then cleaned and lubricated as needed, and the casting cycle is repeated.

Following extraction from the die, parts are often quenched and then trimmed to remove the runners, which were necessary for metal flow during mold filling. Trimming is also necessary to remove the overflow wells and any parting-line flash that is produced. Subsequently, secondary machining and surface finishing operations may be performed.

10.4   DIE CASTING MACHINES

Die casting machines consist of several elements: the die mounting and clamping system, the die, the metal pumping and injection system, the metal melting and storing system, and any auxiliary equipment for mechanization of such operations as part extraction and die lubrication.

10.4.1   Die Mounting and Clamping Systems

The die casting machine must be able to open and close the die and lock it closed with enough force to overcome the pressure of the molten metal in the cavity. The mechanical or hydraulic systems needed to do this are identical to those found on injection molding machines and described in Chapter 8. This fact should not be surprising since injection molding machines were developed from die casting technology.

10.4.2   Metal Pumping and Injection Systems

The two basic types of injection systems are hot-chamber and cold-chamber. Hot-chamber systems, in which the pump is placed in the container of molten metal, are used with alloys having low melting temperatures, such as zinc. Cold-chamber machines must be used for high-melting temperature alloys such as aluminum, copper-based alloys, and the ZA zinc alloys, which contain large amounts of aluminum. The high-melting-temperature alloys used in a hot-chamber machine would erode the ferrous injection pump components, thereby degrading the pump and contaminating the alloy. Magnesium alloys, although they are cast at high temperatures, can be cast in hot-chamber machines as well as in cold-chamber machines because they are inert to the ferrous machine components [1].

10.4.3   Hot-Chamber Machines

A typical hot-chamber injection or shot system, as shown in Fig. 10.1, consists of a cylinder, a plunger, a gooseneck, and a nozzle. The injection cycle begins with the plunger in the up position. The molten metal flows from the metal-holding pot in the furnace, through the intake ports, and into the pressure cylinder. Then, with the dies closed and locked, hydraulic pressure moves the plunger down into the pressure cylinder and seals off the intake ports. The molten metal is forced through the gooseneck channel and the nozzle and into the sprue, feed system, and die cavities. The sprue is a conically expanding flow channel that passes through the cover die half from the nozzle into the feed system. The conical shape provides a smooth transition from the injection point to the feed channels and allows easy extraction from the die after solidification. After a preset dwell time

FIG. 10.1 Hot-chamber injection system.

for metal solidification, the hydraulic system is reversed and the plunger is pulled up. The cycle then repeats. Cycle times range from several seconds for castings weighing a few grams to 30 s or more for large, thick-walled castings weighing over a kilogram [2]. Specifications for a range of hot-chamber machines are given in Table 10.3.

TABLE 10.3 Hot-Chamber Die Casting Machines

Clamping force (kN)	Shot size (cm³)	Operating rate ($/h)	Dry cycle time (s)	Max. die opening (cm)	Platen size (cm)
900	750	58	2.3	20.0	48 × 56
1150	900	60	2.5	23.0	56 × 64
1650	1050	62	2.9	25.0	66 × 70
2200	1300	64	3.3	31.0	70 × 78
4000	1600	70	4.6	38.0	78 × 98
5500	3600	73	5.6	45.7	100 × 120
6000	4000	76	6.2	48.0	120 × 150
8000	4000	86	7.5	53.0	120 × 150

FIG. 10.2 Cold-chamber die casting machine elements.

10.4.4 Cold-Chamber Machines

A typical cold-chamber machine, as shown in Fig. 10.2, consists of a horizontal shot chamber with a pouring hole on the top, a water-cooled plunger, and a pressurized injection cylinder. The sequence of operations is as follows: when the die is closed and locked and the cylinder plunger is retracted, the molten metal is ladled into the shot chamber through the pouring hole. In order to tightly pack the metal in the cavity, the volume of metal poured into the chamber is greater than the combined volume of the cavity, the feed system, and the overflow wells. The injection cylinder is then energized, moving the plunger through the chamber, thereby forcing the molten metal into the die cavity. After the metal has solidified, the die opens and the plunger moves back to its original position. As the die opens, the excess metal at the end of the injection cylinder, called the biscuit, is forced out of the cylinder because it is attached to the casting. Material in the biscuit is required during the die casting cycle in order to maintain liquid metal pressure on the casting while it solidifies and shrinks.

Specifications for a number of cold-chamber machines are given in Table 10.4.

TABLE 10.4 Cold-Chamber Die Casting Machines

Clamping force (kN)	Shot size (cm³)	Operating rate ($/h)	Dry cycle time (s)	Max. die opening (cm)	Platen size (cm)
900	305	66	2.2	24.4	48 × 64
1,800	672	73	2.8	36.0	86 × 90
3,500	1,176	81	3.9	38.0	100 × 108
6,000	1,932	94	5.8	46.0	100 × 120
10,000	5,397	116	8.6	76.0	160 × 160
15,000	11,256	132	10.2	81.0	210 × 240
25,000	11,634	196	19.9	109.0	240 × 240
30,000	13,110	218	23.3	119.0	240 × 240

10.5 DIE CASTING DIES

Die casting dies consist of two major sections—the ejector die half and the cover die half—which meet at the parting line; see Fig. 10.3. The cavities and cores are usually machined into inserts that are fitted into each of these halves. The cover die half is secured to the stationary platen, while the ejector die half is fastened to the movable platen. The cavity and matching core must be designed so that the die halves can be pulled away from the solidified casting.

The construction of die casting dies is almost identical to that of molds for injection molding. In injection molding terminology, the ejector die half comprises the core plate and ejector housing, and the cover die half comprises the cavity plate and backing support plate.

Side-pull mechanisms for casting parts with external cross-features can be found in exactly the same form in die casting dies as in plastic injection molds, described in Chapter 8. However, molten die casting alloys are much less viscous than the polymer melt in injection molding and have a great tendency to flow between the contacting surfaces of the die. This phenomenon, referred to as "flashing," tends to jam mold mechanisms, which must, for this reason, be robust. The combination of flashing with the high core retraction forces due to part shrinkage makes it extremely difficult to produce satisfactory internal core mechanisms. Thus, internal screw threads or other internal undercuts cannot usually be cast and must be produced by expensive additional machining

FIG. 10.3 Die for cold-chamber die casting machine.

operations. Ejection systems found in die casting dies are identical to the ones found in injection molds.

"Flashing" always occurs between the cover die and ejector die halves, leading to a thin, irregular band of metal around the parting line. Occasionally, this parting line flash may escape between the die faces. For this reason, full safety doors must always be fitted to manual die casting machines to contain any such escaping flash material.

One main difference in the die casting process is that overflow wells are usually designed around the perimeter of die casting cavities. As mentioned earlier, they reduce the amount of oxides in the casting, by allowing the first part of the shot, which displaces the air through the escape vents, to pass completely through the cavity. The remaining portion of the shot and the die are then at a higher temperature, thereby reducing the chance of the metal freezing prematurely. Such premature freezing leads to the formation of surface defects called cold shuts, in which streams of metal do not weld together properly because they have partially solidified by the time they meet. Overflow wells are also needed to maintain a more uniform die temperature on small castings, by adding substantially to the mass of molten metal.

10.5.1 Trimming Dies

After extraction from the die casting machines, the sprue or biscuit, runners, gates, overflow wells, and parting-line flash must be removed from the casting. This is done either manually or, if production quantities are larger, with trimming presses. The dies used for trimming operations are similar to blanking and piercing dies used for sheet metal pressworking. They are mounted on mechanical or hydraulic presses, and because the required forces are low, the bed area to tonnage rating ratio is relatively large. The thickness of the metal to be trimmed is usually in the range of 0.75 to 1.5 mm.

It is desirable, when designing a casting, to locate the main gates from the feed channels as well as the gates to the overflow wells around the parting line of the cavity and to design a parting line that is not stepped. This simplifies both the casting die and the trimming die.

10.6 FINISHING

Following trimming, castings are often polished and/or coated to provide corrosion resistance and wear resistance, and to improve aesthetic appearance. Polishing is often the only surface treatment for aluminum castings, or it may be the preparation stage for high-gloss painting or plating of zinc castings. Before coating, parts are put through a series of cleaning operations to remove any contamination that could prevent the adhesion of these applied coatings. The cleaning operations usually performed are degreasing, alkaline cleaning, and acid dipping.

Following cleaning, several coatings are available depending on the type of alloy cast. These coatings may be separated into three groups: electroplating, anodizing, and painting. Electroplating is used mainly for zinc alloy castings because aluminum and magnesium alloys oxidize quickly, which prevents the electroplate layers from adhering properly. Brass castings, although they may be electroplated after removal of oxides, are often used unfinished.

The most common type of electroplating is a decorative chrome finish on zinc die castings, which consists of several layers of applied metal. First, a very thin layer of copper (0.008 mm) is applied to aid in the adhesion of the subsequent layers. A second layer of copper is then sometimes added to improve the final surface finish. Two layers of nickel, 0.025 mm thick, are then applied. These layers aid in corrosion resistance by diverting the corrosion to the outer layer of nickel because of the difference in electrical potential between the two layers. The final layer is a thin coat of chromium (0.003 mm), which also helps to prevent corrosion by serving as a barrier.

Anodizing, used on aluminum, zinc, and magnesium alloy castings, provides corrosion resistance and wear resistance, and it may also serve as a base for painting. Anodizing of aluminum is the formation of a layer, 0.005 to 0.030 mm thick, of stable oxides on the surface of the base metal by making the casting the anode in an electrolytic cell, with separate cathodes of lead, aluminum, or stainless steel. This surface is usually a dull gray and therefore not ordinarily applied for decorative purposes.

The most common form of applied coating for aesthetic appearance and protection is painting. Paint may be applied to bare metal, primed metal, or surfaces that have additional protective coatings. Paint is often applied by electrostatic painting, which uses powdered paint sprayed through a nozzle having an electric potential opposite to that of the castings.

The process of impregnation, while not a surface finishing process, is sometimes performed after the casting and polishing processes have been completed. Impregnation is used on castings where porosity may produce structural problems, as when castings are to be used to hold fluids or to contain fluid pressure. The process of impregnation consists of placing the castings in a vacuum chamber, evacuating the pores, and immersing the castings in a sealant. The sealant is then forced into the pores once the casting is in atmospheric pressure.

The cost of surface treatments is often represented as a simple cost per square area of casting surface. Typical costs for the more common surface treatments and for sealant impregnation are given in Table 10.5.

TABLE 10.5 Costs of Common Finishing Processes

Finishing process	Cost per 50 cm^2 of surface area (cents)
Sealant impregnation	1.8
CU/N$_{i/Cr}$ plate	4.5
Polish	1.3
Anodize	1.6
Prime cost	2.1
Finish paint coat	2.4

10.7 AUXILIARY EQUIPMENT FOR AUTOMATION

Several operations in die casting may be automated in order to reduce cycle times and to produce more consistent quality. These operations, which may utilize mechanized equipment or simple programmable manipulations, are the removal of the casting from the die, transfer of castings to subsequent operations such as trimming, application of die lubricants, and transfer of molten metal to the shot chamber of cold-chamber machines.

Automatic extraction involves the use of a mechanical manipulator that simulates the actions of a human operator in removing the part from the die. The fingers of the manipulator are open upon entry into the die opening; they then close on the casting, which is usually suspended on the ends of the ejector pins, pull it out of the die opening, and drop in onto a conveyor belt or into a trim die. These devices range from simple two-degrees-of-freedom mechanisms to programmable robots that are capable of multiple-axis motions. Small nonpreci-

FIG. 10.4 Simple mechanical ladle for cold-chamber machine.

sion die castings may simply be dropped from the die in the same manner as small injection moldings.

Die lubricants may be applied automatically by stationary spray heads located near the die, or by reciprocating spray heads located near the die, or by reciprocating spray heads that enter the die after the casting has been extracted. These are sometimes mounted on the back of the extractor arm and are sprayed as the arm is retracting from the die.

Automatic metal transfer systems are used to transfer molten metal from the holding furnace to the shot chamber of cold-chamber die casting machines. These systems may be simple mechanical ladles as shown in Fig. 10.4 or a variety of more complex systems, some of which fill at the bottom in order to reduce the transfer of oxides.

10.8 DETERMINATION OF THE OPTIMUM NUMBER OF CAVITIES

Die casting processing cost is the product of the die casting cycle time and the operating rate of the die casting machine and its operator. In order to determine the operating rate, the machine size must be known. This, in turn, can only be determined if the number of die cavities is known. Since the procedures being developed in this work are to be used in early design, the number of cavities that may be used in later manufacturing cannot be determined with certainty. It can only be assumed that the part will be manufactured in an efficient manner. Thus, a value for what is likely to be an optimum number of cavities must be used. The determination of this value is the subject of this section.

The optimum number of die cavities to be used in the die casting die, equal to the number of apertures in the trim die, can be determined for a particular die casting task by first calculating the most economical number of cavities, and then analyzing the physical constraints of the equipment to ensure that the economical number of cavities is practical. The most economical number of cavities can be determined by the following analysis, which is almost identical to the one for injection molding.

$$C_t = C_{dc} + C_{tr} + C_{dn} + C_{tn} + C_{ta} \; \$ \tag{10.1}$$

where
  C_t = total cost for all the components to be manufactured, N_t, $
  C_{dc} = die casting processing cost, $
  C_{tr} = trimming processing cost, $
  C_{dn} = multicavity die casting die cost, $
  C_{tn} = multiaperture trim die cost, $
  C_{ta} = total alloy cost, $

The die casting processing cost, C_{dc}, is the cost of operating the appropriate size die casting machine, and it can be represented by the following equation:

$$C_{dc} = (N_t/n)C_{rd}t_d \ \$ \tag{10.2}$$

where
N_t = total number of components to be cast
n = number of cavities
C_{rd} = die casting machine and operator rate, \$/h
t_d = die casting machine cycle time, h

The hourly operating rate of a die casting machine, including the operator rate, can be approximated by the following linear relationship:

$$C_{rd} = k_1 + m_1 F \ \$/h \tag{10.3}$$

where
F = die casting machine clamp force, kN
k_1, m_1 = machine rate coefficients

This relationship, which is identical in form to the one for injection molding, was arrived at through examination of the machine hourly rate data. Linear regression analysis of the data in Tables 10.2 and 10.3 gives the following values:

Hot chamber : $k_1 = 55.4$, $m_1 = 0.0036$
Cold chamber : $k_1 = 62.0$, $m_1 = 0.0052$

The form of the relationship is supported by the nature of the variation of die casting machine capital costs with rated clamp force values as shown in Fig. 10.5. This machine cost data, obtained from five machine makers, shows a linear relationship between clamp force and machine costs for hot- or cold-chamber machines up to 15 MN. However, it should be noted that very large cold-chamber machines in the range of 15 to 30 MN are associated with greatly increased cost. For these machines, the smooth relationship results obtained in this section should be applied with caution.

The cost of trimming, C_{tr}, can be represented by the following equation:

$$C_{tr} = (N_t/n)C_{rt}t_p \ \$ \tag{10.4}$$

where
C_{rt} = trim press and operator rate, \$/h
t_p = trimming cycle time, h

In the present analysis, the hourly rate for trimming is approximated by a constant value for trim presses of all sizes. This is done because the cost of trim presses is relatively low due to the small forces required, and therefore only small-capacity

FIG. 10.5 Capital costs of die casting machines.

presses are necessary in the trimming of die casting alloys. For this reason, C_{rt} is dominated by the hourly rate of the trim press operator rather than by the cost of the press itself.

The trimming cycle time may be represented by the following equation:

$$t_p = t_{p0} + n\Delta t_p \ \mathrm{h} \tag{10.5}$$

where
t_{p0} = trimming cycle time for a single-aperture trimming operation for a single part, h
Δ_{tp} = additional trimming cycle time for each aperture in a multiaperture trimming die, mainly due to increased loading time of the multicavity casting into the press

The cost of a multicavity die casting die, C_{dn} relative to the cost of a single-cavity die, C_{d1}, follows a relationship similar to that of injection molding dies. Based on data from Reinbacker [3], this relationship can be represented as the following power law:

$$C_{dn} = C_{d1}n^m \ \$ \tag{10.6}$$

where
C_{d1} = cost of a single-cavity die casting die,
m = multicavity die cost exponent
n = number of cavities

The decreased cost per cavity resulting from the manufacture of multiple identical cavities follows the same trend as for the manufacture of injection molds. Thus, as discussed in Chapter 8, a reasonable value for m is 0.7.

The cost of a multiaperture trim die, C_{tn} relative to the cost of a single-aperture trim die, C_{t1}, will be assumed to follow a similar relationship, namely:

$$C_{tn} = C_{t1}n^m \text{ \$} \tag{10.7}$$

where
C_{t1} = cost of a single-aperture trim die, \$
m = multiaperture trim die cost exponent

It is assumed that the cost exponent for multiaperture trim tools is the same as that for multicavity die casting dies.

The equation for the total alloy cost, C_{ta}, is

$$C_{ta} = N_t C_a \text{ \$} \tag{10.8}$$

where

C_a = alloy cost for each casting, \$

Compiling the previous equations gives

$$\begin{aligned} C_t = {} & (N_t/n)(k_1 + m_1 F)t_d \\ & + (N_t/n)t_{p0} + n\Delta t_p)C_{rt} \\ & + (C_d + C_t)n^m + N_t C_a \end{aligned} \tag{10.9}$$

If full die casting machine clamp force utilization is assumed, then

$$F = nf \text{ kN}$$

or

$$n = F/f \tag{10.10}$$

where
F = die casting machine clamp force, kN
f = separating force on one cavity, kN

Substituting Eq. (10.10) into Eq. (10.9) gives

$$\begin{aligned} C_t = {} & N_t(k_1 f/F + m_1 f)t_d \\ & + N_t C_{rt}t_{p0} f/F \\ & + N_t C_{rt}\Delta t_p \\ & + (C_d + C_t)(F/f)^m + N_t C_a \end{aligned} \tag{10.11}$$

In order to find the number of cavities that gives the lowest cost for any given die casting machine size, the derivative of Eq. (10.11) with respect to the clamp force, F, is equated to zero. This gives

$$dC_t/dF = -N_t f(k_1 t_d + C_{rt} t_{p0})/F^2$$
$$+ mF^{(m-1)}(C_d + C_t)/f^m = 0 \tag{10.12}$$

Finally, rearranging Eq. (10.12) gives

$$n^{(m+1)} = N_t(k_1 t_d + C_{rt} t_{p0})/[m(C_d + C_t)] \tag{10.13}$$

as the equation for the most economical number of die cavities for any given die casting task.

Example

An aluminum die cast component has an estimated die casting cycle time of 20 s for a single-cavity die and an estimated 7 s trimming cycle time for a single-aperture trim die. The cost of a single-cavity die for this part has been estimated to be $10,000 and the trim die has been estimated to be $2000. Determine the optimum number of cavities for production volumes of 100,000, 250,000 and 500,000 assuming $k_1 = 62\,\$/h$, $C_{rt} = 35\,\$/h$, and $m = 0.7$.

Using Eq. (10.13), when $N_t = 100,000$ components, gives

$$n_c^{(1.7)} = 100,000(62 \times 20 + 35 \times 7)/(3600 \times 0.7 \times 12,000)$$
$$n_c = 2.6$$

Similarly, for 250,000 components, $n_c = 4.4$, and for 500,000 components, $n_c = 6.6$. These numbers indicate that for production volumes of 100,000, 250,000 and 500,000, dies with cavity numbers of 3, 4, and 7, respectively, would lead to most efficient manufacture. In practice, it is unusual to have an odd number of cavities, so for these three cases, the likely number of cavities would be 2, 4, and 6, respectively.

Once the most economical number of cavities has been determined for a particular die casting task, the physical constraints of the equipment must be examined. The first consideration is the number and position of sliding cores in the die.

As with injection molds, sliding cores must be located in the die such that they may be retracted, and such that there is space for their driving mechanisms. Also, as with injection molds, cavities that require sliding cores on four sides are limited to single-cavity dies, while cavities with cores on three sides are limited to two-cavity dies. Cavities containing core slides on two sides are restricted to either two- or four-cavity dies, depending on the angle between slides; see Fig. 10.6.

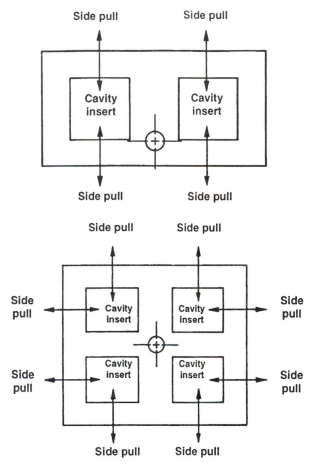

FIG. 10.6   Restricted number of cavities with two side-pulls.

The remaining constraints are on the die casting machine and trim press to be used for the task. The die casting machine must be large enough to provide the required clamp force, as well as to provide a platen area, shot volume, and die opening large enough for the specified casting arrangement. Similarly, the bed area of the trim press must be large enough to accommodate the area of the shot. If the available machines and presses cannot meet all of these constraints, then the number of cavities must be lowered until the corresponding machine size falls within the range of available machines. The process of determining the appropriate machine size will be covered in detail in the next section.

10.9 DETERMINATION OF APPROPRIATE MACHINE SIZE

Several factors must be considered when choosing the appropriate machine size with which to cast a particular die cast component. These factors include the machine performance, as well as the dimensional constraints imposed by the machine. The most important machine performance capability to be considered is the machine clamping force. Dimensional factors that must be considered include the available shot volume capacity, the die opening stroke length (also called clamp stroke), and the platen area.

10.9.1 Required Machine Clamp Force

Die casting machines are primarily specified on the basis of machine clamping force. In order to prevent the die halves from separating, the clamp force, F, exerted by the machine on the die must be greater than the separating force, f, of the molten metal on the die during injection:

$$F > f \qquad (10.14)$$

For a given die casting task, the force exerted by the molten metal may be represented as follows:

$$f = p_m A_{pt}/10 \qquad (10.15)$$

where
f = force of molten metal on die, kN
p_m = molten metal pressure in the die, MPa
A_{pt} = total projected area of molten metal within the die, cm^2

The total projected area, A_{pt}, is the area of the cavities, feed system, and overflow wells, taken normal to the direction of die opening, and can be represented by the following equation:

$$A_{pt} = A_{pc} + A_{po} + A_{pf} \qquad (10.16)$$

where
A_{pc} = projected area of cavities
A_{po} = projected area of overflow wells
A_{pf} = projected area of feed system

Figure 10.7 shows the relative size of a typical casting before and after trimming. The proportions of A_{pf} and A_{po} to the cavity area, A_{pc}, vary with the size of the casting, the wall thickness, and the number of cavities. However, analysis of a wide variety of different castings has failed to establish any logical relationships between the geometry of the cavity and the area of the feed and overflow system. One reason for this situation may be, as stated by Herman [4], that the relationships between casting geometry and overflow size are not well understood. The size of overflow wells is thus a matter of individual diemaker judgment coupled with trial and error modifications during die tryout. The range of variation of $(A_{po} + A_{pf})$, from examination of actual castings, appears to be from 50% of A_{pc} to 100% of A_{pc}. The mean value of total casting projected area can thus be represented approximately by

$$A_{pt} \approx 1.75 A_{pc} \qquad (10.17)$$

Equation (10.17) is intended to be used at the sketch stage of design in order to obtain a first estimate of required clamp force from Eq. (10.15). The pressure at which the molten metal is injected into the die depends primarily on the die casting alloy being used. Typical pressures for the main classes of alloys are given in Table 10.6. It should be noted that the metal pressure is often increased from the instant that the die is filled in order to reduce metal porosity and surface defects which can result from metal shrinkage. However, this intensification of pressure occurs when a skin of solidified metal has already formed from contact with the die surface. This skin acts like a vessel which helps to contain the pressure increase, and for this reason machine builders suggest that the unin-tensified pressure should be used for clamp force calculations. Thus, the values for p_m in Eq. (10.15) may be taken directly from Table 10.6.

10.9.2 Shot Volume

The shot volume required for a particular casting cycle may be represented by

$$V_s = V_c + V_o + V_f \text{ cm}^3 \qquad (10.18)$$

where
V_s = total shot volume
V_c = volume of cavities

FIG. 10.7 Hot-chamber die casting before and after trimming.

TABLE 10.6 Typical Cavity Pressures in Die Casting

Alloys	Cavity pressure (MN/m^2)
Zinc	21
Aluminum	48
ZA	35
Copper	40
Magnesium	48

V_o = volume of overflow wells
V_f = volume of feed system

As with the projected area contributions, the volumes of the overflow wells and the feed system represent a significant portion of the shot volume. The proportion of material in the overflow and runner system is usually considerably greater for relatively thin wall castings. Blum [5] analyzed a number of different castings and has suggested that the volumes of the overflow and feed systems can be represented by the approximate relationships

$$V_o = 0.8V_c/h^{1.25} \text{ cm}^3 \tag{10.19}$$

$$V_f = V_c/h \text{ cm}^3 \tag{10.20}$$

where h is the average wall thickness of the part measured in millimeters. The trend of these relationships is supported in part by Herman [4], who recommends overflow volumes for die design, the average values of which fit almost precisely to the curve

$$V_o = V_c/h^{1.5} \text{ cm}^3 \tag{10.21}$$

For the present early-design assessment purposes, these tentative relationships will be further reduced to the simple expression for shot size:

$$V_s = V_c(1 + 2/h) \tag{10.22}$$

where again h = average wall thickness, mm. The difference between Eq. (10.22) and Eqs. (10.18), (10.19) and (10.20) over the range $h = 1$ mm to 10 mm is only 4 to 7%.

It should be noted that the feed system and overflow wells, which are trimmed from the casting, cannot be reused immediately, as with injection moldings. The scrap material from die casting must be returned to the material supplier, where oxides are removed and the chemical composition is recertified. This "conditioning" process typically costs 15 to 20% of the material purchase cost. Material cost per part should, therefore, be estimated from the weight of the part, plus say

20% of the weight of overflow wells and feed system, using the costs per kilogram given in Table 10.1.

10.9.3 Dimensional Machine Constraints

For a part to be die cast on a particular machine that has sufficient clamp force and shot volume, two further conditions must be satisfied. First, the maximum die opening or clamp stroke must be wide enough so that the part can be extracted without interference. Thus, the required clamp stroke, L_s, for a hollow part of depth D, with a clearance of 12 cm for operator or mechanical extractor, will be

$$L_s = 2D + 12 \text{ cm} \tag{10.23}$$

The factor 2 is required to achieve separation from both the cavity and core.

The second requirement is that the area between the corner tie bars on the clamp unit, sometimes referred to as the platen area, must be sufficient to accommodate the required die. The size of the die can be calculated in the same way as the mold base for injection molding. Thus, the clearance between adjacent cavities or between cavities and plate edge should be a minimum of 7.5 cm with an increase of 0.5 cm for each 100 cm² of cavity area. Reasonable estimates of the required plate size are given by allowing a 20% increase of part width for overflow wells and 12.5 cm of added plate width for the sprue or biscuit.

Example

A 20 cm long by 15 cm wide by 10 cm deep box-shaped die casting is to be made from A360 aluminum alloy. The mean wall thickness of the part is 5 mm and the part volume is 500 cm³. Determine the appropriate machine size if a two-cavity die is to be used.

Projected area of cavities is given by

$$A_{pc} = 2 \times 20 \times 15 = 600 \text{ cm}^2$$

and so estimated shot area is

$$A_{pt} = 1.75 \times 600 = 1050 \text{ cm}^2$$

Thus, the die separating force from Eq. (10.15) and Table 10.6 is

$$F_m = 48 \times 1050/10$$
$$= 5040 \text{ kN}$$

The shot size is given by Eq. (10.22) to be

$$V_s = 2 \times 500(1 + 2/5) = 1400 \text{ cm}^3$$

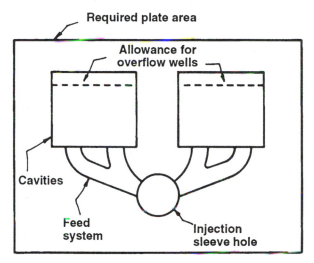

FIG. 10.8 Layout of two-cavity die.

The clamp stroke, L_s, must be at least

$$L_s = 2 \times 10 + 12 = 32 \text{ cm}$$

The clearance between the cavities and with the plate edge is likely to be

$$\text{Clearance} = 7.5 + 0.5 \times (20 \times 15)/100$$
$$= 9.0 \text{ cm}$$

Thus, the two cavities may be arranged end to end with 9.0 cm spacing between them and around the edges and with a 20% width increase to allow for the overflow wells. If an additional increase of 12.5 cm is then applied to the plate width for the biscuit, a final plate size of 67×42.5 cm is obtained. The layout within this plate is shown in Fig. 10.8. The appropriate machine from Table 10.4 would thus be the one with 6000 kN clamp force, which can accommodate plate sizes up to 100 cm × 120 cm.

10.10 DIE CASTING CYCLE TIME ESTIMATION

A die casting machine cycle consists of the following elements:

1. Ladling the molten shot into the shot sleeve (for cold-chamber machines only)
2. Injection of molten metal into the feed system, cavities, and overflow wells
3. Cooling of the metal in the feed system, cavities, and overflow wells
4. Opening of the die and the safety door

5. Extraction of the die casting, which is usually held on the projecting ejector pins
6. Lubrication of the die surfaces
7. Closing of the die for the next cycle

10.10.1 Ladling of Molten Metal

The time for manual ladling of the molten metal shot into a cold-chamber machine has been studied by Ostwald [6], who presents time standards for different shot volumes. This can be represented almost precisely by the linear relationship

$$t_{lm} = 0.0048 V_s \tag{10.24}$$

where
t_{lm} = manual ladling time, s
V_s = total shot volume, cm^3

This time does not include the transfer of the ladle to the machine pouring hole, which occurs while the die and safety door on the machine are closing.

10.10.2 Metal Injection

Metal injection and the resultant filling of feed system, cavities, and overflow wells occurs extremely rapidly in die casting. This is essential to avoid premature solidification, which would prevent complete cavity filling or cause casting defects in which partially solidified streams of metal come together with incomplete bonding taking place. It is clear that the problem of premature solidification will be greater for thinner wall die castings since, during filling, a thinner stream of molten metal, with less heat content, will contact the cooled die walls. The Society of Die Casting Engineers [7] has recommended that fill time should be directly proportional to the average casting wall thickness governed by the following equation:

$$t_f = 0.035h(T_i + T_l + 61)/(T_i - T_m) \tag{10.25}$$

where
t_f = fill time for feed system, cavities and overflow wells, s
T_i = recommended melt injection temperature, °C
T_l = die casting alloy liquidus temperature, °C
T_m = die temperature prior to shot, °C
h = average wall thickness of diecasting, mm

TABLE 10.7 Typical Die Casting Temperatures (°C)

Alloy	Injection temp.	Liquidus temp.	Die temp.	Ejection temp.
Zinc	440	387	175	300
Aluminum	635	585	220	385
ZA	460	432	215	340
Copper	948	927	315	500
Magnesium	655	610	275	430

Typical values of T_i, T_l, and T_m for the different families of die casting alloys are given in Table 10.7. Substitution of these values into Eq. (10.25) yields fill times ranging from 0.005 h to 0.015 h. It can be seen that fill times in die casting will rarely exceed 0.1 s and are usually represented in milliseconds. Thus, for the purposes of estimating cycle times, fill time can simply be neglected.

10.10.3 Metal Cooling

As described briefly above, the casting cycle proceeds when molten metal, at temperature T_i, is injected rapidly into the die, which is at initial temperature T_m. The casting is then allowed to cool to a recommended ejection temperature T_e, while the heat is being removed from the die through the circulation of cooling water.

During solidification of the metal, latent heat of fusion is released as the metal crystallizes. This additional heat can be represented by an equivalent increase in temperature, ΔT, given by the following equation:

$$\Delta T = H_f/H_s \tag{10.26}$$

where
H_f = latent heat of fusion coefficient, J/kg
H_s = specific heat, J/(kg K)

The equivalent injection temperature, T_{ir}, then becomes

$$T_{ir} = T_i + \Delta T \tag{10.27}$$

This approach to the inclusion of heat of fusion in cooling calculations has been used extensively in the literature. The term ΔT is often referred to as "superheat."

It is accepted in the literature [8] that the main resistance to heat flow from the casting is the interface layer between the casting and the die. This is in direct contrast to injection molding, where the resistance to heat flow is provided by the polymer itself. This is because in injection molding, the thermal conductivity coefficient of the thermoplastics is of the order of 0.1 W/(mK) whereas die casting alloys have typical conductivity values of approximately 100 W/(mK).

This leads to a cooling problem in die casting which is entirely opposite to that existing in injection molding. In injection molding, the goal is to cool the polymer as rapidly as possible in order to reduce the major component of cycle times. In die casting, "lubricants" are sprayed onto the die to protect the die surface, but also to provide a heat resistant coating in order to slow cooling for satisfactory die filling.

The resistance of the die interface is represented by its heat transfer coefficient, H_t, which has units $kW/(m^2K)$. The rate of heat flow into the die surface is then given by

$$\dot{W} = H_t A(T - T_m) \tag{10.28}$$

where
T = alloy temperature adjacent to die face, °C
T_m = temperature of die adjacent to die face, °C
A = area of contact with die surface, m^2
H_t = heat transfer coefficient, $kW/(m^2K)$
$\dot{W}$ = heat flow rate, kW

Reynolds [9] has shown that for permanent mold (nonpressurized) casting of aluminum alloy, the heat transfer coefficient with a polished die surface is as high as $13 \, kW/(m^2K)$. However, with a thin coat of amorphous carbon, the heat transfer coefficient varies between 1 and $2 \, kW/(m^2K)$. Sekhar et al. [10] confirmed the pronounced effects on heat transfer of the thin layers of carbon produced from the die lubricants by the hot metal contact. They also showed that the heat transfer coefficient is increased by applied pressure on the metal. For typical die casting pressures between 20 and $50 \, MN/M^2$, and carbon layers between 0.05 and 0.2 mm thick, Sekhar's results show an average heat transfer coefficient value of approximately $5 \, kW/(m^2K)$.

Dewhurst and Blum [11] have shown that, based on the heat transfer coefficient as the principal heat resistance mechanism, the cooling time may be represented by the simple equation

$$t_c = \rho H_s \, \log_e[(T_{ir} - T_m)/(T_e - T_m)]h_{max}/2H_t \tag{10.29}$$

where
ρ = density, Mg/m^3
H_s = specific heat, $J/(kgK)$
T_{ir} = "superheat" injection temperature, °C
T_m = mold temperature, °C
T_e = casting ejection temperature, °C
h_{max} = maximum casting wall thickness, mm
H_t = heat transfer coefficient, $W/(m^2K)$

Example

Determine typical cooling times for zinc die castings. For a typical zinc die casting alloy, the following parameter values can be used:

$$\rho = 6.6 \text{ Mg/m}^3$$
$$H_s = 419 \text{ J/(kgK)}$$
$$T_i = 440°\text{C}$$
$$T_e = 300°\text{C}$$
$$T_m = 175°\text{C}$$
$$H_f = 112 \times 10^3 \text{ J/kg}$$
$$H_t = 5000 \text{ W/(m}^2\text{K)}$$

Thus, from Eq. (10.26) and (10.27)

$$T_{ir} = 440 + (112 \times 10^3)/419$$
$$= 707.3°\text{C} \tag{10.30}$$

Substituting the preceding values into Eq. (10.29) gives an estimate of nominal cooling time as

$$t'_{cz} = 0.4h_{max} \text{ s} \tag{10.31}$$

where

t'_{cz} = nominal cooling time for zinc alloys
h_{max} = maximum wall thickness, mm

Similar substitutions into Eq. (10.29) with appropriate parameter values for other die casting alloys give the following simple expressions for cooling time.

$$t'_c = \beta h_{max} \text{ s} \tag{10.32}$$

where

β = cooling factor

and

$$\beta = 0.4 \text{ for zinc alloys}$$
$$= 0.47 \text{ for aluminum alloys}$$
$$= 0.42 \text{ for ZA alloys}$$
$$= 0.63 \text{ for copper alloys}$$
$$= 0.31 \text{ for magnesium alloys}$$

Equation (10.32) is based on the assumption that there is negligible resistance to heat flow through the steel die and into the cooling channels. This is a good assumption for basically flat castings where cooling channels can be arranged through the cavity and core blocks to cover the casting surfaces. For complex casting shapes, however, cooling of the dies becomes less efficient and the cooling time increases. Herman [4] has suggested that the cooling time increases with casting complexity in proportion to the ratio of the cavity surface area divided by the cavity projected area. However, comparisons with industrial case studies show that this tends to overestimate the cooling time for geometrically complex castings and that the trend represented by

$$t_c = (A_f/A_p)^{1/2} \beta h_{max} \tag{10.33}$$

where
A_f = cavity surface area
A_p = cavity projected area

gives a better fit to actual cooling times.

Note that for thin wall castings, the cooling time for the feed system may be longer than for the casting itself. A rule of thumb, obtained from industrial sources, is that the tooling time will never be less than for a flat 3 mm thick casting. Also, in calculating the values for A_p and A_f, the overflow wells and feed systems are neglected.

Example

Determine the cooling time for a 50 mm diameter by 100 mm deep plain cylindrical cup, with 3 mm wall thickness, which is to be die-cast from aluminum alloy.

Cavity area, $A_f = 17671 \text{ mm}^2$
Cavity projected area, $A_p = 1963 \text{ mm}^2$

Thus, from Eq. (10.33), with the appropriate b value of 0.47,

$$t_c = (17671/1963)^{1/2} \times 0.47 \times 3$$
$$= 4.23 \text{ s}$$

10.10.4 Part Extraction and Die Lubrication

On die opening, the ejector pins protrude through the core and push the casting, with its feed and overflow system, into the gap between the cavity core plate. Small nonprecision castings may then be dropped into the gap below the die, usually into a water tank where a conveyer belt transports the part into a bin. However, die castings always stick onto the ejector pins because of flashing around the pin ends in the die, and a secondary ejection mechanism must be employed to break the casting free. This usually involves putting small rack-and-pinion actuators behind a small proportion of the pins, which move these pins further forward at the end of the main ejector stroke.

For larger or precision parts, the casting must be removed from the ejector pins by the machine operator or by a pick-and-place device on an automatic machine. In this case, the time for casting removal depends principally on casting size. Discussions with die casters suggest that a typical time for unloading a 10 cm × 15 cm casting is 3 s, and that unloading a 20 cm × 30 cm casting will take 5 s. If it is assumed that the unloading time increases linearly with casting size, then these values give the relationship

$$t_x = 1 + 0.08(W + L) \text{ s} \qquad \text{for } W + L > 25 \text{ cm} \qquad (10.34)$$

and

$$t_x = 3 \text{ s} \qquad \text{otherwise}$$

where

t_x = casting extraction time, s
W, L = width and length of the smallest rectangle that will enclose feed system, cavities and overflow wells, cm

Example

A box-shaped aluminum alloy casting, 8 cm wide by 10 cm long by 2 cm deep, is to be cast in a six-cavity die. Estimate the time for extraction of the casting from the die casting machine.

Using the guidelines for casting layout given in Sec. 10.8, the size of each cavity plus overflow wells will be approximately (8 × 1.2) by 10 or 9.6 cm by 10 cm. Assuming a two-by-three pattern of castings with a separation of 8 cm (7.5 plus 0.5 for a cavity area of approximately 100 cm^2), gives a cavity array size equal to 28 cm × 44.8 cm. Finally, allowing a 12.5 cm width increase for the sprue or biscuit gives a total casting size of 40.5 × 44.8 cm.

The time for part extraction, from Eq. (10.34) is thus

$$t_x = 1 + 0.08(40.5 + 44.8) = 7.8 \text{ s}$$

The time for die opening and closing is estimated in the same way as for injection molding. The only difference is that in die casting, the full clamp stroke is commonly utilized to give adequate access for casting removal. As in injection molding, the die must be opened at less than full clamp speed to allow safe separation of casting from the cores. If 40% of full speed is assumed, as discussed in Sec. 8.6.3, then the die opening plus closing time can be given by

$$t_{\text{open}} + t_{\text{close}} = 1.75 t_d \text{ s} \qquad (10.35)$$

TABLE 10.8 Lubricant Application Times for Die Casting (Seconds) and Required Cycles per Lubrication

		Added time	
Part size	Basic time	Per side-pull	Per extra cavity
Small	3	1	1
(10 cm × 10 cm)			
Medium	4.5	1	2
(20 cm × 20 cm)			
Large	6	1	3
(30 cm × 30 cm)			
Number of machine cycles per lubrication			
Aluminum	1		
Copper	1		
Zinc-aluminum	2		
Magnesium	2		
Zinc	3		

where

t_d = machine dry cycle time

Thus, if the six-cavity die above is operated on the 6000 kN cold-chamber machine in Table 10.3, the die opening and closing time will be given by

$$t_{open} + t_{close} = 1.75 \times 5.8 = 10.2 \text{ s}$$

After part extraction, and before the die is closed, the die surfaces are sprayed with an appropriate lubricant. The resulting lubricant film serves two purposes. It forms a barrier to heat flow, as discussed in Sec. 10.10.3, to allow more time for satisfactory cavity filling. It also protects the die surface from erosion by the high-pressure wave of hot metal. The time for application of die lubricant depends on the alloy being cast and on the number of cavities and cavity size. It also increases with the number of side-pulls, since the slideways require additional lubricant concentration. Typical times obtained from industrial contacts are given in Table 10.8.

Example

The box-shaped aluminum castings discussed above have a hole in the side wall that requires one side-pull for each of the six cavities. The average wall thickness is 4 mm and the maximum wall thickness is 10 mm. The projected area, A_p, of each cavity is 80 cm^2 and the cavity surface area, A_f, equals 280 cm^2. The volume of each casting is 85 cm^2. Thus, from Eq. (10.22), the shot volume is given by

$$V_s = 6 \times 85 \times (1 + 2/4) = 765 \text{ cm}^3$$

Referring to Table 10.8, the die lubrication time per cycle is given by

$$t_1 = 3 + 1 \times (n_s \times n_c) + 1 \times (n_c - 1) \tag{10.36}$$

where
n_s = number of side-pulls per cavity
   = 1
n_c = number of cavities
   = 6

Thus

$$t_1 = 14 \text{ s}$$

Note that if the boxes were cast from zinc alloy, then lubrication would occur every three cycles, so the value for t_1 would be 4.67 s.
   The cooling time for the box castings is given by Eq. (10.33) as

$$t_c = (280/80)^{1/2} \times 0.47 \times 10$$
$$= 8.8 \text{ s}$$

   The time for ladling of the molten shot into the cold-chamber machine is given from Eq. (10.24) as

$$t_{lm} = 2 + 0.0048 \times 765 = 5.7 \text{ s}$$

Finally, from Sec. 10.10.2, it can be shown that the fill time will be approximately 0.05 s.
   Thus, the complete cycle time for the six-cavity die casting operation is as follows:

$$\text{Cooling time} = 8.8$$
$$\text{Part extraction time} = 7.8$$
$$\text{Die lubrication time} = 14.0$$
$$\text{Die open/close time} = 10.2$$
$$\text{Metal ladling time} = 5.7$$
$$\text{Total} = 46.5 \text{ s}$$

10.10.5 Trimming Cycle Time

The need for the trimming operation in the die casting process is an important factor distinguishing die casting cost estimation from that of injection molding. The trimming processing cost is the product of the trimming time and the hourly operating rate of the machine and operator. As previously mentioned in the discussion on optimum number of cavities, the hourly trimming rate can be approximated by a constant value for all machine sizes because of the small tonnages of the machines as well as the relatively small range of machine sizes. This small press size requirement means that the hourly rate is dominated by the hourly labor rate of the trim press operator rather than by the cost of the press itself.

The trimming cycle time, including the time to load the shot into the press, is similar to punch press loading and cycle times for sheet metalworking, given by Ostwald [6]. These times vary linearly with the sum of the length and width of the part and can be represented by the following relationship developed from these data:

$$t_s = 3.6 + 0.12(L + W) \text{ s} \tag{10.37}$$

where
t_s = sheet metal press cycle time, s
L = length of rectangular envelope, cm
W = width of rectangular envelope, cm

Discussion with industrial sources indicates that press loading times of die casting shots are generally longer than sheet metal loading times. One reason for this is that die casting shots are oddly shaped and are, therefore, more difficult to align in the die. Additional time is also required for periodic cleaning of the trimming die, as flash and other scrap create a buildup of debris.

It appears from data on a limited number of castings that trim press cycle times are typically 50% higher than manual press operations for sheet metal. Thus, for early costing purposes, the trim press cycle time can be estimated by

$$t_p = 5.4 + 0.18(L + W) \text{ s} \tag{10.38}$$

Note that for castings with side holes produced by side-pulls, trimming is required in both the die opening and side-pull directions. In some cases, these separate trimming tasks will be carried out with separate trim tools on separate presses. In these cases, a multiple of the time estimate from Eq. (10.38) would give the appropriate total trimming cycle time. The cost of trim dies and the effect on die cost of multiple trim directions are discussed in Sec. 10.10.3.

10.11   DIE COST ESTIMATION

The tooling used for die casting is somewhat more expensive than for injection molding. There are three main reasons for this. First, because of the much greater thermal shocks to which a die casting die is subjected, finer steels must be used for the die set and cavity and core inserts than are necessary for injection molding. This gives rise to increased costs for the die set even though it is identical in basic construction to an injection molding mold base. It also results in greater costs for manufacturing the cavity and core inserts from the more difficult to machine material. Second, the overflow wells and larger sprue or biscuit take up more plate area than in injection molding with the requirement for a larger die set than the corresponding mold base for an injection-molded part. Third, for other than the smallest production volumes, a separate trim tool must be manufactured to remove the flash, feed system, and overflow wells from the finished castings.

10.11.1   Die Set Costs

A major supplier of interchangeable die sets and mold bases offers them with three qualities of steel: No. 1 steel (SAE 1030), No. 2 steel (AISI 4130), and No. 3 steel (P-20 AISI 4130). Steels No. 1 and No. 2 are recommended for injection molding, while steels No. 2 and No. 3 are recommended for die casting. For the same plate areas and thicknesses, the average cost of mold bases (steel No. 1 or 2) is compared with the average price of die sets (steel No. 2 or 3) in Fig. 10.9. The line drawn on the graph has a slope equal to 1.25, which means that for the same plate sizes, a die casting die set will typically be 25% more expensive than a mold base for injection molding. Thus, from Sec. 8.7.1 in Chapter 8, the cost of a die set, C_d, can be represented by

$$C_d = 1250 + 0.56 A_c h_p^{0.4} \ \$ \tag{10.39}$$

where
A_c = area of die set cavity plate, cm^2 (which can be estimated from the casting layout rules given in Sec. 10.9.3)
h_p = combined thickness of cavity and core plates in die set, cm.

As with injection molding, typical plate thicknesses should be based on 7.5 cm of plate material separating the casting from the outer plate surfaces. Thus, a 10 cm plain cylindrical cup would typically have the main core mounted onto a 7.5 cm thick core plate and the cavity sunk into a 17.5 cm thick cavity plate.

Also, as for injection molding, the plate area should be increased to allow for any necessary side-pulls; see the description in Sec. 8.7.1.

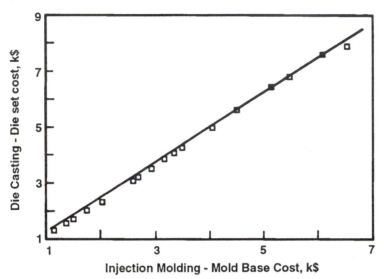

FIG. 10.9   Relative cost of die casting die sets.

10.11.2   Cavity and Core Costs

The equations developed for estimating the costs of cavities and cores for injection molding in Chapter 8 can be applied directly to die casting with only minor changes. The most important change is the use of a factor to allow for the more difficult to machine steels and the machining of overflow wells. A survey of die and mold makers found reasonable agreement that die casting dies are in the range of 20 to 30% more expensive than equivalent molds for injection molding. This also agrees with the comparison of die set to mold base costs in the last section. It seems that typically the cavities and cores will be 25% more expensive for die casting. The equations in Sec. 8.8.2 of Chapter 8 should, therefore, be applied with a multiplying factor of 1.25.

The range of tolerances that can be achieved with die casting is approximately the same as for injection molding, and the effect on cavity and core manufacturing time is as given in Sec. 8.9. However, as opposed to the six different surface finish and appearance factors applicable to injection molding, only three surface finish categories are typically used for die casting. These can be represented as:

1. Minimum finish required to achieve clean separation from the die
2. Medium finish that will allow parts to be buffed or polished
3. Highest-quality finish, which is usually reserved for zinc alloy parts that are to be chrome-plated to mirror standard

The percentage increases that should be used in the point cost system of Sec. 8.9 to account for cavity and core finishing are given in Table 10.9.

TABLE 10.9 Surface Finish Effect
on Point Score

Appearance	Percent increase
Minimum finish	10
Medium finish	18
Highest quality	27

10.11.3 Trim Die Costs

The basic trim die performs essentially the same function as a sheet metal
blanking die. However, the trim die construction is less expensive than for
blanking dies because of the smaller forces that are encountered. Examination of
a range of industrial trim dies indicates that the cost of a die to trim a casting with
a flat parting plane and no internal holes is approximately half the cost of an
equivalent sheet metal blanking die. Moreover, if additional punches are required
to trim internal holes in a casting, then the cost is approximately the same as for
the purchase and fitting of standard punches into a sheet metal die set.

Thus, from the equations developed in Chapter 9, the cost of a trim die can be
estimated as follows. Complexity of the outer profile is defined as for sheet metal
parts by

$$X_p = P^2/(LW) \tag{10.40}$$

where
$P =$ outer perimeter of one cast part, cm
$L, W =$ length and width of smallest rectangle that surrounds outer perimeter of
one cast part, cm

Taking 50% of the basic manufacturing points for blanking dies give

$$M_{t0} = 15 + 0.125 X_p^{0.75} \text{ h} \tag{10.41}$$

Using the average of the curves used in Chapter 9 for area correction of blanking
dies gives

$$f_{lw} = 1 + 0.04(LW)^{0.7} \tag{10.42}$$

Using the sheet metal value of 2.0 equivalent manufacturing hours for the
purchase and fitting of standard punches gives the estimated hours for a basic
trim tool as

$$M_t = f_{lw} M_{t0} + 2N_h \tag{10.43}$$

where

M_t = tool manufacturing time, h
N_h = number of holes to be trimmed

Two additional factors can substantially increase the cost of the basic trim tool. These are the complexity of the parting line and the existence of through holes, produced by side-pulls, which require trimming in a nonaxial direction. Data on trim tool costs obtained from die casters suggest the following approximate relationships for these added cost factors.

(i) Each additional trim direction will require approximately 40 extra hours of tool making. This is the time to produce an extra trim tool with one or more shaving punches, or to incorporate a cam action into the main trim tool for angled punch action.

(ii) Approximately 17 h of additional tool making are associated with each increase in parting line complexity. Parting line complexity is defined according to the levels given in Table 8.8 of Chapter 8. Thus, a casting with parting line complexity 4 will require approximately 68 extra hours of manufacture for the trim tool. This time is required to produce a segmented die with cutting edges on the different levels required to follow the parting line contour.

Example

A die casting has the following defining characteristics:

Outer perimeter, $P = 68.6 \, \text{cm}$
Envelope dimension, $L = 24.0 \, \text{cm}$
Envelope dimension, $W = 13.5 \, \text{cm}$
Number of holes to be trimmed, $N_h = 9$
Number of side pulls per cavity, $n_s = 2$
Number of cavities, $n_c = 2$
Parting line factor $= 2$

Estimate the cost of the required trim tool.

$$\text{Outer profile complexity, } X_p = 68.62/(24 \times 13.5)$$
$$= 14.5$$
$$\text{Area correction factor, } f_{lw} = 1 + 0.04(24 \times 13.5)^{0.7}$$
$$= 3.3$$
$$\text{Basic manufacturing points, } M_{t0} = 15 + 0.125(14.5)^{0.75}$$
$$= 16$$

The base manufacturing hours for the trim tool for a single cast part is given by Eq. (10.43) as

$$M_t = 3.3 \times 16 + 2 \times 9 = 70.8 \, \text{h}$$

Two additional trim directions are required for the side-pull features, which will require 40 added hours of tool making. The parting line has several steps, giving a parting line factor of 2 and, thus, 34 added hours to achieve the stepped trim die surface.

If a typical tool making rate of \$40/h is assumed, then the cost of the trim tool(s) for one cast part would be approximately

$$C_{t1} = (70.8 + 80 + 34) \times 40 = \$7392$$

In this case a two-cavity die is used, so the trim tool must have two shaving dies and two sets of punches to accommodate the two-part casting. To allow for the manufacture of trim tools for multicavity castings, the multicavity cost index is used exactly as for the cost estimating of multicavity dies and molds. Thus, using a multicavity index value of 0.7 (as used in Sec. 10.8), the estimated total cost of the complete trim tool is

$$C_{tn} = 7392 \times 2^{0.7} = \$12,000$$

The actual purchase cost for the tool for trimming this casting was \$14,000.

10.12 ASSEMBLY TECHNIQUES

Die castings can be produced with a variety of features that assist with assembly. Alignment features such as chamfered pins, holes, slots, projecting alignment edges, and so forth, have insignificant cost and yet can ensure frustration-free quality assembly work. Unfortunately, there is no analogy to the injection-molded snap fit elements in die casting. The only integral fastening method available seems to be the cold forming of cast projections to achieve permanent fastening. The cast alloy must, of course, possess sufficient ductility, and this limits the assembly method to zinc and ZA alloys. With these alloys, projecting rivet posts, tabs, or projecting edges can be upset or bent after assembly to achieve strong attachment.

As with injection molding, the die casting process lends itself to the use of inserts, which are simply loaded into the die before injection. This practice is used widely to produce castings with steel screw studs for assembly purposes. However, unlike injection molding, screw thread bushings are not used since cored holes in the casting can be tapped to produce high-strength attachment.

Examples can be found of die castings with spring steel inserts where the spring satisfies a functional requirement in the assembly. This suggests the possible use of spring steel inserts to produce snap fit die cast parts. However, the authors are unaware of any such application.

10.13 DESIGN PRINCIPLES

Die casting and injection molding are closely competing processes, and the detail design principles to ensure efficient manufacture are similar for both. Generally accepted guidelines for die casting design are listed below.

1. Die castings should be thin-walled structures. To ensure smooth metal flow during filling and minimize distortion from cooling and shrinkage, the wall should be uniform. Zinc die castings should typically have wall thicknesses between 1 and 1.5 mm. Similar size castings of aluminum or magnesium should be 30 to 50% thicker than zinc, and copper die castings are usually 2 to 3 mm thick. These thickness ranges result in a fine-grained structure with a minimum amount of porosity and good mechanical properties.

Thicker sections in a casting will have an outer skin of fine metal, with thicknesses about half of the preceding values, with a center section that has a coarser grain structure, some amount of porosity, and poorer mechanical properties. The designer should, therefore, be aware that mechanical strength does not increase in proportion to wall thickness. However, large die castings are often designed with walls as thick as 5 mm and sections up to 10 mm thick. An important consideration in these cases is that, compared to injection molding, little cost penalty is associated with the casting of thick sections. Recall that the cooling time for die castings is proportional to thickness [Eq. (10.33)], while cooling time for injection molding is proportional to thickness squared. Thus, a 5 mm thick injection-molded part will typically take about 60 s to cool, while a 5 mm thick die casting may take only about 4 s. Perhaps of greater significance, a 2 mm thick die casting, which may have the same stiffness as the 5 mm injection-molded part, would only take 2 s to cool. This comparison suggests an economic advantage for die casting where good mechanical properties or heavy walls are required.

2. Features projecting from the main wall of a die casting should not add significantly to the bulk of the wall at the connection point. As with injection molding, this would produce delayed cooling of the localized thickened section of the main wall, resulting in contraction of the surface (sink marks) or internal cavitation. A general rule is that the thickness of projections, where they meet the main wall, should not exceed 80% of the main wall thickness.

3. Features projecting from the side walls of castings should not, if possible, lie behind one another when viewed in the direction of the die opening. In this way, die locking depressions between the features will be avoided which would otherwise require side-pulls in the die. Projections that are isolated when viewed in the direction of the die opening can often be produced by making a step in the parting line to pass over the center of the projection.

4. Internal wall depressions or internal undercuts should be avoided in casting design, since moving internal core mechanisms are virtually impossible to operate

with die casting. Such features must invariably be produced by subsequent machining operations at significant extra cost.

Notwithstanding the preceding guidelines, the power of die casting lies in its ability to produce complex parts with a multitude of features to tight tolerances and with good surface finish. Thus, having made the decision to design for die casting, the most important rule is to get as much from the process as is economically possible. In this way, the structure of the assembly will be simplified with all the resulting cost and quality benefits. The main purpose of the procedures established in this chapter is to help the designer to identify economic applications of the die casting process and to quantify, if necessary, the cost of alternative designs.

REFERENCES

1. Metals Handbook, ASM, Metals Park, OH, 1986.
2. Introduction to Die Casting, American Die Casting Institute, Des Plaines, IL, 1985.
3. Reinbacker, W.R., A Computer Approach to Mold Quotations, PACTEC V, 5th Pacific Technical Conference, Los Angeles, February 1980.
4. Herman, E.A., Die Casting Dies: Designing, Society of Die Casting Engineers, River Grove, IL, 1985.
5. Blum, C., Early Cost Estimation of Die Cast Components, M.S. Thesis, University of Rhode Island, Kingston, 1989.
6. Ostwald, P.F., American Machinist Cost Estimator, American Machinist, New York, 1985.
7. Pokorny, H.H., and Thukkaram, P., Gating Die Casting Dies, Society of Die Casting Engineers, River Grove, IL, 1981.
8. Geiger, G.H., and Poirier, D.R., Transport Phenomena in Metallurgy, Addison-Wesley, Reading, MA, 1973.
9. Reynolds, C.C., Solidification in Die and Permanent Mold Castings, Ph.D. Dissertation, Massachusetts Institute of Technology, Cambridge, 1963.
10. Sekhar, J.A., Abbaschian, G.J., and Mehrahian, R., Effect of Pressure on Metal-Die Heat Transfer Coefficient During Solidification, Mater. Sci. Eng., Vol. 40, p. 105, 1979.
11. Dewhurst, P., and Blum, C., Supporting Analyses for the Economic Assessment of Die Casting in Product Design, Annals CIRP, Vol. 38. p. 161, 1989.

11

Design for Powder Metal Processing

11.1 INTRODUCTION

A variety of structural parts, bearings, gears, and so on, are produced from raw materials in the form of powders. This area of processing is generally called powder metallurgy, although parts can also be made from nonmetallic powders, such as ceramics, by these methods. In the main processing sequence used, raw material powders are mixed and then compressed into the required shape. The compact is then heated in a controlled atmosphere (sintered) to bond the particles together and produce the required properties of the part.

The powder metallurgy process has a number of features that are not found in other metalworking processes, including [1–4]:

1. Precise control of material and product properties. Through careful control of the constituent powder materials, compaction levels, and so on, precise control of the final product properties can achieved.
2. Unique material compositions. Although most powder metallurgy parts are produced with standard material compositions [5,6], a wide range of compositions can be produced by the powder metallurgy process, including combinations impossible by other means. For example, metals and ceramics can be combined by compaction and sintering.
3. Unusual physical properties. Physical properties can be readily varied by powder metallurgy from low-density, highly porous filters to high-density parts with low porosity. Tensile strengths can also be varied from low to very high. It is possible to compact dissimilar materials in layers to obtain different properties at various places in the part. The porosity can be

FIG. 11.1 Material utilization for basic shape-producing processes. (Adapted from Ref. 7.)

controlled so that impregnation with oil or other lubricants can lead to self-lubricating properties for bearings.

4. Net-shape manufacturing. Complex-shaped parts that require no further processing can be produced by the powder metallurgy process. Many gears, cams, and so forth, which can require expensive machining to produce from wrought material stock, can be readily produced from metal powders.

FIG. 11.2 Energy utilization per unit part weight for basic shape-producing processes. (Adapted from Ref. 7.)

Counter bores, holes, flanges, and similar features can be formed when the parts are compacted.

5. Little material wastage and loss. Material wastage from the powder metal process is very low, as illustrated in Fig. 11.1, which compares the overall material wastage for a range of processes [7].

6. High accuracy and repeatability. Close tolerances and a high degree of repeatability, particularly in the transverse direction, can be readily obtained.

7. Low overall energy utilization. The total energy usage by the process is low compared to other forming and shaping processes (Fig. 11.2) [7].

11.2 MAIN STAGES IN THE POWDER METALLURGY PROCESS

Unlike some other material shaping processes, such as injection molding or casting, the powder metallurgy process consists of several independent stages involving different equipment. The basic processing stages are shown in Fig. 11.3.

11.2.1 Mixing

The initial stage in the powder metallurgy process is blending and mixing of the powder materials, together with additives such as lubricants, which are included to aid the compaction process. Metallic stearates, such as zinc stearate, are commonly used as lubricants for compaction and usually form 0.5 to 1.5% of the

FIG. 11.3 Main processing stages for sintered powder parts. (Adapted from Ref. 2.)

mixture. Premixed powders are available direct from suppliers, for some standard materials, but, in general, mixing from constituent elemental or pre-alloyed powders within the plant will be necessary. A variety of mixers are available with different configurations and capacities. The overall purpose of mixing is to obtain as homogeneous a mixture of powders and lubricants as possible.

11.2.2 Compaction

For the majority of engineering powder metallurgy parts, the next stage is cold compaction of the mixed powders. This is most often achieved by die pressing, but other methods are available. Hot compaction, which eliminates the subsequent sintering stage, is used for some specialized applications. In this chapter only cold die pressing is considered in detail as this is the most commonly used method of compaction. A wide variety of different densities, porosities, and tensile strengths can be obtained by using different degrees of compaction or compaction pressures.

During the compaction process, the mixed powders are pressed, usually from both sides, by a number of separately moving punch elements, depending on the complexity of the part. The output of the process is a "green" compact, which has sufficient "green strength" properties to allow handling without damage prior to sintering. The final material properties are directly related to the compaction density achieved, as discussed in more detail in Section 11.4.

11.2.3 Sintering

During sintering the compacted parts are passed through a controlled atmosphere furnace and heated to a temperature below the melting point of the constituent powders. The individual particles become bonded together by diffusion bonding. A variety of different furnace types are available. Most commonly, continuous flow furnaces are used for high productivity, but batch furnaces are also used, particularly if special atmosphere conditions or high temperatures are required.

11.3 SECONDARY MANUFACTURING STAGES

A number of secondary manufacturing processes can be used in conjunction with the main processing steps, usually to refine material properties and to produce features not obtainable from the basic processes. The commonly utilized secondary processes are shown schematically in Fig. 11.4.

11.3.1 Repressing and Resintering

In order to achieve high component densities, partial sintering, followed by repressing and a further sintering stage, is usually necessary. The part is

FIG. 11.4 Secondary processing stages for sintered powder parts. (Adapted from Ref. 2.)

compacted to a moderate density and then presintered. Subsequently, the part is returned to a die set, usually different from the initial compaction die set, and pressed again. A further final sintering stage follows to achieve the required material properties. This combination of processes should be used only when density is critical, because the extra processing stages increase part costs considerably.

11.3.2 Sizing and Coining

Secondary pressing operations (sizing) are used to refine dimensional accuracy or compensate for warpage, and so forth, in the sintered part, with little increase in density occurring. Coining may also be used to engrave or emboss small features in the faces of the part. In each case a suitable die set must be designed and manufactured, in addition to the compaction tools.

11.3.3 Infiltration

For some structural applications it is necessary to eliminate any residual porosity from the powder metallurgy part and to increase strength properties. This can be achieved by infiltrating the sintered part with a lower melting point metal, such as copper, with metal being taken up into the part by capillary action. Infiltration is often carried out by placing a suitable slug of the lower melting point material on the base material compact and then heating to a temperature above the melting point of the infiltrant material. Infiltration can be carried out during the initial sintering process, but is most often done in a second pass through the furnace. In both cases a compact of the infiltrant material must be prepared, in addition to the main part compact, which requires an additional compaction process and corresponding die set. A range of standard copper-infiltrated steels can be produced by this process [5]. These materials can only be produced by the powder metal process in combination with infiltration.

11.3.4 Impregnation

Self-lubricating properties can be achieved by impregnating the porous sintered parts with oil. The sintered parts are usually submerged in a bath of oil for several hours, but for the best results vacuum impregnation should be used. A range of standard self-lubricating bearing materials are used [6], but most moderate density parts can be given self-lubricating properties by impregnation if required.

11.3.5 Resin Impregnation

Parts can also be impregnated with plastics, such as polyester resins, either to improve machinability or to remove porosity, which could adversely affect such finishing operations as plating or interfere with a requirement for gas tightness. The process is similar to oil impregnation, with the interconnected porosity filled with resin by capillary action.

11.3.6   Heat Treatment

Sintered parts may require heat treatment, particularly ferrous-based powder metal parts and some aluminum alloy parts. The heat treatment processes are generally the same as those used for wrought parts.

11.3.7   Machining

Machining may be required, in particular to produce features that cannot be produced, or may be uneconomical to produce, by sintering. However, because powder metallurgy is a net shape process, the amount of machining required is generally small. Such machining is usually restricted to undercuts or threaded holes, for example. The machining properties of sintered materials are similar to the equivalent wrought materials and can be enhanced sometimes by resin impregnation.

11.3.8   Tumbling and Deburring

Burrs produced during the compaction process can be removed by tumbling in a barrel or by vibratory deburring. In this process the parts, after sintering, are loaded into a container, together with some abrasive particles, and then tumbled or vibrated together sufficiently to remove burrs and sharp edges.

11.3.9   Plating and Other Surface Treatments

All of the common plating processes and other surface protection treatments, such as painting, can be used for powder metal parts. Residual porosity can cause plating solution entrapment; consequently, plating is often preceded by resin impregnation.

11.3.10   Steam Treating

Steam oxidizing can be used to increase surface wear and corrosion resistance of iron-based parts. Strength properties and density are also improved. Surfaces are coated with a hard black magnetic iron oxide (Fe_3O_4). The process closes some of the interconnected porosity and all surface porosity. Parts are heated to 480 to 600°C and exposed to superheated steam under pressure. Heat-treated parts cannot be steam-treated since the properties obtained by heat treatment will be altered.

11.3.11   Assembly Processes

Joining powder metal parts is not commonly required because complex shapes can be achieved relative to other forming processes. Should joining of powder

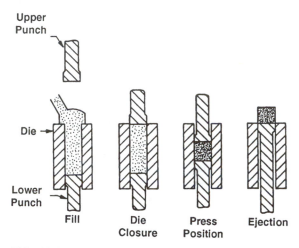

FIG. 11.5   Basic compaction sequence for powder metal parts. (From Ref. 1.)

metal parts be required, many of the commonly used welding processes for wrought parts can be used. As a result of the residual porosity in powder metal parts a unique joining process is possible. Parts are assembled and then infiltrated with a lower melting point metal to effect a bond similar to brazing and soldering.

11.4   COMPACTION CHARACTERISTICS OF POWDERS

During compaction loose powders are poured into a die cavity with the compaction punches retracted (Fig. 11.5). Then the punches are moved relative to the die to compact the powder and increase the density. Subsequently, the "green" compact is ejected from the tooling, prior to transfer to the sintering process.

TABLE 11.1   Extract from Data on Standard Iron Based Materials

Material designation or name	Material condition (AS or HT)	Yield stress (N/mm^2)	Ultimate tensile (N/mm^2)	Part density (g/cc)
F-0000-10	AS	89.6	124.1	6.10
F-0000-15	AS	124.1	172.4	6.70
F-0000-20	AS	172.4	262.0	7.30
F-0005-15	AS	124.1	165.5	6.10
F-0005-20	AS	158.6	220.6	6.60
F-0005-25	AS	193.1	262.0	6.90

The final material properties of the part are largely determined by the compaction density achieved. For example, Table 11.1 shows an extract from the data in the Metal Powder Industries Federation (MPIF) standard on materials [5] and shows the designation and material properties of some standard powder metal (PM) carbon steels. As can be seen, the composition of some of these standard materials is the same, but increased strength and modification of other properties are achieved by compaction to increased densities. For example, see the group of materials designated F-0005-15, F-0005-20, and F-0005-25, which are all low-carbon steel with a 0.05% carbon content; the different properties are achieved by the final density to which the material is compacted. The consequence of this is that for reasonably uniform properties in powder metal parts, it is necessary to achieve relatively uniform density in the "green" compacts. The basic mechanics of the powder compaction process influence the ways in which this can be done.

11.4.1 Powder Compaction Mechanics

The mechanics of powder compaction is governed by friction between the die and the powder and between the individual powder particles. Friction losses cause localized reductions in compaction pressure, and as a result stresses are not distributed uniformly throughout the compact. First, for all but very thin parts (< 6 mm), it is necessary to compact the powder from both sides. Figure 11.6 [8] shows the density distribution in a nickel powder compact pressed only from one side. The highest densities are found in the upper outer circumference, where the wall friction causes the maximum relative motion of the particles. The lowest densities occur in the bottom of the compact remote from the moving punch. Thus, in order to achieve more uniform density, it is necessary to press from both sides with independently moving punches. However, in this case the lowest densities are in the middle of the part due to the effects of container wall friction (Fig. 11.7). Even with double compaction the density gradients impose a limit on the total length of the compact. Successful compaction becomes difficult to achieve when the length-to-diameter ratio of the compact exceeds 5 to 1 [1].

Many powder metal parts consist of a number of levels of different thicknesses in the compaction direction. In order to achieve uniform density in the compact, and hence uniform properties, these different levels must be compacted by separately moving punches. For example, Fig. 11.8 shows a cross section through a part with two levels [8]. If this is compacted with only one lower punch, with a step machined in the upper face, different compression ratios will be achieved in the columns of powder associated with each level, resulting in higher densities in the thinner portion of the compact. In order to achieve more uniform density, it is necessary to separate independently moving lower punches, with the relative movements controlled, to obtain the same compression ratio within each level.

FIG. 11.6 Density distribution in a nickel powder specimen compacted from one side only (densities in gm/cc). (From Ref. 8.)

Pressed from top only

Pressed from top and bottom

FIG. 11.7 Density variations during two-sided compaction.

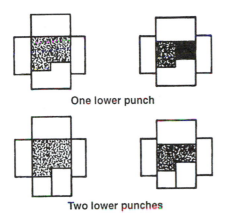

One lower punch

Two lower punches

FIG. 11.8   Density variations in a two-level part. (From Ref. 8.)

From this it can be seen that to achieve uniform properties in a part, the overall complexity of the tooling must increase with the number of different thicknesses or levels in the parts produced.

11.4.2   Compression Characteristics of Metal Powders

The loads required during compaction are determined from the pressures required to achieve a certain density in the parts. Figure 11.9 shows some typical compaction curves for different materials [8]. Such curves are usually obtained

FIG. 11.9   Typical compression curves for metal powders. (From Ref. 8.)

TABLE 11.2 Typical Compaction Pressures for Powder Materials

Material	Tons/in^2	MPa
Aluminum	5–20	69–276
Brass	30–50	414–687
Bronze	15–20	207–276
Carbon	10–12	138–165
Carbides	10–30	138–414
Alumina	8–10	110–138
Steatites	3–5	41–69
Ferrites	8–12	110–165
Iron (low density)	25–30	345–414
Iron (medium density)	30–40	414–552
Iron (high density)	35–60	483–827
Tungsten	5–10	69–138
Tantalum	5–10	69–138

From Ref. 8.

from standard tests using short, cylindrical specimens. As can be seen, as the compaction pressure is increased, the density increases rapidly at first and then slows down, so that the curve eventually becomes asymptotic to a density somewhat below the wrought density of the material. It is difficult to obtain high-density parts because at higher densities the applied loads must be increased by large amounts to obtain even small increases in the part density. For this reason the maximum density obtained using a single compaction operation and sintering is usually around 90% of the equivalent wrought density of the material. Typical compaction pressures for a range of materials are given in Table 11.2.

In order to achieve near full density parts it is necessary to use additional processing steps. In particular, repressing and resintering can be used. This can be illustrated by Fig. 11.10, which shows compaction curves for iron powders for single pressing and for repressing. It can seen that to achieve a part density of, say, 7.3 g/cc by single pressing, it will be necessary to use compaction pressures of 65 tons/in.^{2} (896 MPa), which will produce high loads and require tooling of increased strength. By repressing and resintering, the initial compaction can be reduced to 6.85 g/cc, with moderate pressures of around 35 ton/in.^{2} (483 MPa). Repressing is then carried out at the same compaction pressure to achieve the required density. The total costs of the part will, however, be increased considerably because of the extra processing stages required and a second set of compaction tooling for repressing.

Compaction curves for most materials follow the basic configuration shown in Fig. 11.8, with the precise shape of the curve dependent on the material and shape

FIG. 11.10   Compression curves for single compaction and repressing of iron powder.

of the raw powder particles among other things. Compaction curves can be approximated by a power law relationship such that:

$$P = A\rho^b$$

where P is the compaction pressure and ρ is the part density. The constants A and b can be determined from two values of P and ρ obtained from suitable tests (Fig. 11.11). Compaction curves are usually obtained from tests on cylindrical shapes with the height equal to the diameter. Consequently, for thicker parts the load must be increased for the required density to compensate for the increased container wall friction. This correction should be around 25% for a part length-to-diameter ratio of 4 to 1. Thus the required compacting pressure can be obtained by increasing the value P_1 (Fig. 11.11a), obtained from the basic compaction curve, by a factor K, obtained as illustrated in Fig. 11.11b; i.e.,

$$K = \frac{0.25}{3}(L/D - 1) \qquad \text{for } L/D > 1$$
$$K = 0 \qquad \text{for } L/D < 1$$

Therefore, the total pressure required is given by

$$P = P_1(1 + K)$$

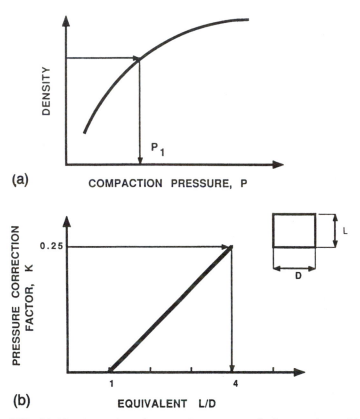

FIG. 11.11   Correction of compaction pressures for increased part thickness.

For parts that are not cylindrical, an equivalent L/D ratio can be used given by

$$L_e/D_e = \frac{V}{2}\sqrt{\frac{\pi}{A^2}}$$

where V is the part volume and A is the projected area in the compaction direction.

11.4.3   Powder Compression Ratio

The depth of loose powder (fill height) required to give the final thickness of the compacted part is determined from the powder compression ratio at the required density (Fig. 11.12). The compression ratio of the powder is given by

$$k_r = \rho/\rho_a$$

FILL

COMPRESSION

FIG. 11.12 Fill height and ejection stroke during powder compaction. (From Ref. 9.)

where ρ_a is the apparent density of the loose powder, which is dependent on the size and shape of the powder particles in the mixture. The compression ratio determines the fill height of powder required for any thickness variations in the final part.

11.5 TOOLING FOR POWDER COMPACTION

The requirement of maintaining relatively uniform densities means that each separate thickness in powder metal parts must usually be compacted by separately moving punch elements. Small thickness changes ($< 15\%$ of the part thickness) can be accommodated by steps in the punch faces with little loss of density uniformity. Various mechanisms are used for achieving the necessary relative motions in the tooling for successful compaction. The main elements for a typical tool set for a multilevel part are shown in Fig. 11.13. The tool set consists of a die, inside which the relative movement of the punch elements to compact the powders takes place. Any through holes in the part are formed by core rods which remain at the same relative position to the upper die surface during the compaction cycle. Other ancillary elements such as punch holder rings, core rod holders, stops, and so on, are required to complete the tool set.

FIG. 11.13 Typical compaction tool elements for a multilevel part. (From Ref. 9.)

During the compaction cycle, the punches are initially retracted to positions to accommodate the fill of loose powder. The retracted positions of the lower punches during filling are determined by the product of the corresponding part thicknesses and the powder compression ratio, k_r. Following filling, the compaction is achieved with the various punches moving relative to each other to give similar compaction densities in the various thicknesses in the part. Finally, after compaction, the "green" compact is ejected and the compression cycle is repeated for the next part.

The complexity and cost of compaction tooling increase as the number of levels or thickness changes in the part increases, since each separate level must be compacted by separately moving punch elements. Many presses utilize standard die sets, with an inserted die held by a suitable tool steel clamping ring. These die sets may be removable or nonremovable. In either case the die sets must be well guided because of the small clearances between tooling elements required. Removable die sets are normally used for press capacities up to around 300 tons, after which the die sets become too large to be readily handled into and out of the press, but occasionally larger removable die sets are used.

11.5.1 Compaction Dies

The die controls the outer peripheral shape of the part, which can contain intricate detail of almost any shape. Compaction dies are usually cylindrical, with the overall thickness dependent on the part thickness and the fill height of powder required. Die surfaces must be highly wear resistant, and the preferred material is tungsten carbide. However, because the cost of carbide is over 10 times that of tool steel, dies are usually constructed with inner carbide inserts, which are shrink-fitted into a tool steel ring. This tool steel ring has a standard outside diameter to suit the recess in the die set or press bed being used. The die insert size that can be accommodated on a particular press is limited.

The required die insert size for a specific part is dependent on the shape being compacted. Preferably any core rods should be as near to the center of the press as possible, and for parts with multiple through holes, positioning the centroid of the projected area of the part at the die center is usually appropriate. A basic rule for determining the carbide insert size required is the diameter of the circle enclosing the part, centered on the centroid, plus an extra amount of material (Fig. 11.14), which is usually not less than 10 mm. The outer die ring, as a practical rule, has a diameter at least three times that of the carbide insert.

11.5.2 Punches for Compaction

Separate punches are required for each level or thickness of the part. These punches move relative to each other during compaction, with the longer punches passing through the shorter punches. Punches must be carefully lapped to ensure close fitting with each other and with the die inner profile. Punches are usually manufactured from cylindrical tool steel stock, by combinations of turning, milling, profile grinding, and lapping.

11.5.3 Core Rods for Through Holes

Through holes of almost any profile can be achieved in powder metal parts, which allows features very difficult or costly to produce by other processes to be readily obtained. Through holes are produced in the part by core rods of the appropriate cross section. Core rods are the longest elements in the tool set. They may have very small cross sections, but such small cross-section rods are relatively difficult and expensive to manufacture. In addition, since core rods are subjected to a cycle of compressive stresses during compaction and tensile stresses during part ejection, the fatigue life of small cross-section rods may be severely reduced. Core rods are produced from tool steel and carbide by similar methods to those used for punches. In all cases, holes of the appropriate shape must be machined in the punches through which the core rods pass.

CENTROID

10mm

MINIMUM   CARBIDE   INSERT   DIAMETER

FIG. 11.14   Compaction die insert dimensions.

11.5.4   Die Accessories

All die sets require a number of additional accessories to be produced, including core rod holders, punch adapters, stops, fasteners and so on. In general, these are produced from tool and other alloy steels.

11.6   PRESSES FOR POWDER COMPACTION

Powder compaction presses may be mechanically or hydraulically driven and differ mainly in the number of actions present in the press mechanisms. Presses are available in a wide range of capacities up to around 25,000 kN (2800 tons). Single-action presses and tooling, in which parts are compacted from one side only, are restricted to relatively thin one-level parts. Double-action presses and tooling systems allow compaction of the part from both sides. Various tool set mechanisms allow the compaction of multilevel parts to be accommodated, although two-level parts are most commonly processed. Multiple-action (adjustable stop) presses, which are capable of producing the most complex parts, are also available.

11.6.1 Factors in Choosing the Appropriate Press

Table 11.3 lists some relevant data for a representative range of compaction presses. The main items that determine the appropriate press for a particular part are as follows:

The number of vertical punch motions possible
Load capacity of the machine
Maximum fill height possible
Maximum ejection stroke possible
Maximum die diameter that can be accommodated

TABLE 11.3 Data on a Range of Typical Compaction Presses

Type of press	Capacity (kN)	Maximum stroke rate (per min)	Minimum stroke rate (per min)	Maximum fill height (cm)	Maximum die insert diameter (cm)
SA	36	150	25	1.524	5.72
DA	45	90	15	3.810	7.30
SA	53	150	20	1.905	9.52
DA	89	60	10	5.080	11.18
DA	134	60	10	6.985	15.24
SA	142	100	15	1.905	13.34
DA	178	50	8	8.255	20.99
DA	267	50	8	8.255	20.32
SA	312	60	10	1.905	15.24
DA	400	40	7	11.430	15.24
DA	534	40	7	15.875	20.32
MA	534	40	7	15.875	20.32
DA	587	34	12	11.430	19.20
DA	890	30	7	15.875	20.32
DA	979	30	12	15.240	21.74
MA	1113	40	7	15.875	21.59
DA	1335	30	7	15.875	21.59
MA	1780	30	7	15.875	22.86
DA	1958	30	10	15.240	26.67
DA	2670	25	7	11.430	25.40
DA	3115	25	7	15.875	25.40
MA	4450	20	7	15.875	25.40
DA	4895	18	6	11.430	50.80

SA, single action; DA, double action; MA, multiple action.

Punch Motions

Some machines are only capable of pressing from one direction and consequently can only be used for relatively thin single-sided parts. Other machines can press from above and below using separate punch actions at the same time, and with suitable tooling they can be used to produce multilevel parts.

Load Required

The total load required for a part is determined by the product of the pressure needed to compact the part to the required density and the projected area of the part in the compaction direction. Determination of the compaction pressure has been described in Sec. 11.4.2.

Fill Height

The fill height or depth of the fill (Fig. 11.14) is the height of the loose powder required to give the part thickness after compaction. The value is determined by the compressibility of the loose powder at the required density. The fill height is obtained by multiplying the finished part height by the compession ratio of the material:

$$\text{Fill height,} \quad h_f = tk_r$$

where t is the part thickness and k_r is the compression ratio.

If the fill height is greater than the maximum that can be accommodated in the press selected on the basis of the compacting load required, a larger-capacity machine must be used. This may be necessary for thick parts with a relatively small cross-sectional area.

Ejection Stroke

The ejection stroke is equivalent to the part thickness plus the penetration of the upper punch into the die (Fig. 11.12). If a greater ejection stroke is required than is available on the press selected on the basis of compacting load, then a large-capacity machine must be used, but often this problem can be overcome by suitable design of the tooling. Consequently, ejection stroke is not a major determining factor in press selection.

Maximum Die Diameter

Presses and tool sets have a maximum die size that can be accommodated. The required die size for a particular part is determined by the procedure described in Sec. 11.5.1. If the required die size is greater than can be used in the press selected from the compaction load, it will be necessary to use a press of larger capacity that can accommodate the required die size. This may be necessary for a part that has a large circumscribing circle diameter, but a relatively small projected area in the compaction direction.

TABLE 11.4 Data on a Range of Typical Coining and Sizing Presses

Type of press	Capacity (kN)	Maximum stroke rate (per min)	Minimum stroke rate (per min)	Maximum clearance height (cm)	Maximum die insert diameter (cm)
DA	30	112	47	1.600	5.08
DA	45	200	50	1.600	5.08
DA	80	90	15	3.000	6.35
DA	134	200	50	2.540	7.62
DA	150	100	17	4.000	8.89
DA	267	60	10	3.810	11.43
DA	300	60	22	6.500	12.70
DA	356	150	30	4.140	15.24
DA	534	100	15	5.080	21.59
DA	890	60	10	5.080	20.32
DA	1000	50	17	7.500	20.32
DA	1780	60	10	5.080	20.32
DA	2500	30	5	10.000	20.32

DA, double action.

11.6.2 Presses for Coining, Sizing, and Repressing

Presses for secondary pressing operations, including coining, sizing, and repressing, are similar to compaction presses, but the powder fill mechanisms are replaced by part loading systems. The shorter strokes required for these operations result in faster cycle times than for compaction presses of similar capacities. Tool designs are similar to compaction, but punch and core rod lengths are considerably smaller as a result of the shorter working strokes required. Table 11.4 lists data on a selected range of secondary operation presses. Selection procedures for these presses are essentially the same as for compaction presses, and it is normally assumed that the loads required are the same as for compaction.

11.7 FORM OF POWDER METAL PARTS

Parts suitable for manufacture by the powder metallurgy process generally consist of one or more levels, with each level being effectively an extruded two-dimensional profile. The profiles at each level may be complex and contain intricate shape features. For example, finished gear profiles can be readily produced. Figure 11.15 shows some typical powder metal parts from industry.

The prime objective in compaction is to compress the powder and achieve relatively uniform density in the compact. Thin (< 6 mm) single-layer parts can be successfully pressed from one side only with a single punch and require a relatively simple single-action press. Steps of less than 15% of the part thickness

FIG. 11.15 Typical powder metal parts.

can be accommodated by steps in the punch faces without too much variation in density. Thicker single-layer parts require pressing from both sides, with two punches moving from above and below at the same time. In multilevel parts each level requires a separate moving punch for successful compaction. Such parts require more complex tooling and presses capable of multiple-action pressing. The number of separate moving punches that can be accommodated in a die set is limited and parts with up to three levels per side of the part can be produced, but most parts have fewer levels than this (Fig. 11.16). Through holes can be formed during compaction and each hole requires a separate core rod, which passes through the punches in the die set.

Tooling costs are increased by the number of levels in the part and the number of through holes. Small-sized through holes require fragile core rods, and their presence tends to increase tool maintenance costs.

The MPIF [9] classifies parts into four complexity levels as follows:

Class I—thin one-level parts of any contour
Class II—thick one-level parts of any contour
Class III—two-level parts of any thickness and contour
Class IV—multilevel parts of any thickness and contour

Table 11.5 indicates the pressing requirements of these classes of parts. However, some consideration of the profile complexity and number of cores is required to indicate increased tool complexity.

ONE LEVEL PARTS

TWO LEVEL PARTS

THREE LEVEL PARTS

FOUR LEVEL PARTS

FIG. 11.16 Levels in powder metal parts.

TABLE 11.5 Press Requirements for MPIF Parts Classification [1]

Part class	Number of levels	Press actions
1	1	Single
2	1	Double
3	2	Double
4	> 2	Double or multiple

11.7.1 Profile Complexity

Intricate profiles of almost any shape can be obtained in the outside profile of the part and similarly in the through holes and individual levels. An indication of increased profile complexity can be determined by a suitable complexity factor. A relation that indicates the profile complexity is the ratio of the perimeter to the perimeter of a circle containing the same area; this is the shortest perimeter that can contain the same area, i.e.:

$$F_c = P_r/2\sqrt{(\pi A)}$$

where F_c is the complexity factor, P_r is the profile perimeter, and A is the area contained within the perimeter.

11.8 SINTERING EQUIPMENT CHARACTERISTICS

Sintering is the process during which the powder particles bond together at temperatures below the melting point of the major constituent elements. In order to achieve the required material properties, the compacts must be heated in a controlled atmosphere to a prescribed temperature (sintering temperature) and held at this temperature for a prescribed time (sintering time). Normally, a preheating phase, mainly to burn off lubricants added to aid compaction, is used prior to the sintering phase. A controlled cooldown period is also required. Table 11.6 shows typical sintering temperatures and times for some of the major material classes processed.

TABLE 11.6 Typical Sintering Temperatures and Times for Different Materials

Material type	Sintering temperature (°C)	Sintering time (min)
Bronze	815	15
Copper	870	25
Aluminum	600	20
Brass	870	20
Iron-based	1150	25
Nickel	1040	38
Stainless steel	1150	40
Ferrites	1370	10–600
Cemented carbides	1460	90
Molybdenum	2050	120
Tungsten	2350	480
Tantalum	2400	480
Titanium	1200	120

TABLE 11.7 Typical Operating Temperatures for Sintering Furnaces

Furnace type	Maximum operating temperature (°C)
Continuous flow	
Belt	1150
Pusher	1150
Roller-hearth	1150
Walking beam	1650
Batch type	
Bell	2800
Elevator	2800
Vacuum	2800

11.8.1 Sintering Equipment

Furnaces for sintering can be divided into two main types: continuous-flow and batch-type furnaces. For high productivity continuous-flow furnaces are preferred, because of the high throughput rates possible. However, when high temperatures or very pure atmosphere conditions are required, batch-type furnaces may be necessary (Table 11.7).

Continuous-Flow Furnaces

Four distinct types of continuous flow furnaces are used (shown schematically in Fig. 11.17): mesh belt furnaces, roller-hearth furnaces, pusher furnaces, and walking-beam furnaces. Table 11.8 lists the basic features of a range of commercially available continuous-flow furnaces. Mesh belt furnaces are the most widely used, particularly for small parts, mainly because of the ease of maintaining a consistent temperature profile throughout the furnace. Parts are placed directly on the belt or onto thin ceramic or graphite plates on the belt. Mesh belt furnaces have limitations on belt load [usually less than $20\,lb/ft^2$ ($73.34\,kg/m^2$)] and in maximum temperature (usually 1150°C).

In roller-hearth furnaces parts are carried on trays driven by rollers. These furnaces can be of large capacity and carry heavier loads than conveyor furnaces. Maximum temperatures are limited to around 1150°C (2100°F). In pusher furnaces parts are loaded onto trays or ceramic plates that are pushed through a stationary-hearth furnace. Throughput rates are generally lower, but these furnaces can operate at higher temperatures [1650°C (3000°F)]. For walking-beam furnaces, parts are placed on ceramic carrier plates and moved through the

MESH-BELT CONVEYOR FURNACE

ROLLER-HEARTH FURNACE

PUSHER FURNACE

WALKING-BEAM FURNACE

FIG. 11.17 Continuous-flow furnaces. (From Ref. 8.)

TABLE 11.8 Data on Typical Continuous Furnaces

Type of furnace	Maximum temp. (°C)	High heat zone length (m)	Overall length (m)	Feeder width (cm)	Throat height (cm)	Load capacity (kg/m²)	Operating cost ($/h)
Belt	1150	0.91	4.57	15.24	7.62	73.24	100
Belt	1150	1.83	9.14	30.48	10.16	73.24	120
Belt	1150	2.44	12.19	45.72	15.24	73.24	144
Belt	1150	3.05	14.78	60.96	15.24	73.24	170
Belt	1150	3.66	14.78	60.96	15.24	73.24	180
Belt	1150	4.57	15.54	66.04	15.24	73.24	216
Belt	1150	5.49	17.17	60.96	15.24	73.24	240
Beam	1204	2.23	8.03	30.48	15.24	244.15	156
Beam	1232	3.05	13.11	60.96	15.24	244.15	192
Beam	1232	3.66	15.54	43.18	12.70	244.15	240
Beam	1371	4.27	18.59	38.10	10.16	244.15	300
Beam	1371	9.14	24.54	76.20	10.16	244.15	300
Pusher	1399	1.52	4.82	20.32	7.62	97.66	120
Pusher	1427	2.44	7.32	30.48	10.16	97.66	168
Pusher	1427	3.05	10.06	30.48	10.16	97.66	240

FIG. 11.18 Temperature profile for continuous-flow furnace. (From Ref. 9.)

furnace by a walking-beam mechanism. These furnaces have high load capacities and can sustain higher maximum temperatures than mesh belt furnaces [1650°C (3000°F)].

Most continuous-flow furnaces are divided into three zones: a burnoff zone, a sintering or high-heat zone, and a cooling zone. A typical temperature profile for parts passing through the furnace is as shown in Fig. 11.18. In the burnoff zone, moderate temperatures of about 650 to 950°C are maintained. The purpose of this zone is to remove any lubricants and so on, before final sintering takes place. In the sintering zone, parts must be maintained at the sintering temperature for a length of time equal to the sintering time appropriate for the part material. These two parameters determine the speed at which parts move through the furnace, which, combined with the number of parts that can be accommodated across the width of the furnace conveyor, determine the overall throughput of parts.

In the cooling zone, parts are maintained in a controlled atmosphere until the temperature is low enough to prevent oxidation occurring if the parts are discharged into the air.

Batch Furnaces

Batch-type furnaces are generally capable of higher maximum temperatures than continuous-flow furnaces (Table 11.7), but production rates are relatively low. The two main types of batch furnace are bell (Fig. 11.19) and elevator furnaces. In both cases the batch of parts is stacked on the furnace base before being loaded into the furnace. Most vacuum furnaces are also batch-type furnaces, although some continuous-flow vacuum furnaces are in use.

FIG. 11.19 Bell-type batch furnace. (From Ref. 9.)

Cycle times for batch furnaces are long and may extend to 10 h or more. Figure 11.20 shows a typical heating and cooling cycle for a batch vacuum furnace. Parts must still be maintained at the sintering temperature and time appropriate for the material of the part. However, the cycle time is dominated by the long heating and cooling times required for batch furnaces. Table 11.9 shows the parameters of a range of commercially available batch furnaces.

FIG. 11.20 Heating and cooling cycle for a vacuum batch furnace. (From Ref. 9.)

TABLE 11.9 Data for Typical Batch Sintering Furnaces

| Type of furnace | Maximum temp. (°C) | Heating space dimensions | | | Operating cost ($/h) |
		Length (m)	Width (m)	Height (m)	
Vacuum	1316	0.36	0.20	0.15	96
Vacuum	1316	0.61	0.46	0.23	108
Vacuum	1316	0.91	0.61	0.51	120
Vacuum	1316	1.22	0.91	0.61	168
Bell	1538	1.22	1.75	1.75	216
Elevator	1528	1.22	1.22	1.40	240
Bell	1649	1.22	1.22	1.22	216
Vacuum	1649	1.22	0.70	0.70	260

11.9 MATERIALS FOR POWDER METAL PROCESSING

Practically all metals and some ceramics can be made into powder and processed into parts by using basic powder metallurgy methods. The most widely used metal material groups are carbon steels, alloy steels, stainless steels, copper alloys, and aluminum alloys. Table 11.10 lists some basic groups of materials that are processed by using powder metallurgy methods. The MPIF has established a range of standard structural materials [5], including iron, carbon steels, copper steels, nickel steels, low-alloy steels, copper-infiltrated steels, stainless steels and

TABLE 11.10 Main Classes of Powder Materials

Iron and carbon steels[a]
Iron copper and copper steels[a]
Iron nickel and nickel steels[a]
Alloy steels[a] and tool steel
Stainless steels[a]
Infiltrated iron, steel, and other metals (with copper,[a] etc.)
Brass, bronze, nickel silver (structural applications)[a]
Aluminum and aluminum alloys
Refractory metals (tungsten, molybdenum, niobium, tantalum, etc.)
High-temperature super alloys (iron, nickel, and cobalt-based)
Self-lubricating bearing materials[a]
Other nonferrous metals
Intermetallic compounds (including mixtures with metals)
Nonmetals and metal/nonmetal mixtures

[a] MPIF Standard Material Classes.

TABLE 11.11   Data for MPIF Standard Iron and Carbon Steel Powder Metal Materials

Material designation	Material condition (AS or HT)	Yield stress (N/mm^2)	Ultimate strength (N/mm^2)	Part density (g/cc)	Cost ($/kg)	Sintering temp. (°C)
F-0000-10	AS	89.6	124.1	6.10	0.84	1121
F-0000-15	AS	124.1	172.4	6.70	0.84	1121
F-0000-20	AS	172.4	262.0	7.30	0.84	1121
F-0005-15	AS	124.1	165.5	6.10	0.84	1121
F-0005-20	AS	158.6	220.6	6.60	0.84	1121
F-0005-25	AS	193.1	262.0	6.90	0.84	1121
F-0005-50HT	HT	413.7	413.7	6.60	0.84	1121
F-0005-60HT	HT	482.6	482.6	6.80	0.84	1121
F-0005-70HT	HT	551.6	551.6	7.00	0.84	1121
F-0008-20	AS	172.4	200.0	5.80	0.84	1121
F-0008-25	AS	206.9	241.3	6.20	0.84	1121
F-0008-30	AS	241.3	289.6	6.60	0.84	1121
F-0008-35	AS	275.8	393.0	7.00	0.84	1121
F-0008-55HT	HT	448.2	448.2	6.30	0.84	1121
F-0008-65HT	HT	517.1	517.1	6.60	0.84	1121
F-0008-75HT	HT	586.1	586.1	6.90	0.84	1121
F-0008-85HT	HT	655.0	655.0	7.10	0.84	1121

Material designation	Sinter time (min)	Theor. density (g/cc)	Apparent density (g/cc)	Density (A) (g/cc)	Pressure (A) (N/mm^2)	Density (B) (g/cc)	Pressure (B) (N/mm^2)
F-0000-10	25	7.87	2.95	6.00	234.4	7.00	551.6
F-0000-15	25	7.87	2.95	6.00	234.4	7.00	551.6
F-0000-20	25	7.87	2.95	6.00	234.4	7.00	551.6
F-0005-15	25	7.87	2.95	6.00	234.4	7.00	551.6
F-0005-20	25	7.87	2.95	6.00	234.4	7.00	551.6
F-0005-25	25	7.87	2.95	6.00	234.4	7.00	551.6
F-0005-50HT	25	7.87	2.95	6.00	234.4	7.00	551.6
F-0005-60HT	25	7.87	2.95	6.00	234.4	7.00	551.6
F-0005-70HT	25	7.87	2.95	6.00	234.4	7.00	551.6
F-0008-20	25	7.87	2.95	6.00	234.4	7.00	551.6
F-0008-25	25	7.87	2.95	6.00	234.4	7.00	551.6
F-0008-30	25	7.87	2.95	6.00	234.4	7.00	551.6
F-0008-35	25	7.87	2.95	6.00	234.4	7.00	551.6
F-0008-55HT	25	7.87	2.95	6.00	234.4	7.00	551.6
F-0008-65HT	25	7.87	2.95	6.00	234.4	7.00	551.6
F-0008-75HT	25	7.87	2.95	6.00	234.4	7.00	551.6
F-0008-85HT	25	7.87	2.95	6.00	234.4	7.00	551.6

From Ref. 5.

TABLE 11.12 Data for MPIF Standard Copper-Infiltrated Steels

Material designation	Material cond. (AS or HT)	Yield stress (N/mm²)	Ultimate strength (N/mm²)	Part density (g/cc)	Material cost ($/kg)	Sintering temp. (°C)	Sintering time (min)
FX1000-25	AS	220.6	351.6	7.30	0.84	1121	25
FX1005-40	AS	334.8	530.9	7.30	0.84	1121	25
FX1005-110HT	HT	827.4	827.4	7.30	0.84	1121	25
FX1006-50	AS	413.7	599.9	7.30	0.84	1121	25
FX1008-110HT	HT	827.4	827.4	7.30	0.84	1121	25
FX2000-25	AS	255.1	317.2	7.30	0.84	1121	25
FX2005-45	AS	413.7	517.1	7.30	0.84	1121	25
FX2005-90HT	HT	689.5	689.5	7.30	0.84	1121	25
FX2008-60	AS	482.6	551.6	7.30	0.84	1121	25
FX2008-90HT	HT	689.5	689.5	7.30	0.84	1121	25

Material designation	Theor. density (g/cc)	Apparent density (g/cc)	Density (A) (g/cc)	Pressure (A) (N/mm²)	Density (B) (g/cc)	Pressure (B) (N/mm²)	Infiltrant quantity (%)	Infiltrant density (g/cc)	Infiltrant cost ($/kg)
FX-1000-25	7.87	2.95	6.00	234.4	7.00	551.6	11.5	8.86	3.53
FX-1005-40	7.87	2.95	6.00	234.4	7.00	551.6	11.5	8.86	3.53
FX-1005-110HT	7.87	2.95	6.00	234.4	7.00	551.6	11.5	8.86	3.53
FX-1006-50	7.87	2.95	6.00	234.4	7.00	551.6	11.5	8.86	3.53
FX-1008-110HT	7.87	2.95	6.00	234.4	7.00	551.6	11.5	8.86	3.53
FX-2000-25	7.87	2.95	6.00	234.4	7.00	551.6	20.0	8.86	3.53
FX-2005-45	7.87	2.95	6.00	234.4	7.00	551.6	20.0	8.86	3.53
FX-2005-90HT	7.87	2.95	6.00	234.4	7.00	551.6	20.0	8.86	3.53
FX-2008-60	7.87	2.95	6.00	234.4	7.00	551.6	20.0	8.86	3.53
FX-2008-90HT	7.87	2.95	6.00	234.4	7.00	551.6	20.0	8.86	3.53

From Ref. 5.

brass, bronze, and nickel silver materials. In addition, there is a range of standard self-lubricating bearing materials [6]. The MPIF has not as yet established standards for aluminum alloys, but industry standards exist to help designers in selecting these materials.

A significant database of material parameters is required for estimating processing costs for powder metal parts. For the purposes of the discussion in this chapter the relevant parameters for three important standard materials classes are given in Tables 11.11 to 11.13. Table 11.11 covers the MPIF standard iron and carbon steels. The two values of density and compaction pressure given in the table enable a power law relationship between compaction pressure and density to be derived, as described in Sec. 11.4.1.

Table 11.12 shows parameters for the standard copper-infiltrated steels designated by MPIF. In this case the part density listed is the infiltrated density, but the compaction data is that corresponding to the iron-based skeleton material prior to infiltration. The percentage of infiltrant material by weight in the final part is included as a parameter for each material.

Table 11.13 lists relevant parameters for some of the standard self-lubricating bearing materials. In this case the density given is the wet density after impregnation with oil, with the oil content by volume also given for each material. The compaction data again corresponds to the base material prior to impregnation.

11.10 CONTRIBUTIONS TO BASIC POWDER METALLURGY MANUFACTURING COSTS

The total manufacturing cost for parts produced by the powder metallurgy process is determined from the material cost, plus the cost associated with the operations in the processing sequence. In general, these processes can be treated independently. Part costs increase with increased geometrical complexity, but also as increased strength properties and densities are required, which may involve secondary processing (Fig. 11.21). Discussion of processing costs will focus initially on basic structural materials, and the modifications necessary for infiltrated and self-lubricating bearing materials will be discussed later.

11.10.1 Material Costs

The material costs for powder metallurgy parts include the cost of the raw powders, the costs of lubricants added to the mixture, and an allowance for losses during processing. Powder losses in industry are generally 2 to 3%. The powder mixtures can be made up from elemental powders or pre-alloyed powders, depending on the application. For some standard materials premixed powders are obtainable from the suppliers. As with most raw materials, the actual costs are

TABLE 11.13 Portion of Data for MPIF Standard Self-Lubricating Bearing Materials

Material designation	Material condition (AS or HT)	Strength constant (N/mm^2)	Part density (g/cc)	Material cost ($/kg)	Sintering temp. (°C)	Sintering time (min)	Theor. density (g/cc)
CT-1000-K19 bronze	AS	131.0	6.2	3.40	827	15	8.71
CT-1000-K26 bronze	AS	179.3	6.6	3.40	827	15	8.71
CT-1000-K37 bronze	AS	255.1	7.0	3.40	827	15	8.71
CTG-1001-K17 bronze	AS	117.2	6.2	3.24	827	15	8.69
CTG-1001-K23 bronze	AS	158.6	6.6	3.24	827	15	8.69
CTG-1001-K33 bronze	AS	227.5	7.0	3.24	827	15	8.69
CTG-1004-K10 bronze	AS	68.9	6.0	3.20	827	15	8.63
CTG-1004-K15 bronze	AS	103.4	6.4	3.20	827	15	8.63
F-0000-K15 iron	AS	103.4	5.8	0.84	1121	25	7.86
F-0000-K23 iron	AS	158.6	6.2	0.84	1121	25	7.86
F-0005-K20 Fe-C	AS	137.9	5.8	0.84	1121	25	7.86
F-0005-K28 Fe-C	AS	193.1	6.2	0.84	1121	25	7.86

TABLE 11.13 (Continued)

Material designation	Apparent density (g/cc)	Density (A) (g/cc)	Pressure (A) (N/mm^2)	Density (B) (g/cc)	Pressure (B) (N/mm^2)	Oil content (%)
CT-1000-K19 bronze	3.40	5.90	68.9	7.65	413.7	24.0
CT-1000-K26 bronze	3.40	5.90	68.9	7.65	413.7	19.0
CT-1000-K37 bronze	3.40	5.90	68.9	7.65	413.7	12.0
CTG-1001-K17 bronze	3.40	5.90	68.9	7.65	413.7	22.0
CTG-1001-K23 bronze	3.40	5.90	68.9	7.65	413.7	17.0
CTG-1001-K33 bronze	3.40	5.90	68.9	7.65	413.7	9.0
CTG-1004-K10 bronze	3.40	5.90	68.9	7.65	413.7	11.0
CTG-1004-K15 bronze	3.40	5.90	68.9	7.65	413.7	7.0
F-0000-K15 iron	2.95	6.00	234.4	7.00	551.6	21.0
F-0000-K23 iron	2.95	6.00	234.4	7.00	551.6	17.0
F-0005-K20 Fe-C	2.95	6.00	234.4	7.00	551.6	21.0
F-0005-K28 Fe-C	2.95	6.00	234.4	7.00	551.6	17.0

From Ref. 6.

					STEAM TREATMENT	HEAT TREATMENT
					RE- SINTERING	RE- SINTERING	RE - SINTERING
		STEAM TREATMENT	HEAT TREATMENT	RE- PRESSING	RE- PRESSING	RE - PRESSING
PRESSING AND SINTERING	PRESSING AND SINTERING	PRESSING AND SINTERING	PRESSING AND SINTERING	PRESSING AND SINTERING	PRESSING AND SINTERING

FUNCTIONAL PROPERTIES (STRENGTH) ⟶

COST

FIG. 11.21 Cost related to increased part material properties. (Adapted from Ref. 2.)

heavily dependent on the quantities ordered. The material cost for each part is determined from the weight of the part, the cost of the material per unit weight, and an allowance of, say, 2% for powder losses during processing. The material costs per unit weight in the database include the cost of mixing and handling. The overall costs for mixing are generally small, with values in the range of $0.09 to $0.13 per kilogram being typical. Thus the material cost for basic powder metallurgy parts is given by

$$C_m = V\rho(1 + l_p)C_p$$

where V is the part volume, ρ is the part density, l_p is the powder loss during processing, typically 0.02, C_p is the cost per unit weight of material, including mixing costs, as given typically in Tables 11.11 through 11.13.

Example

Consider the two-level part shown in Figure 11.22. The approximate volume of this part is $4.97\,\text{cm}^3$. Assuming the material of the part is F-0005-20, then the final density of the part is $6.60\,\text{g/cc}$ and the material cost is $0.84/\text{kg}$ (Table 11.11). Thus the material cost of this part is

$$C_m = 4.97 \times 6.6(1 + 0.02)0.84/10 \text{ cents} = 2.8 \text{ cents}$$

11.10.2 Compacting Costs

The cost of compaction is largely determined by the tooling required to compact the part and the press to be used. For each batch of parts a setup cost is also required. The required press type and capacity are determined by the part complexity and the compacting pressure needed. Tooling costs are also determined by the part complexity.

**lower level
1 - 4 separate
areas**

**lower level
2 - one area**

D

**circumscribing
circle - lower
level 1**

D

**circumscribing
circle - lower
level 2**

FIG. 11.22   Typical two-level powder metal part.

Press Selection

A press must be selected to suit the compaction requirements of the part. The main features that determine the suitability of compaction presses for a particular part are:

1. The number of vertical punch motions possible
2. Load capacity of the machine
3. Maximum fill height possible
4. Maximum die diameter that can be accommodated

   The press load required is obtained by multiplying the required compaction pressure by the projected area of the part. A press with this capacity, plus an excess allowance of, say, 10%, can be used, provided the required fill height and die size can be accommodated. If not, a press of larger load capacity may be required in order to provide the excess fill height or die size requirements.

   The compaction pressure is determined directly from the part density, unless repressing and resintering are required. A basic rule is that repressing should be used if a density of greater than 89% of the equivalent wrought material density is required. Initial compaction at a pressure corresponding to 89% of the wrought density is assumed. This is followed by presintering and repressing. It is generally assumed that the repressing load is the same as for the initial compaction.

   For the selected press, the compaction cost is determined from the product of the cycle time per part and the operating cost per unit time (Table 11.3), which includes labor and overhead costs. The cycle time can be estimated from the following commonly applied rule:

   Stroke rate = maximum press rate minus 10% for every level in the part
      more than one minus 15% for every 13 mm of fill depth more than 13 mm
      minus half of the percentage of the maximum load capacity required,

If this resulting value is less than the minimum stroke rate for the selected press, then the minimum rate is used. One part is produced for each cycle of the press.

Example

For the part shown in Fig. 11.22, which is assumed to be made of F-0005-20 low-carbon steel, the following can be determined.

   The projected area of the part is 7.02 cm^2. Table 11.11 the compaction characteristics of F-0005-20 are defined by the following:

Density 6.00 g/cc, pressure 234.4 N/mm^2
Density 7.00 g/cc, pressure 551.6 N/mm^2

Assuming the compaction pressure $P = A\rho^b$, then for these values $P = 0.01\rho^{5.558}$. Thus for a part density of 6.6 g/cc the corresponding compaction pressure is $0.011 \times 6.6^{5.558}$, which gives 395 N/mm^2. No correction is necessary for part thickness in this case. The compaction load required is $7.02 \times 395/10$, which equals 277 kN. Thus, assuming a safety margin of 15%, the press load required is 320 kN. From Table 11.3 the smallest press to provide this load is rated at 400 kN and has the following charcteristics:

Maximum fill height, 11.43 cm
Maximum die insert diameter, 15.24 cm

These must be checked against the part requirements. The fill height required is given by:

Powder compression ratio $= 6.6/2.95 = 2.24$

Part thickness $= 8.89$ mm and the fill height required

$$= 8.89 \times 2.24/10 = 2.0 \text{ cm}$$

The die insert size required is given by the following calculation: the enclosing circle diameter for the part is 4.76 cm. Thus, the insert diameter required is $(4.76 + 2.0)$ 3, which gives 20.3 cm. Thus, a larger press must be used that will accommodate this size die insert, which from Table 11.3 has the following characteristics:

Load capacity, 534 kN
Maximum stroke rate, 40 per min
Minimum stroke rate, 7 per min
Operating cost, \$63/h

The press stroke rate for this part is then given by

$$40(1 - 0.1 - 0.05 - 0.5 \times 277/534) = 24 \text{ strokes/min}$$

Thus, the press cycle time per part is 60/24, which equals 2.54 s, from which the compaction cost is $63 \times 2.54/3600$, or 0.04 cents per part.

Setup Cost

As the number of levels and complexity of the part increase, the setup time for each batch of parts is increased, due to the need for more complex tools that require more setup time. A rule used for estimating setup time is as follows:

Setup time $= 1$ h $+ 1$ h for every level more than $1 + 1$ hour for thickness

tolerances $\leqslant 0.25$ mm

The setup cost per part is then determined from the product of the setup time and the press cost per unit time divided by the batch quantity of parts produced. As an example, consider the part shown in Fig. 11.22. This part has two levels and, assuming no close tolerances in the compaction direction, the setup time estimate is 2 h. Thus, assuming a batch size of 24,000 parts, the setup cost per part is

$$2 \times 100 \times 63/24,000 = 0.53 \text{ cents per part}$$

11.10.3 Compaction Tooling Costs

The total tooling costs for powder compaction are determined from three elements:

1. The initial cost of the dies, punches, and core rods required for the part
2. The cost of tooling accessories, including punch holders, core rod holders, powder feeding devices, and so on
3. The tool replacement costs, particularly for punches and core rods, which can have relatively short lives in practice

The life of the die is usually high (> 500,000 parts), but cores and punches may require replacement more frequently, particularly if they have small cross-sectional areas. Cores for small-sized through holes and punches that have small width steps will be relatively fragile and will require more frequent replacement and increased maintenance.

Each of these three cost contributions is determined primarily by the size and complexity of the part and secondarily by the press required for compaction of the part.

Initial Tooling Costs

The costs of the basic tooling elements for compacting a part are divided into the tool material costs and the tool manufacturing costs.

Tool Material Costs. The tool material costs can be determined from some general design rules for compacting tooling that have been brought together from a number of sources [10]. These rules are based upon accepted practice in the industry and enable the volumes of tool materials required to be estimated. The basic information required for this determination is as follows:

Fill height, h_f
Enclosing diameter of the whole part, D_o
Enclosing diameter of the separate levels, D_{li}, where $i = 1, 2, 3 \ldots$

The following general rules are then applicable. For dies:

Die thickness $T = h_f + 17.8 \, \text{mm}$
Carbide insert diameter, $D_c = D_o + 20 \, \text{mm}$
Die case diameter $= 3D_c$ or size of corresponding press recess

From these, the volumes of the material required are readily determined.

Carbide insert cost $= \pi D_c^2 T C_c / 4$

where C_c is the cost of tungsten carbide per unit volume, typically $1.22/cm^3$.

Cost of the tool steel die case $= \pi (D_d^2 - D_c^2) T \rho_t C_t / 4$

where D_d is the die case diameter for the selected press, ρ_t is the density of tool steel, 7.86 g/cc (0.283 lb/in.^{3}), C_t is the cost of tool steel, typically $17.6/kg. For the separate lower punches, the punch lengths are given by

$L_1 = h_f + 88.9 \, \text{mm}$

$L_2 = h_f + 119.4 \, \text{mm}$
$L_3 = h_f + 127.0 \, \text{mm}$

The stock diameter for each punch is $D_{pi} = D_{li} + 38.1 \, \text{mm}$, where D_{li} is the punch face enclosing the circle diameter.

Similarly for the separate upper punches (usually up to two) the lengths are given by

$U_1 = 0.5h_f + 68.6 \, \text{mm}$
$U_2 = 0.5h_f + 87.4 \, \text{mm}$

Again the stock diameter for each punch is $D_{li} + 38.1 \, \text{mm}$.

The following basic rules are applied to determine which punch corresponds to each level:

1. The longest punch corresponds to the level with the smallest enclosing diameter and the shortest punch to the level with the largest enclosing diameter.
2. If two levels have the same enclosing diameter, then the level with the smallest enclosed area corresponds to the longer punch.

Again, from these values the volume of the punch materials can be estimated. Punch material costs are given by

$$\pi D_{pi}^2 L_i \rho_t C_t / 4$$

where D_{pi} is the punch stock diameter and L_i is the punch length (L_1 to L_3, U_1, U_2).

Example. For the sample part under consideration one upper and two lower punches are required, with six separate core rods passing through the punches.

Fill height, 20 mm
Enclosing diameter of whole part, 47.6 mm
Enclosing diameter of lower level 1, 47.6 mm
Enclosing diameter of lower level 2, 18.1 mm

From these data the following are determined for the die material costs:

Die thickness, $20 + 17.8 \, \text{mm} = 37.8 \, \text{mm}$
Carbide insert diameter, $47.6 + 20 = 67.6 \, \text{mm}$
Die case diameter corresponding to selected press, 20.32 cm
Cost of the carbide insert material $= \pi \times 67.6^2 \times 37.8 \times 1.22/4000 = \$165.5$
Cost of die case material $= \pi(203.2^2 - 67.6^2)37.8 \times 7.86/(4 \times 10^6) = \$19.18$

The following are determined for the punch material costs:

Length of upper punch, $0.5 \times 20 + 68.6 = 78.6\,\text{mm}$
Length of lower punch 1, $0.5 \times 20 + 88.9 = 98.9\,\text{mm}$
Length of lower punch 2, $0.5 \times 20 + 119.4 = 129.4\,\text{mm}$
Stock diameter for upper punch, $47.6 + 38.1 = 85.7\,\text{mm}$
Stock diameter for lower punch 1, $85.7\,\text{mm}$
Stock diameter for lower punch 2, $38.1 + 38.1 = 76.2\,\text{mm}$

Material costs for these punches are as follows:

Upper punch cost, $\pi 85.7^2 \times 78.6 \times 7.86 \times 17.6/(4 \times 10^6) = \$62.76$
Lower punch 1 cost, $\pi 85.7^2 \times 98.9 \times 7.86 \times 17.6/(4 \times 10^6) = \$78.92$
Lower punch 2 cost, $\pi 76.2^2 \times 129.4 \times 7.86 \times 17.6/(4 \times 10^6) = \$81.63$

Tool Manufacturing Costs

The procedure for estimating the tool element manufacturing costs is to consider the machining operations required to produce each tool element, and then utilize estimating procedures developed for machining (Chapter 7), to develop expressions that give the times to machine each element related to the profile complexities and other geometrical features [10]. The manufacturing costs are then determined by multiplying the manufacturing times by a suitable tool shop cost rate, $/h (say $45/h).

Dies. The die profiles are assumed to be produced by wire electro discharge machining (EDM) and then finished and lapped to the punches. Figure 11.23 shows a typical relationship between cutting time and die thickness for wire

FIG. 11.23 Relationship between cutting time and material thickness for wire EDM processing (●, carbide; ◆, tool steel). (From Ref. 10.)

EDM. The following expressions are used to estimate the machining and finishing times:

EDM cutting time, $1.6 + [e^{(0.5T/25.4+3)}]P_r/(60 \times 25.4)$ h
Finishing time, $1.0 + P_r F_c T/25.4^2$ h

where F_c is the complexity factor for the outside profile of the whole part.

As an example for the sample part the following are determined:

Part projected area, $7.02 \, \text{cm}^2$
Part outside perimeter, $13.4 \, \text{cm}$
Die thickness, $37.8 \, \text{mm}$

Then the part outside profile complexity factor, $F_c = 13.4/2\sqrt{(\pi 7.02)} = 1.43$. The EDM cutting time for the die profile $= 1.6 + e^{(0.5 \times 37.8/25.4+3)}$ $\times 134/(60 \times 25.4) = 5.32$ h. The die finishing time $= 1.0 + 134 \times 1.43 \times$ $37.8/25.4^2 = 12.23$ h. Thus the total die manufacturing time $= 16.55$ h. Therefore, assuming a die shop manufacturing rate of \$45/h, the die manufacturing cost becomes $16.55 \times 45 = \$745$.

Punches. Similar expressions have been developed for the machining and finishing of the external punch profiles and for the holes that must pass through some of the punches. Holes are required for core rods and for the longer punches to pass through the next shorter punches as required. In each case the relationships to determine the manufacturing times utilize the perimeters, enclosed areas, and complexity factors of each separate level and through hole.

The times for machining and finishing the outside profiles of punches are given by

Machining time $(h) = 1.25 + 0.1\Sigma P_r/25.4 + 0.89(0.6L_i + 9.53)$
$$\times (\pi D_{pi}^2/4 - A_t)/25.4^3 + 0.18L_i\Sigma P_r/25.4^2$$
Finishing time $(h) = 0.4L_i\Sigma(F_c P_r)/25.4^2$

In these expressions ΣP_r is the sum of perimeters of all the separate regions at the level on which the punch acts. Similarly, A_t is total area enclosed for all the separate regions on which the punch acts. For one-level parts ΣP_r will be the outside perimeter of the whole part and A_t the total enclosed area. The term $\Sigma(F_c P_r)$ is the sum of the product of the perimeter and complexity factor for all regions at the level on which the punch acts.

Example

For the sample part consider the manufacture of the outside profile of the lower punch corresponding to level 2. In this case one separate region exists at this level, so that ΣP_r is the length of the perimeter for this level, which is 119.1 mm. The following data has been determined previously:

Punch length, $L_i = 129.4 \, \text{mm}$
Enclosing diameter, $D_{li} = 38.1 \, \text{mm}$

The area enclosed by the perimeter at this level is 689 mm^2. From these the complexity factor for this level is $119.1/2\sqrt{(\pi 689)} = 1.29$. Thus, for this level $\Sigma P_r F_c$ is $1.29 \times 119.1 = 153.6$.

The machining time for this punch is given by

$$1.25 + 0.1 \times 119.1/25.4 + 0.89(0.6 \times 129.4 + 9.53)(\pi 38.1^2/4 - 689)/25.4^3$$
$$+ 0.18 \times 129.4 \times 119.1/25.4^2 = 8.16 \text{ h}$$

The finishing time for the punch is $0.4 \times 129.4 \times 153.3/25.4^2 = 12.3$ h. Thus the total punch manufacturing cost is $(12.3 + 8.16) \times 45 = \$920.7$.

Holes are required through the punches for core rods to pass through and for the longer punches to pass through the shorter punches. These holes are produced by a combination of drilling and EDM for nonround holes. In most cases a hole is drilled part way through the punch from the back, and then the actual hole profile is produced by EDM through the remainder of the punch. Finally, these holes must be finished and lapped to the corresponding punch element or core rod.

For circular holes the processing times are given by

Machining time (h) $= 0.006 D_h^2 L_i/25.4^3$, plus a setup time of 15 min/hole

and finishing time $= P_r L_i/25.4^2$

where L_i is the length of the punch, P_r is the hole perimeter, and D_h is the hole diameter.

Example. Consider the cost of producing a circular hole through the long lower punch for the sample part. The hole diameter is 4.76 mm, the hole perimeter 14.95 mm, and the punch length 129.4 mm. Thus the machining time is $0.006 \times 4.76^2 \times 129.4/25.4^3 = 0.001$ h, and the finishing time is $14.95 \times 129.4/25.4^2 = 3$ h.

For noncircular holes the following assumptions are made: The hole profile will be machined by EDM for a length of punch given by $L_h = 0.5$ $(T + 12.7)$ mm. The remainder of the punch length is drilled out to a diameter equal to the circumscribing diameter of the hole profile, D_h. For larger-sized holes, the portion of punch length machined by EDM is initially drilled through with a smaller drill, to reduce the amount of material to be removed by the EDM process. An equivalent hole diameter can be calculated as $D_{eh} = 2(A_h/\pi)^{0.5}$, where A_h is the cross-sectional area of the hole required. If D_{eh} is less than 19 mm, then the following apply:

$A_{hol} = A_h/3$ and $Q = 0$, otherwise $A_{hol} = A_h$ and $Q = 8A_h L_i/(3\pi)$

The machining time (h) for the hole is then given by

$$0.8 + (0.006 D_h^2(L_i - L_h) + 0.006Q + 12.56 A_{hol} L_h)/25.4^3$$
$$+ 0.16(A_{hol} L_h/25.4^3)^{0.5}$$

The finishing time (h) for the hole is given by

$$P_r F_c L_h / 25.4^2$$

These calculations are repeated for all holes required through all punches.

Example. Consider the manufacturing time for one of the crescent-shaped holes for the sample part in Fig. 11.22. The following data applies:

Die thickness, $T = 37.8$ mm
Hole enclosing diameter, $D_h = 15.08$ mm
Cross-sectional area of hole, $A_h = 53.74$ mm^2
Hole perimeter, $P_r = 44.5$ mm
Punch length, $L_i = 129.4$ mm

Thus, $L_h = 0.5(37.8 + 12.7) = 25.25$ mm and $D_{eh} = 2(53.74/\pi)^{0.5} = 8.27$ mm, which is less than 19 mm. Thus, $A_{hol} + 53.74/3 = 17.91$ and $Q = 0$. The hole machining time is then given by

$$0.8 + (0.006 \times 15.08^2(129.4 - 25.25) + 12.56 \times 17.91 \times 25.25)/25.4^3$$
$$+ 0.16 \times (17.91 \times 25.25/25.4^3)^{0.5} = 1.177 \text{ h}$$

The hole complexity factor is $44.5/2(\pi 53.74)^{0.5} = 1.71$, and hence the hole finishing time is given by $44.5 \times 1.71 \times 53.74/25.4^3 = 0.25$ h. Thus the total hole manufacturing time is 1.43 h.

Core Rods. From an examination of the costs of a number of core rods [10], the equivalent processing time (including an allowance for material) can be determined from the following relationship:

$$(4.375 D_{eh}/25.4 + 2.7)F_c,$$

where D_{eh} is the equivalent hole diameter corresponding to the core rod as determined above and F_c is the hole profile complexity. If the hole area is less than 6.45 mm^2, then it is assumed that the processing time is doubled, to account for increased difficulty of manufacture.

Example. The part in Fig. 11.22 requires six core rods in all. Consider first the rods for the circular holes. In this case the complexity factor is 1, the equivalent hole diameter is 4.775 mm, and the cross-sectional area of the hole is 17.42 mm^2. Therefore the core rod manufacturing time is $4.375 \times 4.775/25.4 + 2.7 = 3.52$ h, which at \$45/h gives an equivalent cost of \$158.

Similarly, for the crescent-shaped holes the following data applies:

Hole equivalent diameter $= 8.27$ mm
Cross-sectional area $= 53.74$ mm^2
Hole complexity factor $= 1.71$

Thus the core rod manufacturing time is $(4.375 \times 8.27/25.4 + 2.7)1.71 = 7.05$ h, which gives an equivalent core rod cost of \$317.

Total Tool Manufacturing Costs. At the end of this procedure, the total time involved in machining and fitting all tooling elements is obtained. This result is multiplied by a suitable tool manufacturing rate (\$/h) to determine the total tooling manufacturing costs. These costs are added to the tool material costs, determined as described above, to give the total tool costs.

11.10.4 Tool Accessory Costs

In addition to the dies, punches, and core rods required for a part, various tooling accessories are needed, including filling mechanisms, punch and core rod holders, and so on. Examination of data from powder metal processing companies has shown that tool accessory costs increase with the size of the press required for the part. The procedure adopted in this analysis is to relate the cost of the different tool accessories to the press capacity through the following relationships.

1. Basic cost for one level part without through holes (includes filling mechanism, one upper and one lower punch holder, etc.)

    $$\text{Cost (\$)} = 0.4C_{20}(C_{AP})^{0.3}$$

2. Core rod holder

    $$\text{Cost (\$)} = 0.04C_{20}(C_{AP})^{0.3}$$

3. Additional lower punch holders (each lower level over one)

    $$\text{Cost (\$)} = 0.16C_{20}(C_{AP})^{0.2}$$

4. Additional upper punch holders (each upper level over one)

    $$\text{Cost (\$)} = 0.2C_{20}(C_{AP})^{0.3}$$

In these expressions C_{20} is the cost of the tool accessories for a one-level part for a typical 20 ton (178 kN) compaction press, and this value is supplied by the user to calibrate these relationships. The parameter C_{AP} is the required press capacity in tons (kN/8.896).

Example

For the sample part shown in Fig. 11.22, the press selected has a capacity of 534 kN (60 tons), a second lower punch holder is required and a core rod holder is required also. Thus assuming C_{20} is \$500, then the tool accessory costs become:

$$\text{Basic cost} = 0.4 \times 500(60)^{0.3} = \$683$$

$$\text{Core rod holder} = 0.04 \times 500(60)^{0.3} = \$68$$
$$\text{Lower punch holder} = 0.16 \times 500(60)^{0.2} = \$181$$

giving a total tool accessory cost of \$932.

11.10.5 Tool Replacement Costs

The estimating procedure also allows for tooling replacement costs and main-tenance. In general, the dies for compaction can be expected to last for the order of half a million parts before replacement is necessary [10]. For punches the useful life can be 200,000 to 300,000 parts [10], but this may be severely reduced for punches with sections of reduced area or thickness. Core rods are subject to considerable stress and friction during the compaction cycle, with the stress changing from compressive to tensile during each cycle. Thus thin-section core rods often break before wear becomes significant, whereas thicker core rods have longer lives and replacement is required when excessive wear has occurred. A simple model for tool replacement and maintenance has been built into the estimating procedure, which increases the frequency of punch and core rod replacements for small-section tool elements. The life of the die is usually high ($> 500,000$ parts), but cores and punches may require replacement more frequently, particularly if they are fragile. Cores for small-sized through holes and punches that have small width steps are relatively fragile and require more frequent replacement and increased maintenance.

By assuming a logarithmic relationship for the life of the various tooling components and fitting some life data, the following expressions can be obtained. For core rods $\text{Ln}(l_f) = 14.5 - \text{Ln}(25.4 P_r / A_h)$, where l_f is the life in number of parts produced, P_r is the hole perimeter, and A_h is the area enclosed by the perimeter. For punches, for each separate region on which the punch acts then $\text{Ln}(l_f) = 14.9 - 1.2\text{Ln}(25.4 P_r / A_h)$. When the punch has more than one separate region, the lowest life determined in this manner is taken as the punch life.

Tool replacement costs are obtained by first determining the number of replacement items required as follows:

No. of replacement items, $N_r =$ integer part of (P_V / l_f), where P_V is the production volume required. This is then multiplied by the original tool element costs to determine the tool replacement costs.

Example

For the sample part in Fig. 11.22, the tool life for circular core rods is given by

$$\text{Ln}(l_f) = 14.5 - \text{Ln}(25.4 \times 14.96/17.41)$$

which gives a life of 90,846 parts, and therefore, assuming a required production volume of 200,000 parts, the number of replacement items is given by:

Integer part of $(200,000/90,846)$, or 2 additional rods per hole

Similarly for the punch for lower level 1, the individual regions at this level are considered separately to determine the punch life as follows:

1. Circular central region punch life is given by

    $\text{Ln}(l_f) = 14.9 - 1.2\text{Ln}(25.4 \times 32.41/84.52)$, or 192,630 parts

2. For the three outer regions, the punch life is given by

    $\text{Ln}(l_f) = 14.9 - 1.2\text{Ln}(25.4 \times 29.69/63.23)$, or 151,065 parts

Thus, the lowest life from these separate regions determines the punch life and the number of replacement punches becomes the integer part of (200,000/151,065), or one replacement punch in this case. Similar calculations are carried out on the other punches and core rods to determine the total number of replacement tooling elements.

11.10.6   Validation of the Tool Cost Estimating Procedure

The reliability of this tool cost-estimating procedure has been validated by application to a number of parts manufactured by the powder metallurgy industry. Figure 11.24 compares the tooling costs estimated by this procedure, with estimates of the corresponding costs obtained from industry for the parts shown in Fig. 11.15. Only the die, punches, and core rods are considered. A similar comparison for a wider range of parts is illustrated in Figure 11.25.

11.10.7   Sintering Costs

During sintering the compacted parts are passed through a furnace in a controlled atmosphere. Sintering costs are determined from the operating cost of the furnace used and the throughput rate of parts. The furnace operating costs per unit time

FIG. 11.24 Comparison of tooling cost estimates with industry values for a range of powder metal parts.

Sample Parts

FIG. 11.25 Comparison of tooling cost estimates with industry values for a larger range of powder metal parts.

include the atmosphere costs, labor, and an allowance for maintenance. The rate of throughput of parts is given by the number of parts per unit area of belt and the area of belt passing through the furnace per unit time. The number of parts per unit area for a given furnace is dependent on the part material, the projected area of the part, and the belt width. Several basic rules can be applied:

1. For iron containing $< 4\%$ alloying elements, parts can be stacked up to the recommended load capacity of the belt or furnace handling system.
2. For bronze, brass, and aluminum, a gap of 3 mm must be left between parts.
3. For iron containing over 4% alloying elements, a gap of 3 mm must be left between parts.

Continuous-Flow Furnaces

A suitable continuous-flow furnace is normally selected based on the sintering temperature required for the part material. However, mesh belt furnaces are most commonly used. The selected furnace will have certain dimensional characteristics that determine the sintering costs per part. These items include:

Furnace overall length, L_{FL}
Sintering zone length, L_{HT}
Width of belt or feeder, w_F
Height of furnace opening, H_F
Maximum weight of parts per unit area, W_{max}

The belt or feeder speed, v_F is determined from the sintering time, T_s, appropriate to the material and the length of the sintering zone of the furnace:

$$v_F = L_{HT}/T_s$$

The throughput rate of parts is determined from the manner in which parts can be stacked or placed on the belt or feeder. For pusher, roller-hearth, and walking-beam furnaces, parts must be placed on trays or plates, and some allowance for the thickness of these plates, say about 1 cm, should be made. The number of parts, N_{fw}, which can be placed across the width of the furnace, limited by the height of the furnace opening, is dependent on whether the parts can be stacked or must be spaced out on the belt or feeder. Allowance for any nesting of parts possible should also be made. From N_{fw} the length of furnace that the batch of parts will take up is determined as,

$$B_L = B_s L / N_{fw}$$

where B_s is the batch size and L is the part length. Therefore, the time for the batch of parts to pass through the furnace is $(B_L + L_{FL})/v_F$ and the throughput time per part is given by

$$t_{FL} = (B_L + L_{FL})/(v_F B_s)$$

From this the sintering cost per part is determined as $t_F C_{LF}$, where C_{LF} is the cost of operating the furnace per unit time, including labor and atmosphere costs.

Example. Assume that the following belt furnace is selected from Table 11.8, which has the following characteristics:

High heat zone length, 1.83 m
Overall length, 9.14 m
Furnace width, 30.48 cm
Throat height, 10.16 cm
Load capacity, 73.34 kg/m^2
Operating cost, \$120/h

The sample part is made from the standard material F-0005-20, which has a sintering temperature of 1121°C and a sintering time of 25 min. In addition, the dimensions of the part are as follows:

Part length, 47.625 mm
Part width, 47.625 mm
Part thickness, 8.89 mm
Part weight, 0.033 kg

The belt velocity in this case is given by 1.83/25 m/min = 0.0732 m/min. The material of the part will allow stacking of the parts and therefore the number of parts that can be placed across the furnace width is the integer part of 304.8/47.625 = 6 parts, and the number of parts that can be stacked vertically is the integer part of 101.16/8.89 = 11 parts. Thus the number of parts in each row will be 66, and the load per unit area becomes $66 \times 0.033/ (0.3048 \times 0.047)$ kg/m^2 = 147 kg/m^2, which considerably exceeds the allowable load of 73.34 kg/m^2. Therefore, the number of layers of parts must be reduced to five to meet this constraint. Thus the number of parts that can be placed across the belt is 30.

Assuming a batch quantity of 24,000 parts, then the length of the batch in the furnace is $47.625 \times 24,000/30$ mm $= 38.1$ m. From this the throughput time per part is

$$(30.1 + 9.14)/(0.0732 \times 24,000) = 0.022 \text{ min}$$

From this the sintering cost per part becomes $0.022 \times 120/60 = \$0.04$ per part.

Batch Furnaces

For batch furnaces the number of parts that can be placed inside the furnace is determined from the part dimensions and the dimensions of the heating chamber. Allowance should be made for any plates required between the layers of parts. This number or the batch size, whichever is smaller, will be the furnace load, N_{BF}.

In order to estimate the time necessary to process the parts in the furnace an approximation to the heating and cooling cycle as shown in Fig. 11.26 can be used. Assuming a constant heating rate R_1 (°C/h) and a constant cooling rate R_2, if the sintering temperature is θ_S, the sintering time T_s, the burnoff temperature, θ_b, and the burnoff time, T_B, then the furnace cycle time is given by

$$t_{FB} = \theta_s/R_1 + T_B + T_s + \theta_s/R_2$$

From this the sintering cost per part for batch furnaces is

$$t_{FB}C_{FL}/B_s$$

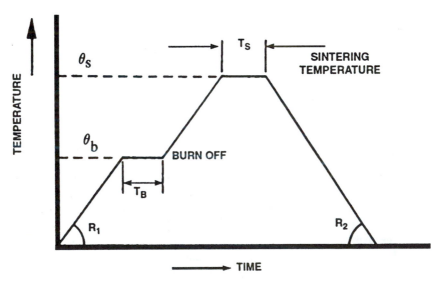

FIG. 11.26 Approximate heating and cooling cycle for batch sintering furnaces.

11.10.8   Repressing, Coining, and Sizing

The secondary pressing operations of repressing, coining, and sizing must be treated similarly to the cold compaction process. The appropriate press load is determined and the corresponding machine selected. From the operating costs of the machine and the estimated cycle time, the cost per piece is determined. For repressing, the same load as for the initial compaction is used, and for sizing, 80% of the compacting load is used. It is reasonable to assume that the tooling costs for these operations are 80% of the initial compaction costs.

11.11   MODIFICATIONS FOR INFILTRATED MATERIALS

Infiltration can be carried out during the initial sintering process or as a separate process after the initial sintering. In either case a compact of the infiltrant material, usually copper, must be prepared. The required amount of copper must be estimated and the compaction cost determined in a similar manner to the compaction of the main part, except that the compact will be a simple one-level part. For the separate infiltration operation, the cost of a second pass through the furnace must be estimated, together with some additional handling costs for the infiltrant compacts.

11.11.1   Material Costs

For infiltrated materials, such as the copper-infiltrated steels listed in Table 11.12, the material costs are determined as follows:

Compaction density of the skeleton material, $\rho = \rho_w (1 - q_i)\rho_p$
Weight of infiltrant, $V\rho_p q_i$

The total material cost for the part is therefore:

$$C_m = V\rho(1 + l_p)C_p + V\rho_p(1 + l_p)q_i C_i,$$

where ρ_w is the equivalent wrought density of the skeleton material, q_i is the percent by weight of the infiltrant material, ρ_p is the part density including infiltrant, ρ_i is the infiltrant material wrought density, and C_i is the cost per unit weight of the infiltrant material. These values can be obtained from Table 11.12.

11.11.2   Compaction Costs

The part prior to infiltration is compacted to the density ρ determined above. The cost of this compaction is determined in the same manner as discussed in Sec. 11.10. An additional compaction operation is required for the compact of infiltrant material. For this purpose the compaction of a one-level part in the infiltrant material is assumed. If it is assumed that the projected area of the infiltrant compact is, say, 80% of the part projected area, then the height of the infiltrant compact can be obtained from the volume determined as described in the previous section. Assuming that the infiltrant compact density is 80% of the equivalent wrought density of the infiltrant raw material, then the procedures for press selection, tool cost calculation, and so on, for processing the one-level compact proceed as described previously for part compaction. Compaction of the main part is determined as for noninfiltrated parts.

11.11.3 Sintering Costs

The sintering operation for infiltrated parts is carried out in a similar manner to that for noninfiltrated parts, with the infiltration step done either as a separate operation after sintering or during the initial sintering process. In both cases, when infiltration is being done, only a single layer of parts can be passed through the furnace and the extra height and weight from the infiltrant compact must be allowed for in any furnace selection procedures. Otherwise, estimates of sintering costs are determined largely as described previously.

11.12 IMPREGNATION, HEAT TREATMENT, TUMBLING, STEAM TREATMENT, AND OTHER SURFACE TREATMENTS

11.12.1 Processing Costs

Costs for secondary processes are largely independent of the shape complexity of the final part. Consequently, for early cost-estimating purposes, a simple cost per unit weight of part, or per unit part surface area, is sufficient to estimate the processing costs. Table 11.14 gives a list of typical processing cost for these operations.

11.12.2 Additional Material Costs

Some parts are impregnated with oil for self-lubricating properties or with polymers or resin to seal the surface-connected porosity in the sintered compact. The costs of the impregnant material must be allowed for, together with the actual cost of the impregnation operation.

Self-Lubricating Bearing Materials

For standard self-lubricating bearing materials, such as those listed in Table 11.13, the following relationships apply:

Part compaction density, $\rho = \rho_p - \rho_o q_o$

Volume of oil in the part, $V_o = q_o V$

TABLE 11.14 Typical Costs for Secondary Processes

Secondary process	Cost ($/kg)
Tumbling	0.22
Impregnation	2.22
Steam treatment	4.41
Quench and tempering	2.31
Carburizing	1.54
Nitriding	2.31
Precipitation hardening	2.31
Annealing	1.54

Secondary process	Cost (cents/50 cm^2)
Cadmium plate	0.54
Hard chrome plate	1.55
Copper plate	0.70
Nickel plate	0.78
Zinc plate	0.39
Anodize	0.93
Chromate	0.39
Prime and paint	1.40

Therefore, total material costs $C_m = V\rho(1 + l_p)C_p + q_o V C_o$ where ρ_o is the density of oil, typically 0.875 g/cc, q_o is the percent by volume of the impregnant oil, C_o is the cost of the oil per unit volume, typically $1.10/liter.

Compaction and sintering costs for the parts are then determined in the same manner as for nonimpregnated parts described previously, but using the compaction density determined as outlined in this section.

Materials Impregnated with Oil or Polymer

For powder metal parts impregnated with other materials, such as oil, resin, or polymer, the basic powder material cost is determined as given in Sec. 11.10.1. The extra material costs are determined as follows:

Volume of polymer or oil, $\quad V_o = V(1 - \rho/\rho_w)$

Hence the impregnant cost, $\quad C_r = V(1 - \rho/\rho_w)C_o$

where C_o is the cost of oil or polymer per unit volume. This cost must be added to the basic powder material cost of the part.

FIG. 11.27   Design recommendations for minimum level widths. (From Ref. 1.)

11.13   SOME DESIGN GUIDELINES FOR POWDER METAL PARTS

The complexity of powder metal parts increases with the number of levels in the part and with the number of through holes in the part, since these require separate tooling elements in the tool set, which increase tooling and other costs. Intricate profiles, including those with significant detail, can be readily produced. However, features that result in thin sections in the compaction tooling elements, which may have a detrimental effect on their life, should be avoided if possible. For example, in multilevel parts, small step widths in the part can result in very thin punch elements that are prone to premature failure and reduced tool life (Fig. 11.27). Thus minimum step widths for all level changes should be specified. Similarly, changes in thickness of the part that result in weak punch sections should be avoided. For example, the compaction of a near spherical shape can be achieved by suitable redesign, as shown in Fig. 11.28. Also the shape of the individual level profiles can result in very fragile punch profiles (Fig. 11.29), particularly where edges meet tangentially. Small changes in the level profiles that avoid very thin punch sections should be made where appropriate.

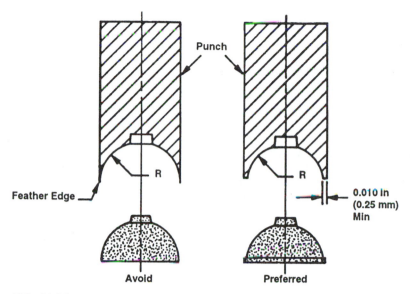

FIG. 11.28 Design modifications to reduce weak punch sections. (From Ref. 1.)

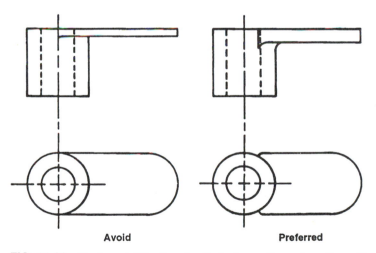

FIG. 11.29 Design modifications to eliminate small punch sections. (From Ref. 1.)

REFERENCES

1. MPIF, Powder Metallurgy Design Manual, Metal Powder Industries Federation, Princeton, NJ, 1989.
2. Mosca, E., Powder Metallurgy: Criteria for Design and Inspection, Associozone Industriali Metallugici Meccanici Affini, Turin, 1984.

3. MPIF, Powder Metallurgy: Principles and Applications, Metal Powder Industries Federation, Princeton, NJ, 1980.
4. Hoeganaes Iron Corp., Iron Powder Handbook, Riverton, NJ, 1962.
5. MPIF Standard No. 35, Material Standards for P/M Structural Parts, Metal Powder Industries Federation, Princeton, NJ, 1990.
6. MPIF Standard No. 35, Material Standards for P/M Self-Lubricating Bearings, Metal Powder Industries Federation, Princeton, NJ, 1991.
7. Kloos, K.H., VDI Berichte, No. 77, p. 193, 1977.
8. American Society of Metals, Metals Handbook, Vol. 7, Powder Metallurgy, ASM, Metals Park, OH, 1984.
9. Bradbury, S., Powder Metallurgy Equipment Manual, Metal Powder Industries Federation, Princeton, NJ, 1986.
10. Fumo, A., Early Cost Estimating for Sintered Powder Metal Components, M.S. Thesis, Department of Industrial and Manufacturing Engineering, University of Rhode Island, Kingston, 1988.

12

Design for Sand Casting

12.1   INTRODUCTION

Sand casting is a process in which molten metal is poured into sand molds, which are broken away from the cast part after the metal has solidified. The sand molds are made of two halves in which impressions are made by a pattern. When the two halves are assembled, the impressions form a cavity into which the molten metal is poured. The pattern resembles the part but is slightly larger to accommodate for the metal shrinkage that occurs during cooling. Patterns may contain multiple impressions, to form multicavity sand molds, for the economical production of large quantities of castings.

A mixture of sand and a binding agent is compacted around the pattern to form the two mold halves in separate metal flasks. Historically, the lower mold half is named the drag, and the upper mold half is referred to as the cope. Channels are formed in the drag to feed the molten metal to the cavity, or cavities in a multicavity mold. The opening of a feed channel to its cavity is the gate, but often the complete set of feed channels is referred to as the gating system. A pouring cup and vertical tapered channel in the cope are used to deliver molten metal to the feed channels. Two additional features may sometimes be required in a sand mold. First, one or more separate sand cores may be placed in formed locations, called core prints, during assembly of the mold halves. Second, an additional cavity, called a riser, may be formed in the cope to supply molten metal pressure during solidification and shrinkage of the casting.

Sand cores are used in sand casting to produce internal cavities or undercuts. Since the cores are broken out of the casting after solidification, complex internal cavities are readily producible. The main requirement is that the internal surfaces

FIG. 12.1   Assembled cope and drag mold.

be accessible for the subsequent cleaning processes to remove adhering sand particles. An assembled cope and drag mold, with both a sand core and a riser, is shown in Fig. 12.1. After metal solidification and cooling in this mold, the casting is broken out of the sand, the gating system is removed, and the cast part is cleaned of adhering sand particles.

Sand castings can be produced singly on the foundry floor with weights measured in tons, or small castings can be produced on high-speed automated lines in quantities measured in hundreds of thousands. The latter type of casting production is a central part of automobile manufacture for such items as pump housings, flywheel blanks, crankshafts, and so on. Small brass castings are produced in a similar manner in large volumes for the plumbing and associated industries.

Sand castings can be made with extremely complex outer shapes. For economical high-volume production the shape should contain a parting line around its perimeter, each side of which has taper, called draft, for separation from the sand mold. This is required not for separation of the finished casting from the mold, but rather for removal of the pattern after the mold sand has been compacted around it. Castings without draft may be produced by the use of multipiece patterns. In these cases the pattern pieces can be disassembled for removal in different directions from the compacted mold. This is a labor-intensive and slow process used only for limited production of very complex parts. When reviewing the economics of sand casting later in this chapter we will focus on economical batch production in modern automated foundries.

TABLE 12.1 Commonly Used Sand Casting Alloys

Generic alloy	Tensile yield strength, MN/m²	Elastic modulus, GN/m²	Castability	Cost, $/kg	Scrap value, $/kg
Ductile iron	515	165	Excellent	0.29	0.07
Gray iron	—	130	Excellent	0.22	0.07
Malleable iron	310	180	Good	0.35	0.07
Carbon steels	345	175	Fair	0.73	0.09
Stainless steel	515	200	Fair	2.53	0.40
Aluminum	95	70	Excellent	1.87	0.50
Magnesium	95	45	Excellent	4.00	1.00
Bronze	345	115	Good	1.65	0.55
Nickel alloys	600	195	Fair	11.00	2.20

12.2 SAND CASTING ALLOYS

Almost any metal alloy can be sand cast, although specific alloying elements are used to increase the metal fluidity and to improve characteristics for feeding to the mold. Table 12.1 gives generic casting alloys. Each of these generic alloys represents a long list of available alloys designed for specific applications. For example, approximately 20 different aluminum alloys are commonly cast for different applications [1].

The castability rating in Table 12.1 is related to fluidity, amount of shrinkage, and the temperature range over which the alloy solidifies. Steels, for example, are not as suitable as irons for casting because of the wider temperature range, called the freezing range, over which they solidify. This results in greater difficulty in mold filling and in avoiding cavities in the solidified casting.

The cost of metal ingots for casting varies substantially with the quantity to be purchased. The typical costs given in Table 12.1 are the costs likely to be paid by a small to medium-sized foundry. Large, automated foundries involved in mass production will obtain lower prices. Also included in Table 12.1 are typical scrap values of the different alloys. These values are needed to calculate the material cost of cast parts, as shown later in this chapter.

12.3 BASIC CHARACTERISTICS AND MOLD PREPARATION

A mold is produced for each casting, or set of castings with multicavity molds. Thus the cost of sand castings is driven, more than by any other factor, by the efficiency of mold preparation. This involves multiple operations from the mixing of mold sand to the inclusion of extra mold features to control cooling and to avoid defects resulting from the metal shrinkage. These will be described in the following sections.

12.3.1 Sand Preparation

The sand must be carefully controlled to ensure consistency, thereby providing each mold with the same properties. Typical molding sand consists of 88% silica, 9% clay, and 3% water. The sand is mixed to evenly disperse the clay additive and to provide uniform coating of the sand grains. This serves to bind the sand grains during compaction, thus providing the required mold strength.

Most foundry sand may be reused. However, the sand must be conditioned before it can be used again, since the sand removed from castings contains impurities and tends to remain in large chunks. Lump breakers, screens, and magnetic separators are used to break up sand clumps and remove metal particles. Some new sand is constantly being added to help maintain the sand quality. This is necessary since the used sand still contains some burned clay even after conditioning. Although only a small portion of sand is lost in the foundry operation, this can still involve the disposal of a large amount of material. In recent years, because of the increasing expense of landfill disposal and environmental concerns, foundries have made increasing efforts to reclaim as much sand for reuse as possible.

12.3.2 The Gating System

The gating system in a sand mold consists of a series of channels and reservoirs designed to feed molten metal to all parts of the mold cavity. The design of the gating system influences a number of factors that can affect the casting's quality and ease of finishing. The common parts of a gating system are shown in Fig. 12.2. Molten metal enters the mold through the pouring cup. It then travels down the tapered sprue and into the sprue well. From there, runners deliver the metal to gates, which open into the mold cavities. Runner extensions and wells capture the early metal flow. This is important because the flow picks up any loose particles in the gating system, and these degrade the metal properties.

The design of the system determines the rate at which the metal will flow and the amount of cooling that will take place before it reaches the mold cavity. The liquid metal may solidify in the feeding passages if it loses too much heat during its journey to the mold cavity. This can block passages (this condition is called a cold shut) and cut off the supply of liquid metal to parts of the mold, resulting in an incomplete casting.

The gating structure must also keep the molten metal from becoming turbulent. Turbulence can lead to excessive absorption of gases into the liquid metal, creating a more porous casting. Turbulence also increases oxidation of the metal and may erode the mold wall.

Another way to prevent mold particles from entering the mold is by using filters. Two types of ceramic filters can be inserted into the runner system. One

FIG. 12.2 Typical gating system.

type of ceramic filter looks like a rectangular sponge. It is used to capture foreign particles in the molten metal passing through it. A simpler, less expensive type of filter consists of a group of holes in a rectangular ceramic brick. Both types of filters reduce erosion of the mold and gas absorption in the metal by creating a more laminar flow.

12.3.3 Mold Risers and Chills

To understand pattern and gating system design, it is important to appreciate the effects on the casting of metal shrinkage. Three types of metal shrinkage may occur during the casting process.

Liquid shrinkage results from the molten metal losing volume as its temperature drops and it approaches the solidification temperature. The shrinkage that occurs in the liquid metal does not affect the process or the design of the feed system. This is because new metal is constantly flowing to fill the mold cavity. Solidification shrinkage occurs as the metal continues to cool and the liquid metal changes to a higher-density solid. This will occur at a single temperature for pure metals but will take place over a range of temperatures (called a freezing interval) for alloys. Metals and alloys with short freezing ranges, like pure metals and eutectic alloys, have a tendency to form large cavities as they shrink. Directional solidification can be used to combat this problem. This is accomplished by designing the feed system so the metal furthest away from the feed gate freezes first and solidification continues toward the feed point, allowing the shrinkage void to be continuously filled as it is formed. The final void ends up in the gating system and not as a defect in the casting.

Alloys with a wide freezing range result in a mixed-state metal (partially solid and partially liquid) that does not flow so readily in the mold. As the cooler areas solidify, it is more difficult for new liquid metal to fill in the voids that form, which results in castings with a large number of small defects. This type of shrinkage is extremely hard to control through the design of the gating system, and a porous casting may be unavoidable. If an air-tight casting is required, then the metal may be impregnated with another material in a secondary operation.

Solid shrinkage, also called patternmaker shrinkage, occurs as the solid casting cools from its solidification temperature to room temperature. It is accounted for by making the pattern for the casting slightly larger than the desired product.

Sometimes, as mentioned in the introduction, an extra cavity is formed in the sand mold to hold a reservoir of molten metal. This cavity, called a riser, is located in the gating system of a casting to feed metal into voids that form during metal shrinkage. In order to accomplish this, the riser must solidify after the casting. Ideally, the mold cavity begins to solidify at the part that is farthest away from the riser and then continues toward the riser, so that liquid metal will always be available to fill in shrinkage voids. However, sometimes this is not possible, and multiple risers are needed. In this case, different sections of the casting solidify in the direction of their respective risers [2–6].

In addition to risers, elements called chills are sometimes used to produce sound castings. Chills are used to assist in directional solidification and may reduce the number of risers required to feed the casting. This is accomplished by increasing the speed at which parts of the casting solidify. External chills are made up of material that has a high heat capacity and high thermal conductivity. When these materials are placed next to the mold cavity, they promote directional solidification and thereby extend the distance that a riser may effectively feed. Internal chills are pieces of metal placed within the mold cavity to absorb heat and assist in increasing the solidification rate. They are usually more effective than external chills, but care must be taken since they will ultimately become part of the casting.

12.3.4 Pattern Types

The patterns used to form sand molds may be made of wood, metal, or plastic. Draft, or taper, must be added to the vertical edges of the pattern so that it may be withdrawn from the sand mold without causing damage. The amount of draft depends on the type of pattern or type of sand mold. Patterns may have core prints (recesses in the mold to accommodate cores), runners, and risers included on the pattern plates so that these will not need to be formed separately.

The material used for patterns depends on the number of molds to be created from the pattern. Hardwood or plastic patterns may be used for a few thousand molds, while cast iron patterns may produce more than 100,000 molds [7]. For the highest-volume production automated casting lines, patterns are often made from stainless steel.

Patterns are commonly made in two pieces, or two halves, split along the parting line of the casting. The parting line is determined by the geometry of the part and must be chosen so that each pattern half may be withdrawn from the mold.

Alternatively, patterns may be mounted to opposite sides of a common board. This arrangement is called a match-plate pattern and provides consistency in mold making from a single former, since the board surfaces define the mold parting surface. A match-plate pattern has the advantage of being a single piece of tooling. It has the disadvantage that the two mold halves cannot be made simultaneously (see Fig. 12.3).

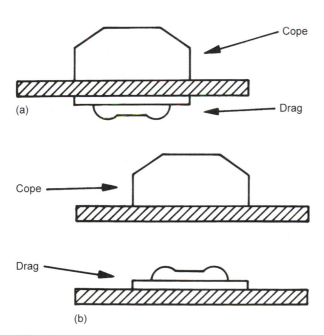

FIG. 12.3   (a) Match-plate pattern; (b) separate cope and drag pattern plates.

12.3.5   Sand Compaction Methods

A sand mold is created by compacting sand around the pattern. It is necessary to pack the sand as densely and homogeneously as possible to obtain good detail and tight tolerances. For this reason, it is very uncommon for a mold to be compacted (rammed up) without some type of machine assistance. Molding machines greatly increase the quality of the mold produced and decrease the skill required of the operator [8]. These machines usually involve a combination of jolting and squeezing actions. Cope and drag machines, for example, begin by pouring sand into flasks mounted over the separate pattern halves. The flasks are then lifted and dropped several times to produce the jolt action. Then a squeeze head is used to further compress the sand. Next, the patterns are removed from the cope and drag. Finally, the cope is lifted, rotated 180 degrees so that the cope impression is facing the drag impression, and then the cope is located onto the drag half by pins and bushings.

Note that the jolt action tends to compress the sand most tightly nearest to the pattern. Squeezing has the opposite effect of compressing the sand most near the outside of the flask. The combination of these two actions thus produces more uniform sand density.

The most modern automated casting systems are referred to as vertically parted flaskless. Vertically parted flaskless molding is a break from traditional mold-making methods. As the name implies, no flask is used to contain the sand during molding. The machine uses hydraulic rams to simultaneously squeeze the cope pattern on one side of a sand block and the drag pattern on the other side. The block is then pushed out of the machine. A new block is then formed and mated to the original block (cores are inserted before closing if needed), creating a complete mold cavity at the interface. The molds form a long, rectangular block of sand coming out of the molding machine. A pouring station is located close to the molding machine. The interconnected molds then travel down a conveyer. The length of the conveyer depends on how long the castings need to cool before being broken out of the molds. A station at the end of the conveyer breaks the mold up, and the castings are separated from the sand. This process is more efficient because a complete mold cavity is formed in each block of sand. Other methods, whether bench or molding transfer line, require separate cope and drag sections to produce one mold. Vertically parted flaskless molding is normally used for relatively simple castings, to be produced in large quantities, with few, if any, cores.

12.4   SAND CORES

One of the most powerful aspects of sand casting is its ability to create internal cavities or exterior undercuts. This is done by using sand shapes, called cores, placed within the mold cavity. Coring allows forms to be produced that cannot be made using any other process. However, a casting designer needs to be aware that cores can significantly increase the cost of a casting, and their use should be avoided if possible.

When developing tooling for a casting, it is always more desirable to create a single large core instead of a number of smaller ones. This is because the labor involved increases proportionally with the number of cores, as does the likelihood of dimensional inaccuracies.

Modern core-making systems use air pressure to blow the sand, mixed with a resin or oil binding agent, from a hopper down into the core-making dies (called core boxes). In a typical core-making process, the cores are removed immediately from the core boxes and put on dryer plates. The cores are then baked in ovens until they attain the required strength. Alternatively, with the so-called hot box process, heated tooling is used to cure the cores in the die cavities. The hot box process may also be combined with subsequent baking in ovens to achieve final cured strength.

12.5 MELTING AND POURING OF METAL

The equipment used for melting the metal ingots varies from large cupolas for producing iron in bulk, to a variety of crucibles, electric furnaces, and induction furnaces [9]. The choice is simply a matter of economics and required production rates. Whichever method is used, the end result is liquid metal at the correct temperature for pouring. The soundness of a casting is largely determined by how the metal enters the mold and solidifies. Therefore, the pouring and handling of the molten metal is an extremely important stage in the casting process. Inefficient handling can lead to excessive heat loss, resulting in premature solidification within the mold and incomplete filling. This is referred to as a cold shut. Faulty pouring can result in air entrapment producing cavities in the casting. This occurs most commonly if the sprue is not kept full during pouring. Short pours are another common example of faulty pouring technique [10].

The pouring operation may be done manually, with mechanical assistance, or automatically. In the case of a manual pour, the worker physically transports a ladle containing liquid metal to the mold and pours it into the mold. The amount of metal poured is limited by the individual's strength. Because of this limitation, hand pouring is usually reserved for low production quantities of small castings.

With mechanical assistance, a much larger cylindrical vessel (still called a ladle) is usually suspended from an overhead monorail system. The metal is poured by turning a handwheel, or by pulling a lever that causes the container to tilt until the metal runs out into the mold. Many more castings may be filled than with the manual method.

High-production foundries pour metal automatically. Instead of transporting the metal to the molds, the molds are transported to the pouring station. A predetermined amount of metal is then poured into the mold. This method allows for the most control over the delivery of the metal into the molds and, if initially designed and executed correctly, it produces castings with a high degree of consistency.

12.6 CLEANING OF CASTINGS

Once the casting has cooled, it must be cleaned before any final operations may be performed. Cleaning consists of the removal of all adhering sand from the casting together with the removal of the gating system and any fins of metal around the parting line. The techniques employed in cleaning, or shakeout, are described in the following section [11,12].

Shakeout is the first stage of cleaning. Shakeout is a foundry term referring to a mechanical operation used to separate the casting from its mold. This should occur without damaging the casting and while keeping noise and dust emissions to a minimal level. An ideal shakeout should achieve the following:

Separate the sand, casting, and flask
Remove as much adhering sand from the casting as possible
Clean the flask of all adhering sand
Break up large mold lumps

It is desirable to remove the casting from the mold as soon as possible, because the sand cores are harder to remove from the casting once they have cooled. However, in some cases the casting is left in the mold to keep it from cooling too quickly. In most modern foundries, shakeout is accomplished by using either vibratory or rotary drum systems.

Shakeout is usually followed by shot blasting. The shot blasting machine is used to provide the casting with a good surface finish and to remove any interior sand that is still adhering to the casting. It is often the most expensive piece of cleaning room equipment in the foundry and is expensive to maintain. For this reason it is important that as much sand be removed from the casting as possible before it is shot blasted. Adhering sand wears the wheel vanes and cups, shot blast liners, and dust collection systems and creates a less pleasant working environment for the workers.

The abrasive metal shot is focused at concentrated areas, using either spinning wheels or concentrated air pressure to project it. The casting is moved through the shot blast streams on a conveyor, rotated on a fixture, or tumbled through the machine.

After the first stage of cleaning, called finishing, the next operation is to remove the gating system and any fins from the castings. Optimizing the finishing process of a casting is very important, because up to 40% of the direct labor content of the casting can be accounted for in these final manual process steps.

Castings sometimes have to go through a grinding operation after being shot blasted. This is done to remove unwanted surface protrusions such as the base of removed fins and gating system contact areas. Grinding operations may be done manually or by using automation. Manual grinding is usually done with hand-held grinders on larger castings. Small castings may be ground using stand grinders and hand manipulation of the castings.

Fixed automation may be possible, depending on the part geometry and the nature of the material being removed. Using fixed automation, the part is held in a fixture and can be touched by several grinding wheels simultaneously. Well-designed coring for the casting can increase the feasibility of using automation. If the flash to be removed grows out from the casting surface, then grinding wheels can easily remove it. Flash growing sideways below the boundaries of the casting surface is very difficult to remove with automation. Such removal is almost always done manually. However, robots with small grinders can be programmed to follow the contour of the pocket being cleaned.

For steel, cast iron, and some copper-based castings, chipping is still a common operation performed to remove the fins and flash. Chipping is performed using pneumatic chisels to quickly cut away unwanted metal. A line of operators equipped with chippers is able to handle a wide variety of castings at the same time. This provides a great deal of flexibility because several different castings can pass down the same line to be cleaned, but this flexibility is gained at the expense of more manual labor.

12.7 COST ESTIMATING

Determining the cost of producing a sand casting is a very complicated procedure, because of the large variations possible in the casting process. Accurate cost estimation is hindered by the amount of equipment involved, and the different combinations in which the equipment and processes may be linked. The cost estimation procedure presented here is divided into three main categories: metal cost, tooling cost, and processing cost.

12.7.1 Metal Cost

The melting department of a foundry represents a very important stage in the casting process. The foundry's ability to manufacture sound castings depends greatly on the chemistry of the metal produced in the melting department. Besides being the cornerstone of the casting process, the melt department also represents a very substantial percentage of the cost of the finished casting. The melt department generates as much as 30 to 50% of the processing costs involved in sand casting.

The first step in determining the total metal cost is to calculate how much material is needed to produce the casting. An examination of a wide variety of castings was carried out [13] to determine the minimum risering and feeding requirements. Variations in casting size were found to occur due to differences in the feeding requirements for different metals and to differences in the arrangement of impressions on the pattern plates.

It is not uncommon to produce siamese castings, where two symmetrical parts are produced as a single piece and are later separated (cut apart). Ideally, the combined dimensions of the siamese casting should be used in the feed calculations. However, this is an example of a foundry technique of which the part designer is unlikely to be aware. For such situations the model used here will overestimate the size of the gating system and thus give a small cost overestimate.

It was found that the total casting weight could be predicted using only the final part envelope dimensions with an accuracy within 15% of actual values [13]. The relationship for the poured casting weight is

$$W_p = \rho V_{fc}[1 + 1.9((L + W)/D)^{-0.701}] \tag{12.1}$$

where

ρ = metal density, kg/m^3
V_{fc} = volume of metal in finished casting, m^3
L = part length, mm
W = part width, mm
D = part depth, mm

Once the poured weight is estimated, the cost of the metal must be determined. The cost of the metal must include the cost of the raw material, the cost of the energy to raise the metal to the required pouring temperature, and the cost of operating the furnace and ancillary equipment. The sum of these costs is often referred to as the cost of metal "at the spout." The steps involved in calculating the cost of metal at the spout are presented below.

Furnace energy cost, measured in dollars per kilogram, is given by

$$C_{en} = E_{ct}M_{me}/F_{ff}(/100) \tag{12.2}$$

where

E_{ct} = cost of electricity, $/(kW\cdot h)$
M_{me} = minimum melt energy, $(kW\cdot h)/kg$
F_{ff} = furnace efficiency

TABLE 12.2 Furnace Cost Data for Sand Casting Alloys

Generic alloy	Density, kg/m^3	Minimum melt energy (kW·h)/kg
Ductile iron	7110	0.391
Gray iron	6920	0.390
Malleable iron	7280	0.395
Carbon steels	7830	0.393
Stainless steel	7750	0.405
Aluminum	2710	0.326
Magnesium	1800	0.332
Bronze	8800	0.185
Nickel alloys	9250	0.341

Values of 0.035 for E_{ct} and 80% for F_{ff} can be used for typical U.S. foundry operations. M_{me} values for a range of typical casting metals are given in Table 12.2.

In addition to the cost of energy, a fixed furnace cost is usually applied to cover the capital investment. This cost can be given by

$$C_{fk} = F_{fc}/\rho \tag{12.3}$$

where

F_{fc} = fixed furnace cost, $/(m^3 of metal)

For a 1000 kg furnace a value for F_{fc} of $284/m^3 represents a reasonable value for current calculations in the United States.

Finally, the labor cost for furnace operation, per kilogram of poured metal (C_{lk}), must be accounted for. This can be estimated for a foundry on a per shift basis—i.e., the cost of furnace workers for one shift divided by the weight of metal poured per shift. A typical value for the furnace labor cost for a midsize automated U.S. foundry is $0.02/kg.

The cost of metal at the spout can thus be represented by

$$C_{ms} = C_{rm} + C_{en} + C_{fk} + C_{lk} \tag{12.4}$$

where

C_{rm} = cost of raw material (given in Table 12.1), $/kg

For example, using the values of C_{rm}, M_{me}, and ρ from Tables 12.1 and 12.2 for gray iron gives

$$C_{ms} = 0.22 + 0.035 \times 0.39/0.8 + 284/6920 + 0.02$$
$$= 0.30 \ \$/kg$$

The cost of metal in a finished casting is higher than this value because of the wasted metal in the feed system as defined by Eq. (12.1), and which has scrap value as given in Table 12.1. With this adjustment the cost of metal in a finished casting is given by

$$C_{mf} = (C_{ms}W_p - C_{cv}(W_p - \rho V_{fc}))/(1 - S_g/100) \tag{12.5}$$

where

C_{ms} = cost of metal at the spout, \$/kg
W_p = weight of poured metal, kg
ρ = density of metal, kg/m^3
V_{fc} = volume of finished casting, m^3
C_{cv} = metal scrap value, \$/kg
S_g = percentage of scrap castings

If, in cost-estimating procedures, the cost of metal ingot, C_{rm}, is used instead of C_{ms} in Eq. (12.5), then the costs of energy and furnace operations are added to the costs of the other casting operations discussed in Section 12.4.7.

12.7.2 Sand Costs

The cost of new mold sand can range from \$13/ton up to \$35/ton depending on the fineness of the grain. This cost does not include the shipping cost, which is the major source of variation between supplier's prices. Transportation cost is important because the characteristics of the sand can vary depending on the site at which it was mined.

Foundries are often willing to pay more for sand from the Midwest because it has better grain characteristics than the angular sand from New Jersey, which requires more binding agents. The need for more binding agents adds expense and creates pollution problems that may justify the higher transportation costs from the midwestern site. Treated sand is considered a hazardous waste and disposal costs can vary from \$300 to \$500 per ton.

Sand must be properly conditioned with the appropriate additives to give it the desired molding properties. The sand must then be transported to the molding area to be used. After use, the sand is separated from the finished casting, transported back to the sand area, is cooled, and any impurities are removed before it can be reconditioned and used again. New grains, as well as binder additives, are added to the recovered sand before it is used again in the system.

Since the influence of the mold sand cost on the casting's manufacturing price is very slight, often less than 3% of the total manufacturing cost, a simple procedure determined from a survey of U.S. foundries will be used. Based on this data [13], it appears that mold sand cost can be approximated as 0.018 \$/kg of metal poured. Thus the cost of mold sand per part can be represented by

$$C_{msd} = 0.018 W_p \tag{12.6}$$

Core sand is considerably more expensive than mold sand. This is because finer sand is necessary to produce cores strong enough to withstand the pressures of pouring metal. From the same foundry data [13], an average cost for core sand was determined to be 0.084 \$/kg.

Cores are frequently damaged during processing, and the scrap rates can vary from 4 to 40%, depending on how delicate the core configuration is and on the type of core-making process being used. For the majority of cases a scrap rate of 8% will provide a reasonable estimate. The cost of the core sand for producing a casting is then

$$C_{csd} = \rho_{cs} V_c C_{cs}/(1 - S_c/100) \tag{12.7}$$

where

ρ_{cs} = density of core sand, kg/m^3 (= 1387)
V_c = volume of core, m^3
S_c = core scrap rate, %
C_{cs} = cost of core sand, \$/kg (= 0.084)

12.7.3 Tooling Costs

The cost of tooling is usually determined in industry through comparisons with similar tools made in the past. The estimator, a tooling expert, then adjusts the cost of the new tooling depending on how much easier or harder the job being estimated is compared to the one previously completed. This is an unacceptable method for the present purposes because the users of the model developed here should not have to be experts in the costing process. Therefore, an alternative method is presented for determining the cost of the tooling.

The first step in determining the cost of the tooling is to determine the equipment on which the tooling will be used. This information is needed to determine the size of the tooling and the number of castings that may be produced per hour. If the casting requires cores, then the cost of core boxes required to support the mold line's core requirements must also be calculated.

To determine the cost of the required pattern it is necessary first to review some of the principles of pattern design. Pattern design is an art and cannot be easily broken down into a few rules of thumb. The rules presented in this section provide a conservative method for locating patterns on a base plate. However, these rules are not absolute and an expert may be able to violate them with impunity.

The first step in pattern layout is to decide where the pattern line will be on the casting. This determines the geometries that will be mounted to the cope and drag plates and is needed to determine the proper spacing between the patterns for multicavity molds. The designer also needs to remember that a feed system is necessary. While casting designers cannot be expected to design this system, they should leave room in areas that they would expect the feed system to pass through. Additional space must also be left for risers when necessary. Finally, core prints should be designed and located so that the cores may be held securely in place. The following rules are suggested in determining the layout of the pattern plates.

For thicker castings, pattern spacing should be equal to the adjacent section heights of the casting. For example, a 20 cm high casting may taper down to 10 cm at the edge adjacent to the location for another pattern. The spacing between the patterns should be 10 cm. For shallower castings, those less than 5 cm high, the spacing rule is increased to 2 times the adjacent section height. The spacing should never become less than 1.5 cm between two sections.

A distance of 2.5 to 10 cm should be kept between the pattern and the edge of the plate. The spacing varies depending on the size and type of equipment being used. For the purposes of this model, a distance of 5 cm is recommended. There should also be approximately 5 cm of spacing between the end of a core print and the edge of the plate.

The main arteries of the gating system should have 2.5 cm of sand surrounding them. However, it is allowable for the runner system to come closer to, or even to contact, a core print.

The layout of the core box cost is simpler than for the pattern plate. The cavities should be laid out so that the shallowest dimension is in the vertical plane. The recommended core box design rules are provided below.

Include the core prints when developing the layout of the die cavities.
Keep a 2 cm wall between cavities.
A border of approximately 5 cm is required on the sides of the box.
A border of 5 cm is required on the top of the core box and 2 cm on the bottom.

For larger-volume production, the plates, on which the pattern impressions are mounted, are usually produced as iron castings. If the parting surface of the mold is flat, then the top and bottom plate surfaces are machined accordingly and flask pins and bushings are added. Anything other than a flat parting line on the cast part requires more complex mounting plate castings. These in turn require more extensive machining, which can significantly increase the cost of the tooling.

The cost of the mounting plates for the pattern impressions, or for the core box cavity inserts, can vary from $1.30 to $4.40 per kilogram depending upon the quantity to be made [13]. For the purpose of this costing model, it will be assumed that the plates are manufactured in large quantities and a figure of $2.00

per kilogram is used. The cost of machining and rigging a simple mounting plate costs about 0.025 $/cm^2 for a medium-sized pattern plate. With these values, the typical cost for a pair of mounting plates can be represented by [13]

$$C_{pm} = 0.58A_{pl} \tag{12.8}$$

where

A_{pl} = plate area, cm^2

Once the cost of the pattern base plate is known, the next step is to calculate the cost of the pattern impressions. This is done by rating the complexity of the casting outer surface, based on the number of surface patches. A surface patch is defined as a segment that is either planar or has a constant or smoothly changing curvature. The intersection of two surface patches appears as a rapid change in slope or curvature. The counting procedure is described below.

Count all of the surface patches that are formed by the pattern but ignore any that are created by a core. When multiple identical features are located on the surface of a part, a power index of 0.7 can be used to estimate the savings in machining identical features. For example, for a boss with 6 surface patches that appears 10 times on a casting, the total number of surface patches for all of the bosses would be calculated as $6 \times 10^{0.7} = 30$ instead of 60. This index is also used in the equation for determining the reduced cost of machining multiple identical impressions. The cost of a set of identical pattern impressions is then given by

$$C_{pi} = R_t(0.313N_{sp}^{1.27} + 0.085A_p^{1.2})N_{pi}^{0.7} \tag{12.9}$$

where

R_t = toolmaking (pattern shop) rate, $/h
N_{sp} = number of surface patches
A_p = projected area of impression, cm^2
N_{pi} = number of identical impressions

Pattern shop rates in the United States are typically between $35 and $40 an hour.

For high-volume production the gating system is formed by additional pattern impressions. The cost of the gating system appears to be based on several factors. The first is the type of metal being fed. For example, gray iron generally does not require risers, so its gating system is easier to design and build than for other metals. Another factor is the number of cavities that must be fed and the distances between the cavities. A survey of pattern costs for an automated casting line [13] revealed that the cost of gating system impressions typically adds 20 to 35% to the cost of the pattern tooling. The more inexpensive the impressions, the higher the percentage of the final cost the gating system will be. Using a mid-range value of 25%, total pattern cost can be represented by

$$C_{pt} = G_f(C_{pm} + C_{pi})$$

(12.10)

where

G_f = gate factor

= 1 for a manual casting process

= 1.25 for an automated casting line

The cost of a core box can be calculated by using a method that is similar to the method for estimating injection molding dies. Like an injection molding die, a core box contains cavities that are formed by the closure of two plates. An extensive amount of machining is required beyond the machining needed to produce the cavities. Ejection pins are located in the lower half to allow for the quick removal of the core. The upper plate contains passages for the sand to flow through as it is blown into the core chamber. Small holes must be drilled into the plate to allow for air to escape the cavity as the sand is blown in to replace it.

The cost of producing a cavity insert is based on the cavity complexity and the projected area (parting line area) of the part. The number of surface patches is determined using the same procedure previously discussed. The cost of the cavity insert for a core box can be represented by Eq. (12.9). The total cost of a core box is thus

$$C_{box} = C_{pi} + C_{pm}$$

(12.11)

where

C_{pm} = the cost of the mounting plates [given by Eq. (12.8)]

Most high-production tooling is made of iron. However, some foundries use stainless steel for producing critical tooling for high-production volumes. For smaller production quantities tools may be made of wood or plastic. The costs for producing tooling from alternative materials can be approximated using the relative tooling costs in Table 12.3. Estimates of tool costs using Eqs. (12.10) and (12.11) are simply multiplied by the appropriate value of the relative tool cost factor R_{tf}.

Table 12.3 provides information on the number of cycles for which a piece of tooling may be used before it is replaced. The data are for one pattern impression or one core box cavity. For example, the predicted production life of a four-cavity cast iron core box is 600,000 cores.

TABLE 12.3 Tool Life and Relative Costs

Pattern or core box material	Relative cost factor, R_{tf}	Tool life per cavity
Wood	0.25	2,500
Plastic	0.40	5,000
Cast iron	1.00	150,000
Stainless steel	1.18	180,000

12.7.4 Processing Costs

Processing costs are broken down into three main categories: the cost of producing the cores, the cost of producing the molds and pouring the metal, and the cost of cleaning the castings.

There are a large number of variables in determining the cost to produce a core. The cost of the core is affected by its size, which determines the number of cavities possible per box and, therefore, the core machine production rate. The production rate of the machine largely depends on the type of core-making process being employed. The type of core-making process and the design of the core influence the scrap rate. The core scrap rate can vary from 4% for a simple core to 40% for a very delicate core.

Some cores call for a refractory coating, which requires equipment and manpower to coat the cores and then bake them to dry the coating. All of these factors combine to form a formidable challenge to obtaining accurate estimates of individual core costs. It is typical, for estimating purposes, for the average weight of cores produced by the core shop per hour to be divided into the number of core process workers. This gives a production rate for the core shop expressed in core kilograms per worker hour. The cost of processing a particular core is then given by

$$C_{core} = \rho_{cs} V_c R_{cm} / [P_{cm}(1 - S_c/100)(P_{ff}/100)] \qquad (12.12)$$

where

ρ_{cs} = density of core sand, kg/m³
V_c = volume of core, m³
P_{ff} = plant efficiency, % (actual production time/total available time) × 100
R_{cm} = worker rate for core making, \$/h
P_{cm} = core production rate, kg/h

For U.S. corporations, a typical value for plant efficiency is 85%.

From a survey of several large foundries [13], an average production rate for cores was found to be approximately 53 kg/worker-hour, with an associated burdened labor rate of 50\$/h. The cost advantage of multicavity core boxes is not as significant as for other molding processes because of the secondary operations involved in removing flash or fins, adding coatings, and curing. For approximate cost estimating, Eq. (12.12) can be used irrespective of the configuration of the core box. As discussed earlier, when considering the cost of core sand, a scrap rate S_c of 8% is reasonable for early costing purposes.

The molding area costs are easier to determine than those of the core area, because there are fewer variations in the processing system. This is because the molding machine can be modeled as a transfer line. The mold area contains a variety of operations. However, since the majority of the operations take place on the mold line, the processing cost is based on the number of castings produced per hour, the number of workers required to support the line, and a burdened labor rate per worker. A typical cope and drag casting line in the Midwest requires 21 workers to achieve a production rate of 285 molds per hour. The burdened rate per worker on this line is $208 per hour. The casting scrap rate, like the core scrap rate, can vary widely depending on the geometry of the part and the type of metal being poured. From a survey of a number of foundries [13], it was found that a value of 2% is reasonable for early cost estimating. The mold line processing cost per casting can be given by

$$C_{mp} = N_{mw}R_{mp}/(N_cP_{mp}(1 - S_m/100)(P_{ff}/100)) \qquad (12.13)$$

where

N_{mw} = number of line workers
R_{mp} = worker rate for molding line production, $/h
N_c = number of mold cavities
P_{mp} = molding line production rate, molds/h
S_m = casting scrap rate, %
P_{ff} = plant efficiency, %

It should be noted that the rate, R_{mp}, of 208$/h is for a large midwestern foundry capable of supplying large castings, such as engine blocks, at high production rates. The large capital equipment to provide this capability results in the very high burdened labor rate. A survey of smaller automated foundries in other parts of the United States [13] suggests that an appropriate rate for R_{mp} for high-volume production of small to medium-sized castings is approximately 95 $/h.

The final operation on the mold line is to remove the castings from the mold. This is usually accomplished by using a ram to push the sand mold out of the flask and to break the mold apart. The mold sand is reclaimed, and the castings are transported to the cleaning department.

The cost of cleaning a casting is mainly based on size and the geometry of the part. However, the type of material cast and the process employed also affect the ease with which the casting may be cleaned.

Cleaning a casting is the most labor-intensive job in the foundry. Even with the aid of automation to remove the gating system and scale from the casting, there can still be a substantial amount of manual work. Chippers and grinders are used to remove the flash and feed system contact area. Of course, the level to which this is done depends on how important the flash removal is. In some cases, the level of finish of the casting is not critical and cost savings can be realized.

The major cost driver in the cleaning process is the casting's geometry. The length of the parting line determines the amount of flash that must be removed. If the part is cored, then the flash formed around the cored openings in the casting must also be removed. However, other factors also affect the cleaning time. Some castings require more reorientations to clean the flash lines than others do, and the time to carry out these reorientations is based on the dimensions and the weight of the casting.

Deep pockets and small openings in a casting also create cleaning problems. A casting with large internal cavities and small holes can create very difficult problems. This is because the sand inside the casting is often not completely broken down and cannot, therefore, be easily removed.

A survey of a large number of castings suggests that the total labor time for cleaning castings (including sand removal, runner system and flash removal, and shot blasting) is approximately proportional to the square root of the casting weight. Regression analysis of casting data from a midsized automated foundry provided the following relationship for cleaning labor time.

$$T_{cl} = 88.4 \times W_p^{0.44} \text{ s/casting} \tag{12.14}$$

The cleaning cost is then given by

$$C_{cl} = R_{cl} T_{cl}/3600 \tag{12.15}$$

where

T_{cl} = total cleaning time, s
R_{cl} = worker rate for cleaning room, \$/h

Cleaning room burdened labor rates vary from approximately \$25/h for small automated foundries in the United States to more than double that value for large foundries supplying major castings to the automobile industries.

12.8 DESIGN RULES FOR SAND CASTINGS

A report written for the North Atlantic Treaty Organization (NATO) [14] states that the primary factor in the selection of sand casting is cost reduction. However, the most important factor in determining the cost of a casting is the design. The study states that the cost of a casting can "easily be doubled by factors controlled by the designer."

The following design rules are taken from industry handbooks [15–17]. Castings certainly can be produced that do not adhere to these design rules. However, this will always be at the expense of increased scrap and possibly large increases in production cost.

Avoid Sharp Angles and Multiple-Section Joints

Metal structure is affected by the shape of the casting section. Solidification of the molten metal begins at the mold face, from which crystals grow into the casting at right angles. A straight section of constant thickness (see Fig. 12.4a) results in uniform cooling, which will in turn produce uniform material properties. On the other hand, sharp angles can cause large temperature variations in the casting, which often lead to casting defects. Hot spots result where the free cooling of the casting is interrupted as parts of the sand are loaded with more energy than other areas. Also, chilled areas arise on external corners from being exposed to two cooling planes (see Fig. 12.4b). The resulting grain structure is not homogeneous and, in particular, weak areas in the casting are created in the areas where the cooling rate is excessive.

A well-designed casting brings the minimum number of sections together at intersections and avoids acute angles. Wherever a number of sections converge, the appropriate solution is to create a large hole like the center of a web. Examples of good and bad section configurations are shown in Fig. 12.5.

Design Sections of Uniform Thickness

Design the casting so that all of the section thicknesses are as consistent as possible. This promotes even cooling of the casting, reducing the likelihood of defects. If larger masses of metal are unavoidable, the designer should make them accessible for feeding either directly or with a riser.

Designing for uniform thickness also reduces the amount of material in a casting, saving weight and reducing machining, and results in a stronger casting. However, if section thicknesses are too small, then feeding problems may occur. The increased cost of scrap caused by incomplete feeding (caused by metal freezing and blocking the section from being completely filled) will normally be higher than the material savings in a lighter casting. The economical minimum section thicknesses of different metals to be sand cast are listed in Table 12.4.

Proportion Inner Wall Thickness

Inner sections in a casting cool more slowly than a section exposed to the mold face. If a complex geometry is necessary, the designer should reduce the inner section thickness to 80% of the outer wall thickness. Also, core section thicknesses should always be greater than the section thickness of the surrounding metal. If the core is too small, it will become overheated and slow down the solidification rate of the surrounding metal, leading to the possibility of defects.

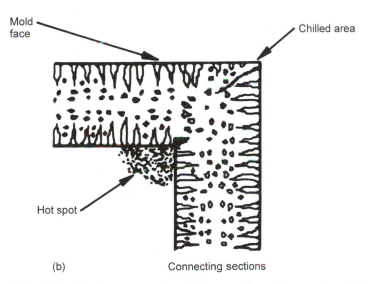

FIG. 12.4 (a) Constant casting thickness with uniform cooling; (b) effects of non-uniform cooling caused by abrupt section changes.

Consider Metal Shrinkage in the Design

Almost all alloys shrink as they solidify. While the patternmaker is the one affected by the shrinkage, the designer must still compensate for it in the design. In a good design, the section thicknesses decrease as the distance from the feed system or riser increases. In order to accomplish this, the designer must be familiar enough with the casting process to be able to visualize how the casting will be fed and adjust the casting's dimensions to assist the metal flow. The greater the shrinkage of the metal, the more the designer must consider it when designing the casting. Table 12.5 lists the shrinkage of several of the commonly cast alloy groups. The amount of shrinkage depends upon the precise carbon content for irons and steels and varies over the ranges shown.

Bad
6 ribs intersect

Better
4 ribs intersect

Best
3 ribs intersect

FIG. 12.5 Examples of good and bad section configurations. (From Ref. 17.)

TABLE 12.4 Minimal Economical Section Thicknesses for Different Section Lengths, S_L

Generic alloy	Thickness for section length 2.5 cm or less (mm)	Thickness for section length between 2.5 and 15 cm (mm)	Thickness for section length greater than 15 cm (mm)
Ductile iron	4.8	12.5	19.0
Gray iron	4.0	9.5	19.0
Malleable iron	3.3	6.4	14.0
Carbon steels	8.0	12.5	25.0
Aluminum	4.0	8.0	16.0
Magnesium	4.0	8.0	16.0

TABLE 12.5 Shrinkage Allowances of Casting Alloys

Generic alloy	Linear shrinkage (%)
Gray iron	0.83–1.3
Malleable iron	0.78–1.0
Carbon steels	1.6–2.6
Aluminum	1.3
Magnesium	1.3

Use a Simple Parting Line

A flat plane, known as a straight parting line, separating the two mold halves, results in more economical casting than a tiered or contoured separating surface. More complex parting lines often result in fewer parts per mold, more costly patterns, less accuracy, and increased scrap. Also, the parting line should be positioned so that it has minimal effect on the functional characteristics of the part.

Locating the parting line in less critical parts of the casting is desirable for two main reasons. First, dimensions around the parting line are the hardest to control. Additionally, flash occurs at the parting line. If the surface around the parting line is not critical, then flash removal costs will be lower.

Define Appropriate Machining Allowances

The machining allowance is material added to the casting to compensate for dimensional and surface variations in the as-cast part. The amount of stock added is a function of the size of the surface to be machined and to a lesser degree the machining method and the final accuracy required. Minimal additional material is needed if only flatness, possibly with some unmachined surface areas, is desired. A larger allowance is required if the full surface is to be machined without any imperfections. Normal machining allowances vary from 0.25 cm for small castings (< 15 cm) to as much as 2.5 cm for large castings (> 250 cm).

Use Economical Tolerances

The tolerances achievable by a foundry vary depending on the types of processes employed at the facility. For example, automated molding machines are capable of producing molds with tighter tolerances than might be produced by hand. Conservative tolerances, which are readily achievable by most foundries and are therefore the most economical, are used in the following discussion.

Tighter tolerances may be obtained by machining, which significantly increases the cost of the casting.

The most basic tolerance is the linear tolerance. It refers to how precisely the distance between two points can be produced. Linear tolerances of ±1.0 mm are readily achievable for small castings. An additional factor of ±0.03 mm should be added for every centimeter over 15 cm for larger parts.

An additional tolerance must be added to the linear tolerance of a dimension that passes through or originates from the parting surface. These additional tolerances reflect variations caused by expansion and contraction of the mold, the metal during solidification, patternmaking tolerances, and vibration of the pattern during removal from the mold. The size of the additional tolerance depends upon the projected area of the casting at the parting surface. The typical tolerance assignment is ±0.25 mm for each 10 cm² of projected area.

Cores create tolerance variation because of the clearance that is necessary for their placement into the mold. The features produced by the core surface can be held to a tighter tolerance than the features produced by the mold surface, because cores are stronger and able to be produced to tighter tolerances than the mold. However, the surface produced by the core may be displaced from the surface created by the mold because of core shift. The additional tolerance for core shift varies with the protected area of the core normal to the dimension being considered. The recommended value is the same as for the additional parting line tolerance given above.

The following list shows the tolerances that must be applied to the different dimensions for the part in Fig. 12.6.

A: linear tolerance
B: linear + core location tolerances
C: linear + core location tolerances
D: linear tolerance
E: linear + parting line tolerances

12.9 EXAMPLE CALCULATIONS

A concept sketch of a pump body is illustrated in Fig. 12.7. It is to be manufactured from cast carbon steel, and a total of 10,000 are required. It is proposed that the bodies be made on a cope and drag casting line using single-cavity molds. The equations in Sec. 12.7 are used to provide initial estimates of the part cost and required tooling investments.

Values of the part and production variables to be used in these calculations are listed below.

Finished part volume $V_{fc} = 438$ cm^3

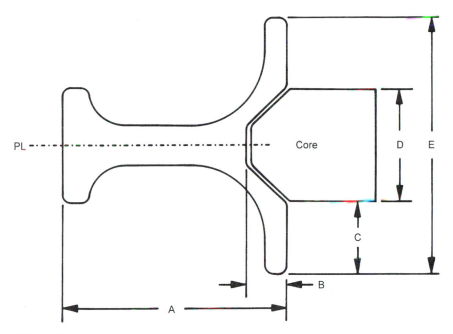

FIG. 12.6 Appropriate tolerances for a cored casting.

FIG. 12.7 (a) Pump body casting; (b) casting core.

Part projected area $A_p = 222$ cm^2
Part length $L = 250$ mm
Part width $W = 150$ mm
Part depth $D = 50$ mm

From Table 12.2, the density of carbon steel is 7833 kg/m^3 and substituting into Eq. (12.1), with appropriate unit conversion, gives the weight of poured metal per part

$$W_p = 7833 \times (438/10^6)(1 + 1.9(400/50)^{-0.701})$$
$$= 4.95 \text{ kg}$$

The minimum melt energy from Table 12.2 is 0.393 (kw·h)/kg, and using the suggested values of $E_{ct} = 0.035\$/$(kw·h) for electricity cost and $E_{ff} = 80\%$ for furnace efficiency, gives furnace energy cost from Eq. (12.2) as

$$C_{en} = 0.035 \times 0.393/0.8 = 0.017 \text{ \$/kg}$$

The fixed furnace cost to cover capital investment is given by Eq. (12.3), using $F_{fc} = 284\$/$m^3,

$$C_{fk} = 284/7833 = 0.036 \text{ \$/kg}$$

Together with an ingot cost $C_{rm} = 0.73\$/$kg (from Table 12.1), and the suggested furnace labor cost $C_{lk} = 0.02\$/$kg, the cost of molten carbon steel at the spout is estimated to be

$$C_{ms} = 0.73 + 0.017 + 0.036 + 0.02 = 0.80 \text{ \$/kg}$$

The weight of the finished casting, after removal of the feed system, is

$$W_{fc} = \rho V_{fc} = 7833 \times (438/10^6) = 3.43 \text{ kg}$$

Thus the cost of metal per finished part, from Eq. (12.5), using steel scrap value $C_{cv} = 0.09\$/$kg and scrap rate $S_g = 2\%$, is

$$C_{mf} = (0.80 \times 4.95 - 0.09 \times (4.95 - 3.43))/(1 - 0.02)$$
$$= 3.90 \text{ \$/part}$$

The estimated cost of mold sand per part is given by Eq. (12.6):

$$C_{msd} = 0.018 \times 4.95 = 0.09 \text{ \$/part}$$

Now turning our attention to the material cost of the cores, the volume of the single core illustrated in Fig. 12.6 is $V_c = 683$ cm^3. Using the cost of core sand

$C_{cs} = 0.084\,\$/\text{kg}$, core sand density $\rho_{os} = 1387\ \text{kg/m}^3$, and a core scrap rate $S_c = 8\%$, then Eq. (12.7) gives the estimated cost of core sand per part to be

$$C_{csd} = 1387 \times (683/10^6) \times 0.084/(1 - 0.08)$$
$$= 0.09\ \$/\text{part}$$

We next estimate the core and molding processing costs in order to determine the likely part cost. The weight of each core to be used in casting a pump body is

$$W_c = 1387 \times (683/10^6) = 0.95\ \text{kg}$$

Using the typical values for core production given in Sec. 12.7.4, the processing cost per core is given by Eq. (12.12):

$$C_{core} = 0.95 \times 50/(53 \times (1 - 0.08) \times 0.85)$$
$$= 1.15\ \$$$

It will be assumed that the pump bodies are to be made at a medium-sized automated foundry in the United States for which an appropriate rate R_{mp} will be $95\,\$/\text{h}$. Equation (12.13) then gives the process cost per casting as

$$C_{mp} = 21 \times 95/(285 \times (1 - 0.02) \times 0.85)$$
$$= 8.40\ \$$$

Finally, the cost of cleaning the casting, and for removal of metal flash and the runner system, is given by Eqs. (12.14) and (12.15). Using the recommended value of $R_{cl} = 25\,\$/\text{h}$

$$C_{cl} = 25 \times 88.4 \times 4.95^{0.44}/3600$$
$$= 1.24\ \$/\text{casting}$$

The piece part cost for the cast part comprises the total material and processing costs. From the preceding calculations these costs have been estimated to be:

$$\text{Material cost} = 3.90 + 0.09 + 0.9$$
$$= 4.08\ \$$$
$$\text{Processing cost} = 1.15 + 8.40 + 1.24$$
$$= 10.79\ \$$$
$$\text{Piece part cost} = 10.79 + 4.08 = 14.87\ \$$$

We now proceed to estimate the cost of the required tooling. Using a spacing of 5 cm between the impressions and the edge of the pattern plates, as recommended in Sec. 12.7.3, the area of the pattern plates is given by

$$A_{pl} = (25 + 2 \times 5)(15 + 2 \times 5) = 875\ \text{cm}^2$$

Thus from Eq. (12.8) the cost of the pattern mounting plates is estimated to be

$$C_{pm} = 0.58 \times 875 = 508 \ \$$$

The number of surface patches, which make up the pump body, equals 12; 5 for the main body, 3 for each side hollow projection, and 1 for the bottom cylindrical hole. Using $A_p = 222 \ cm^2$ (the value given at the beginning of this section), and $R_t = 40 \ \$/h$, the cost of manufacture of the pattern impressions is given by Eq. (12.9) as

$$C_{pi} = 40(0.313 \times 12^{1.27} + 0.085 \times 222^{1.2}) = 2518 \ \$$$

Thus the cost of the two pattern plates, for the cope and drag flasks, manufactured from cast iron, is given by Eq. (12.10) as

$$C_{pt} = 1.25(508 + 2518) = 3783 \ \$$$

The core, illustrated in Fig. 12.7, has a projected area in the vertical direction of $175 \ cm^2$. The length and width of the core are 25 and 13.4 cm, respectively. The area of plates required to mount the cavity and core inserts is given by

$$A_{pl} = (25 + 2 \times 5)(13.4 + 2 \times 5) = 819 \ cm^2$$

Thus from Eq. (12.8) the cost of core box plates is estimated to be

$$C_{pm} = 0.58 \times 819 = 475 \ \$$$

The number of surface patches, which make up the core, equals 6; 3 for the main body and 2 for each side projection. Using $A_p = 175 \ cm^2$ in Eq. (12.9), the cost of cavity and core inserts is estimated to be

$$C_{pi} = 40(0.313 \times 6^{1.27} + 0.085 \times 175^{1.2}) = 1793 \ \$$$

Thus the cost of the core box, manufactured from cast iron, is estimated to be

$$C_{box} = 1793 + 475 = 2268 \ \$$$

Finally, the tooling investment may be amortized over the total number of pump bodies to be produced to give

$$C_{tc} = (3783 + 2268)/10000 = 0.61 \ \$$$

where C_{tc} is the tooling cost per casting.

REFERENCES

1. Brown, J.R., Foseco Foundryman's Handbook, 10th ed. Butterworth-Heinemann, 1996.
2. Wukovich, N., Evaluating Side Risers and Necks, Part 1, Modern Casting, December 1988, p. 42.

3. Wukovich, N., Evaluating Side Risers and Necks, Part 2, Modern Casting, January 1988, p. 49.
4. Wukovich, N., Evaluating Side Risers and Necks, Part 3, Modern Casting, February 1988, p. 56.
5. American Society of Metals Casting, Vol. 15, Metals Handbook, 9th ed. ASM International, Metals Park, OH, 1988, pp. 577–597.
6. Suschil, T., Designing Gates and Risers in an Artful Compromise, Modern Casting, March 1989, pp. 27–29.
7. Wieser, P.F., Steel Castings Handbook, 5th ed., Steel Founders Society of America, 1980.
8. Bralower, P.M., Sand Molding: From Hand Ramming to Near Net Shape Castings, Modern Casting, May 1989, pp. 53–58.
9. Burditt, M.F., Designs and Operation of Melting Furnaces Differ Markedly, Modern Casting, August 1989, pp. 51–55.
10. Burditt M.F. and P.M. Bralower, Good Pouring Practice Contributes to Quality Castings, Modern Casting, 1989, pp. 59–63.
11. Mrdjenovich, R., Shakeout: Separating the Casting from Its Mold, Modern Casting, October 1989, pp. 45–47.
12. Luther, N., Cleaning and Finishing: Getting the Casting Ready for Shipping, Modern Casting, November 1989, pp. 53–58.
13. Kobrak, G., Design and Early Cost Estimation of Sand Castings, M.S. Thesis, University of Rhode Island, Kingston, 1993.
14. Mietrach, D., AGARD Handbook on Advanced Casting, AGARD-AG-299. North Atlantic Treaty Organization Advisory Group for Aerospace Research and Development, Bremen, Germany, p. 9.
15. Casting Engineering and Foundry World, Continental Communications, Inc., Bridgeport, CT.
16. Customers Foundry Orientation Manual, Robinson Foundry, Alexander City, AL.
17. Bralla, G.B., Handbook of Product Design for Manufacturing, McGraw-Hill, New York, 1986.

13

Design for Investment Casting

13.1 INTRODUCTION

The investment casting process is capable of producing complex castings with tight tolerances and superior surface finish and can meet the highest performance standards, such as in aircraft jet engine applications. Other advantages of the process include the ability to cast materials that are impossible to forge and difficult to machine and to produce prototypes through direct machining of wax patterns.

Investment casting is a uniquely flexible process for making parts of high complexity, and it is therefore ideally suited to reduced part count designs—one of the main objectives of design for manufacture and assembly (DFMA).

There are two types of investment casting processes: solid mold investment casting and ceramic shell mold investment casting. The two processes differ primarily in the way the mold is formed. In the "solid" mold process, the wax or plastic pattern is placed into a container, and mold material is poured around the pattern, solidifying into a solid block. In the "shell" process, the pattern is dipped or "invested" in the mold material, leaving a coating of uniform thickness. The coating is allowed to dry and the dipping process is repeated. The multiple coats form a hard ceramic shell mold. The ceramic shell mold process described in this chapter is the major form of investment casting for engineering applications.

13.2 PROCESS OVERVIEW

A simplified representation of the ceramic shell mold investment casting process is shown in Fig. 13.1. The process starts with the fabrication of a wax or plastic

549

FIG. 13.1 Simplified ceramic shell mold investment casting process.

7

8

9

pattern. The pattern has the same shape as the final cast part, but with dimensional allowances for the shrinkage that occurs during the casting process. Usually a number of patterns are then assembled into a cluster. The cluster includes all the necessary gates and runners for providing metal flow to each pattern in the assembly. The completed cluster is invested (dipped) into a ceramic slurry, removed, and then covered with a stucco material. Investing and stuccoing is repeated a number of times. The investing and stuccoing build up a shell around the pattern cluster. Once the ceramic shell mold is completely dry, the pattern is melted out leaving cavities in the shape of the parts. The mold is fired to reach its final strength and, while the mold is still hot, molten metal is poured into the cavity and allowed to solidify. The mold material is then knocked off the casting and the parts are separated.

13.3   PATTERN MATERIALS

Patterns for investment casting are injection-molded of either wax or plastic. Paraffin and microcrystalline waxes are the most common base materials for patterns [1].

Their low melting points and low viscosity make waxes easy to mold into patterns, assemble into clusters, and melt out of shell molds without damaging the mold. Since waxes can be injected at low pressure and temperature, and have low abrasiveness, tooling costs are low.

The strength and toughness of pattern waxes can be improved by the addition of plastics; solidification shrinkage of the patterns is reduced by the addition of resins and powdered solid fillers. Other additives are used in wax blends for various purposes. Dyes are added to allow different wax formulations to be distinguished by color. Antioxidants are added to help reduce thermal degradation. Oils and plasticizers are added to adjust injection properties [1].

After wax, plastic is the next most favored material for patterns—polystyrene being the most common. Plastics patterns are strong and resist damage when extremely thin sections are required in the cast part. However, the tooling for plastic patterns is more expensive than for wax.

13.4   PATTERN INJECTION MACHINES

Investment casting patterns are usually made by injecting wax or plastic into metal molds. Wax is injected in a liquid, slush, paste, or solid form. Typical injection temperatures range from 43 to 77°C (110 to 170°F) and injection pressures from 275 kPa to 10.3 MPa (40 to 1500 psi). Liquid wax injection takes place at the higher temperature and lower pressure ends of the ranges, while solid wax injection is done using lower temperatures and higher pressures. Simple equipment for wax injection consists of a pneumatic unit with a closed, heated tank of wax, as shown in Fig. 13.2. The tank has a thermostat, pressure regulator, heated valve, and nozzle. Shop air provides the pressure needed. A metal mold is held to the nozzle with one hand, while the other hand operates the valve. Units of this type are usually limited to less than 690 kPa (100 psi) and only inject liquid wax. Equipment like this is used on a large variety of small parts [1].

Patterns that cannot be made on simple equipment because of size, complexity, dimensional requirements, or quantity, are made on hydraulic machines (Fig. 13.3). Hydraulic machines can use higher injection pressures that improve injection time and quality. They allow the use of larger molds and higher injection pressures and are readily available in a range from 9.1 to 363 Mg (10 to 400 tons) clamping force [1].

Hydraulic machines with manual or semiautomatic modes of operation typically have horizontal platens. The lower platen is stationary and holds the

FIG. 13.2 Simple investment casting pattern injection equipment.

lower half of the mold. The other platen moves vertically, holding the upper mold half, and is forced down to clamp the mold shut prior to injection of the wax. With semiautomatic machines the operation is automatic except that the pattern is removed manually. Automatic hydraulic machines typically have vertical platens. One platen is fixed while the other slides back and forth to open and close the mold, and the pattern is automatically ejected.

FIG. 13.3 Hydraulic pattern injection machine.

Hydraulic machines usually have a reservoir for liquid and slushy wax. The wax must be continuously mixed to keep it uniform and to keep any solids in suspension. Some facilities have central wax supplies. Machines used for paste wax injection receive their supply of wax from a removable metal cylinder containing preconditioned wax. The wax is injected as a piston forces the wax out of the cylinder. Solid wax injection uses preconditioned wax billets loaded into the machine. Both paste and solid injection machines require separate baths or ovens for preconditioning or pretempering of the wax cylinders or billets.

Polystyrene patterns are produced using standard plastic injection molding machines as described in Chapter 8.

13.5 PATTERN MOLDS

Most molds for molding patterns are manufactured by machining the part form directly into the mold. Aluminum is the most widely used material for wax injection molds, whereas tool steel is most frequently used for plastic injection molds. Molds for wax patterns will last indefinitely but those for plastic patterns will wear and require replacement after 1 million cycles or so.

13.6 PATTERN AND CLUSTER ASSEMBLY

The majority of patterns are injected in one piece with gates attached. Large size or complexity can require a pattern to be made up of several pieces. The pieces may be all wax, all polystyrene, or a combination of the two. When small, delicate details are required on parts too large to be made as one-piece plastic patterns, fine plastic details may be assembled to a larger wax pattern.

Once the individual pattern assembly is complete, the metal feeding system made up of the required runners, sprue, and pouring cup is added. Smaller parts are processed in clusters when a number of patterns share the same feeding system. The number of patterns on a given cluster varies according to part size and process limitations.

Most assembly of patterns, feeding systems, and clusters is done by hand. Wax components are assembled by wax welding. This is done by melting the joining surfaces with a hot iron, spatula, or small gas flame. The two wax pieces are pressed together until the wax solidifies, welding the two components together.

Once the cluster, or individual part, is fully assembled it must be cleaned. The cluster must be free of dust, loose wax, and mold lubricant. The cluster is sometimes lightly etched to promote adhesion of the ceramic mold slurry.

13.7 THE CERAMIC SHELL-MOLD

The cluster or single pattern assembly is dipped into ceramic slurry and then covered with stucco particles. The ceramic slurry contains very small particles and is capable of reproducing the fine detail and smooth surface finish from the pattern. The slurry also contains a binder that gives strength to the mold. The stucco particles prevent the slurry from cracking and pulling away when drying; they also help to provide adhesion between coats and help build mold thickness. Each mold layer, consisting of one slurry application followed by one stucco coat, is allowed to harden prior to adding the next layer. Layers are added until the required thickness or mold strength is reached.

The three most commonly used refractories for ceramic shell molds are silica, zircon, and aluminum silicate, and they are usually used in combination [1].

Silica is used in the form of silica glass or fused silica. It is made by melting natural quartz sand, solidifying it into glass, and then crushing it into various size particles for stucco and into fine powder for slurry mixes. Since it can be readily dissolved, it can be easily cleaned from those cavities of finished castings that cannot be cleaned by mechanical methods. Zircon is natural sand, ground into powder for slurry mixes, and is usually used in combination with fused silica. Where used for stucco, zircon is used only for primer coats. It does not come in sizes large enough to be used as a backup coat stucco. Aluminum silicates contain 42 to 72% alumina, with the balance being silica.

Slurries require the addition of a binder to give them strength. The most common binders are hydrolyzed ethyl silicate, colloidal silica, and sodium silicate [1].

Dipping, draining, and stuccoing of patterns or clusters can be done manually, robotically, with hard automation, or by a combination of the methods. Robots are increasingly being used for making shell molds [1]. They increase productivity, provide the ability to process larger parts or clusters, and produce more uniform coatings.

When dipped into the primer slurry, the cluster is manipulated in the slurry tank to guarantee that all surfaces and cavities are coated with slurry. The cluster is continually rotated and manipulated to make sure the slurry coats the cluster evenly. After the excess slurry has drained off, the cluster is covered with stucco particles to stop any further runoff of slurry.

Stucco particles are applied by either dipping the cluster in a fluidized bed or rotating the cluster under falling particles from an overhead screen. Primer slurries have a finer refractory powder and are stuccoed with finer particles than the following backup layers. This is to provide as smooth and as detailed a representation as possible of the pattern surface and to help resist metal penetration. The primary function of backup coats is to provide the necessary strength in the mold to survive pattern meltout and metal pouring. Usually one or two primer coats in addition to the first coat are applied. The total number of primer and backup coats is within the range of 5 to 15 [1].

13.8 CERAMIC CORES

Cores are widely used in investment casting. The two basic categories of cores are self-forming cores and preformed cores.

Self-forming cores are formed by the mold-making process. The internal passages are already present in the wax pattern. The self-forming core is produced when the pattern is dipped into the ceramic slurry that fills the internal passages. The cavities for self-forming cores are produced by two different methods. One method is the use of core pulls in the pattern mold. The other is by using soluble wax cores placed in the pattern mold prior to wax injection. After wax is injected around the soluble wax core, the pattern is removed from the mold and the soluble core is dissolved out of the pattern. Soluble wax cores are made of solid polyethylene glycol with sodium bicarbonate or calcium carbonate used as a filler powder. The cores are usually dissolved out in an aqueous acid solution that will not attack the base wax used for the pattern body [1]. When openings are large enough to allow ceramic molding slurry complete access to internal cavity features, self-forming cores are preferred. They are used extensively in small hardware castings [1]. When access to internal features is restricted, preformed cores must be used.

Preformed ceramic cores require their own molds and are usually injection-molded in a process in which fine refractory powder is mixed with an organic vehicle and injected into hardened steel molds. The injected mixture is very abrasive and limits mold life. After forming, cores are heat-treated in a two-stage process. In the first stage, the organics are removed and in the second stage the cores are sintered to achieve their final strength and dimensions [1]. Preformed ceramic cores are usually placed in the pattern mold and wax is injected around them.

13.9 PATTERN MELTOUT

The thermal expansion of the waxes used for patterns is many times that of the mold materials. To reduce the tendency for mold cracking, molds are heated very quickly, so that the surface of the pattern melts before the temperature of the main body of the pattern rises appreciably. As the pattern heats up and expands, the melted surface layer is squeezed out of the mold, making room for the expanding pattern and preventing the mold from cracking [1].

The primary methods used for melting patterns from molds are steam autoclave dewaxing and high-temperature flash dewaxing, with steam autoclave dewaxing being the most common method. Saturated steam with rapid pressurization to 550 to 620 kPa (80 to 90 psi) is used. Full pressure is attained in 4 to 7 seconds [1]. Molds are dewaxed in about 15 minutes depending on the size of the mold. The wax can be recovered for reuse.

In flash dewaxing, the mold is placed in a furnace at 870 to 1095°C (1600 to 2000°F). The furnace has an open bottom to allow the wax to collect outside of the furnace as it pours out of the mold. Some of the wax burns as it falls from the furnace. This means that wax reclaimed by this method is somewhat deteriorated. Polystyrene patterns are also removed from molds by this method [1].

13.10 PATTERN BURNOUT AND MOLD FIRING

After pattern meltout, the mold is fired to remove any remaining free or chemically bonded moisture and to burn out any remaining pattern material. Any organic material used in the slurry is burned off and the mold is sintered. Sometimes the mold is allowed to cool for inspection and a separate preheat cycle is done just prior to metal pouring. Batch- or continuous-type gas furnaces are most often used for mold firing, burnout, and preheat. Burnout and sintering temperature is between 870 and 1095°C (1600 and 2000°F) [1].

13.11 KNOCKOUT AND CLEANING

After the casting has cooled, it must be freed from the mold. This is referred to as knockout. The mold is usually broken off using a vibrating pneumatic hammer. Shot blasting is used to remove any mold material stuck to the part surface. The blasting operation is done both manually and on automatic cycles [1].

If cores need to be removed, the entire cluster is dipped into a molten caustic bath where the cores are dissolved out.

13.12 CUTOFF AND FINISHING

After the mold has been cleaned, risers and gates must be removed. For a cluster of parts, each part must be cut from the cluster at its gates; then any separate risers need to be cut off. Materials such as aluminum, magnesium alloys, and some copper alloys are cut off using band saws. Abrasive wheels are used to cut off other copper alloys, steels, ductile iron, and superalloys. Brittle alloys may be knocked off with a hammer if notches were provided in the gates. Cutting torches are used when gates are not accessible to the other methods. After cutoff, abrasive wheels and belts are used to grind the gate stubs flush.

13.13 PATTERN AND CORE MATERIAL COST

There are hundreds of different wax pattern and core materials available. Table 13.1 lists material properties and costs for polystyrene and for some typical wax

TABLE 13.1   Pattern Material Properties

Pattern material	Density (g/cm^3)	Thermal diffusivity (mm^2/s)	Inject temp. $(°C)$	Eject temp. $(°C)$	Mold temp. $(°C)$	Inject pressure (MN/m^2)	Cost $(\$/kg)$
Polystyrene	1.59	0.090	218	77	27	96.5	1.12
Wax 1: liquid	0.99	0.092	65	50	25	1.4–3.4	3.09
Wax 2: liquid	0.97	0.092	67	50	25	1.4–3.4	2.87
paste	0.97	0.092	60	50	25	1.4–3.4	2.87
solid	0.97	0.092	49	48	25	2.7 min	2.87
Wax 3: liquid	1.00	0.092	68	50	25	1.4–3.4	3.17
Wax 4: liquid	1.13	0.092	64	50	25	1.4–3.4	2.45
paste	1.13	0.092	58	50	25	1.4–3.4	2.45
solid	1.13	0.092	51	50	25	2.7 min	2.45
Wax 5: liquid	1.00	0.092	64	50	25	1.4–3.4	4.74

Compiled from an industry source of pattern materials and Chapter 8.

blends used for patterns. Table 13.2 lists the common applications for these pattern materials, and Tables 13.3 and 13.4 present the same information for core materials. The material costs given in these tables include allowances for typical order quantities and for conditioning of waxes before delivery to the wax injection machine.

TABLE 13.2   Pattern Material Applications

Material	Description
Polystyrene	Plastic. For use in very small or fragile parts required in large quantities.
Pattern wax 1	Nonfilled. Liquid injection. For thin-walled parts. Good in applications requiring cores. Good for aircraft castings requiring superior surface finish.
Pattern wax 2	Nonfilled. Liquid, paste, or solid injection. For thin-walled parts. Low brittleness. Good where pattern tooling has many loose or moveable pieces. Commonly used in commercial applications.
Pattern wax 3	Filled. Liquid injection. Good for patterns with fragile cores. Can be used in thin- or thick-section patterns. Good for aircraft castings requiring superior surface finish.
Pattern wax 4	Filled. Liquid, paste, or solid injection. For parts with heavy sections, or thick as well as thin walls. Commonly used in commercial applications.
Pattern wax 5	Nonfilled. Liquid injection jewelry wax. Superior surface finish. Easy to repair. Very flexible. May be milled or filed without chipping.

Compiled from an industry source of pattern materials and Chapter 8.

TABLE 13.3 Core Material Properties

Core material	Density (g/cm^3)	Thermal diffusivity (mm^2/s)	Inject temp. (°C)	Eject temp. (°C)	Mold temp. (°C)	Inject pressure (MN/m^2)	Cost ($/kg)
Soluble: liquid	1.00	0.092	67	50	25	1.4.–3.4	3.66
solid	1.00	0.092	52	51	25	2.7 min	3.66
Silica ceramic	1.60	0.110	232	52	27	140	1.00

Compiled from an industry source of core materials and Chapter 8.

TABLE 13.4 Core Material Applications

Material	Description
Soluble wax	For liquid or solid injection. Dissolved out of cluster prior to investing in ceramic slurry.
Silica ceramic	Fused silica-based. For most applications. Silica can be leached out of the casting. Assumed to have a polyethylene or wax-based organic vehicle for injection.

Compiled from an industry source of core materials.

The material cost for a pattern or core, C_{pm} ($), is given by

$$C_{pm} = D_{pm}M_{cp}V(1 + S_a)/1000 \qquad (13.1)$$

where

D_{pm} = density of pattern or core material, g/cm^3
M_{cp} = pattern or core material cost per unit weight, $/kg
V = volume of the part, cm^3
S_a = volume shrinkage allowance for the cast metal

Typical metal shrinkage allowances, S_a, are presented in Table 13.5. It is assumed that the material in the gate and runner systems is formed from recycled wax.

Example

The part shown in Fig. 13.4 is to be cast in phosphor bronze. It has a volume of 3.326 cm^3 and two holes perpendicular to the direction of forming. These holes could be formed by side-pulls in the mold—one on each side. The cost for the type 2 liquid wax material in the pattern is

$$C_{pm} = 0.97 \times 2.87 \times 3.326(1 + 0.04)/1000 = \$0.00963$$

TABLE 13.5 Volume Allowances for Solid Contraction of Metals

Metal or alloy	Shrinkage (%)	Metal or alloy	Shrinkage (%)
Aluminum alloys	3	Magnesium bronze	6
Aluminum bronze	6	Copper nickel	6
Yellow brass	4	Nickel	6
Grey cast iron	3	Phosphor bronze	4
White cast iron	6	Carbon steel	5
Tin bronze	5	Chromium steel	6
Lead	8	Magnesium steel	8
Magnesium	6	Tin	6
Magnesium alloys (25%)	5	Zinc	8

Adapted from Ref. 4.

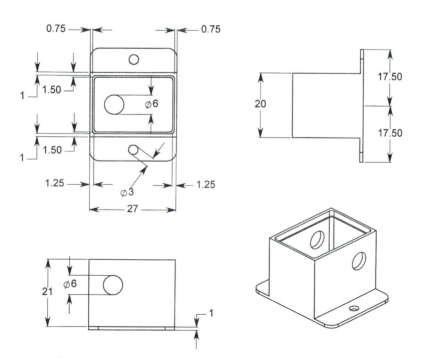

FIG. 13.4 Sample part (dimensions in mm).

13.14 WAX PATTERN INJECTION COST

The first factor to be determined is the size of the injection molding machine and, hence, the rate. The clamp force is a key factor in this determination and its estimation is covered in Chapter 8 for plastic injection molding. The data for wax injection machines is given in Table 13.6.

For plastic injection molding it was assumed that because of the restrictions of runners and gates, the pressure in the mold is about 50% of the injection pressure. With wax injections these restrictions do not cause a significant pressure drop, so the pressure in the mold will be assumed equal to the injection pressure.

The total projected shot area A_s, (cm^2), is given by

$$A_s = n_{pd}A_p(1 + P_{rv}) \tag{13.2}$$

where

n_{pd} = number of patterns (cavities) per mold
A_p = projected area of one part in the molding direction, cm^2
P_{rv} = proportion of runner volume

The shot size and proportion of runner volume will be assumed to be the same as that for plastic injection molding. Approximate values are given in Table 8.2.

Example

(i) The projected area for our example part is 8.88 cm^2 and roughly extrapolating the data in Table 8.2, the proportion of runner volume would be about 60%. Assuming two parts (cavities) per mold, Eq. 13.2 gives a total projected shot area of

$$A_s = 2 \times 8.88 \times (1 + 0.6) = 28.42 \text{ cm}^2$$

TABLE 13.6 Wax Injection Machine Data

Clamping force (kN)	Shot size (cm^3)	Opening speed (cm/s)	Closing speed (cm/s)	Maximum clamp stroke (cm)	Maximum flow rate (cm^3/s)
107	1,885	2.54	2.54	38.1	82
311	1,885	2.54	2.54	51.1	82
445	1,885	2.54	2.54	51.1	82
890	18,275	3.81	3.18	863.6	82
1,334	18,375	2.54	2.54	1,371.6	82
2,670	37,697	2.54	2.54	863.6	82

Compiled from 1994 industry supply literature.

(ii) The recommended injection pressure P_i for liquid wax 2 is, on average, $2.4\,\text{MN/m}^2$ and thus the maximum separating force, F, is given by the shot area times the injection pressure.

$$F = (28.42 \times 10^{-4}) \times 2.4 \times 10^6 \text{ N} = 6.82 \text{ kN}$$

If the available wax injection machines are those listed in Table 13.6, the smallest machine is easily large enough for our example part. This machine would take a shot size of $34\,\text{cm}^3$, has a clamp stroke of 20 cm, and is more than adequate.

13.15   FILL TIME

Liquid wax flows easily under very low pressure; even when solidified in the mold it will flow by shearing with relative ease. Unlike plastic injection, we can assume that the flow rate remains at the maximum value of $82\,\text{cm}^3/\text{s}$ (Table 13.6). The mold fill time, t_f (s), is given by

$$t_f = V_s/Q_{mx} \tag{13.3}$$

where

V_s = required shot size, cm^3
Q_{mx} = maximum wax injection flow rate, cm^3/s

Example

The shot size for our two-cavity mold would be about three times the pattern volume, or roughly $10\,\text{cm}^3$. Thus the estimated fill time would be

$$t_f = 10/82 = 0.122 \text{ s}$$

13.16   COOLING TIME

Unlike plastic injection molded parts, which are generally thin-walled, patterns for investment casting may also be thick-walled or cylindrical. For a thin-walled pattern, the approximate cooling time given by Eq. (8.5) can be used where it is assumed that heat flow is perpendicular to the part wall.

Figure 13.5 shows the three different conditions considered here. In Fig. 13.5a, the thin-wall section, heat flow is perpendicular to the section, in Fig. 13.5b heat flows perpendicular to the length and width of the section, and in Fig. 13.5c heat flows radially from the surfaces of the cylindrical section.

The following equations will be used for the cooling time, t_c (s):

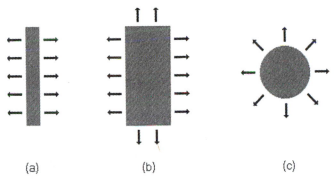

FIG. 13.5 Cooling sections.

for thin-walled sections:

$$t_c = \frac{h_{max}^2}{9.87\alpha} \ln \frac{1.273(T_i - T_m)}{(T_x - T_m)}$$

(13.4)

for thick-walled sections:

$$t_c = \frac{l_s^2 w_s^2}{9.87(l_s^2 + w_s^2)\alpha} \ln \frac{1.62(T_i - T_m)}{(T_x - T_m)}$$

(13.5)

for solid cylindrical sections, where $L/D > 1$

$$t_c = \frac{d_{max}^2}{23.1\alpha} \ln \frac{1.6(T_i - T_m)}{(T_x - T_m)}$$

(13.6)

where

h_{max} = maximum wall thickness, mm
l_s = section length, mm
w_s = section width, mm
d_{max} = maximum section diameter, mm
α = thermal diffusivity, mm^2/s
T_i = injection temperature, °C
T_m = recommended mold temperature, °C
T_x = recommended pattern or core ejection temperature, °C

The cooling times given by these equations can be used without correction for plastic, liquid wax, or paste wax injection. For solid wax injection, the equations will overestimate the cooling time, and correction factors of 0.46, 0.6, and 0.49 should be applied to Eqs. 13.4, 13.5, and 13.6 respectively.

Example

Our example part is thin-walled with a maximum thickness of 1.5 mm. Thus the cooling time is given by

$$t_c = [1.5^2/(9.87 \times 0.092)] \ln(1.273(67 - 25)/(50 - 25)) = 1.884 \text{ s}$$

13.17 EJECTION AND RESET TIME

The ejection and reset time for semiautomatic wax injection is handled differently for wax patterns than for plastic injection molding. The machine open and close times, t_{oc} (s), can be estimated by dividing the required stroke distance by the respective speeds given in Table 13.6 and adding a 1 s dwell time. Thus

$$t_{oc} = (c_f h_d + c_h)(1/v_c + 1/v_o) + 1 \tag{13.7}$$

where

c_f = pattern clearance factor
h_d = pattern depth, cm
c_h = hand clearance, cm
v_c = press closing speed, cm/s
v_o = press opening speed, cm/s

The pattern clearance is from one to two times the thickness of the pattern and the hand clearance is about 10 cm.

For semiautomatic mold operation, the wax pattern is usually removed manually by reaching into the opening mold halves. The ejection time is the time it takes the operator to reach into the mold, pick up the wax pattern, remove the pattern, place it on the work bench, and press buttons to initiate the next cycle. It is assumed that the operator is reaching into the mold as the mold is still opening. It takes about 2 s for an operator to remove the pattern and initiate the machine cycle by pressing the palm buttons. If a core needs to be placed in the mold, it takes about 3 s per core of additional time. About every ten cycles the mold is sprayed with a release agent such as silicon. This operation takes about 4 s, or 0.4 s per cycle.

The total reset time, t_r (s), for semiautomatic pattern injection can now be determined by adding the ejection time to the open and close time.

$$t_r = t_{oc} + 0.4 + (3n_c + 2)n_{pd} \tag{13.8}$$

where

n_c = number of cores per pattern to be placed in the mold
n_{pd} = number of patterns (cavities) per mold

Example

From Table 13.6 the press closing and opening velocities are 2.54 cm/s. We will assume a pattern clearance factor of 2 to remove the 2.2 cm deep pattern from the mold. Thus the machine open and close time is

$$t_{oc} = (2 \times 2.2 + 10)(1/2.54 + 1/2.54) + 1 = 12.34 \text{ s}$$

and since there are two patterns and no cores in the mold, the total reset time is

$$t_r = 12.34 + 0.4 + (3 \times 0 + 2)2 = 16.74 \text{ s}$$

For manual molds the ejection time and reset time is slightly different. The mold is assumed not to be fastened to the injection machine platens. The mold is slid into place on the lower platen and against the injection nozzle. Then the operator hits the buttons to activate the injection cycle. The upper platen need only travel about a centimeter to allow the mold to slide. The open and close time is then

$$t_{oc} = 0.5(1/v_c + 1/v_o) + 1 \tag{13.9}$$

Ejection of the pattern is accomplished manually for manual hand molds. First the mold halves must be separated. Any side-pull-type cores or inserts must be removed and the mold reassembled. The palm buttons can now be hit to initiate the next cycle.

To estimate the time to accomplish manual mold ejection, 15 s is allowed for sliding the mold in and out of the press, separating and replacing the mold halves, removing the part from the mold, and pressing the palm buttons. The time to remove and replace each side-pull insert is estimated at 4 s per insert. If there is a lifter type device, 6 s is allowed. An unscrewing device is given 10 s. Silicon spray mold release is applied about once every five cycles. Equation 13.10 can be used for estimating the total reset time, t_r (s), for a manual mold with one pattern

$$t_r = t_{oc} + 15.8 + 3n_c + 4n_{sp} + 6n_l + 10n_{ud} \tag{13.10}$$

where

n_{sp} = number of side-pulls per pattern
n_l = number of lifters per pattern
n_{ud} = number of unscrewing devices per pattern

Example

For a hand mold, the open and close time will be

$$t_{oc} = 0.5(1/2.54 + 1/2.54) + 1 = 1.4 \text{ s}$$

and the total reset time, for one pattern per mold and two side-pulls, is

$$t_r = 1.4 + 15.8 + 3 \times 0 + 4 \times 2 + 6 \times 0 + 10 \times 0 = 25.2 \text{ s}$$

13.18 PROCESS COST PER PATTERN OR CORE

The total cycle time, t_t (s), can be found by adding together the three times: injection time, cooling time, and reset time. Thus,

$$t_t = t_f + t_c + t_r \tag{13.11}$$

Finally, the process cost per pattern or core, C_{ip} ($), is the cycle cost divided by the number of patterns per mold.

$$C_{ip} = M_i t_t / (3600 n_{pd}) \tag{13.12}$$

where

M_i = machine and operator rate, $/hr (obtained from Tables 13.7 or 13.8)

The combined machine and operator rates in Tables 13.7 and 13.8 were calculated by assuming machine and operator overheads of 100%. The hourly labor rate was taken to be $11.50/hr.

TABLE 13.7 Wax Injection Machine Rates

Clamping force (kN)	Machine cost ($)	Combined operator and machine rate ($/hr)
107	31,000	30
311	43,000	31
445	51,000	33
890	83,000	38
1334	143,000	47
2670	220,000	58

Compiled from 1994 industry supply literature.

TABLE 13.8 High Pressure Ceramic Core Injection Machine Rates

Clamping force (kN)	Machine cost ($)	Combined operator and machine rate ($/hr)
667	90,000	39
1423	160,000	49

Compiled from 1994 industry supply literature.

Example

The total cycle time for semiautomatic operation is

$$t_t = 0.122 + 1.884 + 16.74 = 18.75 \text{ s}$$

and for a machine and operator rate of $30/hr, the process cost per pattern is

$$C_{ip} = 30 \times 18.75/(3600 \times 2) = \$0.078$$

13.19 ESTIMATING CORE INJECTION COST

The cost to inject soluble wax cores is the same as for wax patterns. Ceramic cores are different. They require higher pressures and the machines are more expensive. Machine rates are given in Table 13.8. Reset and ejection time should be handled like wax patterns. Cooling and injection times are similar to those for plastic patterns and should be estimated using the techniques for the injection of plastic parts. Ceramic cores require additional processing after injection before they can be used in a wax pattern. They must undergo a two-stage heat treatment. The first is a potentially lengthy process to remove the organic injection vehicle. The second stage is a sintering process to give the core its final strength. Ceramic cores can also have a high scrap rate. A scrap rate of between 10 and 15% can be assumed. They can break at ejection from the mold, during heat treatment, or during handling.

13.20 PATTERN AND CORE MOLD COST

Methods for the estimation of the cost of plastic injection molds are presented in Chapter 8. Molds for wax pattern and cores are similar to plastic injection molds. However, because of the lower pressures and temperatures required for wax patterns and because aluminum can be used for the molds instead of steel, the cost is very much lower than that for plastic injection molds.

Wax patterns can be made using semiautomatic molds or manual hand molds. For the purposes of estimating the mold base cost for wax pattern injection using semiautomatic molds, it will be assumed that the mold base consists of two sets of plates made of aluminum. The mold base cost C_b ($) can be estimated using the following equation:

$$C_b = C_{vp} + C_{fp} \tag{13.13}$$

where C_{vp} ($) is the cost of the plates containing the core and cavity and is given by

$$C_{vp} = 0.0215h_p A_c + 0.428A_c n_{pl} + 14.27h_p + 32.18n_{pl} \tag{13.14}$$

and where C_{fp} (\$) is the cost of the ejector, riser, and stripper plates, whose thickness does not depend on the depth of the pattern:

$$C_{fp} = 0.66A_c + 366 \qquad (13.15)$$

where

A_c = projected area of the mold base, cm^2
n_{pl} = number of core and cavity plates
h_p = combined thickness of the core and cavity plates, cm

Hand molds require only the variable plates and therefore C_{fp} is zero.

The area of the mold base, A_c, is determined by adding appropriate clearances to the length and width of the part measured in a plane perpendicular to the forming direction. We can use the same clearance value of 7.5 cm used for plastic injection molds. This clearance allows the cavity and core details to be inserted in the mold base plates. However, additional space is not required for side-pulls. These features are usually operated by pneumatic cylinders mounted on the outer mold surface. When a hand mold is used, it will generally have only one cavity. Clearance at the edge of the mold will be assumed to be 2.5 cm. The combined thickness of the core and cavity plates, h_p (cm), is equal to the sum of the depth of the part and the clearance required to the outer surfaces of the plates. The following equation for h_p will account for all types of wax dies.

$$h_p = h_d + n_{cl}h_{cl} \qquad (13.16)$$

where

h_d = depth of the part, cm
n_{cl} = number of clearances required
h_{cl} = minimum clearance, cm

Normally the number of clearances will be 2, and the minimum clearance is 2.5 cm for hand molds and 7.5 cm for semiautomatic wax injection.

Example

The depth of our sample part is 2.2 cm. From Eq. 13.16, the combined thickness of the core and cavity plates for semiautomatic wax injection is

$$h_p = 2.2 + 2 \times 7.5 = 17.2 \text{ cm}$$

The length and width of the part are 3.5 and 2.7 cm respectively. The area of the mold base is therefore

$$A_c = (3.5 + 15)(2.7 + 15) = 327.5 \text{ cm}^2$$

The cost of the ejector, riser, and stripper plates (Eq. 13.17) is

$$C_{fp} = 0.66 \times 327.5 + 366 = \$582$$

and, assuming two cavity and core plates, the cost of these (Eq. 13.16) is

$$C_{vp} = 0.0215 \times 17.2 \times 327.5$$
$$+ 0.428 \times 327.5 \times 2 + 14.27 \times 17.2 + 32.18 \times 2 = \$711$$

The total cost of the mold base (Eq. 13.13) is

$$C_b = 582 + 711 = \$1293$$

The total thickness or height of the mold base, h_{pt} (cm), is equal to the sum of the combined thickness of all of the plates:

$$h_{pt} = h_p + h_{fp} \tag{13.17}$$

where

h_{fp} = thickness of the ejector, riser and stripper plates, cm

The thickness of each ejector, riser, and stripper plate averages about 10 cm for semiautomatic wax dies if an ejection system is required and is zero for hand molds.

Example

For semiautomatic wax injection the ejector, riser, and stripper plates will have a total thickness of 20 cm and the height of the mold base will be

$$h_{pt} = 17.2 + 20 = 37.2 \text{ cm}$$

Once the required size of the mold base and its cost have been established, the custom work cost can be estimated. For semiautomatic wax injection, this includes the preparation of electrical connections, an ejection system, and pneumatic connections. For aluminum semiautomatic dies this cost is only about 10% of the mold base cost. The cost of the mold base with custom work, C_{ab} ($), is now given by

$$C_{ab} = 1.1 C_b \tag{13.18}$$

Example

The cost of the mold base with custom work is

$$C_{ab} = 1.1 \times 1293 = \$1422$$

The cost of forming the mold base into a finished mold depends on the time it will take to perform each of the required operations. The first operation to be considered is making the shape details. This is done in a similar way to that for plastic injection molds. The following equation gives the manufacturing hours M_x required to generate the shape.

$$M_x = 0.1n_s \qquad (13.19)$$

where

n_s = number of surface patches on the part

To account for the size of the part and the removal of material the following equation can be used for the manufacturing hours

$$M_{po} = 5 + 0.014A_{tp}^{1.2} \qquad (13.20)$$

where A_{tp} (cm^2) is the total projected area of all parts in the mold and is given by $A_p n_{pd}$, the projected area for one part multiplied by the number of patterns in the mold.

Additional manufacturing hours are required for retractable side-pulls, internal lifters, and unscrewing devices. For semiautomatic wax molds, side-pulls require an average of about 16 hr each. Since they are actuated by external pneumatic cylinders, construction is simple. There is, however, a cost for the pneumatic cylinder of about $30. For hand molds, the manufacturing time is about 8 hr with no pneumatic actuator.

Internal lifters and unscrewing devices are rarely used in investment casting wax pattern molds. If they are used, it can be estimated that they require about 40 hr for semiautomatic molds and 8 hr for hand molds. It is estimated that actuators for lifters will cost about $30 and for unscrewing devices about $50.

An ejection system is generally used for wax injection molding. When an ejection system is required, the quantity of pins is much less than that used for plastic injection. The aluminum mold material is also easier to prepare for ejector pins. The time in hours, M_e, required to make these pins and fit them to the mold is as follows:

$$M_e = 0.5A_{tp}^{0.5} \qquad (13.21)$$

The adjustment factor of 0.5 is for semiautomatic wax injection.

The manufacturing hours required for different appearances or surface finishes, for required tolerance levels, and for texturing will be assumed to be the same as for plastic injection molding and will be estimated using the factors presented in Tables 8.6 and 8.7 and by adding 5% of the complexity and material removal hours for textured surfaces.

For investment casting patterns, the surface finish required will rarely be better than standard opaque.

The final feature to impact manufacturing time is the shape of the parting surface. The time to make a nonflat parting surface in an aluminum die M_s (hr) is

$$M_s = 0.2 f_p A_{tp}^{0.5} \tag{13.22}$$

where

f_p = parting surface adjustment factor

The total manufacturing hours for mold making, M_{tot}, is now obtained by adding together all of the individual contributing factors. The mold-making cost, C_{dm} ($), can be determined by the following equation:

$$C_{dm} = R_{ds} M_{tot} \tag{13.23}$$

where

R_{ds} = mold-making rate, $/h

and the final cost of the mold, C_d ($) is

$$C_d = C_{dm} + C_{ab} + C_{ac} \tag{13.24}$$

where

C_{dm} = mold manufacturing cost, $
C_{ab} = cost of the mold base, $
C_{ac} = cost of standard mold components or actuators, $

Example

The sample part has 30 surface patches and therefore the complexity manufacturing hours (Eq. 13.19) are estimated to be

$$M_x = 0.1 \times 30 = 3 \text{ h}$$

The total projected area of all parts in the mold is

$$A_{tp} = 8.88 \times 2 = 17.76 \text{ cm}^2$$

and the manufacturing hours for removal of material (Eq. 13.20) are

$$M_{po} = 5 + 0.014 \times 17.76^{1.2} = 5.44 \text{ h}$$

The two side-pulls require 32 hr and, for the ejection system

$$M_e = 0.5 \times 17.76^{0.5} = 2.1 \text{ h}$$

From Tables 8.6 and 8.7, a tolerance of ± 0.25 mm will result in additional hours equal to 10% of $(M_x + M_{po})$ and a standard opaque surface finish will give additional hours equal to 15% of $(M_x + M_{po})$, i.e.,

$$\text{Tolerance hours} = 0.1 \times (3 + 5.44) = 0.84 \text{ h}$$
$$\text{Surface finish hours} = 0.15 \times (3 + 5.44) = 1.27 \text{ h}$$

Finally, since we have a flat parting surface there are no additional hours and the total manufacturing hours are

$$M_{tot} = 3 + 5.44 + 32 + 2.1 + 0.84 + 1.27 = 44.65 \text{ h}$$

with a mold-making rate of 40 $/h this gives a mold-making cost (Eq. 13.23) of

$$C_{dm} = 40 \times 44.65 = \$1786$$

The cost (C_{ac}) of two side-pulls and actuators is $60 and the final cost of the mold (Eq. 13.24) is

$$C_d = 1786 + 1422 + 60 = \$3268$$

13.21 CORE MOLD COST

Soluble cores are basically wax patterns, and the cost of the molds can be estimated by using the preceding equations. Ceramic core molds are similar to plastic injection molds. The ceramic core material is injected under high pressures and temperatures. Ceramic core material is also extremely abrasive and usually requires steel molds. The cost of ceramic core molds may be estimated using the methods described for plastic injection molding in Chapter 8.

13.22 PATTERN AND CLUSTER ASSEMBLY COST

Complex patterns may require assembly of separate pattern pieces before the pattern is complete, and this may require the use of special fixtures. These fixtures vary greatly in size and complexity depending on the pattern to be assembled. The cost of these fixtures is beyond the scope of this text, but a crude estimate can be made. It is estimated that a pattern assembly fixture costs $300 for a small two-piece pattern and $100 per additional piece. Sometimes features can be designed into the pattern pieces that will align them for assembly, and a fixture may not be required.

The assembly operation starts with the placing of the main body of the pattern into the assembly fixture. A hot spatula is applied to the mating surface of the loose piece and the pattern piece, which is then pressed onto the main body of the pattern and held in place until the wax sets. Pattern assembly is delicate precision work. For this reason a time of 20 s per piece is allowed. It is also assumed to take about 10 min to set up the work bench for each different type of pattern.

The pattern assembly cost, C_{tpa} ($), is given by

$$C_{tpa} = (W_{pa}t_{pa}/3600) + W_{pa}S_{pa}/B_s + C_{af}/P_v \qquad (13.25)$$

where

W_{pa} = operator rate for pattern or cluster assembly, $/h
S_{pa} = setup time for pattern assembly, h
B_s = number of parts in production batch
C_{af} = cost of assembly fixture, $
P_v = total production volume for a part

and where the time required for pattern assembly, t_{pa} (s), is given by

$$t_{pa} = 20(n_{pa} - 1) + 5 \qquad (13.26)$$

where n_{pa} is the number of pattern pieces to be assembled.

Occasionally, plastic pattern pieces will need assembly, but this will be assumed to cost the same as a wax welding operation.

Example

In our sample part only one pattern piece would be required. However, we can estimate the pattern assembly time for a three-piece pattern by way of illustration. Using an operator rate of $23.00 per hour ($11.50 plus 100% overhead), a batch size of 1000, and a life volume of 10,000, Eq. 13.26 would give a pattern assembly time of

$$t_{pa} = 20(3 - 1) + 5 = 45 \text{ s}$$

and Eq. 13.26 gives a cost of

$$C_{tpa} = 23 \times 45/3600 + 23(10/60)/1000 + 400/10000 = \$0.3313$$

Once each pattern is complete, it is ready for assembly to the metal feeding system. This includes the gate (if not already part of the pattern), down runners or sprues, and a pouring basin. Large parts are often cast singly. Medium-sized and small parts are typically cast in clusters of two to hundreds of parts, all sharing the same feeding system.

Since the components of the feeding system are made of recycled wax, it will be assumed that there is no raw material cost. There is some cost associated with the recovery and reprocessing of the meltout wax, but it is assumed to be small and can be included in the general operation overhead. In addition, since the feeding system component shapes are standard and wax injection molds last almost indefinitely, it will be assumed that the cost of the molds used to make the standard wax extrusions and cluster bases can be spread over millions of parts and will be negligible when compared to the processing cost.

Assembly fixtures will be assumed standard and used for many different casting patterns. The assembly operation includes the wax welding of all the patterns to the cluster base runners and the wax welding of a ceramic pouring cup or ring to the cluster base pouring basin. A minimum of two reorientations is required during the assembly. It is estimated that a wax welding operation in cluster assembly takes about 10 s and each reorientation of the cluster base about the same. The cost of a cluster base is about \$1 and a ceramic pouring cup \$0.5. The time to set up the workstation for each new type of part is estimated at 7 min. With these assumptions the total cost for cluster assembly, C_{tca} (\$), is given by

$$C_{tca} = (W_{pa}t_{ca}/3600) + 1.5 + W_{pa}S_{ca} \tag{13.27}$$

where

S_{ca} = setup time required for the cluster to be assembled, h

and where t_{ca} (s) is the time to assemble a cluster and is given by

$$t_{ca} = 10(n_{pc} + 3) \tag{13.28}$$

where

n_{pc} = number of parts per cluster

13.23 NUMBER OF PARTS PER CLUSTER

The more patterns or parts that can fit onto a cluster, the less the cluster processing cost per part. The number of parts per cluster is limited by the maximum weight and size of the cluster that can be handled.

When the cluster must be manually handled, the total weight should not exceed 18 kg (40 lb), which is the maximum weight that a person can be required to handle on a regular basis. If robotic handling is used, the weight limit can be exceeded. However, for the purposes of early cost estimating, it will be assumed that the weight of the dry shell mold filled with metal must not exceed 16 kg (35 lb).

In order to estimate the number of parts per cluster it is necessary to have values for two ratios:

1. The casting yield, Y_d, which is the ratio of the weight of finished castings per cluster divided by the total poured weight of metal.
2. The shell mold yield, Y_{sm}, which is the ratio of the weight of finished castings per cluster divided by the weight of the dry shell mold.

Empirical formulas based on industry data for these ratios are

$$Y_d = 0.0483(\ln W) + 0.455 \qquad (13.29)$$
$$Y_{sm} = 0.148(\ln W) + 0.843 \qquad (13.30)$$

where

W = weight of part, kg

The number of parts per cluster, n_{pc}, is now given by

$$n_{pc} = 16/(W/Y_d + W/Y_{sm}) \qquad (13.31)$$

Example

The volume of our sample part is $3.326\,cm^3$. With a density of $8.94\,g/cm^3$ the weight of each part will be 0.0297 kg.
The yields are

$$Y_d = 0.0483 \ln 0.0297 + 0.455 = 0.285$$
$$Y_{sm} = 0.148 \ln 0.0297 + 0.843 = 0.323$$

and the number of parts per cluster is

$$n_{pc} = 16/(0.0297/0.285 + 0.0297/0.323) = 81$$

The time to assemble a cluster, t_{ca}, is given by Eq. 13.28 as

$$t_{ca} = 10(81 + 3) = 840 \text{ s}$$

and the total cluster assembly cost, C_{tca} from Eq. 13.27 is

$$C_{tca} = 23 \times 840/3600 + 1.5 + 23 \times 7/60 = \$9.55$$

Finally, the cost per part for cluster assembly, C_{pca} ($), is

$$C_{pca} = 9.55/81 = \$0.118$$

13.24 PATTERN PIECE COST

The total cost per pattern piece, C_{tp} ($), will be the sum of the cost of injection, pattern material, setup, and the cost of the die amortized over the total product volume.

$$C_{tp} = (C_{ip} + C_{pm})/(1 - P_{psr}) + M_i S_{ds}/B_s + C_d/P_v \qquad (13.32)$$

where

C_{ip} = process cost per pattern piece, $
C_{pm} = cost of wax per pattern piece, $
P_{psr} = pattern piece scrap rate

M_i = injection machine and operator rate, \$/h
S_{ds} = time to set up the mold on the injection machine, h
B_s = batch size
C_d = pattern piece mold cost, \$
P_v = total production volume

The setup time will be estimated at 15 min for a semiautomatic mold and 5 min for a hand mold. These figures include the time to fasten the mold to the injection machine if necessary, adjust the shot size and injection pressure, and remove the mold from the machine.

For estimating purposes an average scrap rate of 5% will be assumed for patterns without cores and 8% for patterns with cores.

Example

Our sample part has only one pattern piece and no cores. With semiautomatic operation the total pattern cost is

$$C_{tp} = (0.078 + 0.00963)/(1 - 0.05) + 30 \times 0.25/1000 + 3268/10000$$
$$= \$0.4265$$

13.25 CLEANING AND ETCHING

Prior to investing the wax cluster in slurry, it must be cleaned and it is usually etched to provide a good surface for the slurry to adhere to. If there are soluble cores, these must be dissolved. For estimating purposes it will be assumed that these operations can be done in one step with a single loading and unloading of the wax cluster. It will be assumed that the costs of the cleaning and etching solutions and equipment are small. With a load and unload time of 8 s and an average labor rate of \$23/hr, including overhead, the cost will be about \$0.06 per cluster.

13.26 SHELL MOLD MATERIAL COST

After cleaning and etching the assembled pattern cluster, it is dipped with the first primer slurry coat, stuccoed, then allowed to dry before application of the next primer coat. For the purposes of cost estimation three zircon-based primer coats will be assumed, followed by four silica-based backup coats and one fused-silica seal coat, all manually applied.

The shell mold yield, Y_{sm}, was defined as the ratio of the weight of the cast parts from the cluster to the weight of the dry shell-mold. Thus, the dry weight of the shell mold, W_{sm} (kg), is given by

$$W_{sm} = n_{pc} W / Y_{sm} \tag{13.33}$$

where

n_{pc} = number of parts per cluster
W = weight of a single part, kg

A detailed estimate of the material cost for a shell mold would involve a knowledge of the number of primary and backup coats, the thickness and area for each, the density of the materials, and their cost per unit weight. However, for quick estimates we can assume a cost of \$1.00/kg for shell mold material.

Example

The weight of our sample part is 0.0297 kg, the shell mold yield is 0.323, and the number of parts per cluster is 81, giving an estimated shell mold weight of

$$W_{sm} = 81 \times 0.0297/0.323 = 7.45 \text{ kg}$$

and the shell mold would cost \$7.45.

13.27   INVESTING THE PATTERN CLUSTER

The cost, C_{pr} (\$), to apply the primer coats to a cluster is given by

$$C_{pr} = M_{pr}[t_{c1} + (n_{cp} - 1)t_{cp}]/3600 \qquad (13.34)$$

where

M_{pr} = machine and operator rate for primer coat application, \$/h
t_{c1} = time to apply the first primer coat, s
t_{cp} = time to apply each subsequent primer coat, s
n_{cp} = number of primer coats

The application of the first coat is done with the most care and is estimated to take about 20 s for slurry and stucco. This includes the time to remove the cluster from a rack, dip it into the slurry tank, remove and manipulate it for even coverage, dip it into the stucco tank, and place it in the drying rack. The following two coats of slurry and stucco take about 15 s each to apply. Using these values and an operator rate of \$31.20 per hour gives a cost to apply three primer coats of

$$C_{pr} = 31.20 \times (20 + (3 - 1) \times 15)/3600 = \$0.43$$

The application of backup coats is assumed to be carried out robotically. The cost, C_{bu} (\$), is given by

$$C_{bu} = n_{cb}M_{bu}t_{cb}/3600 \qquad (13.35)$$

where

n_{cb} = number of backup coats
M_{bu} = machine and operator rate for backup coat application, \$/h

t_{cb} = time to apply a backup coat, s

Assuming a value of $35.15 per hour for the robot, the conveyor and 50% of an operator, and a cycle time for each backup coat application of 10 s, the cost for a single backup coat is about $0.10. If an average of five backup coats is used, the total cost for applying backup coats is

$$C_{bu} = 5 \times 35.15 \times 10/3600 \doteq \$0.49$$

13.28 PATTERN MELTOUT

The cost of wax pattern meltout includes the labor cost of loading and unloading the clusters into the autoclave or furnace, the cost of using the meltout unit, and the energy consumed.

The time for load and unload is estimated at about 8 s. Using an average operator rate including overhead of $23 per hour, gives a cost per cluster for load and unload of $0.05.

The machine rate for a $10,000 dewaxing furnace with an average capacity of ten clusters is about $1.52 per hour, assuming 100% machine overhead, a ten-year depreciation, and a 10% interest rate on the machine purchase. The cost per cluster for a 15 min cycle is, therefore, about $0.04 per cluster. Flash dewaxing with gas furnaces is reported to be only 5 to 10% efficient [2]. Assuming specific heats of 840 J/kgK and 2890 J/kgK for silica and paraffin respectively, melting the wax from an average size cluster will require about 10,200,000 J. With a cost of $0.60 per therm for a high-volume user (1 therm = 105,500,000 J) the average energy cost per cluster will be $0.06.

13.29 BURNOUT, SINTER, AND PREHEAT

Burnout, sinter, and preheat are assumed to form a single operation taking about one hour for a group of ten clusters in a sintering furnace. The machine rate for a sintering furnace will be comparable to that for a dewaxing furnace. Therefore, the cost per cluster will be about $0.15 per cluster.

The operator loads the clusters into the furnace. The unload operation is done in the foundry and is accounted for as part of the metal-pouring operation. The time to load a single cluster is estimated to be 4 s, giving a cost of $0.03 using our typical operator rate.

With a furnace efficiency of 15% and the specific heat of silica, it will cost about $0.43 per cluster to bring an average size shell mold to a temperature of 1090°C. Combining the labor, machine, and energy costs, we have a total cost per cluster of $0.61 for burnout, sinter, and preheat.

13.30 TOTAL SHELL MOLD COST

The following summarizes the costs for the shell mold:

Assemble cluster	9.55
Clean and etch pattern cluster	0.06
Apply three primer coats	0.43
Apply five backup coats	0.49
Pattern meltout (dewax)	0.05
Burnout, sinter, and preheat	0.61
Shell mold material	7.45
Total	$18.64

Example

For our sample part with 81 patterns per cluster, the shell mold cost per part, including cluster assembly, would be

18.64/81 = $0.23

13.31 COST TO MELT METAL

The cost to melt an alloy or metal has three components: energy, equipment, and labor.

Most investment casting facilities use induction furnaces to melt metal. Induction furnaces are relatively environmentally clean and efficient, do not add contaminants from combustion of fuel to the melt, and have low material losses.

For the purposes of estimating, it will be assumed that induction furnaces are being used. The cost of the energy used is the cost of the electricity supplied by the power company. For 1994 in the northeastern United States, the cost of power C_p for users with a demand of 500 kW or more was $0.035 per kilowatt hour ($/kWh) during peak hours. Equation 13.36 gives a relation for determining the cost per kilogram, C_e ($/kg), to melt an alloy.

$$C_e = C_p E_m / n_e \qquad (13.36)$$

where

C_p = cost of electricity, $/kWh
E_m = minimum energy required to melt a metal, kWh/kg
n_e = efficiency of the induction furnace

The efficiency of an induction furnace is dependent on the material to be melted and the frequency at which the furnace operates. For the purposes of cost estimating, the efficiencies given in Table 13.9 will be used. The cost of the power

TABLE 13.9 Approximate Efficiency
of Induction Melting for Various Materials

Material group	Approximate furnace efficiency
Ferrous	0.80
Aluminum	0.60
Other Nonferrous	0.70

Source: Industry supplier of induction furnaces.

to melt the metal forms a relatively small portion of the material cost, so these approximate values will be adequate for estimating purposes.

The minimum energy requirements to bring a material to pouring temperature are given in Table 13.10. These figures are only approximate. Pouring temperatures may be different from one operation to the next, and differences in material composition will also have an impact.

Example

The alloy for our sample part is phosphor bronze. The furnace efficiency, n_e, from Table 13.9 is 0.7 and the minimum energy for melting, E_m, is 0.185 kWh/kg. Thus the cost of energy to melt the metal ready for pouring is

$$C_e = 0.035 \times 0.185/0.7 = 0.00925 \text{ \$/kg}$$

The equipment cost, C_f (\$), for a complete induction melting unit, including furnace, power supply, and installation, can be determined from Eq. 13.37. This equation should only be used for furnaces with a capacity range of from 100 to 4000 kg.

$$C_f = 100,000 \ \log(S_z^{1.64/851}) \tag{13.37}$$

where

S_z = size of the furnace in kilograms of capacity for iron, kg

Example

If the furnace for our sample part has an iron capacity of 250 kg, then the cost of the furnace is

$$C_f = 100,000 \ \log(250^{1.64/851}) = \$100,269$$

The costs given by Eq. 13.37 are for high-quality coreless induction furnaces. They are more expensive than some of the cheaper units used by many small investment casters but they meet strict European guidelines for shielding, have a

TABLE 13.10 Minimum Energy Required to Melt Alloys

Metal	Final temperature (°C)	Energy required (kWh/kg)	Energy required (Btu/lb)
Aluminum bronze	1200	0.237	367
Aluminum alloy (general)	750	0.326	505
Aluminum (pure)	750	0.326	505
Brass (58% copper)	1030	0.173	268
Brass (63% copper)	1030	0.177	274
Brass (73% copper)	1080	0.183	283
Brass (85% copper)	1120	0.170	263
Brass (90% copper)	1140	0.194	300
Bronze	1100	0.185	287
Copper (pure)	1200	0.204	316
Gold (pure)	1150	0.066	102
Gray iron	1500	0.390	604
Iron (pure)	1600	0.391	605
Lead (pure)	450	0.023	35.6
Magnesium (pure)	700	0.332	514
Malleable Iron (black)	1550	0.395	612
Malleable Iron (white)	1550	0.405	627
Cu 70%–Mn 30%	1030	0.193	299
Nickel (pure)	1600	0.341	528
Silver (pure)	1050	0.107	166
Spheroidal iron	1550	0.400	620
Steel (general)	1600	0.393	609
Stainless steel	1650	0.405	627
Zinc (pure)	500	0.093	144
Tin (pure)	400	0.037	57.3

Source: Industry supplier of induction furnaces.

high initial purchase price, but have longer life with lower overall maintenance and operating cost.

It has been estimated that a furnace can be charged manually at about 1500 kg of iron per hour. Induction furnaces can be set with automatic melting cycles and require only about 4 min an hour to set up and monitor.

Some additional labor is required to check and certify the composition of the melt. Many investment casters purchase certified alloy from suppliers in the form of billets, bars, or ingots. Induction melting allows them to melt this metal with little compositional change in the material. However, if scrap runners and gates are to be remelted, the metal composition will change after repeated melting. If the foundry does not have the capacity to test metal composition, readjust it to within specification, and certify it, then they must sell their scrap after a limited number of melts. Sometimes only one melt can be allowed for critical applications. It will be assumed here that a metallurgist is available and can test, adjust, and recertify the metal composition in about 10 min per melt. This assumption allows the recovery of all gates and runner scrap for remelting an unlimited number of times.

The national average labor rate for foundry workers is about $13.50 per hour. The rate for the metallurgist is assumed to be about $18 per hour. Labor overheads are estimated at 100%. The amortization period for the furnace is assumed to be 10 years at an interest rate of 10%. Machine overheads are estimated at 100%. Machine overhead in this case includes maintenance such as changing the refractory liner and power cables. It does not include power, as that is determined separately by material type.

The data in Table 13.11 are for iron. Different materials have different densities, and the weight of each material that a furnace can hold will vary. Each furnace can hold a certain volume. The volume capacity is assumed constant for a given furnace. The weight of a particular material that a furnace can hold can be determined by comparing its density with iron. The machine cost, C_{mf} ($/kg), for a particular alloy can be determined by multiplying the cost per kilogram of iron by the ratio of the density of iron to the desired alloy as shown in Eq. 13.38:

$$C_{mf} = C_{mi}(\rho_i/\rho_a) \tag{13.38}$$

where

C_{mi} = cost of the furnace per kilogram of iron, $/kg
ρ_i = density of iron, g/cm^3
ρ_a = density of alloy, g/cm^3

TABLE 13.11 Machine Rates for Induction Furnaces

Furnace size (kg of iron)	Furnace cost buy and install ($)	Machine rate with overhead ($/hr)	Furnace cycle time (hr)	Machine cost per cycle ($/cycle)	Machine cost per kg iron ($)
2000	250,000	38	1.6	61	0.031
1000	200,000	30	1.4	51	0.041
500	150,000	23	1.2	28	0.056
250	100,000	15	1.2	18	0.072
150	64,000	10	1.1	11	0.073

TABLE 13.12 Operator Rates for Induction Furnaces

Furnace size (kg of iron)	Labor cost per melt cycle ($)	Labor cost per kg iron ($/kg)
2000	34	0.017
1000	21	0.021
500	15	0.030
250	12	0.048
150	11	0.070

Most investment casters use furnaces under 500 kg capacity. If it is assumed that the average furnace size is 250 kg, Eq. 13.38 reduces to Eq. 13.39 for a density of iron of 7.08 g/cm^3.

$$C_{mf} = 0.51/\rho_a \qquad (13.39)$$

Example

If an average-sized furnace of 250 kg capacity for iron is used and the density of phosphor bronze is 8.94 g/cm^3, Eq. 13.39 gives the cost of using the induction furnace per kg of alloy:

$$C_{mf} = 0.51/8.94 = 0.057 \ \$/kg$$

The data in Table 13.12 are also for iron. It is assumed that the pouring rate is relatively constant, so differences in density will have less effect on the labor time required for pouring. Density differences should also not affect the monitoring and certification of the melt. Charge time may be affected. If it is assumed that the same number of equally sized billets of different weights are handled, then the labor time to charge a furnace should not be dramatically different for different alloys. For the purposes of estimating, it is assumed that the labor time to charge a furnace per cycle is the same regardless of the material. The values in Table 13.12 for the cost per cycle for iron can, therefore, be used for all materials. The typical furnace size for investment casting is about 250 kg. The labor cost per kilogram for any material will be estimated as $0.048/kg.

13.32 RAW BASE METAL COST

Because investment casters often handle many different alloys, they often do not have the volume to refine metal from raw scrap. They usually buy their metal in the grade in which they intend to cast the final part. Some investment casters do have the capacity to make adjustments to the melt and bring the melt within specification for certain materials, but for the purposes of estimating it will be assumed that an alloy is purchased from a supplier already certified to the final part material specification. The raw metal cost is therefore the purchase price of the alloy from the supplier.

13.33 READY-TO-POUR LIQUID METAL COST

The cost of liquid metal that is ready to pour may be found by adding together the melting cost and the raw alloy cost. The molten metal cost, C_m ($/kg), is the sum of the cost of energy, the cost of the machine, the cost of the labor, and the cost of raw material:

$$C_m = C_e + C_{mf} + C_{ml} + C_{rm} \qquad (13.40)$$

where

C_e = energy cost per kilogram for melting metal, $/kg
C_{mf} = cost of the furnace per kilogram of alloy, $/kg
C_{ml} = cost of labor per kilogram of alloy, $/kg
C_{rm} = cost of the raw alloy, $/kg

Since labor cost is assumed constant regardless of alloy and the furnace size is estimated as 250 kg, Eq. 13.40 can be written as follows:

$$C_m = C_e + C_{mf} + C_{rm} + 0.048 \qquad (13.41)$$

Example

For a raw material cost of phosphor bronze of 1.764 $/kg, the cost of ready-to-pour liquid metal is

$$C_m = 0.00925 + 0.057 + 1.764 + 0.048 = 1.878 \ \$/kg$$

13.34 POURING COST

For the purposes of estimating, it will be assumed that pouring is a four-worker operation and that they can pour about 1500 kg/hr. Using an average foundry rate for foundry workers of 27 $/hr, including 100% overhead, gives a cost per kilogram for pouring of

$$C_{mp} = 0.072 \ \$/kg$$

13.35 FINAL MATERIAL COST

The final material cost is the cost of the poured material less the value of the scrap recovered after cutoff. Gates and runner are assumed returnable to the melt. When returned in this fashion they replace raw purchased material kilogram for kilogram. When material cannot be returned to the melt and must be sold back to the supplier, it will be sold back at a reduced value. For the purposes of estimating, unless it is known to be otherwise, it will be assumed that scrap is returned to the melt.

Only a portion of the total weight poured goes to making the final part. Some of the material is lost during melting and some during cutoff. During melting with an induction furnace losses are often less than 1%. For the purposes of estimating it will be assumed that total metal losses are about 2%. The remainder of the metal that does not go into the part weight is taken up by the gates and runners for return to the melt. The ratio of part weight to poured metal weight is defined as the casting yield. Casting yield for investment casting can be as low as 10% to as high as 90%.

Equation 13.42 gives the relation for final part material cost per kilogram, C_{mat}, given the preceding assumptions and simplifications.

$$C_{mat} = [C_m + C_{mp} - V_{sc}(1 - P_1 - Y_d)]/Y_d \qquad (13.42)$$

where

$$Y_d = n_{pc}W/W_{pr} \qquad (13.43)$$

and where

C_m = unpoured liquid metal cost, \$/kg
C_{mp} = pouring cost, \$/kg
V_{sc} = value of scrap, \$/kg
P_1 = metal loss rate
Y_d = casting yield
n_{pc} = number of parts per cluster
W = weight of a single part, kg
W_{pr} = weight of material poured into a single mold, kg

Casting yield can vary greatly from part to part. Casting yield varies because of part weight and shape. Two parts of the same weight can have different casting yields due to their shape. One part may require extra gates, longer runners, and multiple risers. In investment casting, the casting yield can also be affected by the number of parts per cluster and the particular type of process used. Material can also have an effect on casting yield. A material such as aluminum with a broad freezing range may require more risering and gating than an identical part made from a ferrous alloy.

Equation 13.29 was developed for estimating casting yield, Y_d, as a function of part weight only. This relation does not take into account the shape of the part. Also, the data available was for steel alloy parts and for traditional shell mold investment casting.

Example

Equation 13.29 gave a casting yield of 0.285, or 28.5%.

If the value of scrap is 0.7 \$/kg, the final part material cost is

$$C_{mat} = (1.878 + 0.072 - 0.7 \times (1 - 0.02 - 0.285))/0.285 = 5.14 \text{ \$/kg}$$

giving a metal cost per part of $5.14 \times 0.0297 = \$0.153$

13.36 BREAKOUT

At this point in the process, the shell mold has lost much of its strength and is often cracked and fragmented. Breaking out the solidified metal cluster from the shell mold is usually accomplished with a pneumatic hammer. The cluster is placed in a fixture that contacts the top of the cluster at the pouring cup and the bottom of the cluster at a central down runner or sprue. The machine cycle is started with the pressing of palm buttons. The pneumatic hammer strikes the cluster at the pouring cup, and the shell mold breaks away.

The cost of pneumatic breakout will be assumed to include a setup cost to adjust the machine to the specific size cluster and a labor cost for the operation of the machine. The machine for this operation is relatively inexpensive, and the cost of the machine will be assumed small when compared to the cost of the operator. The cost of the machine will be included in the general overhead. The cost per cluster, C_{bo} (\$), for pneumatic breakout is given by Eq. 13.44.

$$C_{bo} = C_{nh} + C_{ns}n_{pc}/B_s \tag{13.44}$$

$$C_{nh} = t_{nh}R_{nh}/3600 \tag{13.45}$$

$$C_{ns} = S_{nh}R_{nh} \tag{13.46}$$

where

C_{nh} = operator cost per cluster for a cycle of the pneumatic hammer, \$
C_{ns} = cost of setting up the pneumatic hammer for the correct size of cluster, \$
n_{pc} = number of parts per cluster, \$
B_s = number of parts per production batch
t_{nh} = cycle time for pneumatic hammer breakout, s
R_{nh} = operator rate for breakout, \$/h
S_{nh} = setup time for a pneumatic hammer, h

The setup time for the breakout machine is estimated at about 5 min. The cycle time includes a load and unload time for a cluster of about 10 s and a machine cycle of 3 s for a total cycle time of 13 s. If the burdened operator rate is estimated at \$25.00 per hour, assuming a base wage of \$11.50 per hour with 100%

overhead and accounting separately for vacations, holidays, and sick days, Eq. 13.44 can be reduced to Eq. 13.47:

$$C_{bo} = 0.09 + 2.08 n_{pc}/B_s \tag{13.47}$$

Example

We have 81 parts per cluster and a batch size of 1000. Thus the cost per cluster for breakout is

$$C_{bo} = 0.09 + 2.08 \times 81/1000 = \$0.259$$

13.37 CLEANING

Most of the shell mold is removed during breakout. Some mold material will remain trapped in crevices and holes. This remaining material is removed either by blasting equipment or by being dissolved off by a caustic bath. If ceramic cores were used or deep holes are present, leaching out of the ceramic in a caustic bath will probably be required.

The costs for batch cleaning in a blasting cabinet and cleaning using a caustic bath are handled similarly. The operator is assumed not to have to be present during either cleaning cycle. The machines are assumed to run unattended during their operation, and the cycle is controlled by a simple timer. The material cost for blast media or caustic is assumed small and is included in the general overhead along with the cost of the blasting cabinet and the leaching tanks. No setup of any kind is assumed necessary for these operations. This reduces the cost estimate of these operations to a simple loading and unloading of a cluster. Using the preceding wage rate and a cycle time for loading and unloading of 10 s, the cost for either a blast-cleaning operation or a leaching operation is estimated to be

$$C_{cl} = \$0.07 \text{ per cluster}$$

13.38 CUTOFF

The cutoff cycle is assumed to include a setup cost, a material (or tool wear) cost, and an operator cost. The cost of the machine, usually either a band saw for softer nonferrous materials or an abrasive cutoff wheel for harder materials, is assumed small and will be included in the general overhead. A detailed cost analysis taking into account the life and cost of the specific cutting tool, the part material and size, and the gate cross section should be possible, but the data for such an analysis were not available. Given the lack of available data, only a crude estimate for cutoff cost can be attempted. Equation 13.48 gives an estimate for the cost, C_{co} (\$), of cutoff per cluster. The type of material is not taken into account in the cutoff time estimate even though this would have an effect. The cost of wear on the band saw blade or abrasive cutoff wheel will be roughly estimated as equal to the labor cost of the operation.

$$C_{co} = C_{cf} + C_{tw} + C_{fs} n_{pc} / B_s \tag{13.48}$$

$$C_{cf} = t_{co} R_{co} / 3600 \tag{13.49}$$

$$C_{fs} = S_{co} R_{co} \tag{13.50}$$

$$t_{co} = t_{cl} + (t_{cg} + t_{gp}) n_{gp} n_{pc} + (t_{sc} + t_{so}) n_{so} \tag{13.51}$$

where

C_{co} = total cost per cluster of cutoff, \$
C_{cf} = operator cost per cluster for cutoff, \$
C_{tw} = cost of cutoff tool wear per cluster, \$
C_{fs} = cost of setting up the cutoff machine for a specific cluster, \$
n_{pc} = number of parts per cluster
B_s = number of parts per production batch
t_{co} = cycle time for cutting off all the parts from a cluster, s
R_{co} = operator rate for the cutoff operation, \$/h
S_{co} = setup time for a cutoff operation, h
t_{cl} = time to load a cluster for cutoff, s
t_{cg} = time to cut through a single gate, s
t_{gp} = time to reposition the cluster from one gate to the next, s
n_{gp} = number of gates per part
t_{sc} = time to make a supplemental cut through runner or sprue, s
t_{so} = time to reposition cluster from one supplemental cut to the next, s
n_{so} = number of supplemental cuts per cluster

The same operator rate of \$25.00 per hour used for breakout is assumed for cutoff. The setup time, S_{co}, is also estimated at about 5 min. The time to load the cluster, t_{cl}, is estimated at 4 s. The time to cut through a single gate, t_{cg}, is crudely estimated by Eq. 13.52

$$t_{cg} = w^2 / m_{rc} \tag{13.52}$$

where

m_{rc} = cutoff rate, cm^2/s
w = thickness of the gate, cm

An average removal rate will be estimated at 0.2 cm^2/s.

Although size and weight of the part will probably have an effect, it will be estimated that it takes about 2 s for the operator to position the cluster between gate cuts. This is actually a conservative number, since the parts are often in a straight line making positioning between gate cuts very fast.

Supplemental cuts will be defined as those cuts that do not free a part from the cluster, but make access to the gates for the cutoff saw possible or just easier. These cuts are usually through the runners in the cluster, which are usually thicker than the gates. It will be estimated that a supplemental cut takes three times as long as it takes to cut through a gate stub. Since the cluster will be more difficult to handle before these cuts are made, positioning the cluster between each supplemental cut will also be assumed to take three times as long as positioning the cluster for a gate cut. Equation 13.53 can now be written by combining Eqs. 13.51 and 13.52 and substituting estimated values for cutting and positioning time.

$$t_{co} = 4 + (5w^2 + 2)(n_{gp}n_{pc} + 3n_{so}) \tag{13.53}$$

The number of supplemental cuts can only be crudely estimated since very little data was available. The number of these cuts is in part dependent on the process envelope, size of part, and other process specific factors. It will be assumed here that the number of supplemental cuts can be estimated from the part weight alone and will be between one and six cuts. Equation 13.54 assumes a simple linear relation between part weight and the number of supplemental cuts.

$$n_{so} = INT(-0.7W + 6.5) \qquad \text{if } W > 7, \text{ then } n_{so} = 1 \tag{13.54}$$

where

n_{so} = number of supplemental cuts (rounded off to nearest whole number)
W = weight of a single part, kg

Example

Equation 13.54 gives an estimate for the minimum number of supplemental cuts:

$$n_{so} = INT(-0.7 \times 0.0279 + 6.5) = 6$$

Assuming a single gate thickness of 0.5 cm, Eq. 13.54 gives the cycle time for cutting off all the parts from the cluster:

$$t_{co} = 4 + (5 \times 0.5^2 + 2)(1 \times 81 + 3 \times 6) = 325.8 \text{ s}$$

giving an operator cost per cluster of

$$C_{cf} = 325.8 \times 25/3600 = \$2.26$$

The cost of setup is

$$C_{fs} = (5/60) \times 25 = \$2.08$$

and the total cost per cluster for cutoff is

$$C_{co} = 2.26 + 2.26 + 2.08 \times 81/1000 = \$4.69$$

Summing the costs per cluster for breakout, cleaning, and cutoff we get
$0.259 + 0.07 + 4.69 = \$5.02$, with a cost per part of $5.02/81 = \$0.062$.

Finally, the total cost of the part, rounded to three decimal places, can be
summarized as follows:

Metal	0.153
Wax	0.010
Shell mold	0.230
Pattern processing	0.078
Breakout, clean, cut off	0.062

and the piece part cost is $0.533

The cost of the mold is $3268 and for a total production quantity of 100,000,
this would result in a mold cost per part of $0.033, giving a total part cost of
$0.566.

It should be noted that this analysis has not included the costs of grinding the
gate stubs flush, straightening (if required), inspection, or heat treatment.

13.39   DESIGN GUIDELINES

Investment casting offers greater freedom in design than any other metal forming
operation. Accurate and intricate castings can be made from alloys that melt at
high temperature. Parts can be cast to such close tolerance that little or no
machining is required [3].

The investment casting process was previously known as the "lost wax"
process and has been in use for thousands of years. Its advantage over other
casting processes is its ability to produce very complex castings with fine details.
However, except for very small production quantities, the wax or plastic patterns
will be produced by injection molding. Therefore, guidelines similar to those for
injection molding will apply. These guidelines apply to the pieces of the pattern
that will be used to make up the final pattern shape. Thus the pattern piece must
be easily removable from its mold and the main wall should have a uniform
thickness, which will minimize distortion by facilitating even cooling throughout
the pattern piece.

In addition, the principles of good sand casting design also apply to investment
casting. For example, nonfunctional mass should be minimized to assist in
providing sufficient gating to "feed" the part. Also, uniform wall sections with
generous radii and fillets will assist in metal flow and reduce stress concentrations.

Separate cores can add significantly to the cost, as well as bosses and undercuts. The minimum section ranges from 0.25 to 1 mm depending upon the metal to be cast. The maximum section is approximately 75 mm.

Castings range from 0.5 g to 100 kg in weight, but the investment casting process is best for parts that weigh less than 5 kg.

A flat parting plane for each pattern piece will help to minimize the cost.

REFERENCES

1. Horton, R.A., Investment Casting, ASM Handbook, Vol. 15, Casting. ASM International, Metals Park, OH, 1998.
2. McCloskey, J.C., Productivity in Investment Casting, American Jewelry Manufacturing, Vol. 38, No. 6, June 1990, p. 32.
3. Niebel, B.W., Draper, A.B., and Wysk, R.A., Modern Manufacturing Process Engineering, McGraw-Hill, New York, 1989.
4. Royer, A., and Vasseur, S., Centrifugal Casting, ASM Handbook, Vol. 15, Casting, ASM International, Metals Park, OH, 1988.

14

Design for Hot Forging

14.1 INTRODUCTION

Hot forging, also referred to as drop forging, is a process that can be used to produce a wide variety of parts in most metals. Forgings are produced in sizes ranging from a few millimeters maximum dimension up to 3 m or more in some cases. The principles and practices of hot forging have been established since the last century, but improvements have obviously been made in equipment, lubricants, and the ability to process the more difficult to forge materials since that time. The basic procedure for hot forging is relatively straightforward. Metal stock in the form of either a bar or a billet is first heated into the hot working temperature range to improve ductility. Then the material is squeezed or hammered in a series of tool steel dies to convert the stock into the finished shape. Excess material in the form of flash is produced as a necessary part of forging, and the final processing stage is to remove the flash to yield the finish forged part. Hot forging is a near net shape process, but all forgings require some subsequent machining, in particular for surfaces that must locate with other surfaces during the final assembly of a product.

14.2 CHARACTERISTICS OF THE FORGING PROCESS

Most forgings require a series of forming stages, called preforms, to convert the initial stock material into the finish-forged shape. The number of preforms required depends on several factors, including the overall shape, shape complexity, and material of the part. Forging complexity is increased by several features, including:

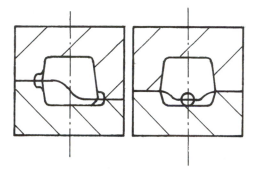

FIG. 14.1   Forging requiring a cranked parting line.

The presence of thin sections in the part
Large changes in cross-sectional area of the part
Part shapes that require the die parting line to be cranked (Fig. 14.1)

14.2.1   Types of Forging Processes

The main types of basic forging processes are referred to as open-die and closed-die forging. In open-die forging a series of relatively simple dies is used to form the final forging incrementally with a large number of blows. This process is largely a more automated version of the old blacksmith-type operations that have been used for centuries. The discussion in this chapter will not include open-die forging, since the process is used to form relatively crude final shapes, but discussion will be devoted to closed-die forging, which is used for the manufacture of a wide range of part shapes.

In closed-die forging a series of shaped dies is used to convert the initial stock into the finish-forged shape. The term "closed-die" forging is something of a misnomer, as the die cavities are not completely closed and material in the form of flash flows out at the die parting line during the final stages of forging. This flash is a critical part of the forging process, and proper control of the flash is essential to ensure die filling. Within closed-die forging two other terms are used: blocker forgings and precision forgings. Blocker forgings, compared to conventional forgings, have thicker sections and more generous radii. They are termed blocker forgings because the performing shape prior to the finishing impression is traditionally called a blocker.

Blocker forgings are easier to form than equivalent conventional forgings, requiring fewer forming stages and lower loads. They are used sometimes when small quantities of parts are required, to reduce die costs, or in difficult-to-form materials, when it is hard to obtain thin sections or there are other problems.

Blocker forgings require more subsequent machining to reach the final part shape than conventional forgings.

Precision forgings are parts formed with thinner sections and closer tolerances than equivalent conventional forgings, i.e., nearer to net shape. Such forgings require careful processing, and peak loads during the final forming stages are 2.5 to 3 times higher than those experienced for equivalent conventional forging (see Sec. 14.7). Thus larger equipment and more precise die-to-die positioning is required. Although the term precision forging implies closer precision than is normally obtained for any material, in practice precision forgings are more often produced in light alloys (aluminum alloys, magnesium alloys, etc.) than in other materials.

14.3 THE ROLE OF FLASH IN FORGING

The flash produced during closed-die forging is scrap material and may in many cases have a volume that is more than 50% of the final part volume. The amount of flash produced increases with the complexity of the part. However, the production of flash is a necessary part of the process, and its control is essential to ensure good die filling, particularly for tall, thin shape features. Figure 14.2 shows the deformation that takes place during the forging of a relatively simple axisymmetrical forging. At the start of deformation the initial stock material

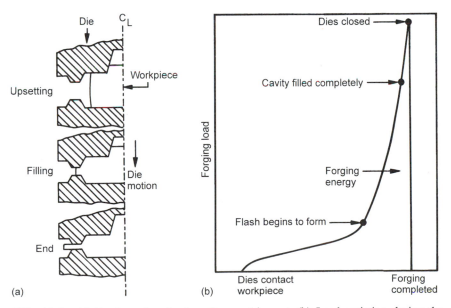

FIG. 14.2 (a) Forging of a simple axisymmetric part. (b) Load variation during the stroke for forging the part. (From Ref. 1.)

(billet) is being upset, and the corresponding forging load is relatively low. Upsetting-type deformation is the most natural form of deformation between dies and the material flows sideways, to form a flattened shape. However, if material is to be forced to move into the extremities of the die cavity, this sideways material flow must be restricted. This is the role of flash formation. A narrow flash land around the split line of the dies restricts the sideways flow of the material. In the final stages of die closure material is extruded through the flash land into the flash gutter around the forging cavity. As the deformation proceeds, the narrowing gap between the flash lands begins to restrict the sideways flow of material, through increased friction and other forces. The forging load begins to rise and the pressure inside the die increases. This increased pressure causes material to flow backwards in the direction of die closure and into the extremities of the die cavity. At the final stage of die closure, the forging load reaches its peak and this corresponds to complete die filling. At this point the last part of the flash is being squeezed through the flash land. The selection of appropriate values for the flash land geometry (gap and width) is critical to good die filling during forging, without excessive forging loads and cavity pressures.

14.3.1 Determination of the Flash Land Geometry

Figure 14.3 shows a typical arrangement for the flash land and the flash gutter on a forging. The gutter must be large enough to accommodate the flash produced. The choice of the appropriate width and thickness of the flash land is an important part of the forging process design. If the geometry is wrong, the dies may not fill completely or the forging loads may become excessive. In addition, the projected area of the flash in the flash lands is usually included in the total projected area of the part for estimation of the forging loads required and

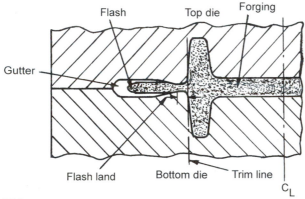

FIG. 14.3   Flash land and flash gutter configuration.

TABLE 14.1 Selected Empirical Formulas for Flash Land Geometry

Reference	Flash thickness, T_f (mm)	Flash land ratio, W_f/T_f
Brachanov and Rebelskii [3]	$0.015A_p^{0.5}$	—
Voiglander [4]	$0.016D + 0.018A_p^{0.5}$	$63D^{0.5}$
Vierrege [5]	$0.017D + 1/(D+5)^{0.5}$	$30/[D\{1 + 2D^2/(h(2r+D))\}]^{0.33}$
Neuberger and Mockel [6]	$1.13 + 0.89W^{0.5} - 0.017W$	$3 + 1.2e^{-1.09W}$
Teterein and Tarnovski [7]	$2W^{0.33} - 0.01W - 0.09$	$0.0038ZD/T_f + 4.93/W^{0.2} - 0.2$

A_p, forging projected area (mm^2); W, forging weight (kg); D, forging diameter (mm); Z, forging complexity factor.

therefore is a determining factor in equipment selection for processing. Determination of the flash land dimensions has been based on experience with forgings of a similar type. As a result there are a number of empirical formulas available for the flash land geometry, and a selection of these is given in Table 14.1 [2].

The first two formulas take no account of the forging complexity and the third formula is based on a limited number of axisymmetric forgings. The fourth and fifth formulas are based on statistical analysis for a large number of forgings and have been shown to be reliable [2, 8], each giving similar results. The fourth formula is used here for the cost estimation procedures described below, because it is simpler to evaluate. This formula is based on data for steel forgings, but it is assumed to be applicable to all materials and is used in the following form in which the main input variable is part volume, V, as opposed to part weight.

$$\text{Flash thickness, } T_f = 1.13 + 0.0789\,V^{0.5} - 0.000134V \qquad (14.1)$$

$$\text{Flash land ratio, } W_f/T_f = 3 + 1.2e^{-0.00857V} \qquad (14.2)$$

This formula is used to determine the area of the flash during forging. The land width, W_f is multiplied by the length of the flash line of the finish forging die (perimeter of the part, P_r).

Example

Figure 14.4 shows a simple steel forging that will be used to illustrate the subsequent calculations in this Chapter. The basic data for this part are as follows:

Part volume $V = 49.9$ cm^3
Projected area $A_p = 78.6$ cm^2
Perimeter $P_r = 31.4$ cm

FIG. 14.4   Steel forging for sample calculations.

For this part the flash parameters can be obtained from Eqs, 14.1 and 14.2.

$$T_f = 1.13 + 0.0789 \, (49.9)^{0.5} - 0.000134 \, (49.9) = 1.68 \text{ mm}$$

$$W_f/T_f = 3 + 1.2e^{-0.00857(49.9)} = 3.78$$

From this W_f is $3.78 \times 1.68 = 6.35$ mm, and the projected area of the flash land is $0.635 \times 31.4 = 19.9$ cm^2.

14.3.2   Amount of Flash

Costs for the material in forging are determined by the weight of the finished forging and any material wasted in processing the part. Material losses result mainly from the flash produced during forging, but further losses may occur due to scale formation for those materials that oxidize significantly during heating and, for hammer forgings, due to bar ends, and so on. Estimation of the flash for a particular forging is difficult and is usually based upon experience with the manufacture of forgings of a similar type. The amount of flash produced varies with the shape of the part, and there are two basic systematic approaches to estimating the amount of flash that have been utilized.

1.   Statistical data giving average ratios of the gross to net weight of forgings for different classes of part and weight are used. This approach has been utilized in different forms by Morgeroth [9], Kruse [10], and the FIA [11] for steel forgings.
2.   The use of average values of the flash amount per unit length of the flash line for different weights of forging (e.g., Refs. 8 and 12).

For the estimating procedures described in this chapter, the second of these approaches has been used. Table 14.2 shows data relating the flash weight per unit length of flash line for different weights of forging. This data has been

TABLE 14.2 Flash Weight per Unit Length of Flash Line
for Steel Forgings

Forging weight (kg)	Flash weight (kg/cm of periphery)
Less than 0.450	0.0047
0.450–2.273	0.0063
2.273–4.545	0.0098
4.545–6.818	0.013
6.818–11.364	0.0168
11.364–22.727	0.0223
22.727–45.455	0.0324
Above 45.455	0.0477

Source: Ref. 8.

recommended by the United Kingdom National Association of Drop Forgers and
Stampers (NADFS) and has been found to be reliable by companies who use it
for flash estimation [7]. This data is for the forging of steel and it has been
assumed that the equivalent volume is produced in other materials. This
equivalent volume of flash can be obtained by dividing by the density of steel.
An expression has been fitted to this data to enable the volume of flash per unit
length of flash line to be estimated, and this relationship is as follows:

The volume of flash per unit length of flash line, V_{fl}, is given by

$$V_{fl} = 0.1234 V^{0.5} \text{ cm}^3/\text{cm} \tag{14.3}$$

Example

For the part shown in Fig. 14.4 the volume of flash per centimetre of flash line
from Eq. 14.3 is $V_{fl} = 0.1234 (49.9)^{0.5} = 0.87 \text{ cm}^3/\text{cm}$ or the total volume of
flash generated is $0.87 \times 31.4 = 27.3 \text{ cm}^3$.

14.3.3   Webs in Forgings

Webs are thin sections with a large projected area in the direction of die closure.
Webs are often designed into the parts for strength and other reasons, often
accompanied by peripheral ribs. These webs add considerably to the load
requirements during forging operations because of the large die contact areas,
which increase cooling rates, friction, and so on. If the finished part has through
holes to be forged in, then these must be filled with webs at the die parting line
and then these webs are removed by shearing (piercing) during the flash removal
process. The material in these webs is additional waste material and add to the
material cost per part.

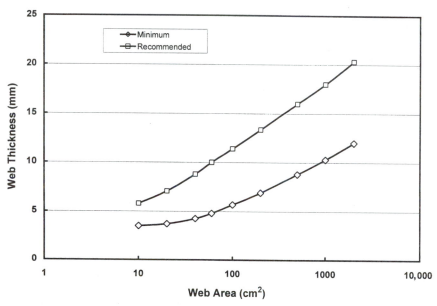

FIG. 14.5 Web thickness related to projected area. (From Ref. 2.)

The appropriate thickness of the webs is dependent on the projected area of the holes to be filled, as shown in Fig. 14.5, from which the following relationship is obtained:

$$\text{Web thickness } T_W \text{ (mm)} = 3.54\, A_H^{0.227}, \tag{14.4}$$

where A_H is the area of the holes in square centimeters.

14.4 FORGING ALLOWANCES

Parts produced by hot forging require machining on surfaces that will locate with other parts in a final product. Thus the detailed shape features of a forging are developed from the required-machined part by adding various allowances to the machined surfaces, although some of these allowances also form part of the forging design for surfaces that will not be machined. Figure 14.6 shows the cross section of a simple forging, which is assumed machined all over. The first allowance added to the machined surface is a finish or machining allowance. This amount is in addition to any dimensional tolerances and must be sufficient to result in a clean surface after finish machining. The allowance for machining is dependent on several factors, but particularly on the amount of oxidation that will result from heating the part up to the forging temperature. The level of oxidation will be dependent on the material type and on the overall size of the forging. Figure 14.7 shows typical finish allowances for different materials [13].

FIG. 14.6 Forging allowances for finish machining and draft.

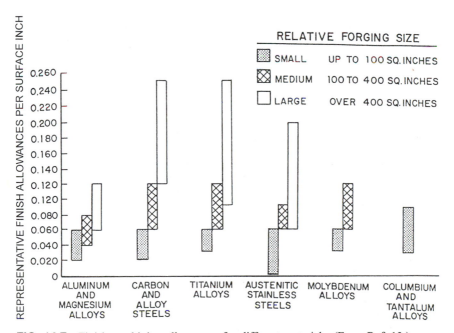

FIG. 14.7 Finish machining allowances for different materials. (From Ref. 13.)

TABLE 14.3 Draft Allowances for Forgings

	Hammer dies		Press dies	
Materials	External	Internal	External	Internal
Steels				
Aluminum alloys	5–7°	7–10°	3–5°	5–7°
Titanium alloys				
Ni-based alloys				
Tolerances in all cases	±1°	±1°	±1°	±1°

Source: Ref. 2.

Draft is an angle allowance added to surfaces parallel to the direction of die closure to facilitate release of the part from the die after forging. In general, draft allowances on inside surfaces are greater than those on outside surfaces, because of the tendency of the part to shrink onto projections in the die as cooling takes place. Table 14.3 gives recommended values of draft angles for both presses and hammers [2].

Finally, all edges and corners in the part must have radii added. These radii are necessary to aid material flow and ensure good die filling. In addition, sharp corners in dies can lead to premature die failure due to fracture as a result of associated stress concentrations, high stresses and so on. Table 14.4 shows typical recommendations for edge and fillet radii for different materials. In general, larger radii are recommended for the more difficult-to-forge materials.

TABLE 14.4 Typical Minimum Edge and Fillet Radii for Rib/Web Type Forgings

Material	Corner radius (mm)	Fillet radius (mm)
Aluminum alloys	2.3	9.7
Low alloy steels	3.0	6.4
Titanium alloys	4.8	12.7
Nickel-based superalloys	6.4	19.0
Iron-based superalloys	4.8	17.0
Molybdenum	4.8	12.7

Source: Adapted from Ref. 13.

14.5 PREFORMING DURING FORGING

In practice very few forgings are produced in the one stage indicated in Fig. 14.2. This will usually result in excessive amounts of flash to ensure die filling and/or large die loads. Thus in most cases a series of preforming operations are necessary to gradually bring the stock material closer to the finished shape before the last forming stage in the finishing die cavity (finisher). The number and type of preforming operations depend largely on the finished forging shape. Figure 14.8 shows a typical sequence for a simple connecting-rod forging [1].

In most cases the starting point of forging is a simple shape—either a length of round or square section bar or a billet cut off from bar stock. The object of

Blank

Edging

Blocking

Finishing

Trimming

FIG. 14.8 Typical forging sequence for a connecting rod. (From Ref. 1.)

preforming is to redistribute the stock material to correspond more closely to the finished shape. The design of preforms is still something of a "black art," relying heavily on the skills of experienced personnel. Progress has been made in the application of finite element and upper-bound plasticity analyses to the design of preforms, but this is still the subject of some research [14, 15].

Most flat or compact forgings start from billets and can usually be produced in two to four forming stages. The first forging stage is usually a simple die, which may just be flat faces, called a buster or scale-breaking die. The purpose of this die is to do some initial flattening of the billet, largely to remove the scale produced by oxidation during heating. For simple shapes the material may then be forged in the finishing die. However, for most parts one or two more preforming stages will be necessary. The preform prior to finishing is called a blocker (sometimes called a semifinisher or in the United Kingdom a molding impression). The blocker is essentially a smoothed-out version of the finisher with thicker sections and larger radii. There are some well-accepted design rules for blocker cross sections [16]. Figure 14.9 shows some typical blocker sections relative to finisher cross sections. If the final part has thin and/or tall features (thin ribs and webs), then a preblocker may also be required and this will have thicker sections and larger radii than the blocker.

For long parts, the starting point is usually the heated end of a bar of material of constant cross section. The initial preforming stages are relatively simple open-die forging operations, the purpose of which is to distribute the material along the length of the forging to correspond more closely to the mass distribution of the finished part. This is achieved by using relatively simple dies called fullers, followed by a die called an edger (or in the United Kingdom a roller die). Figure 14.10 shows a typical sequence for forging a connecting rod. Fullers are used to elongate and draw down the bar stock as appropriate. Fullers have crowned faces and the stock is placed between the dies, with one or two blows taking place. The stock is then rotated through 90 degrees and the process is repeated. Usually only one fuller stage is used, but if there are two or more major changes in cross-sectional area along the length of the part, more than one fuller may be used. After fullering, the edger or roller die is used to smooth out the stock material and to further elongate it somewhat.

For the connecting-rod example in Figure 14.10, after two fullering stages and one edger die, the result is a dumbbell shape, with approximately round cross sections and with an axial mass distribution similar to the finished shape. These initial mass-distribution-preforming stages can be done on reducer rolls, which use a series of shaped rolls to elongate and draw down the bar stock. Reducer rolls are sometimes used in conjunction with mechanical presses for higher productivity.

Following these initial mass-distribution-preforming stages, the cross sections of the part are then formed to correspond to the finished shape. For simple shapes this may be done directly in the finishing cavity, but usually a blocker die is used,

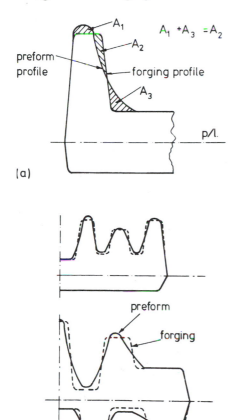

FIG. 14.9 Typical blocker cross sections compared to the finish forging cross sections. (a) General design procedure. (b) Sample cross sections. (From Ref. 17.)

and for forgings with very thin sections a preblocker may also be required. The usual reason to include a blocker forging stage is to increase the life of the finishing-die impression before resinking is necessary. Whether a blocker die is required is usually decided based on experience with similar parts and on the total quantity of forgings required. For simple forgings no blocker impression may be needed. Chamouard [16] gives recommendations for the use of blocker impressions for rib/web type forgings, based on the rib height to rib thickness ratio. A blocker impression is recommended when this ratio exceeds 2.5.

Blocker forging cross sections are essentially smoothed-out versions of corresponding sections in the finished forging, with thicker sections and larger radii. The

FIG. 14.10 Forging sequence design for a connecting rod. (a) Mass distribution stages. (b) Blocker cross sections. (From Ref. 17.)

blocker sections are designed by modifying the corresponding finished sections, using empirically established design rules. For a connecting-rod forging several transverse sections along the length of the part would be selected, together with radial sections at the ends (Fig. 14.10). From these sections corresponding blocker sections are developed, and these define the shape of the blocker die impression. For rib/web-type cross sections, blocker sections have been developed using logarithmic curves, based on the recommendations of Chamouard [16].

If the final forging has bends in the longitudinal axis, a bending operation will be added to the sequence. In this case the mass-distribution-preforming stages (fullers and edgers) will be developed with the axis of the part straightened out, followed by the bending impression. Any blocker stages will be carried out on the bent stock, and again the corresponding die will be developed from cross sections of the finishing impression.

14.5.1 Die Layout

As seen above, several die impressions will be needed to process a hot forging completely. For small and medium-sized hammer forgings these impressions will be laid out on a single die block. Figure 14.11 shows two typical examples [18].

(a)

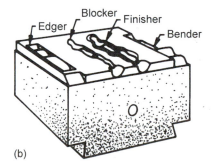

(b)

FIG. 14.11 Typical multi-impression hammer forging dies. (From Ref. 18.)

For larger forgings the various stages may be carried out on separate machines with reheating of the forging stock between stages. For press forgings the various die impressions may be machined into one die block or into separate die inserts attached to the machine bed.

For multiple impression dies the various impressions must be laid out on the die surface to enable successful forging with a minimum-sized die block. The die block depth should be sufficient to enable several resinks of the cavities as wear occurs. A number of factors must be taken into account in the layout of die impressions, including the minimum spacing between cavities, which depends among other things on the cavity depth. In general, the finisher and blocker impressions are placed in the center of the die block, with the fullers to one side and the edger and/or bending die to the other (Fig. 14.12). The finisher is positioned such that the center of loading corresponds to the dowel pin that is used to position the dovetails on the back of the die on the hammer bed. If more than one forging is to be made at once, the finisher and blocker impressions can sometimes be nested to conserve space. The fuller dies are usually inclined at 10 to 15 degrees across the left-hand corner of the die block, again to conserve space. For the purposes of estimating die block size the following cavity-spacing rules, derived from data provided by Thomas [2], can be used.

FIG. 14.12 Die layout for hammer forging die. (From Ref. 17.)

FIG. 14.13 Typical die lock configuration. (From Ref. 2.)

Cavity depth $d_c = 0.5\,T$, where T is the part thickness
Cavity spacing $S_d = 3.1(d_c)^{0.7}$
Cavity edge distance $S_e = 3.4(d_c)^{0.76}$
Die block depth $= 5\,d_c$

Die locks or registers are provided on some dies to prevent mismatch during forging for parts with cranked parting lines. Die locks absorb the side loads produced, but add to the size of the die block and increase the machining costs of the die blocks. Figure 14.13 shows a typical die lock configuration. To be effective the die lock must engage just before the top die comes into contact with the forging stock. An overlap of 10 to 12 mm is recommended, and to allow adequate strength the width of the lock should be at least 1.5 times the depth [1]. If the die parting line is only marginally cranked, a die lock may not be required, but for the purposes of the classification of forgings below it will be assumed that locked dies will be necessary for forgings with cranked parting lines.

14.6 FLASH REMOVAL

The final stage in hot forging is the removal of the flash to yield the finish forging. The flash removed is scrap material and can be more than 50% of the material used for some forgings. The flash is usually removed with a trimming die, which shears the flash off at the parting line of the forging. The webs in any through holes will also be pierced out at the same time. Flash trimming will usually be done on a mechanical press adjacent to the main forging machine, with the forging still hot. In some cases flash trimming may be done later when the part is cold. The operation and the dies used are similar for both hot and cold flash trimming, but the press loads are higher for cold flash trimming. The flash may also be removed by a machining operation, such as band sawing, but this is slow and relatively expensive. Consequently, band sawing should only be considered for small quantities of parts or for some larger forgings.

Trimming and piercing dies have a shearing edge corresponding to the parting line of the forging. The complexity is therefore increased by the need for a cranked parting line. The corresponding punch forces the forging through the trimming die to remove the flash, and the design of the punch must be such that this can be achieved without distortion or damage to the forged part.

14.7 CLASSIFICATION OF FORGINGS

A number of classification schemes for forgings have been developed over the years [19]. These range from relatively simple pictorial systems to quite complex numerical coding schemes. The general objective has been to indicate forging complexity in some way in order to relate this difficulty to different aspects of the forging process design. Several early systems were proposed to systematically provide data on typical gross to net weights for different forging types for estimating purposes [9–11].

A relatively complex classification and coding scheme was used in a *Design for Forging Handbook* [20] in order to indicate general forging costs [21]. This classification scheme covered the presence of individual shape features of the forging such as holes, depressions, bosses, ribs, and so on. Parts were allocated to different classes dependent on the presence of these features. For the purposes of the current procedure a more simplified approach based not on the presence of specific features, but on numerical evaluations of complexity, is used [12]. Parts are first divided into main classes determined by the overall dimensions of the rectangular envelope that encloses the part (Fig. 14.14). For long forgings with a bent axis this envelope is determined after the part has been straightened out. This

First Digit	Description	
0	Compact Parts, $L/W \le 2$, $L/T \le 2$	
1	Flat Parts, $L/W \le 2$, $L/T > 2$	
2	Long Parts, $L/W > 2$	Main Axis Straight
3		Main Axis Bent

L = Envelope Length
W = Envelope Width
T = Envelope Thickness
L >= W >= T

FIG. 14.14 Forging classification, allocation of first digit.

initial broad classification divides parts according to the basic sequence of operations required for processing.

Class 0: Compact Parts $(L/W \leq 2.0, L/T \leq 2.0)$

The basic sequence of operations for this class is

Scale break (buster)
Blockers (one or two)
Finisher
Clip and pierce to remove flash and webs for through holes

The forging complexity for this class is increased by the presence of

Thin sections
Cranked die split lines
Forged in side depressions

Class 1: Flat Parts $(L/W \leq 2.0, L/T > 2.0)$

The basic sequence of operations for this class is

Scale break (buster)
Blockers (one or two)
Finisher
Clip and pierce to remove flash and webs in through holes

The forging complexity for this class is increased by the presence of

Thin sections
Ribs and webs
Cranked die split lines
Forged in side depressions

Second Digit	Description	
0	Parting Line Flat	No Side Depressions
1	Parting Line Not Flat	
2	Parting Line Flat	Side Depressions
3	Parting Line Not Flat	

FIG. 14.15 Allocation of second digit for compact and flat parts.

Second Digit	Description
0	Parting Line Flat
1	Parting Line Not Flat

FIG. 14.16   Allocation of second digit for long parts.

Classes 2 and 3: Long Parts ($L/W > 2.0$)

The basic sequence of operations for these classes is

Fullers (one or two)
Edger (roller)
Bender (for bent parts)
Blockers (one or two)
Finisher
Clip and pierce to remove flash

In some cases, passes through reducer rolls may replace the first two stages.
The forging complexity for these classes is increased by the presence of

Large changes in cross-sectional area
Thin sections
Ribs and webs
Cranked die split lines

Example
For the part shown in Fig. 14.4, $L = W = 100$ mm and $T = 20$ mm. Thus $L/W = 1$ and $L/T = 5$. The first digit is 1 (flat part). There are no side depressions and the parting line is flat, so the second digit is zero.

14.7.1   Forging Complexity

Two numerical indications of forging complexity are used: the shape complexity factor and the number of surface patches in the part.

Shape Complexity Factor
This factor is a modification of that used in the European tolerancing standards for forgings to indicate complexity [23], i.e.

$$\text{Complexity factor } F_{fc} = \frac{\text{volume of rectangular envelope for part}}{\text{part volume}} = \frac{\text{LWT}}{V}$$

For bent parts, the axis is straightened before this complexity factor is calculated. This factor indicates in a general way the amount of deformation necessary within specific classes of forging, as the presence of thin sections and large changes in cross-sectional area result in increased values of this complexity rating. For the example part shown in Fig. 14.4, $F_{fc} = 4.00$.

Number of Surface Patches in the Part

This rating is similar to the counting procedure for surface patches used for cost estimation for injection molded parts as outlined in Chapter 8. The number of surface patches that make up the shape in the upper and lower cavities are counted up. All standard surface elements, such as planes, cylinders, cones, and so on, are given equal rating, but free form or sculptured surface elements are counted equal to four standard surface patches. This number is a measure of the forging complexity that indicates the presence of more complex shape features. For example, a multiribbed part will have increased numbers of surface elements present relative to a simpler forging.

14.8 FORGING EQUIPMENT

Hot forgings can be produced on a variety of equipment, including mechanical and hydraulic presses, friction screw presses, and hammers. This forging equipment can be divided into two basic types: work-restricted and stroke-restricted machines. In work-restricted machines the amount of deformation that can be achieved during each stroke or blow of the machine is limited by the energy or maximum force available. If the energy or force capacity is less than is required to deform the part, then more than one stroke or blow is needed. Machines that fall into this category are hammers, friction screw presses, and hydraulic presses. In stroke-restricted machines the amount of deformation that can be done is fixed by the stroke of the machine. If sufficient force or energy to carry out the operation is not available, then the machine will stall and a larger machine should be used. Mechanical presses fall into this category, as a crank or eccentric determines the amount of ram movement.

Hammers are the most common types of machine used [23], and the basic technology of forging hammers was developed in the last century. The choice of forging equipment depends on a number of factors, including part size and complexity, material, and the quality of the parts to be produced. Hammers are often preferred for small to medium batches because of quicker tool setups and lower overheads. They are also used for elongated and branch-type forgings because die areas can be provided for the larger number of preform dies required for such shapes. Additionally, mechanical presses are not available in very large load capacities, so, for large forgings, hammers or large hydraulic presses must be used.

14.8.1 Gravity Drop Hammers

Gravity drop hammers [18] are the oldest type of forging equipment available. The principle of operation is that the moving die block (tup) is raised by a lifting mechanism and then released, so that it falls onto the fixed die attached to the anvil. The amount of deformation that can be carried out is determined by the potential energy of the moving die block at its maximum height. This potential energy is converted into kinetic energy as the die block falls and is then dissipated in deformation of the workpiece. Various lifting mechanisms are used, including frictional means with boards, band brakes or belts, or a lifting cylinder employing steam, compressed air, or hydraulic fluid (Fig. 14.17). These machines are available in a range of blow energies from 0.6 kNm (60 kg-m) to 400 kNm (40,000 kg-m).

14.8.2 Double Acting or Power Hammers

These machines [18] are similar to gravity hammers in that a lifting cylinder raises the moving tup, but power is also applied to the downward-moving tup to increase the energy capacity. Energy ratings for similar tup weights are considerably more than for gravity hammers, and the die closing speeds are higher also. Power comes from double-acting steam, compressed air, or hydraulic cylinders. Double-acting hammers are manufactured in a range of energy ratings from 3 kNm (300 kg-m) to 825 kNm (82,500 kg-m).

FIG. 14.17 Schematic of various types of drop hammer. (a) Board hammer. (b) Belt hammer. (c) Chain hammer. (d) Airlift hammer. (From Ref. 1.)

14.8.3 Vertical Counterblow Hammers

In these machines two tups with nearly equal masses are driven by double-acting cylinders toward each other and impact in the center of the machine. More energy is dissipated in the workpiece than in the foundations and subsoil compared to single-acting hammers. Very high energy capacities are available in the largest machines, with ranges from 30 kNm (30,000 kg-m) to 2000 kNm (200,000 kg-m).

14.8.4 Horizontal Counterblow Hammers

These machines are also called impacters and two rams are actuated by double-acting cylinders. Heated stock is positioned vertically between the dies by an automatic transfer mechanism. Energy ranges from 4 kNm (400 kg-m) to 54 kNm (5400 kg-m) are typical.

14.8.5 Mechanical Presses

In mechanical presses [18] a crank, knuckle joint, scotch yoke, or moving-wedge mechanism is used to apply a vertical squeezing motion between the upper moving die and a lower fixed die. Figures 14.18a and b show typical mechanical

FIG. 14.18 Schematic of various types of forging press mechanisms. (a) Crank press. (b) Knuckle joint press. (c) Friction screw press. (d) Hydraulic press. (From Ref. 1.)

press mechanisms. In general, guidance of the two dies is better than in hammers, so improved die matching is possible. Press ratings from 3 MN to 140 MN (300–14,000 tons) are available. Thus, mechanical presses are not available for the largest of parts processed by forging.

14.8.6   Screw Presses

In screw presses [18] the upper ram and die are connected to a large vertical screw that can be rotated by a flywheel, so that the ram can move up and down relative to the fixed die in the bed of the machine (Fig. 14.18c). The ram has a limited amount of energy for each stroke; thus multiple blows are usually employed similar to hammers. Screw presses are available in ratings from 0.63 MN to 63 MN (63–6300 tons).

14.8.7   Hydraulic Presses

Hydraulic presses [18] are available in a wide range of sizes up to the largest at 50,000 tons or more capacity. The moving die is attached to a ram actuated by a large hydraulic cylinder (Fig. 14.18d). Various strokes, forces, and closing speeds can be obtained on hydraulic presses. In some cases hydraulic presses are fitted with auxiliary horizontally moving rams, and these enable side depressions to be forged into some parts, although this is not done to a great extent.

14.8.8   Choice of Forging Machine Type

In general, whether to use hammers or presses depends on a number of factors, but some guidelines exist [23].

Circular or ring forgings in steels are particularly suited to crank presses until the dimensions of the forgings result in loads in excess of the range of mechanical presses, when power or counterblow hammers become necessary.

TABLE 14.5 Some Comparative Data for Forging Equipment

Hammers	Maximum blow energy (kJ)	Die closing speed (m/s)
Gravity drop	47–120	3–5
Power drop	1150	4.6–9
Counterblow	1220	4.6–9
Impacter	34	10–17
Presses	Force (MN)	—
Mechanical	2.2–143	0.06–1.5
Screw	1.3–280	0.5–1.2
Hydraulic	2.2–623	0.03–0.8

Source: Ref. 24.

Asymmetrical and branched forgings tend to be produced on hammers, as more total die area is required than is available on mechanical presses of appropriate load capacity.

Large circular forgings are manufactured using hammers because of the large force requirements.

Close-to-form blade-type forgings tend to be made using screw presses or with wider tolerances on hammers.

Thin-plate-type forgings with edge ribs, together with light alloy precision forgings are usually produced on hydraulic presses.

Smaller batch sizes tend to be made on hammers rather than presses.

14.8.9   Comparisons of Forging Equipment

Table 14.5 [24] shows some comparative data for the different types of forging equipment. As can be seen, die-closing speeds are significantly higher for hammers compared to presses. This has relevance for the forging of strain rate sensitive materials. Table 14.6 shows the ranges of the different types of forging equipment available. The different machines are compared based on their maximum deformation energy capacity. However, this comparison is not related to the energy available per blow or stroke of the machine, but is based on the energy requirements for parts typically processed on the size of machine indicated. The relative comparisons between machines are derived from ratios presented in Dallas [24] and other reports. The ratings at the top of each column are those normally used in industry, with hammers usually rated by the weight of the tup and presses by the load capacity in tons. As can be seen, mechanical presses are not usually available above 14,000 tons capacity; thus large forgings must be produced on either hammers or hydraulic presses, with the very largest restricted to counterblow hammers or large hydraulic presses.

Figure 14.19 shows typical average usable blow or stroke rates for forging equipment. This data is largely based on data from Scott and Wilson [23], with some supplementary calculations from Dallas [24]. These blow rates are not the maximum obtainable from the equipment, but are typical usable blow rates, which reflect such factors as the time taken to manipulate the part between impressions. As can be seen, mechanical presses are capable of the highest usable blow rates for smaller parts, which is why they are preferred for higher productivity situations for parts that are appropriate. For large parts the usable blow rates become comparable for the various types of equipment, because the time taken to manipulate the forgings between blows becomes dominant rather than the blow rate capabilities of the machines. These curves enable estimates of the forging cycle times to be made once an appropriate machine has been chosen.

TABLE 14.6 Equivalent Capacities of Presses and Hammers

Maximum energy		Hydraulic presses, rating (tons)	Counterblow hammers, rating (kg-m)	Power hammers, rating (lb)	Mechanical presses, rating (tons)	Gravity drop hammers, rating (lb)	Friction screw presses, rating (tons)
ft-lb	kg-m						
3,850	533	170	450	455	210	1,000	170
5,870	812	260	650	680	320	1,500	250
8,830	1,200	400	1,000	910	475	2,000	390
11,000	1,540	400		1,000	600	2,200	490
11,320	1,570	550		1,130	660	2,500	500
14,200	1,960	680		1,360	820	3,000	620
16,700	2,310	750	2,000	1,500	900	3,300	730
19,400	2,690	900		1,800	1,080	4,000	850
21,700	3,000	935	2,500	1,870	1,120	4,100	950
22,500	3,120	1,000		2,000	1,200	4,400	990
24,700	3,420	1,125		2,250	1,350	5,000	1,050
26,200	3,630		3,000				1,150
28,500	3,950	1,250		2,500	1,500	5,550	1,250
30,000	4,160	1,350	3,500	2,700	1,620	6,000	1,520
34,400	4,770	1,500	4,000	3,000	1,800	6,600	1,600
41,600	5,770	1,800		3,600	2,160	8,000	1,830
46,000	6,370	2,000	5,500	4,000	2,400	8,800	2,020
52,000	7,200	2,250	6,000	4,500	2,700	10,000	2,290
58,000	8,030	2,500		5,000	3,000		2,590
70,000	9,700	3,000	8,000	6,000	3,600		3,080
86,800	12,000	3,700	10,000	7,400	4,400		3,800
94,000	13,000	4,000		8,000	4,800		4,180
118,000	16,000	5,000	13,000	10,000	6,000		5,000
138,000	19,100	5,850	16,000	11,700	7,000		6,060
142,000	19,600	6,000		12,000	7,200		6,220
173,000	23,900	7,300	20,000	14,600	8,700		7,590
220,000	27,120	8,300	25,000	16,600	10,000		9,650
240,000	32,200	10,000	32,000	20,000	10,600		10,220
300,000	41,500	12,500	13,650	25,000	13,650		
361,500	50,070	15,000	40,000	30,000			
425,000	59,000	17,500		35,000			
	60,330	18,000		36,000			
484,000	67,036	20,000	50,000	40,000			
546,800	75,500		63,000				
610,000	84,500	25,000		50,000			
694,400	95,800		80,000				
738,000	102, 00	30,000					
868,000	120,200	35,000	100,000				
1,280,000	172,000	50,000	144,000				

From Ref. 20, with data from several sources (Refs. 24, 25).

FIG. 14.19 Average usable blow rates for forging equipment. (From Ref. 12.)

FIG. 14.20 Operating cost per operation relative to 1000-lb power hammer. (From Ref. 25.)

Figure 14.20 shows data on the cost per operation for different types of forging equipment relative to the cost of a 1000 lb power hammer [25]. In this context, operation means forging stage (i.e. blocker, finisher, etc.). The data have been collected from a variety of industry sources and also reflect the fact that on average hammers utilize two or three times the number of blows per forming stage as is used on presses, for which only one per operation is assumed. Figure 14.21 [26] shows some combined curves for the data in Fig. 14.20. The curve for hammers is a combination of the data for all types of hammers. In the upper range of energy capacities the curve corresponds to counterblow hammers and at the

FIG. 14.21   Relative operating cost per operation for forging equipment. (From Ref. 26.)

lower ranges to power hammers. The curve for mechanical presses ends at the point of intersection with the curve for hydraulic presses. These curves can be used to estimate forging processing costs once the number of forging stages has been established.

TABLE 14.7 Classification of Forging Materials

Class	Materials	Load factor, α_m(kg-m/mm^2)	Die life factor, β_m
0	Aluminum alloys		
	Group A	0.054	1.0
	Group B	0.087	1.0
	Magnesium alloys	0.065	1.0
1	Copper and copper alloys	0.065	1.0
2	Carbon and alloy steels		
	Group A	0.065	1.0
	Group B	0.076	0.8
	Group C	0.087	0.7
	Group D	0.098	0.6
3	Ferritic and martensitic stainless steels	0.109	0.3
	Tool steels	0.109	0.3
	Maraging steels	0.109	0.3
4	Austenitic stainless steels	0.130	0.3
	PH stainless steels	0.141	0.3
	Nickel alloys of iron	0.130	0.3
5	Titanium and titanium alloys		
	α and α/β alloys	0.163	0.2
	β alloys	0.195	0.2
6	Iron-based superalloys	0.152	0.2
7	Cobalt-based superalloys	0.195	0.2
8	Nickel-based superalloys	0.220	0.1
9	Niobium (Columbium) alloys	0.195	0.1
	Molybdenum alloys	0.217	0.1
	Tantalum alloys	0.124	0.1
	Tungsten alloys	0.260	0.1
	Beryllium	0.065	0.5

Source: Ref. 20.

14.9 CLASSIFICATION OF MATERIALS

A wide variety of materials can be used for forging, but by far the largest proportion of parts produced is from carbon and alloy steels, with a significant number also made from light alloys. Table 14.7 shows a general classification of materials used in hot forging [11, 20]. These material classes are arranged roughly in increasing order of forging difficulty. However, there is considerable overlap in these class. For example, many high-strength aluminum alloys are more difficult to process than steels. Increase in forging difficulty is represented by increased forging load requirements and usually reduced die life. In addition, for the more difficult to deform materials it may be impossible to obtain very thin sections (ribs and webs), and consequently the end product must be less close to net shape than for the easier to forge materials. Thus so-called precision forgings are usually associated only with light alloy materials.

Table 14.7 contains two factors that are used in cost estimating procedures described later. The first of these is a load factor α_m that gives the approximate deformation energy per unit area required for a relatively simple shape in the specific material. The second factor is a material die life factor β_m, which represents the approximate reduction in die life for the forging of similar shapes in different materials.

14.10 FORGING COSTS

Figure 14.22 shows the breakdown of the average forging costs for hot forgings found in industry [27]. Material costs usually make up around 50% of forging costs, and of this material, a significant proportion is waste material in the form of flash, scale losses, and so on. Die costs represent about 10% of forging costs and the remainder includes direct labor, equipment operating costs and overhead costs. For the purposes of early cost estimating, three main cost elements are considered.

1. Material costs, including flash and scale losses
2. Equipment operating costs, including labor, heating costs, ancillary equipment, and overhead
3. Die costs, including initial tooling costs and maintenance and resinking costs

Each of these will be considered in more detail in the following discussion. The early costing procedure described for hot forging is currently restricted to parts produced on hammers and forging presses using conventional dies. Preforming carried out on forging rolls is not considered, nor are such processes as ring rolling and hot upset forging.

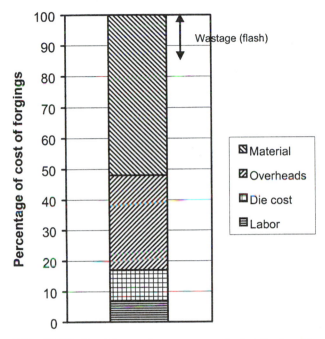

FIG. 14.22 Breakdown of average costs for hot forging. (From Ref. 27.)

14.10.1 Material Costs

The material costs for a forged part are determined as follows:

$$\text{Material cost } C_{\text{mat}} = C_{\text{m}}\rho[(V + P_{\text{r}}V_{\text{fl}} + A_{\text{H}}T_{\text{w}})(1 + S_{\text{sl}}/100)], \tag{14.5}$$

where

C_{m} = cost per unit weight of material
ρ = part density
V = the part volume
P_{r} = length of the flash line (perimeter of part)
V_{fl} = flash volume per unit length of flash line (Eq. 14.3)
A_{H} = area of through holes
T_{w} = web thickness recommended (Eq. 14.4)
S_{sl} = scale loss for material, %

Example

For the part shown in Fig. 14.4 the following data are available:

Part volume $V = 49.9$ cc

Flash volume per length of flash line $V_{fl} = 0.87$ cc/cm
Length of flash line $P_r = 31.4$ cm
There are no through holes

Assuming that $C_m = 1.1\$/kg$, $\rho = 7.86$ gm/cc, and $S_{sl} = 5\%$, then

$C_{mat} = 1.1 \times 7.86[(49.9 + 31.4 \times 0.87)1.05]/1000 = \$0.43$ per forging

14.10.2 Equipment Operating Costs

The production cost per piece is basically expressed as

$$C_{pr} = M/R_p \tag{14.6}$$

where M is the operating cost per unit time of the equipment and R_p is the production rate (pieces per unit time). The operating cost per unit time, M, includes direct labor, ancillary equipment for handling and heating, and not just the operating cost of the main press or hammer used for the forging. It is usual for a separate press to be used for the trimming of the flash and piercing, and this is usually a mechanical press. The costs for this are determined separately. Thus the operating cost of the equipment used is given by

$$C_{pr} = MN_b/B_r \tag{14.7}$$

where N_b is the number of blows or strokes required to form the part and B_r is the effective blow or stroke rate of the press or hammer used. It should be noted that the number of strokes is usually equal to the number of operations if a mechanical or hydraulic press is used. For work-restricted machines such as hammers, several blows per operation are usually used, the average being between two and three blows per operation in general. This assumption has been used for generating the data in Fig. 14.21.

Thus, the operating costs of the equipment are dependent on the equipment selected for the forging and its effective stroke or blow rate. The required equipment must be selected based on the load or energy required for shaping the part. A variety of methods are available for estimating the load needed for a specific forging [18, 24, 28–32], including some relatively complex plasticity analyses. For the purposes of early cost estimation, it has been decided to modify the procedure outlined in the ASM handbook [18]. The plan area of the forging (including area of the flash in the flash lands) is multiplied by a material factor to give the energy capacity of the hammer required to forge the part. Variations of this approach are generally used by estimators in industry to determine the appropriate equipment to be used. However, the load or energy required for a forging is influenced by the shape complexity of the part in addition to the

Forging Data:

Second Digit	Forging Shape Complexity Factor, F_{fc}							
	<= 1.5		>1.5 and <= 2.5		> 2.5 and <= 5.0		>5.0	
0	1.6	0.95	1.7	0.9	1.9	0.75	2.2	0.55
	2		3		3		4	
1	1.6	0.7	1.7	0.6	1.9	0.5	2.2	0.3
	2		3		3		4	
2	1.6	2	1.7	0.6	1.9	0.5	2.2	0.3
	2		3		3		4	
3	1.6	0.65	1.7	0.5	1.9	0.45	2.2	0.25
	2		3		3		4	

Key: Shape Load Factor, α_s, upper left
Shape Die Life Factor, β_s, upper right
Number of Forging Operations, N_{op}, lower left

Forging Operations Required:

	Number of Forging Operations, N_{op}		
	2	3	4
Scale break, n_{sb}	1	1	1
Blocker, n_{bk}	0	1	1
Semifinisher, n_{sf}	0	0	1
Finisher, n_{fn}	1	1	1

Bender, $n_{bnd} = 0$; Edger, $n_{edg} = 0$; Fuller stage 1, $n_{f1} = 0$; Fuller stage 2, $n_{f2} = 0$

FIG. 14.23   Data for compact forgings, Shape Class 0.

material being forged. Thus, for the purpose of this early cost-estimating system, the following is used:

$$\text{Energy capacity } E_f = A_p \alpha_m \alpha_s \qquad (14.8)$$

where
A_p = the projected part area including flash
α_m = material load factor (Fig. 14.7)
α_s = shape load factor

Once the energy capacity of the required equipment is known, a specific piece of equipment can be selected from Table 14.6 and the relative operating cost per operation determined from Fig. 14.21.

Forging Data:

Second Digit	Forging Shape Complexity Factor, F_{fc}							
	<= 1.5		>1.5 and <= 3.0		> 3.0 and <= 6.0		>6.0	
0	1.0 2	1.0	1.25 3	0.75	1.4 3	0.45	1.6 4	0.3
1	1.05 2	0.9	1.3 3	0.7	1.45 3	0.4	1.65 4	0.3
2	1.0 2	1.0	1.25 3	0.75	1.4 3	0.45	1.6 4	0.3
3	1.05 2	0.9	1.3 3	0.7	1.45 3	0.4	1.65 4	0.3

Key: Shape Load Factor, α_s, upper left
Shape Die Life Factor, β_s, upper right
Number of Forging Operations, N_{op}, lower left

Forging Operations Required:

	Number of Forging Operations, N_{op}		
	2	3	4
Scale break, n_{sb}	1	1	1
Blocker, n_{bk}	0	1	1
Semifinisher, n_{sf}	0	0	1
Finisher, n_{fn}	1	1	1

Bender, $n_{bnd} = 0$; Edger, $n_{edg} = 0$; Fuller stage 1, $n_{f1} = 0$; Fuller stage 2, $n_{f2} = 0$

FIG. 14.24 Data for flat forgings, Shape Class 1.

The material load factor is obtained from the material classification given in Table 14.7. The shape load factor reflects the increased load required to finish-forge more complex shapes. This factor is obtained from the classification scheme described above (Figs. 14.14 to 14.16). Figures 14.22 to 14.26 show the data associated with the classification. The appropriate shape load factor, α_s, is given in the upper left of each classification block. The other data include the shape die life factor, α_s (upper right), and the number of forging operations required, N_{op} (lower left). Shown in the lower part of these figures are the specific operations required for each value of N_{op}.

Forging Data:

Second Digit	Forging Shape Complexity Factor, F_{fc}			
	<= 2.0	>2.0 and <= 5.0	> 5.0 and <= 10	>10
0	1.0 0.9 3	1.1 0.85 4	1.2 0.75 5	1.3 0.6 6
1	1.2 0.65 2	1.3 0.6 3	1.45 0.5 3	1.7 0.35 4

Key: Shape Load Factor, α_s, upper left
       Shape Die Life Factor, β_s, upper right
       Number of Forging Operations, N_{op}, lower left

Forging Operations Required:

	Number of Forging Operations, N_{op}			
	3	4	5	6
Fuller stage 1, n_{f1}	0	1	1	1
Fuller stage 2, n_{f2}	0	0	1	1
Edger, n_{edg}	1	1	1	1
Blocker, n_{bk}	1	1	1	1
Semifinisher, n_{sf}	0	0	0	1
Finisher, n_{fn}	1	1	1	1

Bender, $n_{bnd} = 0$; Scale break, $n_{sb} = 0$

FIG. 14.25 Data for long straight forgings, Shape Class 2.

14.10.3   Examples of Equipment Selection

Figures 14.27, 14.28, and 14.29 [12] show comparisons of the forging equipment capacities determined by this procedure, compared to the actual equipment used for a range of forgings in different classes. As can be seen, the correlation is generally good, except for the case of precision forgings in light alloy materials, where a factor of 2 to 2.5 is necessary to give the required result (parts F and M in Fig. 14.27) . Thus, for the purposes of early cost estimating, the basic calculation is applied to conventional forgings and this value increased by 2.5 for precision forging and multiplied by a factor of 0.9 for blocker-type forgings, to reflect the lower loads required in this case.

Forging Data:

Second Digit	Forging Shape Complexity Factor, F_{fc}			
	<= 2.0	>2.0 and <= 5.0	> 5.0 and <= 10	>10
0	1.05 0.9 4	1.15 0.85 5	1.25 0.75 6	1.4 0.5 7
1	1.25 0.65 4	1.35 0.6 5	1.5 0.5 6	1.7 0.35 7

Key: Shape Load Factor, α_s, upper left
        Shape Die Life Factor, β_s, upper right
        Number of Forging Operations, N_{op}, lower left

Forging Operations Required:

	Number of Forging Operations, N_{op}			
	4	**5**	**6**	**7**
Fuller stage 1, n_{f1}	0	1	1	1
Fuller stage 2, n_{f2}	0	0	1	1
Bender, n_{bnd}	1	1	1	1
Edger, n_{edg}	1	1	1	1
Blocker, n_{bk}	1	1	1	1
Semifinisher, n_{sf}	0	0	0	1
Finisher, n_{fn}	1	1	1	1

Scale break, $n_{sb} = 0$

FIG. 14.26 Data for long bent forgings, Shape Class 3.

14.10.4 Forging Processing Costs

Once the appropriate forging equipment type and size and the forging complexity have been determined, the processing cost for the forging can be estimated. Equation 14.7 can be modified to the form:

$$C_{pr} = C_{op} N_{op} / N_c \qquad (14.9)$$

where
C_{op} = the forging equipment cost per operation
N_{op} = the number of operations required
N_c = the number of identical forgings per cycle

The forging equipment operating cost per operation, C_{op}, is obtained from

$$C_{op} = C_{ro} C_{1000} \qquad (14.10)$$

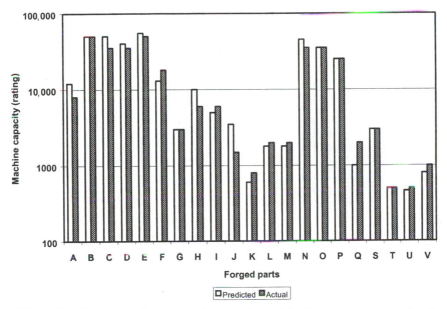

FIG. 14.27 Comparison between predicted and actual forging equipment capacities for flat, round forgings. (From Ref. 12.)

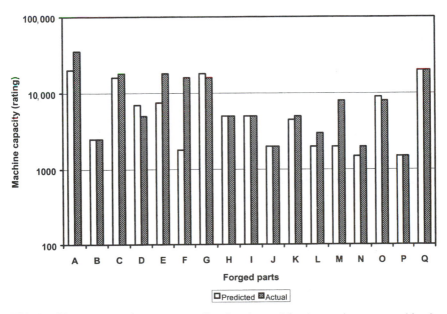

FIG. 14.28 Comparison between predicted and actual forging equipment capacities for flat, nonround forging. (Forgings F and M are precision forgings in aluminum alloy.) (From Ref. 12.)

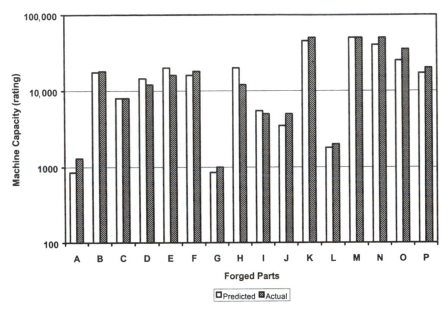

FIG. 14.29 Comparison between predicted and actual forging equipment capacities for long forging. (From Ref. 12.)

where C_{ro} is the relative cost per operation compared to a 1000 lb power hammer, obtained from Fig. 14.21, and C_{1000} is the operating cost per operation of a 1000 lb power hammer.

Example

For the sample part shown in Fig. 14.4 the following data are available:

Projected area $= 78.6$ cm^2
Projected area of flash land $= 19.9$ cm^2
Material load factor $= 0.065$ kg-m/mm^2 (Table 14.7)
Shape classification number $= 10$
Forging complexity factor $= 4$
Shape load factor $= 1.4$ (Fig. 14.23)
Number of operations $= 3$ (Fig. 14.23)

The equipment energy capacity required, E_f, is given by Eq. 14.8:

$$E_f = (78.6 + 19.9) \times 100 \times 0.065 \times 1.4 = 896 \text{ kg-m}$$

From Table 14.6 this part will require a power hammer rated at just below 1000 lb or a mechanical press of about 450 tons. From Fig. 14.21, the relative operating cost per operation for a hammer of this size is 0.95. Thus, assuming the processing cost per operation for a 1000 lb power hammer is $0.15, then the processing cost per forging will be

$$C_{pr} = 0.95 \times 0.15 \times 3 = \$0.42 \text{ per forging}$$

14.10.5 Forging Machine Setup Costs

The forging press or hammer must be set up prior to each batch of forgings produced. The setup cost per forging is given by

$$C_{set} = T_{set}M/B_s \qquad (14.11)$$

where T_{set} is the setup time and B_s is the batch size. Based on data available for average setup times for forging equipment [31] , a relationship between setup time and forging equipment capacity of the following form can be found:

$$T_{set} = 0.3925(E_f)^{0.28} \text{ h} \qquad (14.12)$$

where E_f is the energy capacity in kg-in determined from Eq. (14.8).

Example

For the sample part in Fig. 14.4 the required energy capacity has been determined to be 896 kg-m. Thus the corresponding setup time from Eq. (14.12) is $0.3925(896)^{0.28} = 2.6$ h. Assuming an operating cost rate, M, of $85/hr and a batch size of 10,000 forgings, the setup cost per forging becomes $85 \times 2.6/10000$, which gives $0.02. This is small, but will increase significantly for small batch quantities.

14.11 FORGING DIE COSTS

Several factors contribute to the die cost per piece, including:

1. The initial tooling costs
2. The total number of forgings to be produced or the expected life of the tools
3. The costs of tool refurbishing and/or resinking, together with the number of resinks possible.

   These factors are influenced by the complexity of the forging and the material used.

14.11.1 Initial Die Costs

The approach adopted for estimating the initial die cost is as follows:

1. Determine the die material costs.
2. Estimate the number of hours required for machining the tooling and dies, including all hand finishing and other factors.

There is one die cavity for each operation in the forging sequence, but some of the initial dies are relatively simple in form. There is a difference in practice between the use of presses and hammers in forging. For hammer forgings it is most usual to utilize multi-impression die blocks manufactured as one piece. In some circumstances the finishing cavity may be inserted into the die block, but this is often a result of resinking after the initial cavity becomes worn. Hammer forging preforming operations include fullering, edging, bending, etc. On presses, it is more normal to use separate die inserts for each operation, which are mounted into standard die containers on the press bed. The preforming stages are busters, preblockers, etc. Thus, the tool material requirements are different for presses and hammers, but the complexity considerations for machining the main cavities are similar.

14.11.2 Estimation of Costs for Multi-Impression Forging Dies

The procedure adopted for determining the initial die costs is a modification of that given by the Forging Industry Association (FIA) for forging estimating [11]. Based on the plan area of the forging, the type of forging, and complexity as expressed as the number of surface patches in the cavity shape, the number of hours to machine each die type required is determined. The size of the die blocks can be estimated based upon the guidelines used in industry.

Die Material Costs

The die material costs are determined from the number of die cavities required and the dimensions of the part, using industry guidelines for such factors as the spacing between die impressions [2]. The number of forging impressions N_{imp} is given by

$$N_{imp} = n_{bd} + n_{bk} + n_{sf} + n_{fin} + n_{sb} \qquad (14.13)$$

where the appropriate values are obtained from the forging classification data given in Fig. 14.23 to 14.26.

Similarly, the number of fuller dies

$$N_{fl} = n_{fl} + n_{f2}. \qquad (14.14)$$

The equivalent bar diameter for long parts or for other types of part when more than one forging per cycle are produced, D_{bar}, is given by

$$D_{bar} = (4d_{ave}W_{plt}/\pi)^{0.5} \qquad (14.15)$$

where d_{ave} is the average cavity depth and W_{plt} is the platter width. The average cavity depth, d_{ave}, is given by $d_{ave} = V/A_p$. The platter width is equal to the width of the part if only one forging per cycle is to be produced. Otherwise, the platter width and length need to be determined taking into account any nesting of the part shapes that may be possible.

The width of the die block, W_{blk}, is given by

$$W_{blk} = n_{edg}(D_{bar} + 19) + (N_{imp} - 1)S_d$$
$$+ N_{imp}W_{plt} + 2S_e + N_{fl}(D_{bar} + 19)\cos\psi_{fl} \qquad (14.16)$$

where W_{plt} is the platter width and ψ_{fl} is the angle of the fullers to the die face, usually 15°. The length of the die block, L_{blk}, is given by

$$L_{blk} = L_{plt} + 2S_e \qquad (14.17)$$

but if locked dies are required to counteract lateral forces from cranked parting lines, then a further S_e is added to the die block length to accommodate the counter lock. The length and width of the platter, L_{plt} and W_{plt} are equal to the part length, L, and the part width, W, respectively, if only one part per cycle is produced. However, if more than one part per cycle is produced, then appropriate values of L_{plt} and W_{plt} should be assigned, taking into account any nesting of the part profiles that may be possible.

The die block depth, T_{blk}, is assumed equal to 5 times the cavity depth, d_c. Then the cost of the die block material, C_{dmat}, is given by

$$C_{dmat} = 2C_t L_{blk} W_{blk} T_{blk} \rho_t \qquad (14.18)$$

where C_t is the cost of tool steel per unit weight and ρ_t is the density of tool steel.

Example

For the part shown in Fig. 14.4, the calculations of die material costs are as follows. For this part, $n_{bd} = 0$, $n_{bl} = 1$, $n_{sf} = 0$, $n_{fin} = 1$, $n_{edg} = 0$, $n_{sb} = 1$, and $n_{fl1} = n_{nfl2} = 0$ (Fig. 14.24). Therefore, $N_{imp} = 3$ and $N_{fl} = 0$. The width of the platter is equal to the width of the part $= 100$ mm, if only one part per cycle is produced. For the part, the cavity spacing $S_d = 3.1(10)^{0.7} = 15.5$ mm and the cavity edge distance, $S_e = 3.4(10)^{0.76} = 19.6$ mm (Sec. 14.5). Consequently, the width of the die block is given by Eq. (14.16) as

$$W_{blk} = 2 \times 15.5 + 3 \times 100 + 2 \times 19.6 = 370 \text{ mm}$$

The length of the die block is $L_{blk} = 100 + 2 \times 19.6 = 140$ mm and the block thickness $T_{blk} = 100$ mm. Assuming the cost of tool steel is $20 per kilogram and the density of tool steel is 7.9 gm/cc, then the die material cost for this part is

$$C_{dmat} = 2 \times 20 \times 7.9(5 \times 14 \times 37)/1000 = \$818.$$

Multi-Impression Die Manufacturing Costs

The cost of manufacturing multi-impression dies is determined by estimating the number of hours required to manufacture the dies and then multiplying this by a die manufacturing hourly rate. The procedure used is modification of the FIA estimating procedure [11]. The total time to manufacture the die set is the sum of the times corresponding to the various steps in the procedure.

i. Block Preparation Time.   The time for initial preparation of the die block is given by

$$T_{\text{prep}} = T_{\text{bt}} + 0.0078 W_{\text{blk}} L_{\text{blk}} \text{ h} \tag{14.19}$$

where T_{bt} is the base time, which is equal to 4 h if the forging complexity factor F_{fc} is less than 2.0, 5 h for F_{fc} between 2.0 and 6.0, and 6 h for F_{fc} greater than 6.0. The die block width and length are given in centimeters.

Example: For the sample part in Fig. 14.4 $T_{\text{bt}} = 5$ h as $F_{\text{fc}} = 4$. Then T_{prep} is determined as $5 + 0.0078 \times 37 \times 14 = 9$ h.

ii. Layout Time.   The time for laying out the die block is given by

$$T_{\text{lay}} = 0.008 N_c^m A_p F_{\text{fc}} S_c S_{\text{lk}} \text{ h} \tag{14.20}$$

where N_c is the number of forgings per cycle, S_c is the cavity standard, S_{lk} is the lock standard, and m is the multicavity index—usually taken as 0.7. The lock standard, S_{lk}, is 1.0 for parts for which the die split line is in one plane and 1.5 if the split line is not in one plane. The cavity standard, S_c, is

$$S_c = 0.6(n_{\text{fn}} + n_{\text{sf}}) + 0.4(n_{\text{sb}} + n_{\text{blk}} + n_{\text{bnd}} + n_{\text{edg}} + n_{\text{f1}} + n_{\text{f2}}) \tag{14.21}$$

Example: For the sample part $S_{\text{lk}} = 1$, as the split line will be flat. In Eq. (14.21), $n_{\text{fn}} = n_{\text{sb}} = n_{\text{blk}} = 1$ and all other terms are zero; thus S_c becomes 1.4. The projected area of the part is 78.6 cm². Therefore, the layout time is given as

$$T_{\text{lay}} = 0.008 \times 78.6 \times 4 \times 1.4 = 3.52 \text{ h}$$

iii. Milling Time.   The time for milling the die cavities is obtained from

$$T_{\text{mill}} = 0.155 N_c^m A_p S_{\text{ml}} S_c S_{\text{lk}} \tag{14.22}$$

where S_{ml} is the milling standard given by

$$S_{\text{ml}} = K(6.45 M_s)^b \quad \text{or} \quad 0.2, \quad \text{whichever is greater} \tag{14.23}$$

Here M_s is the number of surface patches per unit projected area, N_{sp}/A_p and

$$K = 0.9(1 - \exp(-0.0098 d_{\text{ave}}))$$
$$b = 0.4 + 0.7 \exp(-0.0039 d_{\text{ave}})$$

Example: For the sample part the following data are available: Number of surface patches, N_s, is 7 and therefore $M_s = 7/78.6 = 0.089$. The average depth, d_{ave}, is given by V/A_p and equals 49.9/78.6 or 6.35 mm. Thus $K = 0.9[1 - \exp(-0.0098 \times 6.35)] = 0.0543$ and $b = 0.4 + 0.7\exp(-0.0039 \times 6.35) = 1.083$. These give a milling standard of 0.03, which is less than 0.2; thus $S_{ml} = 0.2$. Thus, the milling time $T_{mill} = 0.155 \times 78.6 \times 0.2 \times 1.4 = 3.41$ h.

iv. Bench Work Time. The bench work time, t_{bw}, on the dies is given by

$$T_{bw} = N_c^m S_{bn} S_c S_{lk}, \tag{14.24}$$

where S_{bn} is the bench standard, which depends on the forging complexity and average cavity depth. The benchwork factor, $F_{ins} = (A_p/6.54 + 0.5N_s)$ and $S_{bn} = B_0 + 0.26(F_{ins} - 15)$. The constant B_0 depends on the average depth, d_{ave}, as follows:

$d_{ave} \leq 12.7$ mm $\qquad\qquad B_0 = 0.056 d_{ave}$
$d_{ave} > 22.86$ mm $\qquad\qquad B_0 = 4.5 + (0.04 d_{ave} - 0.9)2.19$
$d_{ave} > 12.7$ and ≤ 22.86 $\qquad B_0 = 0.5 + (0.04 d_{ave} - 0.35)7.27$

Example: For the example part $d_{ave} = 6.35$ mm. Thus $B_0 = 0.056 \times 6.35 = 0.356$. Also, $A_p = 78.6$ cm^2 and $N_s = 7$. From this $F_{ins} = (78.6/6.6.54 + 0.5 \times 7) = 15.52$ and $S_{bn} = 0.356 + 0.26(15.52 - 15) = 0.49$. Thus the benchmark time is $0.49 \times 1.4 = 0.69$ hours.

v. Planing Time. The block planning time

$$T_{pl} = 0.008 T_{cav}^{1.5} \tag{14.25}$$

where the cavity time $T_{cav} = T_{lay} + T_{mill} + T_{bw}$.

Example: The block planing time for the example part $T_{pl} = 0.008(5.3 + 3.41 + 0.69)^{1.5}$, or 0.23 h.

vi. Dowel Time. The dowel time, T_{dl}, is 3 h if the die material volume is less than 4260 cm^2, or else is 4 h.

Example: For the example part the total volume of the die material is 10,350 cm^3, thus $T_{dl} = 4$ h.

vii. Flash Gutter Time. The time to machine flash gutters on the die cavities,

$$T_{fl}, = N_c P_r/635 \quad \text{or } 0.8 \text{ h} \quad \text{whichever is the larger} \tag{14.26}$$

In this expression P_r is the forging outside perimeter in millimeters.

Example: For the example part, $T_{fl} = 314/635 = 0.49$ h and therefore T_{fl} is 0.8 h.

viii. Edger Time. If an edger die is required, the time for manufacture is T_{edg} and is given by

$$T_{edg} = n_{edg}L(D_{bar}/25.4 + 1)0.005 \tag{14.27}$$

No edger is required for the example part.

ix. Finish-Polish Time. The time to finish-polish the dies cavities, T_{pol}, is given by

$$T_{pol} = N_c[1 + (F_{fc} - 1)0.6] \tag{14.28}$$

Example: For the example part $F_c = 4$, and therefore $T_{pol} = (1 + (4 - 1)0.6) = 2.8$ h.

The total die manufacturing time is the sum of the above times and consequently the die manufacturing cost is

$$C_{dman} = C_{man}(T_{prep} + T_{lay} + T_{mill} + T_{bw} + T_{pl} + T_{dl} + T_{fl} + T_{edg} + T_{pol}) \tag{14.29}$$

The total initial die manufacturing costs are thus

$$C_{DIE} = C_{dmat} + C_{dman} \tag{14.30}$$

Example: Assuming a die manufacturing cost rate of \$45 per hour, then for the example part $C_{dman} = 45(9 + 3.51 + 3.41 + 0.69 + 0.23 + 4 + 0.8 + 0 + 2.8) = \$1100$.

Thus the total initial die cost is $C_{DIE} = \$1100 = \$1918$.

14.12 DIE LIFE AND TOOL REPLACEMENT COSTS

The life of hot forging dies is relatively short, and therefore it is necessary to include in the estimating procedure a consideration of the expected tool life and a general strategy for tool replacement and refurbishing. The life of the finishing cavity particularly is determined by the complexity of the forging and the material being forged. For example, the same shape produced by different materials will result in different die lives. This can be accommodated by the use of the material die life factor, β_m, and typical values for these are given by the FIA [12], as shown in Table 14.7.

Similarly, the life of the tools is reduced as the complexity of the forging (thinner sections, ribs, etc.) is increased. This reduction is accommodated by the shape die life factor, β_s, which is obtained from the classification of forgings described above. Using these factors, die costs are determined by specifying a typical die replacement and refurbishing strategy for a relatively simple forging in low carbon steel. For such a part, the die life would be, say, 40,000 pieces, with up to five resinks of the main die cavity possible. Subsequent to this, a new tool set is assumed necessary. For other parts it is assumed that this basic tool life and the periods before refurbishing will be reduced by the die life factors for material and shape complexity. The die resink quantity, Q_{rs}, is determined by

$$Q_{rs} = Q_{rb}\beta_s\beta_m \tag{14.31}$$

where Q_{rb} is the basic resink quantity (say 40,000), β_s is the shape die life factor, and β_m is the material die life factor. Thus the total die life, L_D, is given by

$$L_D = (N_{rs} + 1)Q_{rs}N_c, \tag{14.32}$$

where N_{rs} is the number of resinks possible, say 5.

The cost of each resink, C_{rs}, is assumed to be equal to

$$C_{man}[0.9(T_{mill} + T_{bw} + T_{fl}) + T_{pol}] \tag{14.33}$$

If the total number of forgings required (life volume), Q_{lv}, is greater than the total die life, then the forging die cost per part, C_D, is given by

$$C_D = (C_{DIE} + (N_{rs} + 1)C_{rs})/L_D \tag{14.34}$$

However, if the life volume is less than the total die life, then the number of resinks required, n_{rs}, will be $Q_{lv}/(Q_{rs}N_c)$ and then the forging die cost per part will be

$$C_D = (C_{DIE} + n_{rs}C_{rs})/Q_{lv} \tag{14.35}$$

Example

For the part $\beta_s = 0.45$ (Fig. 14.24) and $\beta_m = 1.0$ (Table 14.7). The resink quantity, $Q_{rs} = 40,000 \times 0.45 \times 1.0 = 18,000$. From this the total die life will be, from Eq. (14.32), $L_D = (5 + 1)\,18,000 = 108,000$ parts. The cost of each resink is $C_{rs} = 45[0.9(3.41 + 0.69 + 0.23) + 0.2.8] = \$301$. Thus, assuming a total required quantity of 25,000 forgings, then from Eq. (14.35), the die cost per part is $(1918 + 301 \times 25,000/18,000)/25,000 = \$0.09$ per forging.

14.13 COSTS OF FLASH REMOVAL

14.13.1 Flash Removal Processing Costs

After the main forging processes the flash must be removed to yield the finished forging. This is usually carried out on a mechanical press, with a purpose-built trimming die and punch. If possible, any webs in through holes are pierced out at the same time. For small quantities of parts and for large forgings the flash may be removed by a bandsaw or a similar machining method, but this is generally time-consuming and more expensive. In order to determine the processing costs for flash trimming, the required size of the trimming press must be estimated, and this is determined from the load needed to shear the flash at the parting line. The trimming load, F_{trm}, is given by

$$F_{trm} = (T_f P_r + T_w P_w) Y_s N_c 1.15 \tag{14.36}$$

where P_w is the perimeter of the through holes and Y_s is the equivalent shear stress of the material of the forging. The constant of 1.15 is used to give a factor of safety of 15%. The equivalent yield stress is usually taken to be 70% of the UTS of the material. For cold trimming the room temperature UTS should be used and for hot forming the UTS should be assumed to be the material flow stress at an appropriate strain rate.

Once the trimming load is known, the energy equivalent for the mechanical press can be determined as

$$E_f = 0.096 F_{trm}^{0.98} \text{ kg-m} \tag{14.37}$$

From this the relative cost per operation can be determined from Fig. 14.20, and multiplying this by the operating cost per operation for a 1000 lb hammer gives the trimming cost, which should be divided by the number of forgings per cycle to result in the trimming cost per part.

There is a small setup cost for flash trimming. This can be determined in the same manner as for the main forging equipment, following the procedure outlined in Sec. 14.10.5.

Example

For the example part the perimeter of the part is 31.4 cm and the flash thickness is 1.68 mm. For hot trimming of mild steel the material flow stress is approximately 97 MPa. Thus from (14.36) the trimming load, F_{trm}, is given by

$$F_{trm} = 314 \times 1.68 \times 0.7 \times 97 \times 1.15 = 41,191 \text{ N}$$

From this equation, (14.37) gives $E_f = 3197$ kg-m, which from Fig. 14.21 results in a trimming cost relative to the cost per operation of a 1000 lb hammer of 0.73. Thus, assuming that the cost per operation of a 1000 lb hammer is $0.15, then the trimming cost per part is $0.73 \times 0.15 = \$0.11$.

14.3.2   Tooling Costs for Flash Removal

The procedure for determining the tooling costs for trimming is modified from the FIA estimating procedure [11]. First, the material required for the tools is estimated. The trimming die material volume, V_{trd}, is given by

$$V_{trd} = 1.2 L_{plt} W_{plt} T / 2 \tag{14.38}$$

The trimming punch material volume, V_{trp}, is given by

$$V_{trp} = L_{plt} W_{plt} T \tag{14.39}$$

From these the trimming tool material cost, C_{trm}, is determined as

$$C_{trm} = (V_{trd} + V_{trp})\rho_t C_t \tag{14.40}$$

Example

For the example part $L_{plt} = W_{plt} = 10$ cm and the part thickness is 2 cm. From these $V_{trd} = 120 \text{ cm}^3$ and $V_{trp} = 200 \text{ cm}^3$. Assuming the cost of tool steel is \$20/kg and the density is 7.9 gm/cc, then the material cost of the trimming tools is

$$C_{trm} = 20 \times 7.9 \times 320/1000 = \$50.6$$

The time to manufacture the trimming die, T_{td}, is given by

$$T_{trd} = T_{int} + (A_0 + M_p A_{tb} + T_{lk})N_c \text{ h} \tag{14.41}$$

where

$T_{int} =$ initial time allowance hours (4 for cold trim and 5 for hot trim)
$A_0 =$ base time, h
$M_p =$ block area factor, cm^2
$A_{tb} =$ die block area, cm^2
$T_{lk} =$ addition for locked forging dies, h

The values of A_0, M_p, and T_{lk} are obtained from Table 14.8, for different values of the parting line profile complexity factor, F_c, where this is determined by

$$F_c = P_r/2(\pi A_p)^{0.5} \tag{14.42}$$

The number of hours required for the trim punch manufacture is estimated from

$$\begin{aligned} T_{tp} &= (0.004A_{pb} + 0.33) + 0.05 + [(A_{pb} - A_p N_c)/6.56] \\ &\quad + [(P_r - P_w)/2.54 + 14F_c - 13]N_c F_{lck} + 0.005N_c A_p F_{fc} \end{aligned} \tag{14.43}$$

TABLE 14.8 Data for Estimating Trim Die Manufacturing Times

Profile complexity, F_c	Base time, A_0 (h)	Block area factor, M_p (h/cm²)	Locked die addition, T_{lck} (h)
1.0 to 1.5	0.62	0.0143	2
1.5 to 1.8	2.52	0.0146	3
>1.8	5.08	0.0168	0
>1.8 + die lock	8.86	0.0203	0

where A_{pb} is the punch block area, given by L_{plt} W_{plt}, and F_{lck} is a lock factor equal to 0.06 unless the die split line is cranked, in which case the factor is 0.065.

The total initial trim tooling cost is

$$C_{trim} = (T_{td} + T_{tp})C_{man} + C_{trm} \qquad (14.44)$$

The trim die life, L_{trm}, is given by $L_{trm} = L_{tbas}\beta_m N_c$, and from this the trim tool cost per part is C_{trim}/L_{trm} or C_{trim}/Q_{lv} if the life volume required, Q_{lv}, is less than L_{trm}.

Example

For the example part the profile complexity factor, F_c, is 1.0 as the forging profile is circular. Thus, from Table 14.8, $A_0 = 0.62$ h, $M_p = 0.0143$ h square centimeters and the forgoing dies are not locked so $T_{lck} = 0$. From these, assuming hot trimming, the manufacturing time for the trimming die is

$$T_{td} = 5 + (0.62 + 0.0143 \times 78.6) = 6.74 \text{ h}$$

The value of F_{lck} is 0.06 and hence the time for manufacturing the trimming punch becomes

$$T_{tp} = (0.004 \times 100 + 0.33) + 0.05 + [(100 - 78.6)/6.56 + 14 - 13]0.06$$
$$+ 0.005 \times 78.6 \times 4 = 2.61 \text{ h}$$

Thus the total cost of the flash trimming tools is

$$C_{trim} = 45[6.74 + 2.61] + 50.6 = \$472$$

Assuming a lifetime production quantity of forgings, Q_{lv}, of 25,000, then the trimming tool cost per part $\$472/25,000 = \$0.02$ per part.

14.14   OTHER FORGING COSTS

There are some additional small costs associated with forging.

14.14.1   Billet Preparation

For most forging operations there will be a small additional cost for billet preparation. For long parts the forging may be made directly from the heated end of an appropriate length of bar stock. For other types an appropriate billet must be cut off from bar stock. This can be achieved by several methods, including sawing, abrasive cutoff, and cold shearing, which involves the least waste material. Cold shearing will usually be carried out on a mechanical press, and the cost of processing can be calculated following a procedure similar to that for flash trimming described in Sec. 14.13.1. Since standard tools will be used the tool cost per part is very small.

14.14.2 Billet Heating Costs

The costs for heating the billet or bar end to the appropriate forging temperature can be estimated by determining the energy costs for heating. These can be obtained by multiplying the billet weight by the material specific heat and the required temperature rise. For small forgings this cost will be relatively small, but for large forgings that may require several reheats between operations the heating costs may become significant.

REFERENCES

1. Kalpakjian, S., Manufacturing Processes for Engineering Materials, 2nd ed. Addison-Wesley, Reading, MA, 1991.
2. Thomas, A., Die Design, Drop Forging Research Association, Sheffield, UK, 1980.
3. Bruchanov, A.N., and Rebelskii, S.K., Gesenkschmieden und Warmpressen, Verlag-Technik, Berlin, 1955.
4. Voiglander, O., Discussion on "Der Schmiedevorgang in Hammer und Pressen, insbesondere hinsichtlich des Steigen," Werkstattstechnik, Vol. 49, p. 775, 1951.
5. Vierrege, K., Die Gestaltung der Gratspalt am Schmiedegesenk, Industrie Anzeiger, Vol. 90, p. 1561, 1970.
6. Neuberger, F., and Mockel, L., Riehtwerte zur Ermittlung der Gratdicke und der Gratbahnverhaltnisses beim Gesenkschmieden von Stahl, Werkstattstechnik, Vol. 51, p. 725, 1961.
7. Teterin, G.P., and Tarnovski J.J., Calculation of Flash Gap Dimensions in Forging of Axisymmetric Parts Under Hammers, Kuznechno-shtamp. Proizvod., Vol. 10, p. 6, 1968.
8. Choi, S.H. and Dean, T.A., Computer Aids to Data Preparation for Cost Estimation, International Journal of Machine Tool Design and Research, Vol. 24, pp. 105–119, 1984.
9. Morgenroth, E., Ermittlung des Einstaz- und Kontingentgewichtes von Gesenksch-miedestucken aus Stahl, Werkstattblatt, 180–82, Carl Hanser-Verlag, Munich, August 1951.
10. Kruse, O., Uber den Einfluss des Gratgewichtes auf die technischwirtschaftlichen Kennziffern und Materialverbrauchsnormen von Gesenkschmiedeteilen aus Stahl, Fertigungstechnik, Vol. 4, p. 126, 1954.
11. Forging Industry Association, Estimator's Handbook, FIA, Cleveland, 1962.
12. Knight, W.A., Simplified Early Cost Estimating for Hot-Forged Parts, International Journal of Advanced Manufacturing Technology, Vol. 7, pp. 159–167, 1992.
13. Jensen, J.E., Forging Industry Handbook, FIA, Cleveland, 1970.
14. Osman, F.H., and Bramley, A.N., Preform Design for Forging Rotationally Symmetric Parts, Annals of CIRP, Vol. 44, No.1, p. 227, 1995.
15. Yang, D.Y. et al Development of Integrated and Intelligent Design and Analysis System for Forging Processes, Annals of CIRP, Vol. 49, No. 1, p. 177, 2000.
16. Chamouard, A.N., Estampage et Forge, Dunod, Paris, 1964.
17. Biswas, S.K., and Knight, W.A., Towards an Integrated Design and Production System for Hot Forging Dies, International Journal of Production Research, Vol. 14, pp. 23–49, 1976.
18. American Society of Metals, Metals Handbook, Vol. 5, Part A, ASM, Cleveland, 1972.

19. Gallagher, C.C., and Knight, W.A., Group Technology, Butterworths Press, London, 1974.
20. Knight, W.A., and Poli, C., Design for Forging Handbook, University of Massachusetts, Amherst, 1982.
21. Knight, W.A., and Poli, C., Design for Economical Use of Forging: Indication of General Relative Forging Costs, Annals of CIRP, Vol. 31, No. 1, 1982.
22. British Standards Institution, BS 4144, Dimensional Tolerances for Steel Drop and Press Forgings, 1967.
23. Scott, K., and Wilson, A., Hammers vs. Presses, Metallurgia and Metal Forming, July, p. 198, 1975.
24. Dallas, D.A., ed., Tool and Manufacturing Engineers Handbook, McGraw-Hill, New York, 1976.
25. Leone, J.L., Relative Forging Costs Analysis and Estimation, MS Thesis, University of Massachusetts, Amherst, 1983.
26. Knight, W.A., and Poli, C.R., A Systematic Approach to Forging Design, Machine Design, January 24, pp. 94–99, 1985.
27. Hobdell, A.C., and Thomas, A., Approaches to Cheaper Forgings, Metal Forming, Vol. 36, No. 1, p. 17, 1969.
28. Drabing, L.A., Guide to Die Making for Multiple-Impression Drop Forging, Chambersburg Bulletin, 157-L-7, 1966.
29. Altan, T., and Akgerman, N., Modular Analysis of Geometry and Stress in closed Die Forging: Application to a Structural Part, ASME Transactions, J. of Engineering for Industry, Vol. 94B, p. 1025, 1972.
30. Altan, T., et al, Forging Equipment, Materials and Practices, Metals and Ceramics Information Center, Columbus, Ohio, 1973.
31. American Society of Metals, Forging Design Handbook, ASM, Cleveland, 1972.
32. Schey, J.A. Introduction to Manufacturing Processes, McGraw-Hill, New York, 1989.

15

Design for Manufacture and Computer-Aided Design

15.1 INTRODUCTION

The use of computer-aided design (CAD) systems, computer graphics, and computer-aided drafting has become widespread in industry, to the extent that they are now an integral part of the product design process in most companies. As a result, there is obvious interest in closer integration of the design for manufacturing and assembly (DFMA) analysis procedures described in earlier chapters into this CAD-based design environment. In particular, some of the information required for DFMA analysis is geometric and may be available directly from the CAD part data. However, even with current CAD systems, considerable time and effort is still required to enter and design all parts and subassemblies of a product. If the initial DFMA analysis is left until all or most of these data have been entered into the CAD system, it may be too late, as there will then be a considerable reluctance to implement any substantial design changes in the product. An important consideration is how to integrate the DFMA analysis early enough in the design process to have the most benefit.

15.2 GENERAL CONSIDERATIONS FOR LINKING CAD AND DFMA ANALYSIS

In considering the possible links between CAD systems and DFMA analysis it is useful first to review the manner in which these systems currently operate, in particular if the purpose is to impact the concept stages of the design of new, rather than modified, products.

15.2.1   DFMA Analysis

Analysis of a product for manufacture and assembly consists of two main stages: design for assembly (DFA) and early assessment of component manufacturing costs (DFM).

Design for Assembly Analysis

DFA analysis is aimed at:

Product structure simplification and part count reduction
Detailed analysis for ease of assembly

An essential feature of DFA methodology is analysis of the proposed product by multidisciplinary teams. The analysis process is used as a means for promoting discussion and the generation of alternative design concepts and product structures. Product simplification and part count reduction are achieved by analyzing the existing or proposed product structure and applying simple criteria for the necessary existence of separate parts in the assembly to meet the basic functional requirements of the product. The detailed geometry of each item is a somewhat secondary consideration in this context.

The analysis of detailed DFA is realized by considering certain geometric features of each part, together with a subjective assessment of assembly difficulty. Although not a strict requirement, detailed DFA analysis is most conveniently applied by analyzing a prototype or existing product, with each item actually assembled during the analysis process. In this case the objectives of product structure simplification and detailed DFA are considered at the same time. This is a very beneficial process, but in considering the application of DFA at the early concept design of a new product a modified procedure may be more applicable.

Early Assessment of Component Manufacturing Costs

DFM is aimed at:

Appropriate process/material selection based on realistic cost estimates
Establishing or highlighting the relationships between part features and manufacturing costs for a given process

The early cost estimation procedures are perhaps more logically used by individuals, to justify and evaluate the alternative design concepts developed as a result of DFA analysis and also when more detailed features of the individual parts are being considered. Most of the information required for these procedures is geometric, and the potential for integration into a CAD environment is easier to visualize and developments in this direction are described later.

15.2.2   Computer-Aided Design Systems

It is important to note that all current CAD systems have been devised basically to be used by an individual designer operating at his own workstation, as opposed to being used by groups of people. Each workstation largely takes the place of the drafting machines used previously. Such CAD workstations represent a more efficient way of creating geometric and other information, including drawings, of the individual parts of a product. The archiving of the data on each item has essentially been done in the same way as the drawing files used previously, but the images and data are stored electronically rather than physically. However, integrated product data management (PDM) systems have been developed for the organization and sharing of all product data.

The way in which a CAD workstation is used is somewhat at odds with the emphasis in DFA on analysis by multidisciplinary groups, and it is important to consider this aspect when links between CAD systems and DFMA software are proposed. There is a more obvious relationship between the geometry creation process carried out in a CAD system and detailed DFA or the early estimation of part manufacturing costs. Considerations of product structure simplification and part count reduction can to some extent be treated independently of the detailed geometry of the items being analysed. The manner in which the parts and subassemblies are related to each other in the CAD system data structure is of more significance. Much of this information is available in a bill of materials and/or assembly structure chart of the product.

15.3   GEOMETRIC REPRESENTATION SCHEMES IN CAD SYSTEMS

There are a number of different geometry representation schemes or models used in CAD systems [1–4] (Fig. 15.1), the main schemes being:

1. Wire frame representation
2. Various surface modelling schemes
3. Solid modelling schemes, including:
   a. Constructive solid geometry (CSG) modelling
   b. Boundary representation solid modeling
   c. Sweep representations

Recently, feature-based models and symbolic or object-oriented representations have received considerable attention.

9 TOP POLYGONS

9 SIDE POLYGONS

9 BOTTOM POLYGONS

S_2

S_1

S_3

SOLID $= S_2 \cup (S_1 \lfloor S_3)$

(a)

(b)

(c)

(d)

(e)

SWEPT SOLID

GENERATOR
SURFACE

GENERATOR
SURFACE

SWEPT SOLID

AXIS &
ROTATION

FIG. 15.1 Geometric modeling schemes. (a) Wire frame model. (b) Surface model. (c) Constructive solid geometry (CSG) model. (d) Boundary representation (B-rep) model. (e) Sweep representations. (Adapted from Refs. 3 and 4.)

Representation in this context refers to the data used to model or represent the object in the system and the manner in which these data are organized. Each of the alternative geometry representation schemes has advantages in certain situations, and most CAD systems support more than one of these modeling schemes. Conversion from one scheme to another may not always be possible. For example, deriving a unique CSG model from a given boundary representation is generally not possible. However, it is straightforward to obtain a wire frame representation from the other geometric models in common use, and this is generally done as a matter of course in most systems for rapid visualization of the design during the geometry creation process.

Which representation scheme is supported in any CAD system is an important consideration, because this has an effect on how each item can be visualized and, more important perhaps, determines which information can be derived automatically from the CAD system data. It is usually important for a given CAD system to support several different models, since each may be the most suitable for describing particular items or tasks. For example, there is little point in trying to represent a sheet metal part as a solid model, when a surface model, with a specified thickness, is more suitable. However, a solid model representation will be the most useful for many mechanical parts, particularly as it usually allows the direct determination of such attributes as part volume, weight, and so on, if required.

An important consideration in the use of a combined CAD/DFMA system at the concept stages of design is the scale of the task or ease of use of the system in creating the geometry of each item. Specifically, it is desirable to have a rapid means of initially capturing the rough geometry of an item—in particular, if this can readily be more fully developed at a later stage. A significant problem in the use of current CAD systems is the total effort required to capture the complete geometry of a product, which tends to limit the desire to make significant changes in product structure once this has been fully carried out. However, modern CAD systems usually allow approximate geometry to be created initially and refined subsequently.

Some further details of the different modeling schemes will now be discussed.

15.3.1 Wire Frame Models

Many CAD systems use wire frame models to define geometry [1–4] (Fig. 15.la), and even those systems that use other methods for defining the basic geometric entities utilize wire frames for rapid visualization, in particular as wire frames are readily derivable from other representations. Wire frame representations contain only vertices (points) and edges (lines), which are the intersections of the surfaces of the object. No other information about the surfaces of the object is carried, and thus wire frame models lack the surface definition for many of the analyses that might be required on the defined objects. Visualization of complex wire frame models can be confusing, as automatic hidden line removal and surface rendering are not possible. A major disadvantage is that a unique object representation is not achieved, and wire frame models can often be interpreted in a number of different ways [1–4]. Thus the identification of manufacturing-related features is usually not possible and interpretation of the model into its manufacturing requirements is generally impossible to do automatically.

15.3.2 Surface Models

With surface models (Fig. 15.1b) the geometry of the object is represented by its bounding surfaces. Several different types of surface representation are available. In general, the problem in using surface models is that often different parts of the overall surface of an object may be modeled separately, sometimes by different methods. Thus a surface model may be a collection of surfaces that do not completely define a physical object. Information concerning the inside or outside of the object may not be available; thus attributes such as the object's volume or mass properties cannot be determined. Two main types of surface model are commonly used:

1. Face or tessellated models
2. Sculptured surface models

With face models [1–4] (Fig. 15.2), the surface of the object represented is approximated by a number of faces (usually planar). This is a very simple representation, and the data used to define the object is a set of face, edge, and vertex tables. This type of model is very commonly used for visualization of the object described, since this approximation enables fairly efficient algorithms for hidden detail removal and surface rendering to be used, which are the basis for realistic computer-generated images of objects. However, such face models may

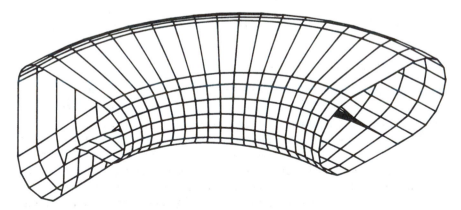

FIG. 15.2 Face model of an object. (From Ref. 2.)

be inadequate for other purposes, such as the automatic generation of numerically controlled (NC) programs, unless a large number of faces are used to represent the object's surfaces. A specific form of face model is now usually available as output from CAD systems and these are called STL files, which stands for stereolithography tessellation language. In this case the geometry is represented by a set of triangular faces, together with the direction of the outwardly facing normal to each face. This very simple representation is utilized as input to rapid prototyping equipment such as for stereolithography and selective laser sintering. The widespread availability of this form of output has led to the development of software for object visualization from STL files, together with estimates of parameters such as part volume, surface area, and so on. Files for STL representations are large for complex objects if good accuracy is required. There is also a significant amount of redundant information since each triangle is represented separately; thus each corner (vertex) will be stated two, three, or more times on adjacent triangles. The STL file format is also sensitive to topological corruption, leading to such errors as gaps between triangles, overlap of triangles, and so on. Some software is available for the repair of STL files that become corrupted in this manner.

In sculptured surface models (Fig. 15.3) various mathematical surfaces, usually based around three-dimensional parametric curve definitions, are blended

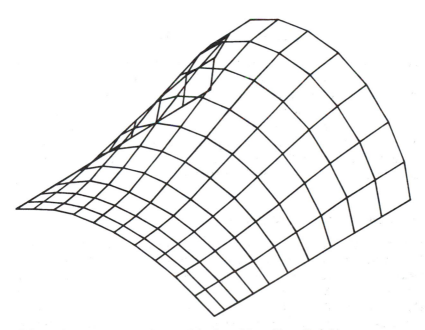

FIG. 15.3 Sculptured surface model of an object. (From Ref. 2.)

to conform to specified boundary curves. Several different mathematical representations have been used. Systems based around these modeling schemes are very suitable for free-form surfaces such as those found in automobile body structures, ship's hulls, aircraft skin structures, and so on. They may be difficult to use for objects that have abrupt changes of surface direction. Since a mathematical description of every point within a surface is readily available, integration with NC machining is usually possible with relative ease.

15.3.3 Constructive Solid Geometry Models

In constructive solid geometry (CSG) models [1–4], the geometry (Fig. 15.1c) is stored as a binary tree of Boolean operations (union, difference, and intersection) applied to a limited set of fundamental shape primitives. The initial range of primitives available limits the range of objects that can be described. A distinct advantage is that impossible objects cannot be created by this modeling scheme, and consequently CSG modeling is well suited to the initial definition of the object geometry. This form of representation is also well suited to certain analytical operations, but considerable computational complexity is required for graphics, NC processing, and so on. Thus CSG models must usually be supported by a boundary evaluator to facilitate these processes.

15.3.4 Boundary Representation Models

In boundary representation (B-rep) (Fig. 15.1d) models, the object geometry is represented by a collection of faces, together with the connectivity (topology) between them. Such representations are not well suited to analytical operations, such as center of gravity determination or mass properties calculation. However, processes involving the surfaces of the object are readily carried out. For example, high-quality surface rendering is comparatively straightforward. Design changes may be difficult to accommodate, and careful control of the part data is necessary to avoid creating impossible objects.

15.3.5 Sweep Representations

Sweep representations (Fig. 15.1e) use an entity of lower order, such as a closed profile or curve, plus sweeping information (rotation or translation) to describe a volumetric object. Sweep models are easily stored and are particularly useful for symmetrical objects, but are less useful for asymmetrical geometries. Sweep models are not generally used for internal representations, but may be part of the procedures available for initial description of the geometry. Boundary representation models are readily derived from sweep models.

15.3.6 Feature-Based Models

The terms "feature" or "feature-based" have become commonly used in relation to CAD systems recently. This stems from a desire to be able to consider an object in the CAD environment in terms of something more immediately meaningful or useful, often in a manufacturing sense, than the points, lines, circles, surfaces or solid primitives that are currently the basis of geometry definition within most CAD systems.

It is unfortunate that, in common with most new developments, the terms "feature" and "feature-based" have different meanings to different people. In addition, what is meant by these terms in the marketplace may in fact be even more confused, as CAD vendors feel the necessity to introduce this terminology to describe the latest developments of their particular systems.

A number of significant questions can be raised in connection with so-called feature-based systems, including (1) What is meant by a feature, (2) Does all the geometry of an object appear as features, and (3) Is all of an object going to be described using features?

1. *What is meant by a feature?* Many different definitions of the term "feature" exist, and Shah [5] gives a comprehensive discussion of this topic. Some definitions of the term, include the following:

Features represent shapes and technological attributes associated with manufacturing operations and tools [5].
Features are groupings of geometric or topological entities that need to be referenced together [5].
Features are elements used in generating, analyzing, and evaluating design [5].
A feature is a classification of object characteristics, which has a significance in a domain [6].

A common thread is often that features represent the engineering meaning of the geometry of a part or assembly. Feature-based models possess additional information levels not found in conventional geometric models. Thus features have a specific meaning in connection with a particular technology and as such become technology specific. For example, it may be useful to be able to describe parts of an injection molding in terms of ribs, bosses, gussets, and so forth; the direct relationship to mold design and manufacturing cost estimates is readily obvious. This certainly has more apparent meaning than the point, lines, circles, and so forth, that generally make up the underlying geometry. However, these features then have much less meaning in the context of, say, machining, when items such as grooves, holes, pockets, slots, and so forth, would appear to be more useful. This is well illustrated by Fig. 15.4, which shows four different possible feature representations of the same object for different technological domains [6].

(a)

FIG. 15.4   Alternative feature models of the same object. (a) Example part.

2. *Does all of the geometry of an object appear as features?* That is, is the entire object covered with features? Can a particular piece of geometry appear in more than one feature? This implies that a feature model and a geometric model, in the conventional sense, do not completely overlap in what they describe. Feature models and a geometric model will both be necessary. Then features may describe or connect to only some or all of this underlying geometry model. Certainly, if not all of the geometry is described as features, or geometry appears in more then one feature, then an underlying conventional geometry model will be required to enable such things as overall size, volume, and so forth to be determined. Several different application specific feature models may be required.

Three main alternatives for creating feature models in geometric modeling exist [5,7]:

A. *Interactive feature definition* (Fig. 15.5): A geometric model is created first and then features are defined by the user by picking entities from a displayed image of the part.
B. *Automatic feature recognition or extraction* (Fig. 15.6): A geometric model is created and then a computer program processes the database to automatically extract features. This aspect of feature modeling has received

FIG. 15.4 (*continued*) Alternative feature models of the same object. (b) Design features.

considerable attention in relation to computer-aided process planning (CAPP). The ease with which this may be achieved is influenced directly by the type of geometric modeling scheme used for the initial object definition.

C. *Design by features* (Fig. 15.7): The part geometry is defined directly in terms of features; i.e., the geometric model is created from features. Features are

FIG. 15.4 (*continued*) Alternative feature models of the same object. (c) Machining features.

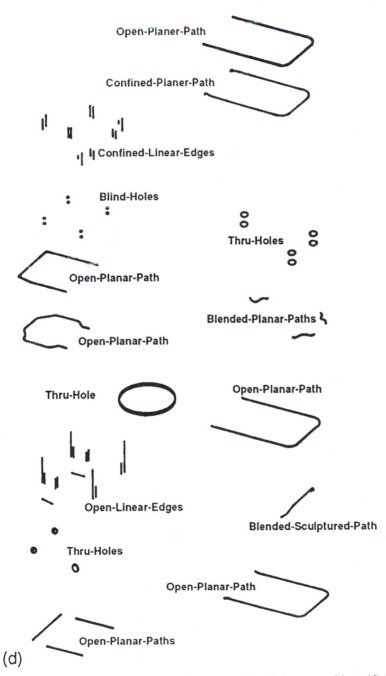

FIG. 15.4 (*continued*) Alternative feature models of the same object. (d) Deburring features.

Planar Faces

Holes

Planar Faces

Holes

Profile Group

Planar Faces

Profile Group

Cylindrical Faces

Datum B

Profile Group

Hole

Datum C

Planar Faces

Holes

Cylindrical & Planar Faces

Datum A

(e)

FIG. 15.4 (*continued*) Alternative feature models of the same object. (e) Inspection features. (Adapted from Ref. 6.)

FIG. 15.5   Interactive feature definition. (From Ref. 5.)

FIG. 15.6   Feature extraction. (From Ref. 5.)

incorporated in the part models at the creation stage. Parameterized generic features from a library are utilized to define the geometry. This enables a richer definition of products to be made at the design stage and facilitates the automation of downstream activities, in particular CAPP.

3. *Is all of an object going to be described using features?* Will a feature-based language replace current geometry-based descriptions for designing parts? There is certainly not yet available a universally applicable feature language or description system. In view of question (1) above, this may be difficult to conceive. When dealing with a specific manufacturing process, the availability of a suitable feature description language will be of benefit to the type of DFM analysis described previously, particularly if this allows data required for this analysis to be captured at the initial input stage, rather than having to develop

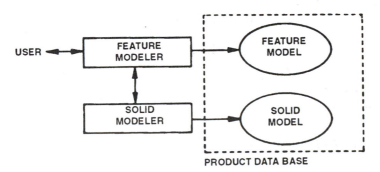

FIG. 15.7   Design by features. (From Ref. 5.)

specific procedures to interpret this information from an existing CAD database describing the parts.

15.3.7 Object-Oriented Programming

In an object-oriented programming environment [8], the basic unit of information is the object, which is defined by a name and a set of attributes that describe the object. The object can be a physical entity, such as a tube with inner diameter, outer diameter, and height (Fig. 15.8).

FIG. 15.8 Object-oriented description of a tube. (From Ref. 8.)

A useful aspect of this environment is that one object can be related to or "inherit" the attributes of another. For example, a colored tube can inherit the attributes and properties of another object that is of the type tube. The values of the colored tube would be the same as the object of the type "tube" unless it is desired to override these values. By allowing inheritance from one object to another, the amount of information needed to describe an object is minimized.

The values of the attributes of an object can be functional relationships, such as arithmetic equations, database look-up values, or the if–then relationships characteristic of expert systems.

One main advantage of using object-oriented programming is that the knowledge about the part is easy to maintain. The information is not scattered around the program structure, but can be stored in objects that can be inherited numerous times. Another advantage is that with objects inheriting from one another, a change in a value of one object will change all other objects and parameters relating to the first object. This is similar to the way one cell can inherit from another cell in a spreadsheet program such as Excel.

15.3.8   Data Transfer Between and from CAD Systems

From the preceding discussion it can be seen that there are a number of ways of representing or modeling the geometry of objects in CAD systems. In addition, different systems will support different sets of geometric entities for the modeling objects. Thus there is no standard way to represent the geometry. This means that the transferring of data between dissimilar CAD systems and to other applications packages, such as those for NC processing, can be problematic. The usual approach to this is through a neutral file format, so that the developers of different CAD systems need only provide translators to process the data into and out of this neutral file format. Several different neutral file formats have been developed, but the format most commonly used is the Initial Graphics Exchange Specification (IGES) [3,4]. This format is particularly used for 3-D line and surface data translation. More recently, exchange formats based on solid model representations have been developed, including Product Data Exchange Specification (PDES), which has been further subsumed into the international Standard for the Exchange of Product Model Data (STEP). While these neutral file formats partially solve the problems of data transfer between systems, it is still not possible to fully transfer data between systems that support different geometric entitles in their modeling schemes.

Other standard formats have been developed to transfer geometric data between CAD systems and other applications. Included in this category are STE files for rapid prototyping systems, as described in Section 15.3.2.

15.4 DESIGN PROCESS IN A LINKED CAD/DFMA ENVIRONMENT

The overall conclusion from the preceding discussion is that there are different methods for the description and geometry creation of objects. The main consideration is the manner in which CAD/DFMA analysis should be incorporated into the design sequence for a new product, with the way in which the part geometry is handled being a somewhat secondary consideration.

A considerable amount of design is an evolution or modification of existing designs, and thus the original design may already have been captured in a CAD system database. In addition, actual products and prototypes will be available for analysis. It is relatively easy to envisage ways in which this information can be accessed by DFMA analysis procedures, although this may not be easy to achieve. The more interesting situation to consider is how the design process for the conception of a new product may be influenced in the CAD environment—in effect, starting from a blank sheet of paper or workstation screen.

It is still improbable that a CAD system will be utilized significantly during the discussions at the very early stages of product conception. Initial discussions will focus largely on function, layout, and so forth, and will most likely still be done by sketching on pieces of paper. The first thing to be established will probably be an initial product structure—i.e., a tentative listing of subassemblies, parts, and so on, with only an approximate consideration of the geometry of each item, perhaps limited to the overall dimensions and approximate shapes. However, the modeling interface for many CAD systems has now been developed in such a way that approximate geometry can be put in initially and then easily modified as the design develops. In addition, there have been considerable developments in product data management (PDM) systems that are aimed at capturing and storing all data on a product as the design develops.

It is at the early stage of product design that a DFA analysis should first be applied and the product structure simplified as much as possible, before great effort has been expended in generating all of the geometry of the proposed design. The current DFA analysis software [9] incorporates a facility for capturing the product structure, i.e., the relationship between subassemblies, parts, and so on. Thus a logical way to Integrate DFA analysis in the CAD environment is as a front end to a CAD system into which the product structure is initially entered, and the DFA system essentially drives the CAD system. The product structure is then used as the basis for the CAD file structure for creating the geometry of each item in the assembly. This will allow simplification of the initial product structure to be done before too much detailed geometry creation has occurred. It is envisaged that the geometry of each item will be created by selection from the DFA product structure in turn, which then gives access to the

CAD system geometry creation window. In this manner different ways of creating the geometry of each part can be readily accommodated. For example:

1. For standard parts, direct retrieval from a database of parameterized standard parts
2. For some families of parts, from a parametric geometry model
3. Using appropriate modeling systems—solid, surface, features, and so on

As the geometry of a particular part is created with a specific process in mind, integration with early cost estimation procedures that utilize the geometric information being generated directly should also be considered.

15.4.1 Example Scenario of CAD/DFMA Integrated System Utilization

The integration of DFA/CAD could have the following basic features: A DFA analysis program is utilized along the lines described in previous chapters, but with the CAD program driven from the DFA program in a separate application window. All graphics and geometry creation facilities should reside in the CAD program (they already exist there). Creation of geometry, drawings, and so on, in the CAD program should be driven and accessed from the product structure charts in the DFA application window.

In order to illustrate how a combined system would operate, consider the following example:

1. Initial concept discussions lead to proposals for a motor drive assembly, as sketched in Fig. 1.8.
2. This proposed product is captured in the DFA analysis software applications window, as shown in Fig. 15.9.
3. Minimum part count criteria can be applied to this product structure and the subsequent discussions lead to several simpler product structure concepts, such as that shown in Fig. 1.9.
4. These modified product structures are captured in the DFA analysis software window, as shown, for example, in Fig. 15.10.
5. These product structures then form the basis for building up (and subsequently accessing) the geometry of each item in the CAD system window. The structure diagram in the DFA window effectively becomes the menu for the file structure set up in the CAD system. Selecting each item in turn allows the geometry to be created by an appropriate method within the CAD application window (Fig. 15.11).
6. As the geometry of each part is created, integration with cost estimation for a selected process could also occur, with the estimators utilizing the geometric data as it is created, although this latter stage is not easy to achieve.

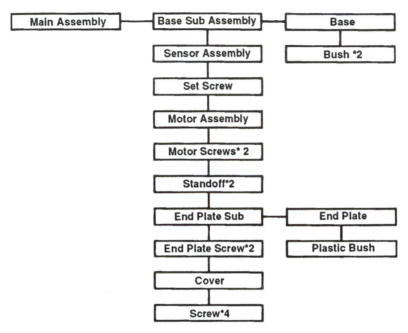

FIG. 15.9   Structure chart of proposed design of motor drive assembly.

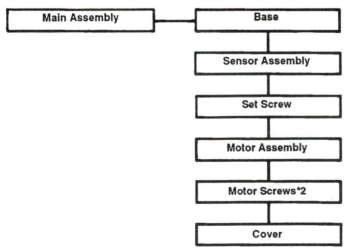

FIG. 15.10   Structure chart of modified design of motor drive assembly.

FIG. 15.11   Use of product structure to access CAD system.

15.5   EXTRACTION OF DFMA DATA FROM CAD SYSTEM DATABASE

Some of the information required for DFA analysis for each part is geometric and as such may be derivable from the CAD system part data if suitable algorithms for determination of the required attributes are available. Consequently, there is an obvious interest in the possibility of extracting the required information directly from a CAD system database, although it should be stressed that if DFA analysis is initially delayed until a design has been completely modeled in a CAD system it may be too late to overcome the natural resistance to implement design changes. However, extraction of the required data from a CAD system database is not straightforward and requires specific software routines developed for this purpose. This overall problem is a specific application of automatic feature recognition because assembly-related features must be determined from the CAD system database.

The overall difficulties with feature recognition procedures discussed previously apply equally to this case. During design for assembly analysis, the designer is asked to provide information relevant to the handling and assembly of each item in the assembly. For manual handling analysis the geometric information required includes:

Certain dimensions (size, thickness, etc.)
The axis of insertion of the part.

The rotational symmetry of the part or subassembly about and perpendicular to the axis of insertion (referred to as alpha and beta symmetry).
For large items, the weight

In addition, information of a more subjective nature is required, including, for example,

Can the part be grasped in one hand?
Do parts nest or tangle?
Are parts easy or difficult to grasp and manipulate?
Are handling tools required?

Some of these items are determined by geometric features of the parts. It may be possible, although very difficult, to formulate specific rules so that these questions may be answered automatically from information contained in a geometric database, although the complexity of achieving this may make this task not worthwhile. The user must still supply many of the more subjective answers.

For manual insertion, additional subjective information is required, including, for example,

Is access for part, tool or hands obstructed?
Is vision of the mating surfaces restricted?
Is holding down required to maintain the part orientation or location during subsequent operations?
Is the part easy to align and position?
Is the resistance to insertion sufficient to make manual assembly difficult?

Again, several of these items could be related to geometric features of the assembly, but complex rule-based decision routines and geometry modeling will be required for these questions to be answered automatically. For example, obstructed access could be related to the closeness of items in an assembly modeled in the CAD system, but automatic determination of this degree of closeness would be very difficult in most systems. Some attention has been paid to problems of this type in relation to the determination of disassembly sequences from CAD models (e.g. [10]), but overall this remains a largely unsolved problem.

Some attention has been given to extracting from CAD system databases some of the information required for DFA analysis, for example [11–13]. Algorithms can be made to work that evaluate such features as part size and symmetry, but the computational efficiency of these procedures is generally low. In addition, since DFA analysis is a concurrent engineering activity there is considerable value in having system users evaluate such items as part symmetry because of the discussion that is generated, which may be lost through automatic evaluation of these features.

Algorithms have been developed that determine the size and symmetry of objects from a wire frame geometric database, by Rosario [11,12], who worked specifically with the AutoCad database. The main algorithms work on the wire frame model stored in the parts database, as described in detail elsewhere [11, 12]. The first of these algorithms determines the smallest rectangular envelope that encloses the part when viewed from each of the orthogonal directions. Evaluation of the rotational symmetry involves modified orthogonal view of the part, which are modeled as graphs. All closed paths on each view are found and then three-dimensional "palindrome" checks of these closed paths and the arrangements between them are made. The word palindrome here means a string of symbols with the property that, regardless of the direction from which it is read, yields the same string. The evaluation of the rotational symmetries of a part consists of calculating the rotational symmetry from each plane view. The whole symmetry evaluation procedure is repeated for the three views of the object, and values of the end-to-end symmetry, alpha, and about axis of insertion symmetry, beta, are assigned according to these values and the insertion axis. For example, if the Z-axis is the insertion axis, then beta is the rotational symmetry about the XY plane. The value of alpha will be the smaller of the two symmetries in the other two planes, since the usual procedure is to reorient the object by rotation about the axis that involves the least effort.

Some success has also been reported by Tapadia [13] in determining alpha and beta symmetries and overall dimensions from the database of a B-rep solid modeler, Romulus. In this case algorithms written in Prolog were used to evaluate these parameters from the part data. One specific link between DFA software and the ProEngineer CAD package has been developed [14]. This package is called ProDFA and extracts some data from the ProEngineer database into the DFA analysis. Specifically, the product structure (list of items) and overall item dimensions are determined from the CAD system database. The order of items in the resulting DFA analysis is initially in the order in which items appear in the ProEngineer assembly lists. The users of the DFA package must still consider the more subjective responses and the minimum parts criteria to complete the analysis. However, much of the repetitive information such as item names and dimensions is transferred automatically.

15.6 EXPERT DESIGN AND COST ESTIMATING PROCEDURES

Expert systems are being applied to CAD. By utilizing an on-line expert system in a CAD environment, major judgments and decisions in the design process can be automated. One of the most important tasks of an intelligent CAD system would be to make useful and relevant manufacturing knowledge available to designers early in the design process. This would allow on-line design checking and analysis that could warn the user that the design does not conform to manufacturability specifications and perhaps suggest suitable changes to improve manufacturability. For example, some developments in this connection have been made for injection molding design [15]. For a limited class of box-like moldings, with bosses, ribs, and holes, a procedure was developed with design rules incorporated so that feature dimensions and other characteristics appropriate to the selected polymer are added automatically as the part design is built up on a graphics screen. At the same time, cost estimates for the part are determined automatically as the design progresses.

Integration of the type of early cost estimation procedures described in earlier chapters into the CAD process for component parts is a very powerful concept, since this will provide immediate feedback to the designer of the cost impact of component design features. Achieving this by interrogating the CAD system database is possible, but this involves the use of algorithms that identify the part features that influence costs from the CAD system part data. An initial procedure for sheet metal stampings produced on turret presses and press brakes has been developed [16]. For this system the number of punch hits, bending operations, and so on, is determined from the neutral file of the ProEngineer CAD system [17]. From this information the part costs are estimated using the methods outlined in Chapter 9.

A more powerful approach is to determine costs as the part geometry is built up in the CAD system, using a design-with-features approach. This may, however, require the availability of a number of process-specific feature-based CAD modules such that the part design is achieved through features that relate directly to the manufacturing cost contributions for the specific process. The basic ProEngineer system is well suited to this approach for machined parts, because part design is basically achieved by defining a base part and then subtracting user-defined features such as holes, grooves, slots, and so on.

More recently, the availability of STL file outputs from CAD systems has enabled some of the geometric data required for various models for early cost estimation to be obtained automatically [18]. The link between CAD systems and the DFM software is the SolidView program that has been developed for the visualization and measurement of objects defined in STL and other formats [19]. Specific routines can be used to determine the following:

Overall envelope dimensions
Part volume (weight)
Projected area in the direction the die or mold closure where appropriate
Number of surface elements or patches.

This means that cost estimates for manufacture by various manufacturing processes can be obtained from an initial CAD description of an object. Figure

FIG. 15.12   Thin-walled part visualized in SolidView. (From Ref. 19.)

15.12 shows a typical part visualized from a STL file representation using Solid-View. Table 15.1 shows the cost breakdown for manufacturing this part by injection molding and die casting, obtained directly from this model using the cost-estimating methods described in Chapters 8 and 10. Since STL files are available as output from all of the major solid modeling CAD systems in common use, this link allows the initial cost estimates for processing by a variety of processes to be readily obtained, independent of the CAD system used to initially define the geometry. In addition, programs such as SolidView will accept other standard input formats, such as IGES.

TABLE 15.1   DFM Cost Estimates for Part Shown in Figure 15.12

Life volume = 100,000	Cost per Part, $						Initial tooling investment
	Material	Setup	Process	Piece Part	Tooling	Total	
CAD Part—Injection Molding in ABS	0.23	0.01	0.62	0.86	0.08	0.94	$8,993
CAD Part—Hot Chamber Die Casting in ZA-8 Zinc Alloy	1.24	0.02	0.37	1.63	0.16	1.79	$15,561
CAD Part—Cold Chamber Die Casting in A356 Cast Aluminum	0.79	0.05	0.62	1.45	0.16	1.61	$15,561

REFERENCES

1. Taylor, D.A., Computer Aided Design, Addison-Wesley, Reading, MA, 1991.
2. Woodark J., Computing Shape, Butterworth Press, London, 1986.
3. Chang, T.-C., Wysk, R.A., and Wang, H.-P., Computer Aided Manufacturing, Prentice-Hall, Englewood Cliffs, NJ, 1991.
4. Bedworth, D.D., Henderson, M.R., and Wolfe, P.M., Computer-Integrated Design and Manufacturing, McGraw-Hill, New York, 1991.
5. Shah, J.J., An Assessment of Features Technology, Proc. CAM-1 Features Symposium, Boston, MA, August 9–10, 1990, p. 55.
6. Hummell, K.E., The Role of Features in Computer Aided Process Planning, Proc. CAM-I Features Symposium, Boston, MA, August 9–10, 1990, p. 55.
7. Shah, J.J., and Mathew, A., Experimental Investigation of the STEP Form-Feature Information Model, Computer Aided Design, Vol. 23, No. 4, pp. 437–464, 1991.
8. Miller, R.C., and Walker, T.C., Artificial Intelligence in Manufacturing Applications, SEAI Technical Publications, Madison, GA, 1986.
9. Boothroyd Dewhurst, Inc., Design for Assembly Software Manual, Version 8.0, Wakefield, RI, 1998.
10. Laperriere L., and El Maraghy, H.A., Planning of Products Assembly and Disassembly. Annals of CIRP, Vol. 41, p. 5, 1992.
11. Rosario, L. Automatic Geometric Part Feature Calculation for Design for Assembly Analysis, Ph.D. Thesis, University of Rhode Island, Kingston 1988.
12. Rosario L., and Knight, W.A., Implementation of Algorithms that Extract Geometric Features from CAD System's Part Data, Proc. Manufacturing International '90, Atlanta, March 1990.
13. Tapadia, R.J., The Use of Feature Information to Generate Assembly Handling Part Codes, M.S. Thesis Arizona State University, Tenpe, AZ., 1989.
14. Boothroyd Dewhurst, Inc., ProDFA Software Manual, Version 1.0, Wakefield, RI, 1998.
15. Cafone, J. and Dewhurst, P., Development of an Intelligent Injection Molding Design Tool, Report No. 37, Department of Industrial and Manufacturing Engineering, University of Rhode Island, Kingston, 1990.
16. Donovan, J.R., A Computer-Aided Cost Estimating System for Sheet Metal Parts, M.S. Thesis, University of Rhode Island, Kingston, 1992.
17. Roberts, A.F., New Directions in Feature-Based Solid Modeling, Proc. International Forum on Design for Manufacture and Assembly, Newport, RI, June 15–17, 1992.
18. Boothroyd Dewhurst, Inc., Concept Costing Software, Version 1.1, Wakefield, RI, 2000.
19. Solid Concepts, Inc., SolidView Software, 2000.

Nomenclature

A	area contained within perimeter; length of the rectangular envelope enclosing a nonrotational machined component
A_0	base time allowance for trim die manufacture
A_c	area of cavity plate; projected area of mold base; cross-sectional area of the undeformed chip
A_f	cavity surface area
A_H	area of through holes in forging
A_h	cross-sectional area of hole
A_{hol}	hole-modified area $= A_h/3$
A_m	area of machined surface
A_p	projected area of one part or pattern piece
A_{pb}	trim punch block area
A_{pl}	area of sand casting pattern plate
A_s	area of sheet metal used for each part; projected shot area
A_t	total area enclosed for all regions at a level in a powder metal part
A_{tb}	trim die block area
A_{tp}	total projected area of all parts or pattern pieces in the mold
A_u	usable die set plate area
a_d	depth of groove to be machined
a_e	depth of cut in horizontal milling; width of cut in vertical milling
a_p	depth of cut in turning, vertical milling, and grinding; width of cut in horizontal milling
a_r	rough grinding stock on radius of a rotational workpiece
a_t	total depth of material to be removed

B small batch size; width of the rectangular prismatic envelope enclosing a nonrotational machined component

B_0 basic bench standard value for forging dies

B_L batch length in furnace

B_r effective blow rate of forging equipment

B_s batch size of parts

b reduction exponent; index for cavity milling standard equation

b_w width of surface to be machined

C thickness of the rectangular prismatic envelope enclosing a non-rotational machined component

C_1 cost of one pair of cavity and core inserts

C_{1000} cost per operation for a 1000 lb power hammer

C_{20} tool accessory cost for 20 ton (178 kN) press for one-level part

C_{ab} cost of mold base with custom work

C_{ac} cost of standard mold components or actuators

C_{af} cost of pattern assembly fixture

C_{AP} press capacity

C_b cost of mold base

C_{bo} cost per cluster for breakout

C_{box} cost of core box for sand casting

C_{bu} cost to apply backup coasts

C_c cost of tungsten carbide per unit volume; grinding cost when recommended conditions are used

C_{c1} cost of single-cavity mold

C_{cf} operator cost per cluster for cutoff

C_{cl} cost per cluster for cleaning or leaching; sand casting cleaning cost

C_{cn} cost of n-cavity mold

C_{co} cost per cluster for cutoff

C_{core} core processing cost

C_{csd} cost of core sand for one casting

C_D forging die cost per part

C_d mold cost

C_{DIE} total forging die cost

C_{d1} cost of single-cavity die

C_{dm} mold-making cost

C_{dman} forging-die manufacturing cost

C_{dmat} material cost for forging die block

C_{dn} cost of n-cavity die

C_{ds} cost of die set

C_e cost of energy to melt metal

C_{en} furnace energy cost

C_f cost of feeding each part; cost of induction melting unit

C_F	cost of feeder
C_{fk}	fixed furnace cost per weight of metal
C_{FL}	furnace operating cost per unit time
C_{fp}	cost of mold base fixed plates
C_{fs}	cost of setup for cutoff
C_g	production cost for a grinding operation
C_i	cost of automatic insertion per part; cost per unit weight of infiltrant material
C_{ip}	process cost per pattern or core
C_{lk}	furnace labor cost per weight of metal
C_m	cost of polymer material per part; material cost; cost of metal ready to pour
C_{man}	forging die manufacturing cost rate
C_{mat}	cost of part material
C_{mf}	cost of furnace for alloy; cost of metal in finished sand casting
C_{mi}	cost of furnace for iron
C_{min}	minimum production cost (minimum value of C_{pr})
C_{ml}	cost of labor to melt metal
C_{mp}	pouring cost; process cost for sand casting
C_{ms}	total cost of metal at the furnace spout
C_n	cost of n identical pairs of cavity and core inserts
C_{nh}	operator cost per cluster for pneumatic hammer
C_{ns}	cost of setup for pneumatic hammer
C_o	cost per unit volume of impregnated oil; resin or polymer
C_{op}	cost per operation of forging equipment
C_p	cost of powder per unit weight; grinding cost when maximum power is used; processing cost of one part; cost of electricity
C_{pca}	cost per part to assemble cluster
C_{pi}	cost of set of pattern impressions
C_{pm}	cost of wax material; cost of pattern mounting plates
C_{po}	production cost when maximum power is used
C_{pr}	production cost per machined component; cost to apply primer coats to cluster; production cost per piece in forging
C_{pt}	cost of sand casting pattern
C_{px}	complexity factor for sheet metal stamping
C_r	relative feeder cost
C_{rm}	cost of raw alloy; cost of raw material for sand casting
C_{rs}	cost of a resink for forging dies
C_{ro}	cost per operation of forging equipment relative to 1000 lb power hammer
C_{set}	setup cost per part for forging

C_t	cost of providing a new or freshly ground tool; cost of tool steel per unit weight; total handling and insertion cost per part
C_{tca}	cost to assemble one cluster
C_{tl}	cost of single-aperture trim tool
C_{tn}	cost of multiaperture trim tool
C_{tp}	total cost of pattern piece
C_{tpa}	pattern assembly cost
C_{trim}	total cost of trimming tools
C_{trm}	material cost for flash trimming tools
C_{tw}	cost of cutoff tool wear per cluster
C_{vp}	cost of mold base variable plates
C_w	wheel wear and wheel changing costs in grinding
c	dimensionless diametral clearance between peg and hole; "individual" reduction exponent
c_f	pattern clearance factor
c_h	hand clearance
D	diameter of hole; diameter of the circular cylinder enclosing a rotational machined component; part diameter; depth of part
D_a	probability of a defective product
D_{bar}	equivalent bar diameter for forging
D_c	carbide insert diameter
D_d	die case diameter
D_e	equivalent part diameter
D_{eh}	equivalent hole diameter
D_h	hole diameter or circumscribing circle diameter for hole
D_i	average probability of assembly defects per operation
D_{li}	circumscribing circle diameter for level $i, i = 1, 2, 3, \ldots$
D_o	circumscribing circle diameter for whole part
D_{pi}	punch stock material diameter, punch $i = 1, 2, 3, \ldots$
D_{pm}	density of pattern material
d	diameter of peg; depth
d_a	outer diameter of surface machined by facing
d_{ave}	average cavity depth for forging
d_b	inner diameter of surface machined by facing
d_c	cavity depth in forging
d_g	grip size
d_m	diameter of the machined surface
d_{max}	maximum section diameter
d_t	diameter of cutting tool
d_w	diameter of the work surface
E	orienting efficiency of a part
E_{ct}	cost of electricity

E_f	required energy capacity of forging equipment
E_m	overall efficiency of machine-tool motor and drive systems; minimum energy to melt a metal
E_{ma}	manual assembly efficiency
E_o	equipment factory overhead ratio
e	eccentricity of force on peg; strain in sheet metal forming
F	press force; maximum separating force
F_c	profile complexity factor
F_{fc}	fixed furnace cost per volume of metal; forging complexity factor
F_{ff}	furnace efficiency
F_{ins}	benchwork factor for forging dies
F_{lck}	trim punch lock factor
F_{lw}	plan area correction factor
F_m	maximum feed rate for standard feeder
F_{trm}	load required for flash removal by trimming press
F_r	required feed rate
f	displacement of the tool relative to the workpiece, in the direction of feed motion per stroke or per revolution of the workpiece or tool; separating force on die or mold; factor increase in output
f_d	die plate thickness correction factor
f_p	parting surface adjustment factor
G_f	gate factor for sand casting pattern
H	height of feature
H_f	latent heat of fusion
H_F	height of furnace opening
H_s	specific heat
H_t	heat transfer coefficient
h	wall thickness or gage thickness
h_{cl}	minimum clearance
h_d	die plate thickness; depth of part or pattern piece
h_f	fill height of powder
h_{fp}	thickness of ejector, riser, and stripper plates
h_{max}	maximum wall thickness
h_p	combined thickness of core and cavity plates
h_{pt}	height or thickness of mold base
K	compaction pressure correction factor; distance traveled by a point on the tool cutting edge relative to the workpiece during the machining time; coefficient for cavity milling standard equation
k_1	constant representing wheel wear and wheel-changing costs per unit metal removal rate in grinding; coefficient of machine hourly rate
k_2	constant representing rough grinding time multiplied by the metal removal rate

k_r	powder compression ratio
L	length of part or feature; length of peg in section in hole; depth of insertion; length of the circular cylinder enclosing a rotational machined component
L_{blk}	length of forging die block
L_b	total length of bend lines
L_D	total life for forging dies
L_e	equivalent part length
L_{FL}	furnace overall length
L_h	length of hole machined by EDM
L_{HT}	sintering zone length of furnace
L_i	lower punch length for punch i, $i = 1, 2, 3, \ldots$
L_{plt}	platter length for forging
L_s	clamp stroke
L_{tbas}	basic trimming tool life quantity
L_{trm}	trimming tool life quantity
L_w	length of wire
l	overall length of part in direction of feeding
l_f	life of tool element
l_p	length of pathway between machine tools; powder loss during PM processing
l_{rd}	distance forklift truck travels to respond to request
l_s	section length
l_t	length of broach
l_w	length of machined surface
M	total machine tool and operator rate; equipment operating cost per unit time
M_{bu}	machine and operator rate for backup coat application
M_{cp}	pattern or core material cost
M_e	manufacturing point score or hours for the ejector pin system
M_i	machine and operator rate
M_{me}	minimum melt energy
M_p	sheet metal die manufacturing points; block area factor for trimming die
M_{pc}	manufacturing points for custom punches
M_{pn}	manufacturing die points for number and length of bends
M_{po}	basic sheet metal die manufacturing points; manufacturing hours for part or pattern piece size
M_{pr}	machine and operator rate for primer coat application
M_{ps}	manufacturing points for standard punches
M_{px}	manufacturing points for geometrical complexity

M_s	manufacturing hours for nonflat parting surface; surface patches per unit projected area for forging dies
M_t	manufacturing points for trim tool
M_{to}	basic manufacturing points for trim tool
M_{tot}	total manufacturing hours for mold making
M_x	manufacturing hours for geometrical complexity
m	multicavity cost index, usually 0.7
m_1	coefficient of machine hourly rate
m_{rc}	cutoff rate
N_1	normal force at point 1
N_2	normal force at point 2
N_b	number of bends to be formed in one die; number of blows or strokes required for a forging
N_c	number of contacts in a connector; number of impressions or cavities in sand casting; number of identical forgings produced per cycle
N_d	number of different punch shapes or sizes; number of assembly errors in one product
N_e	number of ejector pins
N_{fl}	number of fuller dies for a forging
N_{fw}	number of parts across furnace width
N_h	number of hits with turret press
N_{hd}	number of holes or depressions
N_{imp}	number of forging impressions
N_{min}	theoretical minimum number of parts
N_{mw}	number of line workers for sand casting
N_{op}	number of operations required for a forging
N_p	number of custom punches
N_{pi}	number of identical impressions on pattern plate
N_r	number of replacement tool items
N_{rs}	number of resinks possible for forging dies
N_{sp}	number of surface patches to be machined
N_{st}	number of stitches in the lacing of a wire harness
N_w	number of wires assembled simultaneously onto a wire harness jig
n	number of workpieces; number of cavities; Taylor tool life index (or exponent) in machining; number of assembly operations
n_{bd}	indicator for bending die required (1 or 0)
n_{bk}	indicator for blocker die required (1 or 0)
n_c	number of cores per pattern piece
n_{cb}	number of backup coats
n_{cl}	number of clearances required
n_{cp}	number of primer coats
n_e	efficiency of induction furnace

n_{edg}	indicator for edger die required (1 or 0)
n_{fin}	indicator for finishing die required (1 or 0)
n_{fl1}	indicator for first fuller die required (1 or 0)
n_{fl2}	indicator for second fuller die required (1 or 0)
n_{gp}	number of gates per casting
n_l	number of lifters per pattern piece
n_{pa}	number of pieces per pattern
n_{pc}	number of parts per cluster
n_{pd}	number of patterns (cavities) per mold
n_{pl}	number of core and cavity plates
n_r	frequency of cutting strokes
n_{rs}	number of resinks required for forging dies
n_s	number of surface patches on the part or pattern piece
n_{sb}	indicator for scale-breaking die required (1 or 0)
n_{sf}	indicator for semifinishing die required (1 or 0)
n_{so}	number of supplemental cuts per cluster
n_{sp}	number of side pulls per pattern piece
n_t	rotational speed of cutting tool
n_{ud}	number of unscrewing devices per pattern piece
n_w	rotational speed of workpiece on worktable
P	compaction pressure; force on peg; perimeter length to be sheared on sheet metal part
P_b	payback period
P_{cm}	core production rate for sand casting
P_e	electrical power required for machining
P_{ff}	sand casting plant efficiency
P_i	recommended injection pressure; uncorrected compaction pressure
P_j	injection power
P_l	metal loss rate
P_m	power required in machining
P_{mp}	sand casting production rate
P_p	perimeter of custom punches
P_{psr}	pattern piece setup rate
P_r	perimeter of projected area
P_{rv}	proportion of runner volume
P_s	specific cutting energy—power required to remove unit volume of material in unit time
P_v	production volume
P_W	perimeter of through holes in forging
Q	hole machining factor
Q	flow rate
Q_{lv}	life production volume for forgings

Q_{mx}	maximum wax injection flow rate
Q_{rb}	basic resink quantity for forging dies
Q_{rs}	resink quantity for forging dies
q_c	proportion by weight of infiltrant material
q_o	proportion by volume of impregnated oil
R_1	batch furnace heating rate
R_2	batch furnace cooling rate
R_a	arithmetical mean surface roughness
R_{cl}	worker rate for cleaning sand castings
R_{co}	operator rate for cutoff
R_{ds}	mold-making rate
R_f	cost of using feeding equipment
R_i	cost of using automatic workhead per part
R_{mp}	worker rate for sand casting line
R_{nh}	operator rate for breakout
R_p	production rate (parts per unit time)
R_t	tool making rate for sand casting
r	inside bend radius; tool profile radius; average time for a factor increase in output divided by the time for the first output expressed as a percentage
r_ϵ	cutting tool corner radius
S	profile cutting speed
S_a	volume shrinkage for metal
S_{bw}	bench standard for forging dies
S_c	cavity standard for forging dies
S_{ca}	setup time for cluster assembly
S_{co}	setup time for cutoff
S_d	cavity spacing in forging
S_{ds}	time to set up the mold on the injection machine
S_e	cavity edge distance in forging
S_{lk}	die lock standard for forging dies
S_g	percentage of scrap for sand castings
S_m	scrap rate for sand casting
S_{ml}	milling standard for forging dies
S_n	number of shifts worked per day
S_{nh}	setup time for pneumatic hammer
S_{pa}	setup time for pattern assembly
S_{sl}	scale loss in forging (%)
S_z	capacity of furnace
T	part thickness; die thickness; temperature
T_1	time of production for the first unit
$T_{1,100}$	assumed basic DFA time value

$T_{1,B}$	adjusted DFA time for batch size B
$T_{1,x}$	average time of production for x units
T_B	burn-off time
T_{blk}	thickness of forging die block
T_{bt}	base time for forging die block preparation time
T_{bw}	bench-working time for forging dies
T_{cav}	cavity time for forging dies
T_{cl}	sand casting cleaning time
T_{dl}	dowel machining time for forging dies
T_{edg}	edger machining time for forging dies
T_f	flash land thickness in hot forging
T_{fl}	flash gutter time for forging dies
T_{int}	initial time allowance for trim die manufacture
T_i	injection temperature
T_{lay}	forging die layout time
T_{lk}	additional time for trimming die manufacture for locked dies
T_m	mold temperature
T_{mill}	cavity milling time for forging dies
T_{pl}	block planing time for forging dies
T_{pol}	cavity polishing time for forging dies
T_{prep}	forging die block preparation time
T_s	sintering time
T_{set}	setup time for forging equipment
T_{tp}	time to manufacture flash trimming punch
T_{trd}	manufacturing time for flash trimming die
T_W	web thickness in forging
T_x	ejection temperature; time to produce the xth unit
t	machine cycle time; part thickness; tool life (machining time between regrinds or between cutting edge replacements)
t_a	basic assembly time for one part
t_b	basic time to insert peg through stack of parts
t_c	nonproductive time in grinding including wheel dressing time and time to load and unload the workpiece; tool life giving minimum cost machining; cooling time
t_{c1}	time to apply first primer coat
t_{ca}	time to assemble cluster
t_{cb}	time to apply one backup coat
t_{cg}	time to cut through single gate
t_{cl}	time to load cluster for cutoff
t_{co}	time to cut off all parts from cluster
t_{cp}	time to apply each subsequent primer coat
t_{ct}	tool changing time

t_d	time to dress a wire
t_f	mold fill time; transportation time for a roundtrip by a forklift truck
t_{FB}	batch furnace batch sintering time
t_{FL}	furnace throughput time per part
t_{gc}	grinding time for recommended conditions
t_{gf}	finish grinding time
t_{gp}	grinding time when maximum power is used; time to reposition cluster from one gate to next
t'_{gp}	corrected value of t_{gp} to allow for wheel costs
t_{gr}	time for rough grinding
t_h	basic time for handling a "light" part
t_i	manual insertion time for peg in hole; average assembly time per operation
t_l	nonproductive time incurred each time the workpiece is clamped and unclamped or loaded and unloaded in the machine tool
t_m	machining time—time between engagement and disengagement of feed motion; time to assemble N_w wires attached to a connector onto a wire harness jig
t_{ma}	total assembly time for product
t_{mc}	machining time when cutting speed for minimum cost is used
t_{mp}	machining time when maximum power is used
t_n	time to assemble N_w wires simultaneously onto a wire harness jig
t_{nh}	breakout cycle time for pneumatic hammer
t_{oc}	open and close time
t_p	time penalty for insertion of peg through stack of parts; time to assemble wires with crimped contacts into a connector
t_{pa}	time for pattern assembly
t_{pw}	additional part handling time due to weight
t_r	total reset time; tool life for a cutting speed of v_r
t_s	spark-out time in grinding; time to assembly a connector having soldered contacts
t_{sc}	time for supplemental cut through runner or sprue
t_{so}	time to reposition cluster for supplemental cuts
t_{st}	time to lace a wire harness
t_t	total cycle time
t_{tr}	transportation time per workpiece
U	ultimate tensile strength
U_i	upper punch length; punch $i, i = 1, 2, 3, \ldots$
V	part volume; required production volume
V_{fc}	volume of metal in finished casting
V_{fl}	volume of flash per unit length of flash line
V_m	volume of material to be removed by machining

V_s	shot size
V_{sc}	value of scrap
V_{trd}	material volume for flash trimming die
V_{trp}	material volume for flash trimming punch
v	cutting speed—relative velocity of the tool relative to the workpiece
v_{av}	average cutting speed
v_c	cutting speed giving minimum cost machining; press closing speed
v_f	feed speed in milling
v_F	belt or feed speed for furnace
v_{max}	maximum cutting speed
v_o	press opening speed
v_{po}	cutting speed giving maximum power
v_r	cutting speed for a tool life of t_r
v_{trav}	traverse speed in grinding
W	weight of part; width of part or feature
$\dot{W}$	heat flow rate
W_{blk}	width of forging die block
W_c	workhead cost
W_f	width of flash land in hot forging
W_{max}	maximum weight per unit area for furnace
W_p	poured weight of casting
W_{pa}	operator rate for pattern assembly
W_{plt}	platter width for hot forging
W_{pr}	weight of material poured into a single mold
W_r	relative workhead cost
W_{sm}	dry weight of the shell mold
w	thickness of gate
w_1	width of chamfer on peg
w_2	width of chamfer on hole
w_F	belt or feeder width for furnace
w_s	section width
w_t	width of grinding wheel
X_i	inner complexity value
X_o	outer complexity value
x	number of identical units
Y_d	casting yield
Y_s	equivalent shear yield stress of material
Y_{sm}	shell mold yield
Z_{pw}	metal removal rate when maximum power is used in grinding
Z_w	metal removal rate
Z_{wc}	metal removal rate for recommended conditions in grinding
Z_{wmax}	maximum metal removal rate

α	thermal diffusivity; alpha symmetry of part
α_m	material load factor for hot forging
α_s	shape load factor for forging
β	beta symmetry of part
β_m	material die life factor for hot forging
β_s	shape die life factor for hot forging
θ	angle of force on peg; angle of bend of sheet metal part
θ_b	burn-off temperature
θ_s	sintering temperature
θ_1	semiconical angle of chamfer on peg
θ_2	semiconical angle of chamfer on hole
μ	coefficient of friction
ρ	part density
ρ_a	density of alloy; powder apparent density
ρ_i	density of iron; infiltrant material wrought density
ρ_o	density of impregnated oil
ρ_p	part density, including infiltrant
ρ_t	density of tool steel
ρ_w	equivalent wrought density of material
ψ_{fl}	angle of fuller dies on die block

Index